Water
and Wastewater
Technology

Water
and Wastewater
Technology

SECOND EDITION

MARK J. HAMMER
Professor of Civil Engineering

JOHN WILEY & SONS
New York
Chichester
Brisbane
Toronto
Singapore

To my mother
Bertha Grundahl Hammer
and in memory of my father
Herbert Hammer

Cover: Photograph by Mark Hammer.
Northeast Wastewater Treatment Plant,
Lincoln, Nebraska.

Library of Congress Cataloging in Publication Data:

Hammer, Mark J., 1931–
  Water and Wastewater technology.

  Includes index.
  1. Water-supply engineering.   2. Sewage disposal.
  I. Title.
TD345.H25   1986   628.1   85-6291
ISBN 0-471-05650-2

Printed in the United States of America

10 9 8 7 6 5 4 3 2

Preface

This book provides a comprehensive understanding of the technology of municipal water processing, water distribution, wastewater collection, wastewater treatment, and sludge disposal. This second edition updates the subjects presented in the first edition and broadens the coverage to include additional design concepts and data on operations and maintenance. The objective is to provide the fundamental knowledge in preparation for practice and continuation of studies in design-oriented courses in sanitary engineering programs and graduate education.

Based on my experience in education, I believe students benefit from a review of the disciplines that have specific applications in water supply and pollution control. Therefore, the introductory chapters cover fundamentals of chemistry, biology, hydraulics, and hydrology that are unique to sanitary studies. The book is organized in a traditional manner with water distribution and processing separated from wastewater collection and treatment. The final chapters give an overview of advanced wastewater treatment, water reuse, and land disposal techniques. I have carefully integrated the subject matter so that students can clearly understand the interrelationships between individual unit operations and integration of systems as a whole. In discussing various topics, I specifically included the latest technology: for example, the use of lasers in laying sewer pipe, the phenomenon of eutrophication, the use of high-purity oxygen in wastewater treatment, belt filters for dewatering sludges, the procedures for determining BOD of industrial wastewaters, and land disposal. A thorough discussion of the book's content is given in the introduction of Chapter 1.

Illustrations help to explain fundamental concepts and show modern equipment and facilities. Also, numerous sample calculations help the reader to understand the unit processes being described. Answers are given for some of the homework problems, mainly to help students who are interested in individual study. Finally, the essential resource material is included in an appendix.

Audrey Hammer typed the manuscript and gave invaluable assistance in assembling and proofreading the final copy of this book. I also wish to thank the following two instructors for their evaluations of the revised manuscript: Leo A. Ebel, Washington University, St. Louis, Missouri; and Jerry A. Nathanson, Union County College, Cranford, New Jersey.

Contents

chapter 1

Introduction

The hydrologic cycle describes the movement of water in nature. Evaporation from the ocean is carried over land areas by maritime air masses. Vapor from inland waters and transpiration from plants add to atmospheric moisture that eventually precipitates as rain or snow. Rainfall may percolate into the ground, join surface watercourses, be taken up by plants, or reevaporate. Groundwater and surface flows drain toward the ocean for recycling.

Humans intervene in the hydrologic cycle, generating artificial water cycles (Figure 1-1). Some communities withdraw groundwater for public supply, but the majority rely on surface sources. After processing, water is distributed to households and industries. Wastewater is collected in a sewer system and transported to a plant for treatment prior to disposal. Conventional methods provide only partial recovery of the original water quality. Dilution into a surface watercourse and purification by nature yield additional quality improvement. However, the next city downstream is likely to withdraw the water for a municipal supply before complete rejuvenation. This city in turn treats and disposes of its wastewater by dilution. This process of withdrawal and return by successive municipalities in a river basin results in indirect water reuse. During dry weather, maintaining minimum flow in many small rivers relies on the return of upstream wastewater discharges. Thus, an artificial water cycle within the natural hydrologic scheme involves (1) surface-water withdrawal, processing, and distribution; (2) wastewater collection, treatment, and disposal back to surface

water by dilution; (3) natural purification in a river; and (4) repetition of this scheme by cities downstream.

Discharge of conventionally treated wastewaters to lakes, reservoirs, and estuaries, which act like lakes, accelerates eutrophication. Resulting deterioration of water quality interferes with indirect reuse for public supply and water-based recreational activities. Consequently, advanced wastewater treatment by either mechanical plants or land disposal techniques has been introduced into the artificial water cycle involving inland lakes and reservoirs.

Installation of advanced treatment systems that reclaim wastewater to nearly its original quality has encouraged several cities to consider direct reuse of water for industrial processing, recreational lakes, irrigation, groundwater recharge, and other applications. However, direct return for a potable water supply is not being encouraged at present because of potential health hazards from viruses and traces of toxic substances that are difficult to detect and may not be removed in water reclamation. Another problem is the buildup of dissolved salts that can be removed only by costly demineralization processes. Nevertheless, it is anticipated, with the increase in demand for fresh water, that direct water reuse by some metropolitan areas may be realistic by the year 2000.

The basic sciences of chemistry, biology, hydraulics, and hydrology are the foundation for understanding water supply and pollution control. Chemical principles find greatest application in water processing, while

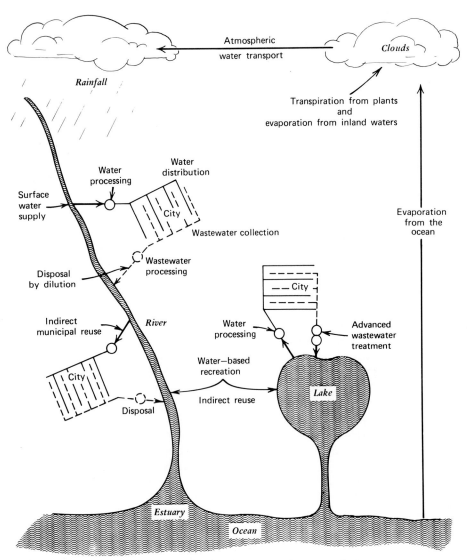

Figure 1-1 Integration of natural and human-generated water cycles.

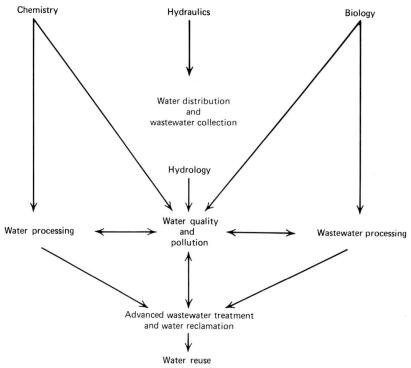

Figure 1-2   Flow diagram relating basic science areas to the disciplines of water and wastewater technology and the integration of various aspects of water supply and pollution control.

wastewater treatment relies on biological systems. Knowledge of hydraulics is the key to water distribution and wastewater collection. Quality of the water environment is the focus of the indirect water reuse cycle. As illustrated in Figure 1-2, chemistry, hydrology, biology, water processing, and wastewater treatment converge to give a perception of water quality and pollution. Finally, insights to future direct water reuse are provided through the technology of advanced wastewater treatment and water reclamation.

In this book the fundamental principles of science are integrated with different aspects of sanitary technology. Table 1-1 lists prerequisite readings included in this book for various subject areas. The purpose of this table is to help the student correlate information from the text on a particular subject of interest. For example, prior to

studying water quality and pollution in Chapter 5, a student should review chemistry fundamentals, aquatic organisms from bacteria through fishes, the aquatic food chain, waterborne diseases, and the hydrology of rivers and lakes. Prerequisite readings for water distribution systems are selections from Chapter 4. The instructor who is preparing a lecture sequence on this topic may prefer to integrate the principles of hydraulics with the descriptive material from the chapter on water distribution. The advantage is that students can read about real pipe networks while solving simplified pipe flow problems. Conversely, when water processing is taught, it may be most advantageous for the student to review the portions on chemistry listed in Table 1-1 prior to doing the reading assignments in Chapter 7.

Wastewater flows and characteristics are discussed

Table 1-1. Listing of Prerequisite Readings for Various Subject Areas in Water and Wastewater Technology

Water Quality and Pollution, Chapter 5

2-1 Elements, Radicals, and Compounds
2-9 Organic Compounds
2-11 Laboratory Chemical Analyses
3-1 Bacteria and Fungi
3-2 Viruses
3-3 Algae
3-4 Protozoa and Multicellular Animals
3-5 Fishes
3-6 Aquatic Food Chain
3-7 Waterborne Diseases
3-8 Indicator Organisms for Water Quality
4-12 Flow in Streams and Rivers
4-13 Hydrology of Lakes and Reservoirs
4-14 Groundwater Hydrology

Water Distribution Systems, Chapter 6

4-1 Water Pressure
4-2 Pressure-Velocity-Head Relationships
4-3 Flow in Pipes under Pressure
4-4 Centrifugal Pump Characteristics
4-5 System Characteristics
4-6 Flow in Pipe Networks
4-9 Flow Measurement in Pipes

Water Processing, Chapter 7

2-1 Elements, Radicals, and Compounds
2-2 Chemical Water Analysis
2-3 Hydrogen Ion Concentration and pH
2-4 Chemical Equilibria
2-5 Chemical Kinetics
2-6 Gas Solubility
2-7 Alkalinity
2-8 Colloids and Coagulation
2-11 Laboratory Chemical Analyses
4-1 Water Pressure
4-2 Pressure-Velocity-Head Relationships
4-9 Flow Measurement in Pipes

Wastewater Collection Systems, Chapter 10

4-1 Water Pressure
4-2 Pressure-Velocity-Head Relationships

4-4 Centrifugal Pump Characteristics
4-5 System Characteristics
4-8 Gravity Flow in Circular Pipes
4-10 Flow Measurement in Open Channels
4-11 Amount of Storm Runoff
Chapter 9 Wastewater Flows and Characteristics

Wastewater Processing, Chapter 11

2-9 Organic Compounds
2-10 Organic Matter in Wastewater
3-1 Bacteria and Fungi
3-3 Algae
3-4 Protozoa and Multicellular Animals
3-11 Biochemical Oxygen Demand
3-12 Biological Treatment Systems
3-13 Biological Kinetics
4-1 Water Pressure
4-2 Pressure-Velocity-Head Relationships
4-10 Flow Measurement in Open Channels
Chapter 9 Wastewater Flows and Characteristics

Operation of Waterworks, Chapter 8

Chapter 6 Water Distribution Systems
Chapter 7 Water Processing
Chapter 5 Water Quality and Pollution

Operation of Wastewater Systems, Chapter 12

Chapter 9 Wastewater Flows and Characteristics
Chapter 10 Wastewater Collection Systems
Chapter 11 Wastewater Processing
Chapter 5 Water Quality and Pollution

Advanced Wastewater Treatment, Chapter 13

Wastewater Processing Plus Prerequisites
Water Processing Plus Prerequisites

Water Reuse and Land Disposal, Chapter 14

Water Quality and Pollution Plus Prerequisites
Wastewater Processing Plus Prerequisites
Water Processing Plus Prerequisites
Advanced Wastewater Treatment

separately in Chapter 9, since this information is needed for both collection systems and wastewater processing. The listing under collection systems also includes applied hydraulics. Conventional municipal wastewater treatment relies on biological processing, and therefore an understanding of living systems is indispensable. The chapters on operation of systems are not intended to give all-inclusive subject coverage. Since this book deals primarily with art and practice, a great deal of information relative to operation is presented throughout in the book's descriptive material.

Advanced wastewater treatment incorporates both biological unit operations and chemical processes that are similar to those applied in water treatment. Therefore, to understand the concepts described in Chapter 13, the reader must have adequate knowledge about the handling of both wastewater and water. Reuse of reclaimed water employs the most recent treatment technology and requires a comprehensive understanding of both treatment processes and water quality.

chapter 2
Chemistry

This chapter provides basic information about chemistry as it applies to water and wastewater technology. Selected data are compiled and presented as an introduction to the chapters dealing with water quality, pollution, and chemical-treatment processes. For example, the characteristics of common elements, radicals, and inorganic compounds are tabulated. The usual method for presenting chemical water analysis is described, since it is not normally presented in general chemistry textbooks. Sections on chemical reactions, alkalinity, and coagulation emphasize important aspects of applied water chemistry. Since organic chemistry traditionally has not been included in introductory courses, persons practicing in water supply and pollution control are not generally exposed to this area of chemistry. For this reason an introduction on the nomenclature of organic compounds and a brief description of the organic matter in wastewater are provided. Finally, the importance, technique, and equipment used in selected laboratory analyses are discussed. Water-quality parameters and their characteristics can be understood better when testing procedures are known.

2-1 ELEMENTS, RADICALS, AND COMPOUNDS

The fundamental chemical identities that form all substances are referred to as elements. Each element differs from any other in weight, size, and chemical properties. Names of elements common to water and wastewater technology along with their symbols, atomic weights, common valence, and equivalent weights are given in Table 2-1. Symbols for elements are used in writing chemical formulas and equations.

Atomic weight is the weight of an element relative to that of hydrogen, which has an atomic weight of unity. This weight expressed in grams is called one gram atomic weight of the element. For example, one gram atomic weight of aluminum Al is 27.0 grams. Equivalent or combining weight of an element is equal to atomic weight divided by the valence.

Some elements appear in nature as gases, for example, hydrogen, oxygen, and nitrogen; mercury appears as a liquid; others appear as pure solids, for instance, carbon, sulfur, phosphorus, calcium, copper, and zinc; and many occur in chemical combination with each other in compounds. Atoms of one element unite with those of another in a definite ratio defined by their valence. Valence is the combining power of an element based on that of the hydrogen atom, which has an assigned value of 1. Thus, an element with a valence of $2+$ can replace two hydrogen atoms in a compound, or in the case of $2-$ can react with two hydrogen atoms. Sodium has a valence of $1+$ while chlorine has a valence of $1-$, therefore, one sodium atom combines with one chlorine atom to form sodium chloride ($NaCl$), common salt. Nitrogen at a valence of $3-$ can combine with three hydrogen atoms to form ammonia gas (NH_3). The weight of a compound, equal to the sum of the weights of the combined elements, is referred to as molecular weight, or simply mole. The molecular weight of $NaCl$ is 58.4 grams, while one mole of ammonia gas is 17.0 grams.

Table 2-1. Basic Information on Common Elements

Name	Symbol	Atomic Weight	Common Valence	Equivalent Weight[a]
Aluminum	Al	27.0	3+	9.0
Arsenic	As	74.9	3+	25.0
Barium	Ba	137.3	2+	68.7
Boron	B	10.8	3+	3.6
Bromine	Br	79.9	1−	79.9
Cadmium	Cd	112.4	2+	56.2
Calcium	Ca	40.1	2+	20.0
Carbon	C	12.0	4−	
Chlorine	Cl	35.5	1−	35.5
Chromium	Cr	52.0	3+	17.3
			6+	
Copper	Cu	63.5	2+	31.8
Fluorine	F	19.0	1−	19.0
Hydrogen	H	1.0	1+	1.0
Iodine	I	126.9	1−	126.9
Iron	Fe	55.8	2+	27.9
			3+	
Lead	Pb	207.2	2+	103.6
Magnesium	Mg	24.3	2+	12.2
Manganese	Mn	54.9	2+	27.5
			4+	
			7+	
Mercury	Hg	200.6	2+	100.3
Nickel	Ni	58.7	2+	29.4
Nitrogen	N	14.0	3−	
			5+	
Oxygen	O	16.0	2−	8.0
Phosphorus	P	31.0	5+	6.0
Potassium	K	39.1	1+	39.1
Selenium	Se	79.0	6+	13.1
Silicon	Si	28.1	4+	6.5
Silver	Ag	107.9	1+	107.9
Sodium	Na	23.0	1+	23.0
Sulphur	S	32.1	2−	16.0
Zinc	Zn	65.4	2+	32.7

[a] Equivalent weight (combining weight) equals atomic weight divided by valence.

Table 2-2. Common Radicals Encountered in Water

Name	Formula	Molecular Weight	Electrical Charge	Equivalent Weight
Ammonium	NH_4^+	18.0	$1+$	18.0
Hydroxyl	OH^-	17.0	$1-$	17.0
Bicarbonate	HCO_3^-	61.0	$1-$	61.0
Carbonate	$CO_3^=$	60.0	$2-$	30.0
Orthophosphate	$PO_4^\equiv$	95.0	$3-$	31.7
Orthophosphate, mono-hydrogen	$HPO_4^=$	96.0	$2-$	48.0
Orthophosphate, di-hydrogen	$H_2PO_4^-$	97.0	$1-$	97.0
Bisulfate	HSO_4^-	97.0	$1-$	97.0
Sulfate	$SO_4^=$	96.0	$2-$	48.0
Bisulfite	HSO_3^-	81.0	$1-$	81.0
Sulfite	$SO_3^=$	80.0	$2-$	40.0
Nitrite	NO_2^-	46.0	$1-$	46.0
Nitrate	NO_3^-	62.0	$1-$	62.0
Hypochlorite	OCl^-	51.5	$1-$	51.5

Certain groupings of atoms act together as a unit in a large number of different molecules. These, referred to as radicals, are given special names, such as the hydroxyl group (OH^-). The most common radicals in ionized form are listed in Table 2-2. Radicals themselves are not compounds but join with other elements to form compounds. Data on inorganic compounds common to water and wastewater chemistry are given in Table 2-3. The proper name, formula, and molecular weight are included for all of the chemicals listed. Popular names, for example, alum for aluminum sulfate, are included in brackets. For chemicals used in water treatment, one common use is given; many have other applications not included. Equivalent weights for compounds and hypothetical combinations, for example, $Ca(HCO_3)_2$, involved in treatment are provided.

EXAMPLE 2-1

Calculate the molecular and equivalent weights of ferric sulfate.

Solution

Formula from Table 2-3 is $Fe_2(SO_4)_3$.
Using atomic weight data from Table 2-1,

Fe	$2 \times 55.8 =$	111.6
S	$3 \times 32.1 =$	96.3
O	$12 \times 16.0 =$	192.0

Molecular weight = 399.9 or 400 grams

The ferric (oxide iron) atom has a valence of $3+$, thus a compound with 2 ferric atoms has a total electrical charge of $6+$. (Three sulfate radicals have a total of $6-$ charges).

$$\text{Equivalent weight} = \frac{\text{molecular weight}}{\text{electrical charge}}$$

$$= \frac{400}{6}$$

$$= 66.7 \text{ grams per equivalent weight}$$

2-2 CHEMICAL WATER ANALYSIS

When placed in water, inorganic compounds dissociate into electrically charged atoms and radicals referred to as ions. This breakdown of substances into their constituent ions is called ionization. An ion is represented by the chemical symbol of the element, or radical, followed by superscript + or − signs to indicate the number of unit

Table 2-3. Basic Information on Common Inorganic Chemicals

Name	Formula	Common Usage	Molecular Weight	Equivalent Weight
Activated carbon	C	Taste and odor control	12.0	n.a.[a]
Aluminum sulfate (filter alum)	$Al_2(SO_4)_3 \cdot 14.3H_2O$	Coagulation	600	100
Aluminum hydroxide	$Al(OH)_3$	(Hypothetical combination)	78.0	26.0
Ammonia	NH_3	Chloramine disinfection	17.0	n.a.
Ammonium fluosilicate	$(NH_4)_2SiF_6$	Fluoridation	178	n.a.
Ammonium sulfate	$(NH_4)_2SO_4$	Coagulation	132	66.1
Calcium bicarbonate	$Ca(HCO_3)_2$	(Hypothetical combination)	162	81.0
Calcium carbonate	$CaCO_3$	Corrosion control	100	50.0
Calcium fluoride	CaF_2	Fluoridation	78.1	n.a.
Calcium hydroxide	$Ca(OH)_2$	Softening	74.1	37.0
Calcium hypochlorite	$Ca(OCl)_2 \cdot 2H_2O$	Disinfection	179	n.a.
Calcium oxide (lime)	CaO	Softening	56.1	28.0
Carbon dioxide	CO_2	Recarbonation	44.0	22.0
Chlorine	Cl_2	Disinfection	71.0	n.a.
Chlorine dioxide	ClO_2	Taste and odor control	67.0	n.a.
Copper sulfate	$CuSO_4$	Algae control	160	79.8
Ferric chloride	$FeCl_3$	Coagulation	162	54.1
Ferric hydroxide	$Fe(OH)_3$	(Hypothetical combination)	107	35.6
Ferric sulfate	$Fe_2(SO_4)_3$	Coagulation	400	66.7
Ferrous sulfate (copperas)	$FeSO_4 \cdot 7H_2O$	Coagulation	278	139
Fluosilicic acid	H_2SiF_6	Fluoridation	144	n.a.
Hydrochloric acid	HCl	n.a.	36.5	36.5
Magnesium hydroxide	$Mg(OH)_2$	Defluoridation	58.3	29.2
Oxygen	O_2	Aeration	32.0	16.0
Potassium permanganate	$KMnO_4$	Oxidation	158	n.a.
Sodium aluminate	$NaAlO_2$	Coagulation	82.0	n.a.
Sodium bicarbonate (baking soda)	$NaHCO_3$	pH adjustment	84.0	84.0
Sodium carbonate (soda ash)	Na_2CO_3	Softening	106	53.0
Sodium chloride (common salt)	$NaCl$	Ion-exchanger regeneration	58.4	58.4
Sodium fluoride	NaF	Fluoridation	42.0	n.a.
Sodium hexametaphosphate	$(NaPO_3)_n$	Corrosion control	n.a.	n.a.
Sodium hydroxide	$NaOH$	pH adjustment	40.0	40.0
Sodium hypochlorite	$NaOCl$	Disinfection	74.4	n.a.
Sodium silicate	Na_4SiO_4	Coagulation aid	184	n.a.
Sodium fluosilicate	Na_2SiF_6	Fluoridation	188	n.a.
Sodium thiosulfate	$Na_2S_2O_3$	Dechlorination	158	n.a.
Sulphur dioxide	SO_2	Dechlorination	64.1	n.a.
Sulfuric acid	H_2SO_4	pH adjustment	98.1	49.0
Water	H_2O	n.a.	18.0	n.a.

[a] n.a. = not applicable.

charges on the ion. Consider the following: sodium, Na^+, chloride, Cl^-, aluminum, Al^{+++}, ammonium, NH_4^+, and sulfate, $SO_4^=$.

Laboratory tests on water, such as those outlined in Section 2-11, determine concentrations of particular ions in solution. Test results are normally expressed as weight of the element or radical in milligrams per liter of water, abbreviated as mg/l. Some books use the term parts per million (ppm), which is for practical purposes identical in meaning to mg/l, since 1 liter of water weighs 1,000,000 milligrams. In other words, 1 mg per liter (mg/l) equals 1 mg in 1,000,000 mg, which is the same as 1 part by weight in 1 million parts by weight (1 ppm). The concentration of a substance in solution can also be expressed in milliequivalents per liter (meq/l), representing the combining weight of the ion, radical, or compound. Milliequivalents can be calculated from milligrams per liter by

$$meq/l = mg/l \times \frac{valence}{atomic\ weight}$$

$$= \frac{mg/l}{equivalent\ weight} \qquad (2\text{-}1)$$

or in the case of a radical or compound the equation reads

$$meq/l = mg/l \times \frac{electrical\ charge}{molecular\ weight}$$

$$= \frac{mg/l}{equivalent\ weight} \qquad (2\text{-}2)$$

Equivalent weights for selected elements, radicals, and inorganic compounds are given in Tables 2-1, 2-2, and 2-3, respectively.

A typical chemical water analysis is in Table 2-4. These data can be compared against the chemical characteristics specified by the drinking water standards[1] to determine whether the water is safe for human consumption, or if treatment is required before domestic or industrial use.

Reporting results in milligrams per liter in tabular form is not convenient for visualizing the chemical composition of a water. Therefore, results are often expressed in milliequivalents per liter, which permits graphical presentation and a quick check on the accuracy of the analyses for major ions. The sum of the milliequivalents per liter of the cations (positive radicals) must equal the sum of the anions (negative radicals). In a perfect evaluation they would be exactly the same, since a water in equilibrium is electrically balanced. Graphical presentation of a water analysis using milliequivalents is performed by plotting

Table 2-4.   Chemical Analysis of a Surface Water (Values in mg/l)

Alkalinity (as $CaCO_3$)	108	Nitrate	2.2
Arsenic	0	pH	7.6
Barium	0	Phosphorus (total inorganic)	0.5
Bicarbonate	131	Potassium	3.9
Cadmium	0	Selenium	0
Calcium	35.8	Silver	0
Chloride	7.1	Sodium	4.6
Chromium	0	Sulfate	26.4
Copper	0.10	Total dissolved solids	220
Fluoride	0.7	Zinc	0
Foaming agents	0.1	Turbidity	5
Iron plus manganese	0.13	Corrosivity	0
Iron	0.10	Threshold odor number	1
Lead	0	Units of color	5
Magnesium	9.9		

the milliequivalents per liter values to scale, for example, by letting 1 meq/l equal 1 inch. The bar graph shown in Figure 2-1 is based on the water data in Table 2-4. The top row consists of the major cations arranged in the order of calcium, magnesium, sodium, and potassium. The under row is arranged in the sequence of bicarbonate, sulfate, and chloride.

Hypothetical combinations of the positive and negative ions can be developed from a bar graph as in Figure 2-1. These combinations are particularly helpful in considering lime water softening. In Figure 2-1, the carbonate hardness $[Ca(HCO_3)_2 + Mg(HCO_3)_2]$ is 2.15 meq/l and the noncarbonate is 0.45 meq/l $(MgSO_4)$.

EXAMPLE 2-2

The results of a water analysis are calcium 29.0 mg/l, magnesium 16.4 mg/l, sodium 23.0 mg/l, potassium 17.5 mg/l, bicarbonate 171 mg/l (as HCO_3), sulfate 36.0 mg/l, and chloride 24.0 mg/l. Convert milligrams per liter concentrations to milliequivalents per liter, list hypothetical combinations, and calculate hardness as mg/l of $CaCO_3$ for this water.

Solution

Using Eq. 2-1,

Component	mg/l	Equivalent Weight	meq/l
Ca^{++}	29.0	20.0	1.45
Mg^{++}	16.4	12.2	1.34
Na^+	23.0	23.0	1.00
K^+	17.5	39.1	0.45
		Total Cations	4.24
HCO_3^-	171	61.0	2.81
$SO_4^=$	36.0	48.0	0.75
Cl^-	24.0	35.5	0.68
		Total Anions	4.24

Hypothetical combinations: $Ca(HCO_3)_2$, 1.45 meq/l; $Mg(HCO_3)_2$, 1.34; $NaHCO_3$, 0.02; Na_2SO_4, 0.75; $NaCl$, 0.23; and KCl, 0.45 meq/l.

Hardness $(Ca^{++} + Mg^{++}) = 2.79$ meq/l

Equivalent weight $CaCO_3 = 50.0$ (Table 2-3)

By Eq. 2-2,

Hardness $= 2.79 \times 50.0 = 140$ mg/l

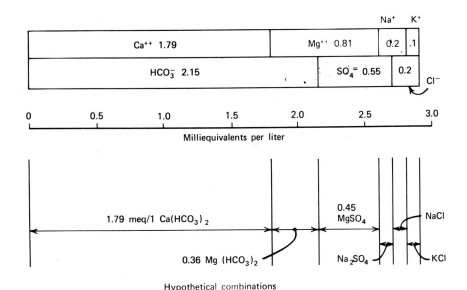

Hypothetical combinations

Figure 2-1 Milliequivalents-per-liter bar graph for water analysis given in Table 2-4.

2-3   HYDROGEN ION CONCENTRATION AND pH

Water (H_2O) dissociates to only a slight degree yielding a concentration of hydrogen ions equal to 10^{-7} mole per liter. Because water yields one hydroxyl (basic) ion for each hydrogen (acid) ion, pure water is considered neutral.

$$H_2O \rightleftharpoons H^+ + OH^- \tag{2-3}$$

The acidic nature of a water is related to the concentration of hydrogen ions in water solution by use of the symbol pH, where

$$pH = \log \frac{1}{[H^+]} \tag{2-4}$$

Since the logarithm of 1 over 10^{-7} is 7, the pH at neutrality is 7.

When an acid is added to water, the hydrogen ion concentration increases, resulting in a lower pH number. Conversely, when an alkaline substance is added, the OH^- ions unite with the free H^+, lowering the hydrogen ion concentration and causing a higher pH. The pH scale, ranging from 0 to 14, is acid from 0 to 7 and basic from 7 to 14.

Acid range	Basic range
0	7	14

2-4   CHEMICAL EQUILIBRIA

Most chemical reactions are reversible to some degree with the final state of equilibrium determining the concentration of reactants and products. For the general reaction expressed by Eq. 2-5, an increase in either A or B shifts the equilibrium to the right, while an increase in the concentration of C or D drives the reaction to the left. A chemical reaction in true equilibrium can be expressed by the mass-action relationship in Eq. 2-6.

$$a\text{A} + b\text{B} \rightleftharpoons c\text{C} + d\text{D} \tag{2-5}$$

$$\frac{[C]^c[D]^d}{[A]^a[B]^b} = K \tag{2-6}$$

where   A and B = reactants
C and D = products
[ ] = molar concentrations
K = equilibrium constant

Homogeneous Equilibria

In homogeneous chemical equilibria, all reactants and products occur in the same physical state, for instance, all are in solution. Inorganic strong acids and bases approach 100 percent ionization in dilute solutions. In contrast, weak inorganic and organic acids and bases are poorly ionized. The degree of ionization of a weak compound is expressed by the mass-action relationship. For example, Eqs. 2-7 through 2-10 characterize the equilibria of carbonic acid.

$$H_2CO_3 \rightleftharpoons H^+ + HCO_3^- \tag{2-7}$$

$$\frac{[H^+][HCO_3^-]}{[H_2CO_3]} = K_1 = 4.45 \times 10^{-7} \text{ at } 25^\circ C \tag{2-8}$$

$$HCO_3^- \rightleftharpoons H^+ + CO_3^= \tag{2-9}$$

$$\frac{[H^+][CO_3^=]}{[HCO_3^-]} = K_2 = 4.69 \times 10^{-11} \text{ at } 25^\circ C \tag{2-10}$$

Section 2-7 discusses bicarbonate-carbonate equilibrium relative to the alkalinity of a water, and Section 2-9 discusses ionization of organic acids.

Heterogeneous Equilibria

These refer to equilibria existing between substances in two or more physical states. For example, at greater than pH 10, solid calcium carbonate in water reaches a stability with the calcium and carbonate ions in solution.

$$CaCO_3 \rightleftharpoons Ca^{++} + CO_3^= \tag{2-11}$$

This kind of equilibrium, between crystals of a substance in a solid state and its ions in solution, can be expressed mathematically as homogeneous equilibrium as follows:

$$\frac{[Ca^{++}][CO_3^=]}{[CaCO_3]} = K \tag{2-12}$$

The concentration of the solid substance can be treated as a constant K_S in mass-action equilibrium. Hence, $[CaCO_3]$ can be assumed equal to KK_S such that Eq. 2-12 becomes

$$[Ca^{++}][CO_3^=] = KK_S = K_{SP}$$
$$= 5.0 \times 10^{-9} \text{ at } 25° \text{ C} \quad (2\text{-}13)$$

where K_{SP} = solubility-product constant

The solution is unsaturated if the product of the ionic molar concentrations is less than the solubility-product constant. Conversely, the solution is supersaturated if the product of the molar concentrations is greater than K_{SP}. In this case, crystals form and precipitation progresses until the ionic concentrations are reduced to equal those of a saturated solution. Based on the K_{SP} given in Eq. 2-13, the theoretical solubility of $CaCO_3$ in pure water at 25° C is 0.000,071 moles per liter, or 7.1 mg/l.

Shifting of Chemical Equilibria

Chemical reactions in water and wastewater processing are used to shift chemical equilibria in order to remove substances from solution. The most common methods for completing reactions are formation of insoluble precipitates, weakly ionized compounds, and gases that can be stripped from solution.

Precipitation Reactions. Forming insoluble precipitates of undesirable substances by the addition of treatment chemicals is the most common method of shifting equilibrium. Examples are the use of coagulants such as aluminum sulfate to remove turbidity and the use of lime to precipitate the multivalent ions of hardness. After formation by chemical reactions, the precipitates are usually removed from suspension by gravity sedimentation followed by filtration of the water through granular media.

An illustration of a precipitation reaction is lime softening. Water containing a high concentration of calcium (that is, hard water) tends to deposit scale on the inside of pipes and on appliances during water use. In the process of precipitation softening, calcium ions are removed from solution by raising the pH of the water by addition of lime (CaO). This results in a shift of the equilibrium in Eq. 2-14 to the right. (The arrow pointing down on the right side of the equation indicates that the $CaCO_3$ is a solid precipitate.)

$$CaO + Ca(HCO_3)_2 = 2CaCO_3{\downarrow} + H_2O \quad (2\text{-}14)$$

A chemical equation, like Eq. 2-14, gives the formulas for reactants and products and the number of moles of each. It is said to be balanced when the number of atoms of each element on the left side equals the number on the right side. By using a balanced equation, one can use molecular relationships to calculate the quantities of reactants and products. The process of using a balanced chemical equation for making calculations is called stoichiometry. In some reactions, equivalent weights can be used to determine amounts of reactants and products. (Refer to Examples 2-3 and 2-4.)

Oxidation and Reduction Reactions. Many chemical changes involve the addition of oxygen and/or the change of valence of one of the reacting substances. Oxidation is the addition of oxygen or removal of electrons, and reduction is the removal of oxygen or addition of electrons. A classic oxidation reaction is the rusting of iron by oxygen.

$$4Fe + 3O_2 = 2Fe_2O_3 \quad (2\text{-}15)$$

The iron is oxidized from Fe to Fe^{+++}, while the oxygen is reduced from O to $O^=$.

An oxidation-reduction reaction in water treatment is the removal of soluble ferrous iron from solution by oxidation using potassium permanganate. In this reaction, Eq. 2-16, the iron gains one positive charge while the manganese in the permanganate ion is reduced from a valence of 7+ to a valence of 4+ in manganese dioxide. (Note that an arrow is used instead of an equal sign in Eq. 2-16 since the equation is not balanced.)

$$3Fe^{2+} + Mn^{(7+)}O_4^- \rightarrow 3Fe^{(3+)}O_x{\downarrow} + Mn^{(4+)}O_2{\downarrow} \quad (2\text{-}16)$$

The precipitates of iron oxides and manganese dioxide are removed from suspension by sedimentation and filtration of the water.

Acid-Base Reactions (Neutralization). Acid added to water dissociates into hydrogen ions and anion radicals resulting in an acid solution. For example, sulfuric acid,

H_2SO_4, forms H^+ and $SO_4^=$ ions in solution. A basic solution is formed by adding an alkali such as NaOH to water. If both an acid and a base are put in the same water, the H^+ ions combine with the OH^- ions to form water, and, if equivalent amounts are added, they neutralize each other forming a salt solution as shown in the following equation:

$$4H^+ + SO_4^= + 2Na^+ + 2OH^-$$
$$= 2H_2O + 2Na^+ + SO_4^= \quad (2\text{-}17)$$

Gas-Producing Reactions.   Chemical processes involving formation of a gaseous product approach completion as the gas escapes from solution. An example is ammonia stripping to remove nitrogen from wastewater. Raising the pH above 10.8 by lime addition converts the ammonia nitrogen in solution from NH_4^+ ions to NH_3 gas.

$$NH_4^+ + OH^- \rightarrow NH_3 + H_2O \quad (2\text{-}18)$$

By the application of wastewater to a countercurrent tower, where air enters at the bottom and exhausts from the top while the wastewater sprinkles down over the tower packing, ammonia gas is stripped out of solution and discharged with the air.

EXAMPLE 2-3

The chemical reaction for removal of calcium hardness by lime precipitation softening is given by Eq. 2-14. What dosage of lime with a purity of 78 percent CaO is required to combine with 70 mg/l of calcium?

Solution

1 mole of $Ca(HCO_3)_2$ (162 g) contains 40.1 g of calcium. Therefore, 70 mg/l of calcium is equivalent to

$$\frac{162}{40.1} \times 70 = 283 \text{ mg/l } Ca(HCO_3)_2$$

56.1 g of CaO combines with 162 g of $Ca(HCO_3)_2$. Therefore, 283 mg/l $Ca(HCO_3)_2$ reacts with

$$\frac{56.1}{162} \times 283 = 98.0 \text{ mg/l CaO}$$

For a purity of 78 percent, the dosage of commercial lime required is

$$\frac{98.0 \text{ CaO}}{0.78} = 126 \text{ mg/l} = 1050 \text{ lb/mil gal}$$

EXAMPLE 2-4

How many pounds of pure sodium hydroxide are required to neutralize an industrial waste with an acidity equivalent to 50 lb of sulfuric acid per million gallons of wastewater?

Solution

Using Eq. 2-17,
2 moles NaOH neutralize 1 mole H_2SO_4
2×40.0 lb NaOH react with 1×98.1 lb H_2SO_4

$$\frac{\text{lb NaOH}}{\text{required}} = \frac{80.0 \text{ lb NaOH}}{98.1 \text{ lb } H_2SO_4} \times 50 \text{ lb } H_2SO_4$$
$$= 40.8 \text{ lb NaOH/mil gal}$$

Alternate Solution

In neutralization reactions one equivalent of base reacts with one of acid. Therefore, by using equivalent weights (Table 2-3) rather than molecular values, the calculation is

$$\frac{\text{lb NaOH}}{\text{required}} = \frac{40.0 \text{ NaOH}}{49.0 \text{ } H_2SO_4} \times 50 \text{ lb } H_2SO_4$$
$$= 40.8 \text{ lb NaOH/mil gal}$$

EXAMPLE 2-5

Theoretically, how many milligrams per liter of reduced iron can be oxidized by one milligram per liter of potassium permanganate?

Solution

From Eq. 2-16, $3Fe^{++}$ ($3 \times 55.8 = 167$) are oxidized for each MnO_4^- reduced ($KMnO_4$ weight is 158)

$$\frac{158 \text{ } KMnO_4}{167 \text{ Fe}} = \frac{1 \text{ mg/l}}{X \text{ mg/l}} \qquad X = 1.06 \text{ mg/l}$$

2-5 CHEMICAL KINETICS

Process kinetics define the rate of a reaction. The kinetics discussed in this section are for irreversible reactions occurring in one phase with uniform distribution of the reactants throughout the liquid. In irreversible reactions, the combination of reactants leads to nearly complete conversion to products, which is true for most processes in water and wastewater treatment. Hence, this assumption is valid for the purpose of interpretation of process kinetics data. Of the reaction types discussed below, first-order kinetics most commonly define the reactions that occur in chemical and biological processing of water and wastewater.

Zero-order reactions proceed at a rate independent of the concentration of any reactant or product; therefore, the disappearance of reactants and appearance of products are linear. Consider the following conversion of a single reactant to a single product:

$$A(\text{reactant}) \rightarrow P(\text{product}) \qquad (2\text{-}19)$$

Using C to represent the concentration of A at any time t, the disappearance of A with respect to time is

$$\frac{dC}{dt} = -k \qquad (2\text{-}20)$$

where $\dfrac{dC}{dt}$ = rate of change in the concentration of A with time

k = reaction-rate constant, per day

By integrating Eq. 2-20 and rearranging the terms,

$$C = C_0 - kt \qquad (2\text{-}21)$$

where C = concentration of A at any time t, milligrams per liter

C_0 = initial concentration of A, milligrams per liter

A plot of the zero-order reaction, represented by Eq. 2-21, is shown by the straight line in Figure 2-2. The negative slope $-k$ means that the concentration of A decreases with time.

First-order reactions proceed at a rate directly proportional to the concentration of one reactant. If one

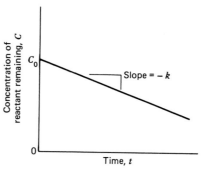

Figure 2-2 A plot of concentration of reactant remaining for a zero-order kinetics reaction, Eq. 2-21.

considers a single reactant converting to a single product (A $\rightarrow$ P), the change in concentration C of A with respect to time is

$$\frac{dC}{dt} = -kC \qquad (2\text{-}22)$$

By integrating Eq. 2-22 and letting C equal C_0 when t equals 0,

$$\log_e \frac{C_0}{C} = kt \text{ or } C = C_0 e^{-kt} \qquad (2\text{-}23)$$

A plot of the first-order reaction, represented by Eq. 2-23, on arithmetic paper is a curve as shown in Figure 2-3a. The slope of a tangent at any point along the curve is equal to $-kC$. Thus, with passage of time from the start of the reaction, the rate decreases with decreasing concentration of A remaining. The first-order reaction can be linearized by graphing on semilogarithmetic paper as shown in Figure 2-3b.

Second-order reactions proceed at a rate proportional to the second power of a single reactant being converted to a single product (2A $\rightarrow$ P). The reaction rate is

$$\frac{dC}{dt} = -kC^2 \qquad (2\text{-}24)$$

The integrated form for Eq. 2-24 is

$$\frac{1}{C} - \frac{1}{C_0} = kt \qquad (2\text{-}25)$$

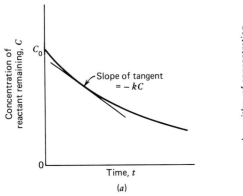
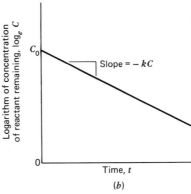

Figure 2-3 Arithmetic and semilogarithmic plots of concentration of reactant remaining for a first-order kinetics reaction, Eq. 2-23.

A plot of the second-order reaction, represented by Eq. 2-25, on arithmetic paper is shown in Figure 2-4. The reciprocal of the concentration of reactant remaining versus time graphs as a straight line with a slope equal to k.

Affect of Temperature on Reaction Rate

The rate of chemical reactions increases with increasing temperature, provided the higher temperature does not affect a reactant or catalyst. The basic relationship between temperature and the reaction-rate constant for chemical reactions was formulated by Arrhenius in 1889. When his original equation is modified and simplified,[2]

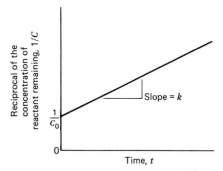

Figure 2-4 A plot of reciprocal of the concentration of reactant remaining for a second-order kinetics reaction, Eq. 2-25.

the common expression for temperature dependence of reaction rates becomes

$$k_2 = k_1 \Theta^{T_2 - T_1} \tag{2-26}$$

where k = reaction-rate constant at respective temperature T, per day
   Θ = temperature coefficient, dimensionless
   T = temperature, $^\circ$C

This equation allows adjustment of the value of a rate constant for a temperature change based on an assumed or experimentally determined temperature coefficient.

Temperature coefficients are relatively constant within the narrow range of temperatures normally associated with operation of water and wastewater processes. Prudent use of a Θ coefficient limits its application to a few degrees (approximately $\pm 5^\circ$C) from the experimental temperature used to establish its value. A commonly used temperature coefficient for both biological and chemical reactions is 1.072, which doubles or halves the reaction rate for a 10°C temperature change. A value of 1.047 doubles or halves the k-rate for a 15°C temperature change.

EXAMPLE 2-6

A chemical reaction that follows first-order kinetics has a measured reaction rate of 50 per day at 20°C. Using a temperature coefficient of 1.072, calculate the k-rate at

$18°$ C for the reactant to decrease in concentration from 100 mg/l to 10 mg/l.

Solution

From Eq. 2-26, the reaction rate at $18°$ C is

$$k_2 = 50(1.072)^{18-20} = 43.2 \text{ per day}$$

Using Eq. 2-23,

$$\log_e \frac{100}{10} = 43.2t$$

$$t = 0.023 \text{ days} = 0.56 \text{ hr}$$

2-6 GAS SOLUBILITY

Most gases either are soluble in water to some degree or react chemically with water. An exception is methane gas (CH_4), commonly referred to as illuminating gas, which does not interact with water to a measurable extent. (Methane is produced in anaerobic decomposition of waste sludge and may be collected and burned for its heat value.) The two major atmospheric gases, nitrogen and oxygen, while not reacting chemically with water, dissolve to a limited degree. The solubility of each gas is directly proportional to pressure it exerts on the water. At a given pressure, solubility of oxygen varies greatly with water temperature and to a lesser degree with salinity (chloride concentration). Dissolved oxygen saturation values, based on normal atmosphere consisting of 21 percent oxygen, for various temperatures and chloride concentrations are listed in Table 2-5. The table footnote explains how to correct oxygen solubility values for barometric pressure. The change of barometric pressure with altitude depends on the rate of decrease in density of the air. Near the earth's surface, it is sufficiently accurate for most engineering computations to say that pressure decreases at a rate of 25 mm (1.0 in.) of mercury per 1000 ft of increased elevation.

Gases may be formed in solution by the decomposition of organic substances in water. Ammonia biochemically released from nitrogenous matter appears in solution as the ammonium radical if the solution is acidic; however, if the water is basic, it remains as ammonia gas. One technique for removing ammonia nitrogen from wastewater uses the principle of raising the pH and then air stripping the ammonia gas. Another gas released from septic wastewater is hydrogen sulfide (H_2S), identifiable by its rotten egg odor. The SH^- group is biochemically produced in solution, converted to H_2S under reduced conditions, and released from solution as a gas. In sewers this can lead to crown corrosion of the pipe by having the H_2S oxidized to sulfuric acid (H_2SO_4) in the condensation moisture hanging on the interior of the pipe.

Chlorine, the most common oxidizing agent used for disinfection of municipal water supplies in the United States, reacts with water as shown in Eq. 2-27. Liquid chlorine is shipped to treatment plants in pressurized steel cylinders. On pressure release, the liquid converts to chlorine gas and is applied to the water. In dilute solution and at pH levels above 4, chlorine combines with water molecules to form hydrochloric acid plus hypochlorous acid which, in turn, yields the hypochlorite ion. The degree of ionization is dependent on pH, with HOCl occurring below 6 and almost complete ionization taking place above pH 9. The pH of a water during disinfection is important, since HOCl is more effective than OCl^- in killing harmful organisms.

$$Cl_2 + H_2O \underset{pH<4}{\overset{pH>4}{\rightleftharpoons}} HCl + HOCl$$

$$\underset{pH<6}{\overset{pH>9}{\rightleftharpoons}} H^+ + OCl^- \qquad (2\text{-}27)$$

Carbon dioxide, although only about 0.03 percent of atmospheric air, plays a major role in water chemistry, since it reacts readily with water, forming bicarbonate and carbonate radicals. CO_2 is absorbed from the air or can be produced by bacterial decomposition of organic matter in the water. Once in solution, it reacts to form carbonic acid.

$$CO_2 + H_2O \rightleftharpoons H_2CO_3$$

$$\underset{pH\ 4.5}{\rightleftharpoons} H^+ + HCO_3^-$$

$$\underset{pH\ 8.3}{\rightleftharpoons} H^+ + CO_3^= \qquad (2\text{-}28)$$

When the pH of the water is greater than 4.5, carbonic

Table 2-5. Saturation Values of Dissolved Oxygen in Water Exposed to Water-Saturated Air Containing 20.90 Percent Oxygen under a Pressure of 760 mm of Mercury[a]

Temperature in° C	Chloride Concentration in Water (mg/l)			Difference per 100 mg Chloride	Temperature in° C	Vapor Pressure (mm)
	0	5000	10,000			
	Dissolved Oxygen (mg/l)					
0	14.6	13.8	13.0	0.017	0	5
1	14.2	13.4	12.6	0.016	1	5
2	13.8	13.1	12.3	0.015	2	5
3	13.5	12.7	12.0	0.015	3	6
4	13.1	12.4	11.7	0.014	4	6
5	12.8	12.1	11.4	0.014	5	7
6	12.5	11.8	11.1	0.014	6	7
7	12.2	11.5	10.9	0.013	7	8
8	11.9	11.2	10.6	0.013	8	8
9	11.6	11.0	10.4	0.012	9	9
10	11.3	10.7	10.1	0.012	10	9
11	11.1	10.5	9.9	0.011	11	10
12	10.8	10.3	9.7	0.011	12	11
13	10.6	10.1	9.5	0.011	13	11
14	10.4	9.9	9.3	0.010	14	12
15	10.2	9.7	9.1	0.010	15	13
16	10.0	9.5	9.0	0.010	16	14
17	9.7	9.3	8.8	0.010	17	15
18	9.5	9.1	8.6	0.009	18	16
19	9.4	8.9	8.5	0.009	19	17
20	9.2	8.7	8.3	0.009	20	18
21	9.0	8.6	8.1	0.009	21	19
22	8.8	8.4	8.0	0.008	22	20
23	8.7	8.3	7.9	0.008	23	21
24	8.5	8.1	7.7	0.008	24	22
25	8.4	8.0	7.6	0.008	25	24
26	8.2	7.8	7.4	0.008	26	25
27	8.1	7.7	7.3	0.008	27	27
28	7.9	7.5	7.1	0.008	28	28
29	7.8	7.4	7.0	0.008	29	30
30	7.6	7.3	6.9	0.008	30	32

[a] Saturation at barometric pressures other than 760 mm (29.92 in.), C'_s is related to the corresponding tabulated values, C_s, by the equation:

$$C'_s = C_s \frac{P - p}{760 - p}$$

where C'_s = solubility at barometric pressure P and given temperature, milligrams per liter
  C_s = saturation at given temperature from table, milligrams per liter
  P = barometric pressure, millimeters
  p = pressure of saturated water vapor at temperature of the water selected from table, millimeters

acid ionizes to form bicarbonate which, in turn, is transformed to the carbonate radical if the pH is above approximately 8.3. Carbon dioxide is very aggressive and leads to corrosion of water pipes; therefore, water supplies with low pH are frequently neutralized with a base to reduce pipe corrosion. On the other hand, alkaline waters with $CO_3^=$ ions are hard and form scale by precipitation of $CaCO_3$. Water treatment practice to lower the pH or soften such a water may be advantageous.

EXAMPLE 2-7

What is the dissolved oxygen concentration in water at 18° C, 800 mg/l chloride concentration, and barometric pressure of 660 mm (4000-ft altitude)?

Solution

From Table 2-5, saturation at 18°C with 800 mg/l chloride concentration is

$$9.5 - 8 \times 0.009 = 9.4 \text{ mg/l}$$

Correcting for a barometric pressure of 660 mm by using the equation in the footnote to Table 2-5,

$$C_s' = 9.4 \frac{660 - 16}{760 - 16} = 8.1 \text{ mg/l}$$

2-7 ALKALINITY

The bicarbonate-carbonate character of a water can be analyzed by slowly adding a strong acid solution to a sample of water and reading resultant changes in pH. This process, called titration, is used to measure the alkalinity of a water. Acidity is measured by titrating with a strong basic solution. If two samples of pure water (pH 7) are titrated with sulfuric acid and sodium hydroxide solutions, respectively, the composite titration curve is as shown by Figure 2-5a. Very small initial additions of either titrant result in significant changes in pH, since addition or withdrawal of hydrogen ions is reflected immediately in changing pH readings. Curve b is a titration curve of a water containing a high initial concentration of carbonate ions prepared by adding sodium carbonate to distilled water. When acid is added,

the majority of the hydrogen ions from acid combine with the carbonate ions to form bicarbonates (Eq. 2-28). The excess hydrogen ions lower the pH gradually until at pH 8.3 all carbonate radicals have been converted to bicarbonates. Additional hydrogen ions reduce the bicarbonates to carbonic acid below pH 4.5. Stirring of the sample at this time results in release of carbon dioxide gas formed from the original carbonates. Figure 2-5c is the reverse titration of curve b where the addition of a strong base results in conversion of carbonic acid to bicarbonate and, in turn, to carbonate; this follows Eq. 2-28 from left to right. Substances in solution, such as the various ionic forms of carbon dioxide that offer resistance to change in pH as acids or bases are added, are referred to as buffers. An understanding of buffering action in water and wastewater chemistry is essential, since many chemical reactions in water treatment and biological reactions in wastewater treatment are pH dependent and rely on pH control. Curve d is a typical titration curve of a relatively hard well water having an initial pH of 7.8. Inflections of the curve near pH values 4.5 and 8.3 indicate that the primary buffer is the bicarbonate-carbonate system. Irregularities and deviations from the ideal bicarbonate buffer curve, Figure 2-5b, are frequently observed in actual titrations of water and wastewater samples; these may be caused by any number of buffering compounds frequently found in low concentrations, such as phosphates ($H_2PO_4^-$, $HPO_4^=$, $PO_4^\equiv$).

Alkalinity of a water is a measure of its capacity to neutralize acids, in other words, to absorb hydrogen ions without significant pH change. Alkalinity is measured by titrating a given sample with 0.02 normal (N) sulfuric acid. (Normality is a method of expressing the strength of a chemical solution. A 1.00 N solution contains one gram of available hydrogen ions, or its equivalent, per liter of solution. A 0.02 N H_2SO_4 solution contains 0.98 gram of pure sulfuric acid, which is 0.02 times the equivalent weight of sulfuric acid.) For highly alkaline samples, the first step is titrating to a pH of 8.3. The second phase, or first in the case of a water with an initial pH of less than 8.3, is titrating to an indicated pH of 4.5. Colorimetric indicators, chemicals that change color at specific pH values, can be used to determine the end points of these titrations if a pH meter is not used. Phenolphthalein turns

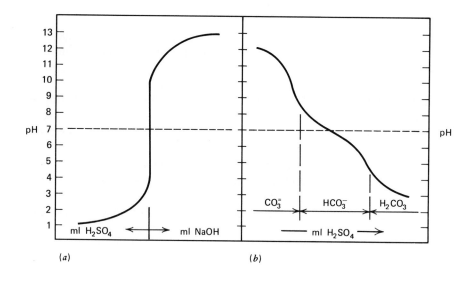

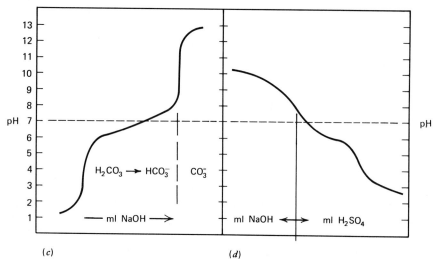

Figure 2-5 Titration curves for pure water (strong acids and bases), carbonate (salt of a weak acid), carbonic acid (weak acid), and typical well water. (*a*) Pure water. (*b*) Carbonate in water. (*c*) Carbonic acid. (*d*) Well water.

from pink to colorless at pH 8.3, and mixed bromocresol green-methyl red indicator changes from bluish-gray at pH 4.8 to light pink at pH 4.6. Methyl orange indicator, a substitute for mixed bromocresol green-methyl red solution, changes from orange at pH 4.6 to pink at 4.0. That part of the total alkalinity above 8.3 is referred to as phenolphthalein alkalinity. Alkalinity is conventionally expressed in terms of milligrams $CaCO_3$ per liter and is calculated by Eq. 2-29. Note, if the water sample is 100 ml and normality of acid 0.02 N, 1 ml of titrant represents 10 mg/1 alkalinity.

Alkalinity as mg/1 $CaCO_3$

$$= \frac{\text{ml titrant} \times \text{normality of acid} \times 50,000}{\text{ml sample}} \quad (2\text{-}29)$$

Figure 2-6 is a graphical representation of the various forms of alkalinity found in water samples. This diagram can be related to Eq. 2-28. If pH of a sample is less than 8.3, the alkalinity is all in the form of bicarbonate. Samples containing carbonate and bicarbonate alkalinity have a pH greater than 8.3, and titration to the phenol-

phthalein end point represents one half of the carbonate alkalinity. If the milliliters of phenolphthalein titrant equal the milliliters of titrant to the mixed bromcresol-methyl red end point, all of the alkalinity is in the form of carbonate. Any alkalinity in excess of this amount is due to hydroxide (OH^-). If abbreviations of P for the measured phenolphthalein alkalinity and T for the total alkalinity are used, the following relationships define the bar graphs in Figure 2-6: (a) Hydroxide $= 2P - T$ and carbonate $= 2(T - P)$; (b) Carbonate $= 2P = T$; (c) Carbonate $= 2P$ and bicarbonate $= T - 2P$; and (d) Bicarbonate $= T$.

EXAMPLE 2-8

A 100-ml water sample is titrated for alkalinity by using 0.02 N sulfuric acid. To reach the phenolphthalein end point requires 3.0 ml, and an additional 12.0 ml is added for the mixed bromocresol green-methyl red color change. Calculate the phenolphthalein and total alkalinities. Give the ionic forms of the alkalinity present.

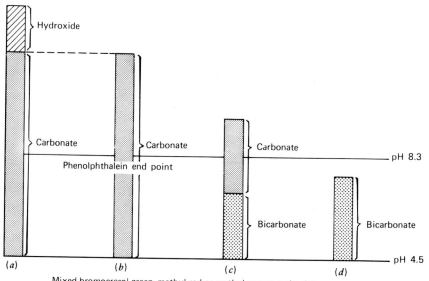

Figure 2-6 Graphical representation of various forms of alkalinity in water samples related to titration end points.

Solution

Using Eq. 2-29,

$$\frac{\text{Phenolphthalein}}{\text{alkalinity}} = \frac{3.0 \times 0.02 \times 50,000}{100}$$

$$= 30 \text{ mg/l (as CaCO}_3)$$

$$\frac{\text{Total}}{\text{alkalinity}} = \frac{15.0 \times 0.02 \times 50,000}{100}$$

$$= 150 \text{ mg/l (as CaCO}_3)$$

Referring to Figure 2-6,

Alkalinity as carbonate $(CO_3^=) = 2 \times 30 = 60$ mg/l

Alkalinity as bicarbonate $(HCO_3^-) = 150 - 60$

$$= 90 \text{ mg/l}$$

2-8 COLLOIDS AND COAGULATION

A large variety of turbidity-producing substances found in polluted waters do not settle out of solution, for example, color compounds, clay particles, microscopic organisms, and organic matter from decaying vegetation or municipal wastes. These particles, referred to as colloids, range in size from 1 to 500 millimicrons (nanometers) and are not visible when using an ordinary microscope. A colloidal dispersion formed by these particles is stable in quiescent water, since the individual particles have such a large surface area relative to their weight that gravity forces do not influence their suspension. Colloids are classified as hydrophobic (water hating) and hydrophilic (water loving).

Hydrophilic colloids are stable because of their attraction for water molecules, rather than because of the slight charge that they might possess. Typical examples are soap, soluble starch, synthetic detergents, and blood serum. Because of their affinity for water, they are not as easy to remove from suspension as hydrophobic colloids, and coagulant dosages of 10 to 20 times the amount used in conventional water treatment are frequently necessary to coagulate hydrophilic materials.

Hydrophobic particles, possessing no affinity for water, depend on electrical charge for their stability in suspension. The bulk of inorganic and organic matter in a turbid natural water is of this type. The forces acting on hydrophobic colloids are illustrated in Figure 2-7a. Individual particles are held apart by electrostatic repulsion forces developed by positive ions adsorbed onto their surfaces from solution. An analogy is the repulsive force that exists between the common poles of two bar magnets. The magnitude of the repulsive force developed by the charged double layer of ions attracted to a particle is referred to as the zeta potential.

A natural force of attraction exists between any two masses (Van der Waals force). Random motion of colloids (Brownian movement), caused by bombardment of water molecules, tends to enhance this physical force of attraction in pulling the particles together. However, a colloidal suspension remains dispersed indefinitely when the forces of repulsion exceed those of attraction and the particles are not allowed to contact.

The purpose of chemical coagulation in water treatment is to destabilize suspended contaminants such that the particles contact and agglomerate, forming flocs that drop out of solution by sedimentation. Destabilization of hydrophobic colloids is accomplished by addition of chemical coagulants, for instance, salts of aluminium and iron. Highly charged hydrolyzed metal ions produced by these salts in solution reduce the repulsive forces between colloids by compressing the diffuse double layer surrounding individual particles (Figure 2-7b). With the forces of repulsion suppressed, gentle mixing results in particle contact, and the forces of attraction cause particles to stick to each other, producing progressive agglomeration. Coagulant aids may be used to enhance the process of flocculation. For example, organic polymers provide bridging between particles by attaching themselves to the absorbant surfaces of colloids, building larger flocculated masses (Figure 2-7c).

Removal of turbidity by coagulation depends on the type of colloids in suspension; the temperature, pH, and chemical composition of the water; the type and dosage of coagulants and aids; and the degree and time of mixing provided for chemical dispersion and floc formation. Although in chemistry the term coagulation means the destabilization of a colloidal dispersion by suppressing the double layer (Figure 2-7a), and flocculation refers to aggregation of the particles, engineers have not

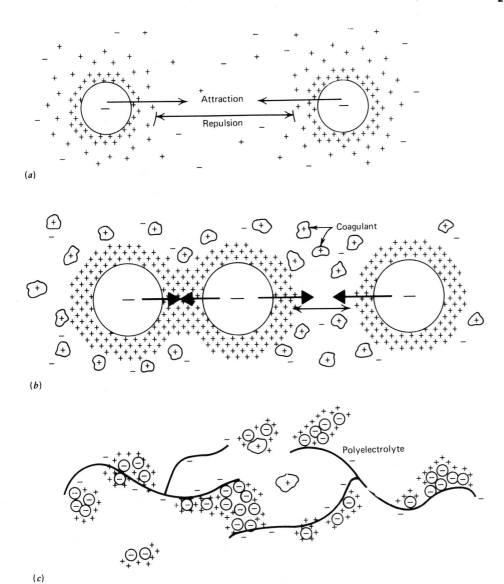

Figure 2-7 Schematic actions of colloids and coagulation. (*a*) Forces acting on hydrophobic colloids in stable suspension. (*b*) Compression of the double-layer charge on colloids (destabilization) by addition of chemical coagulants. (*c*) Agglomeration resulting from coagulation with metal salt and polymer aid.

traditionally restricted the use of these terms to describe the chemical mechanisms only. More frequently, coagulation and flocculation are associated with the physical processes used in chemical treatment. Mixing, involving violent agitation, is used to dissolve and disperse coagulant chemicals throughout the water being treated. Flocculation is a slow mixing process, following chemical dispersion, during which the destabilized particles form into well-developed flocs of sufficient size to settle. The word coagulation is commonly used to describe the entire mixing and flocculation operation. Actual units used in chemical treatment may be constructed in series (mixing—flocculation—sedimentation) or may be housed in a single compartmented tank. An in-line system, as shown in Figure 7-8, normally provides about 1 min for flash mixing, 30 min flocculation, and 2 to 4 hr sedimentation, followed by filtration to remove non-settleable matter. A flocculator-clarifier (Figure 7-10) mixes raw water and applied chemicals with previously flocculating solids in the center mixing compartment, thus contacting settleable solids with untreated water and fresh chemicals. Solids settled in the peripheral zone automatically return to the mixing area; excess solids buildup is withdrawn from the bottom for disposal.

2-9 ORGANIC COMPOUNDS

All organic compounds contain carbon atoms connected to each other in chain or ring structures with other elements attached. Major components are carbon, hydro-

gen, and oxygen; minor elements include nitrogen, phosphorus, sulfur, and certain metals. Each carbon exhibits four connecting bonds. Organics are derived from nature, for example, plant fibers and animal tissues, produced by synthesis reactions producing rubber, plastics, and the like, and through fermentation processes, for instance, alcohols, acids, antibiotics, and others. In contrast to inorganic compounds, organic substances are usually combustible, high in molecular weight, only sparingly soluble in water reacting as molecules rather than ions, and a source of food for animal consumers and microbial decomposers.

The molecular formula for an inorganic compound is specific; however, in organic chemistry an empirical formula may represent more than one compound. For example, the formula C_2H_6O may be arranged to represent the following two substances:

$$\begin{array}{cc}
\begin{array}{ccc}
 & H & H \\
 & | & | \\
H - & C - & C - OH \\
 & | & | \\
 & H & H
\end{array}
&
\begin{array}{ccc}
H & & H \\
| & & | \\
H - C - & O & - C - H \\
| & & | \\
H & & H
\end{array}
\end{array}$$

The one on the left is ethyl alcohol while the one on the right is dimethyl ether. Compounds having the same molecular formula but that differ in structural formulas are called isomers. Because of isomerism and the frequency of ring structures, organic compounds are normally given as graphic formulas rather than as molecular ones. For ease of printing, formulas not containing rings are frequently written on a single line. For example, ethyl alcohol and dimethyl ether become

$$CH_3CH_2OH \quad \text{and} \quad CH_3 - O - CH_3$$

Hydrocarbons

Saturated hydrocarbons (paraffins and alkanes) contain only the elements of carbon and hydrogen with single bonds between carbon atoms. Methane (CH_4), the simplest hydrocarbon, is a gas produced in the anaerobic decomposition of waste sludge. When mixed with approximately 90 percent air, it is highly explosive and can be used as a fuel for gas engines or sludge heaters.

Table 2-6. Saturated Hydrocarbons

Name	Formula	Radical	Formula
Methane	CH_4	Methyl	CH_3-
Ethane	CH_3CH_3	Ethyl	CH_3CH_2-
Propane	$CH_3CH_2CH_3$	n-Propyl	$CH_3CH_2CH_2-$
		Isopropyl	$CH_3\diagdown\!\!\!\diagup CH-$ CH_3
n-Butane	$CH_3CH_2CH_2CH_3$	n-Butyl	$CH_3CH_2CH_2CH_2-$

Short-chain paraffin hydrocarbons can be obtained from petroleum by fractional distillation. Propane is available in high pressure cylinders for a heating fuel, Eq. 2-30.

$$CH_3CH_2CH_3 + 5O_2 \rightarrow 3CO_2 + 4H_2O + \text{energy} \quad (2\text{-}30)$$

Names and formulas of one to four carbon saturated hydrocarbons along with the radicals of these parent compounds are given in Table 2-6.

Unsaturated hydrocarbons (olefins) are distinguished from paraffins by the presence of multiple bonds between some carbon atoms. For example, ethene (ethylene) and acetylene are, respectively,

$$H-\underset{\underset{H}{|}}{C}=\underset{\underset{H}{|}}{C}-H \qquad H-C\equiv C-H$$

The multiple bonds between carbons displace hydrogen atoms, creating units containing fewer hydrogens than could be attached. Hence, the term "unsaturated" is used. Natural vegetable oils containing a large number of unsaturated bonds are liquids at room temperature. Popular vegetable shortenings available as solid fats are commercially produced from oils through the process of hydrogenation—the addition of hydrogen gas under controlled conditions. Reducing the number of unsaturated bonds increases the melting point, converting an oil to solid fat.

The parent compound of aromatic hydrocarbons is benzene. It is a six-carbon ring with double bonds between alternate atoms. Frequently the carbon atoms

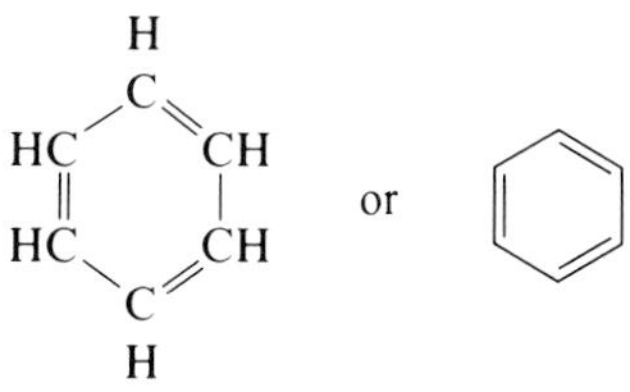

Benzene

are not shown in writing the graphic formula. Benzene is used in the manufacture of a wide variety of commercial products including explosives, insecticides, certain plastics, solvents, and dyes.

Alcohols

These are formed from hydrocarbons by replacing one or more hydrogen atoms by hydroxyl groups (—OH). Names and formulas of three primary alcohols are listed in Table 2-7. Methanol is manufactured synthetically by a catalytic process from carbon monoxide and hydrogen. It is used extensively in manufacturing organic compounds, such as formaldehyde, solvents, and fuel additives. Ethyl alcohol for beverage purposes is produced by fermentation of a variety of natural materials—corn, wheat, rice, and potatoes. Industrial ethanol is produced from fermentation of waste solutions containing sugars, for example, blackstrap molasses, a residue resulting from the purification of cane sugar. Propanol has two isomers, the more common being isopropyl alcohol widely used by industry and sold as a medicinal rubbing alcohol. The

Table 2-7. Primary Alcohols

Common Name	Proper Name	Formula
Methyl alcohol	Methanol	CH_3OH
Ethyl alcohol	Ethanol	CH_3CH_2OH
n-Propyl alcohol	1-Propanol	$CH_3CH_2CH_2OH$
Isopropyl alcohol	2-Propanol	$\begin{matrix} CH_3 \\ \\ CH_3 \end{matrix}\!\!>\!CHOH$

three primary alcohols listed in Table 2-7 have boiling points less than 100° C, and they are miscible with water.

The derivative of benzene containing one hydroxyl group, known as phenol, has a molecular formula of C_6H_5OH and a graphic representation of

H
O

The formula and -ol name ending indicate the characteristics of an alcohol, but phenol, commonly known as carbolic acid, ionizes in water-yielding hydrogen ions and exhibits features of an acid. It occurs as a natural component in wastes from coal-gas petroleum industries and in a wide variety of industrial wastes where phenol is used as a raw material. Phenol is a strong toxin that makes these wastes particularly difficult to treat in biological systems. Phenols also impart undesirable taste to water at extremely low concentrations.

Aldehydes and Ketones

These are compounds containing the carbonyl group. Formaldehyde, in addition to being a preservative for biological specimens, is used in producing a variety of plastics and resins. Acetone (dimethyl ketone) is a good solvent of fats and is a common cleaning agent for laboratory glassware.

Carbonyl group   Formaldehyde   Acetone

Carboxylic Acids

All organic acids contain the carboxyl group written as

$$\overset{\displaystyle O}{\overset{\|}{-C}}-OH \qquad or \qquad -COOH$$

This is the highest state of oxidation that an organic radical can achieve. Further oxidation results in the formation of carbon dioxide and water.

$$CH_4 \xrightarrow{+O} CH_3OH \xrightarrow{-2H} H_2C{=}O \xrightarrow{+O}$$

Hydrocarbon   Alcohol   Aldehyde

$$HCOOH \xrightarrow{+O} CO_2 + H_2O$$

Acid

Names, formulas, and ionization constants for the five simplest acids are given in Table 2-8. Acids through nine carbon are liquids but those with longer chains are greasy solids and, hence, the common name fatty acid. Since

Table 2-8. Carboxylic Acids

Common Name	Proper Name	Formula	Ionization Constant, K
Formic	Methanoic	$HCOOH$	0.000,214
Acetic	Ethanoic	CH_3COOH.	0.000,018
Propionic	Propanoic	CH_3CH_2COOH	0.000,014
n-Butyric	Butanoic	$CH_3CH_2CH_2COOH$	0.000,015
Valeric	Pentanoic	C_4H_9COOH	0.000,016
Caproic	Hexanoic	$C_5H_{11}COOH$	Almost insoluble

organic acids are weak and ionize poorly, equilibrium constants are used to express their degrees of ionization. (Refer to Section 2-4.) The values of the ionization constants in Table 2-8 are for dilute solutions. For a typical organic acid, consider the dissociation of acetic acid.

$$CH_3-\overset{\overset{\displaystyle O}{\|}}{C}-OH \rightleftarrows CH_3-\overset{\overset{\displaystyle O}{\|}}{C}-O^- + H^+ \quad (2\text{-}31)$$

The ionization (equilibrium) constant is equal to the molar concentration of acetate ion times that of hydrogen ion divided by the molar quantity of molecular acetic acid in solution, Eq. 2-32. The K-value for acetic acid is 0.000,018 at 25° C, which means that in moderately dilute solution only 18 out of every million acetic acid molecules in water dissociate releasing hydrogen ions.

$$K = \frac{[CH_3-\overset{\overset{\displaystyle O}{\|}}{C}-O^-][H^+]}{[CH_3-\overset{\overset{\displaystyle O}{\|}}{C}-OH]} = 0.000,018 \quad (2\text{-}32)$$

Formic, acetic, and propionic acids have sharp penetrating odors, while butyric and valeric have extremely disagreeable odors associated with rancid fats and oils. Anaerobic decomposition of long-chain fatty acids results in production of two- and three-carbon acids that are then converted to methane and carbon dioxide gas in digestion of waste sludge.

Basic compounds react with acids to produce salts. For example, NaOH reacts with acetic acid to produce sodium acetate. Soaps are salts of long-chain fatty acids.

Table 2-9. Functional Groups of Organic Compounds

Functional Group	Type of Compound	Proper Name Ending	Example	Common Name	Proper Name
$-\overset{\overset{\displaystyle \|}{}}{\underset{\underset{\displaystyle \|}{}}{C}}-\overset{\overset{\displaystyle \|}{}}{\underset{\underset{\displaystyle \|}{}}{C}}-$	Saturated hydrocarbon	-ane	CH_3CH_3	Ethane	Ethane
$-C=C-$	Olefin	-ene	$H_2C=CH_2$	Ethylene	Ethene
$-C\equiv C-$	Acetylene	-yne	$H-C\equiv C-H$	Acetylene	Ethyne
$-OH$	Alcohol	-ol	CH_3CH_3-OH	Ethyl alcohol	Ethanol
$-O-$	Ether	—	$CH_3CH_2-O-CH_2CH_3$	Ethyl ether	Ethoxyethane
$-\overset{\overset{\displaystyle O}{\|}}{C}-H$	Aldehyde	-al	$CH_3-\overset{\overset{\displaystyle O}{\|}}{C}-H$	Acetaldehyde	Ethanal
$-\overset{\overset{\displaystyle O}{\|}}{C}-$	Ketone	-one	$CH_3-\overset{\overset{\displaystyle O}{\|}}{C}-CH_3$	Acetone	Propanone
$-\overset{\overset{\displaystyle O}{\|}}{C}-OH$	Carboxylic acid	-oic acid	$CH_3-\overset{\overset{\displaystyle O}{\|}}{C}-OH$	Acetic acid	Ethanoic acid
$-N\overset{\diagup H}{\diagdown H}$	Amine	—	CH_3-NH_2	Methylamine	Aminomethane

Sodium acetate   Sodium palmitate (a soap)

Other derivatives of carboxylic acids include esters, such as ethyl acetate, and amides, for example, urea.

Ethyl acetate   Urea (diamide)

Table 2-9 is a summary of functional groups of organic compounds. These data are useful in identifying an organic compound based on a formula or name. Amines, shown at the bottom of the table, may be regarded as derivatives of ammonia in which hydrogen atoms have been replaced by carbon chains or rings.

2-10 ORGANIC MATTER IN WASTEWATER

Biodegradable organic matter in municipal wastewater is classified into three major categories: carbohydrates, proteins, and fats. Carbohydrates consist of sugar units containing the elements of carbon, hydrogen, and oxygen. A single sugar ring is known as a monosaccharide; few of these occur naturally. Disaccharides are composed of two monosaccharide units. Sucrose, common table sugar, is glucose plus fructose, while the most prevalent sugar in milk is lactose consisting of glucose plus galactose. Polysaccharides, long chains of sugar units, can be divided into two groups: readily degradable starches abundant in potatoes, rice, corn, and other edible plants; and cellulose found in wood, cotton, paper, and similar plant tissues. Cellulose compounds degrade biologically at a much slower rate than starches.

Proteins in the simple form are long strings of amino acids containing carbon, hydrogen, oxygen, nitrogen, and phosphorus. They form an essential part of all living tissue, and constitute a diet necessity for all higher animals. The following is a small section of a protein showing four amino acids connected together.

Sucrose   Lactose

Section of a cellulose molecule

Fats refer to a variety of biochemical substances that have the common property of being soluble to varying degrees in organic solvents (ether, ethanol, acetone, and hexane) while being only sparingly soluble in water. Because of their limited solubility, degradation by microorganisms is at a very slow rate. A simple fat is a triglyceride composed of a glycerol unit with short- or long-chain fatty acids attached. The formula for glycerol oleobutyropalmitate, found in butter fat, is

$$H_2C-O-\overset{\overset{O}{\|}}{C}-C_3H_7$$
$$H-\overset{|}{\underset{|}{C}}-O-\overset{\overset{O}{\|}}{C}-C_{15}H_{31}$$
$$H_2C-O-\overset{\overset{O}{\|}}{C}(CH_2)_7C\underset{H}{=}\overset{}{C}(CH_2)_7CH_3$$

The majority of carbohydrates, fats, and proteins in wastewater are in the form of large molecules that cannot penetrate the cell membrane of microorganisms. Bacteria, in order to metabolize high-molecular-weight substances, must be capable of breaking down the large molecules into diffusible fractions for assimilation into the cell. The first step in bacterial decomposition of organic compounds is hydrolysis of carbohydrates into soluble sugars, proteins into amino acids, and fats to short fatty acids. Further aerobic biodegradation results in the formation of carbon dioxide and water. By anaerobic digestion, decomposition in the absence of oxygen, the end products are organic acids, alcohols, and other liquid intermediates as well as gaseous entities of carbon dioxide, methane, and hydrogen sulfide.

Of the organic matter in wastewater, 60 to 80 percent is readily available for biodegradation. Several organic compounds, such as cellulose, long-chain saturated hydrocarbons, and other complex compounds, although available as a bacterial substrate, are considered nonbiodegradable because of the time and environmental limitations of biological wastewater treatment systems. Detergents and pesticides are also a part of the 20 to 40 percent inert fraction of wastes, since they contain organic ring structures that are difficult to decompose. A common

group of surfactants are the sulfonated alkyl benzenes (ABS). In the past, ABS units were derived largely from polymers of propylene, and the alkyl group was highly branched. This structure, very resistant to biological attack, resulted in contamination of both surface- and groundwater supplies with an objectionable foaming property. These materials are now made largely from normal (straight-chain) hydrocarbons, and thus the unbranched carbon chains allow ease in bacterial decomposition.

One of the most versatile and effective chlorinated pesticides for the control of objectionable insects is DDT, a chlorinated aromatic compound. An example of a more recent insecticide is endrin. Both of these highly toxic compounds may appear in waters through direct application or through percolation and runoff from treated areas, as well as through industrial wastes from their manufacture. In general, the chlorinated pesticides are the most resistant to biological degradation and may persist for months or years in soils and water, and may tend to concentrate in aquatic plants and animals. They are also resistant to removal by conventional wastewater and water treatment processes.

DDT

Endrin

Although some waste odors are inorganic compounds, for example, hydrogen sulfide gas, many are caused by volatile organic compounds, such as mercaptans (organics with SH groups) and butyric acid. Industries may produce a variety of medicinal odors in the processing of raw materials. Surface-water supplies plagued with blooms of blue-green algae have fishy or pigpen odors. Actually very little is known about the origin or characteristics of the organic compounds that produce disagreeable odors. Each case of a smelly wastewater treatment plant or disagreeable taste in a water supply must be investigated separately, keeping in mind that the cause may be anaerobic decomposition, industrial chemicals, or growths of obnoxious microorganisms.

2-11 LABORATORY CHEMICAL ANALYSES

Standard Solutions

Solutions of chemical reagents with known concentrations are commonly used in laboratory testing. A 1 molar (1 M) solution contains 1 gram molecular weight of chemical per liter of solution. For example, a molar solution of NaOH would contain 40.0 grams of the pure salt dissolved in a total solution volume of 1 liter. A 1 normal (1 N) solution contains 1 gram equivalent weight of reagent per liter of solution. A normal NaOH solution would contain the same amount of salt as the 1 molar solution, since the molecular weight and equivalent weight are identical (See Table 2-3). However, in the case of sulfuric acid, a 1 M solution would contain 98.1 g/l whereas a 1 N solution would contain 49.0 g/l.

Solution concentrations can also be expressed in terms of milligrams per liter (mg/l), milliequivalents per liter (meq/l), grains per gallon (gpg), pounds per million gallons (lb/mil gal), and pounds per million cubic feet (lb/1,000,000 cu ft). The relationship between milliequivalents per liter and milligrams per liter is given in Eq. 2-1. One mg/l, being 1 part by weight per 1,000,000 parts by weight, is equivalent to 8.34 lb/mil gal, since the weight of 1 gal of water is 8.34 lb. Also, since 1 cu ft of water weighs 62.4 lb, 1 mg/l equals 62.4 lb/1,000,000 cu ft. By definition, 1 lb contains 7000 grains; hence, 1 mg/l equals 0.0584 gpg, or 1 gpg equals 17.1 mg/l. Equation 2-33 summarizes these concentration factors.

$$1.00 \text{ mg/l} = 8.34 \text{ lb/mil gal}$$

$$= \frac{62.4}{1,000,000 \text{ cu ft}} = 0.0584 \text{ gpg} \quad (2\text{-}33)$$

Hydrogen Ion Concentration

The term pH, defined by Eq. 2-4, is used to express the intensity of an acid or alkaline solution. Measurement of hydrogen ion concentration is most frequently accomplished by using a meter that reads directly in pH units, as shown in Figure 2-8. A glass electrode, in association with a calomel electrode, dipped into the solution detects hydrogen ions. Prepared standard solutions are used to calibrate the meter.

Colorimetric techniques are available for determining pH. Special indicator solutions when added to water reveal pH by color development. A viewer equipped with a disk of standard colors can be used to estimate pH by the color of a water sample. Alternately, special paper can be purchased that varies in color according to pH when dipped into water. Unfortunately, although these techniques require less expensive equipment, their application is limited to clear water samples where high accuracy is not needed. Color and turbidity in the sample can interfere with colorimetric comparisons, leading to erroneous readings.

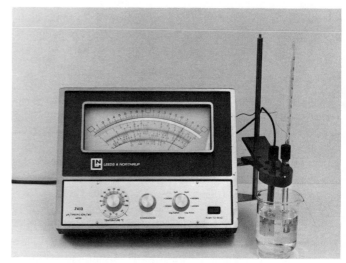

Figure 2-8 A pH meter provided with glass and calomel electrodes for hydrogen ion measurements.

Alkalinity and Acidity

Section 2-7 discusses alkalinity and gives the common forms (Figure 2-6), equation for calculation (Eq. 2-29), and sample computation (Example 2-8). A typical laboratory apparatus for titration analysis is shown in Figure 2-9. The container with a measured sample volume is stirred while a standardized titrant solution in a calibrated burette is dispensed into it. End point of titration is determined either by a colorimetric indicator added to the sample or by use of a glass pH electrode and meter.

The primary need for alkalinity measurements is related to water processing, although this measurement is routinely included in any water analysis. Since lime in water softening and coagulants for turbidity removal react with alkalinity, it is essential that this parameter be monitored in both raw and treated water to ensure optimum dosages of treatment chemicals. Buffering action for control of pH in biological systems is the carbon dioxide-bicarbonate system; therefore, alkalinity measurements are performed on aerating wastewaters and digesting sludges in evaluating environmental conditions.

Acidity is a measure of carbon dioxide and other acids in solution. Analysis is by titration, similar to that used in the determination of alkalinity. Strong inorganic-acid acidity exists below pH 4.5, while carbon dioxide acidity (carbonic acid) is between pH 4.5 and 8.3 (Figure 2-5c). A measured volume of water sample is titrated with 0.02 N NaOH from the existing pH to a pH of 8.3, noting the amount of milliliters of titrant used below pH 3.7 and that neutralized between 3.7 and 8.3. Acidity, conventionally expressed in terms of milligrams $CaCO_3$ per liter, is calculated by Eq. 2-34. Methyl orange acidity is that below pH 4.5, phenolphthalein exists in the pH range of 3.7 to 8.3, and total acidity is the sum of these two values.

Acidity as mg/l $CaCO_3$

$$= \frac{\text{ml titrant} \times \text{normality of base} \times 50,000}{\text{ml sample}} \quad (2\text{-}34)$$

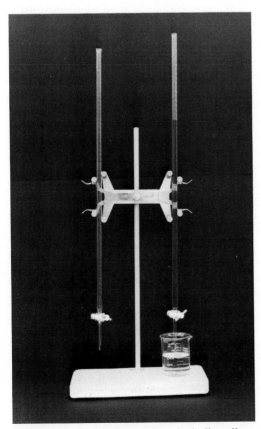

Figure 2-9 Titration apparatus including dispensing burette stand and sample container.

Hardness

Hardness is caused by multivalent metallic cations; those most abundant in natural waters are calcium and magnesium. Hard waters from both underground and surface supplies are most common in areas having extensive geological formations of limestone. Although satisfactory for human consumption, Ca^{++} and Mg^{++} precipitate soap, reducing its cleansing action, and cause scale [$CaCO_3$ and $Mg(OH)_2$] in water distribution mains and hot-water heaters. Even with the advent of synthetic detergents and methods for dealing with scaling problems, partial softening of extremely hard waters by municipal treatment plants is desirable. Waters with less than 50 mg/l hardness are considered soft, up to 150 mg/l moderately hard, and in excess of 300 mg/l very hard.

The most common testing method for hardness is the EDTA titrimetric method. Disodium ethylenediaminetetraacetate (Na_2EDTA) forms stable complex ions with Ca^{++}, Mg^{++}, and other divalent cations causing hardness, thus removing them from solution. If a small amount of dye, Eriochrome Black T, is added to the water containing hardness ions at pH 10, the solution becomes wine red; in absence of hardness the color is blue. The test procedure uses a titration apparatus like that shown in Figure 2-9. The burette is filled with a standardized EDTA solution for dispensing into a measured water sample containing indicator dye. EDTA added complexes hardness ions until all have been removed from solution and the water color changes from wine red to blue, indicating end of titration, Eq. 2-35.

$$Ca^{++} + Mg^{++} + EDTA$$
$$\text{Wine red color}$$

$$\xrightarrow{pH = 10} Ca \cdot EDTA + Mg \cdot EDTA \quad (2\text{-}35)$$
$$\text{Blue color}$$

Hardness is conventionally expressed in the units of mg/l as $CaCO_3$ and is calculated from the laboratory data by the following:

Hardness as mg/l $CaCO_3$

$$= \frac{\text{ml titrant} \times CaCO_3 \text{ equivalent of EDTA} \times 1000}{\text{ml sample}}$$

$$(2\text{-}36)$$

Iron and Manganese

These metals at very low concentrations are highly objectionable in water supplies for domestic or industrial use. Traces of iron and manganese can cause staining of bathroom fixtures, can impart brownish color to laundered clothing, and can affect the taste of water. Groundwaters devoid of dissolved oxygen can contain appreciable amounts of ferrous iron (Fe^{++}) and manganous manganese (Mn^{++}), which are soluble (invisible) forms. When exposed to oxidation, they are transformed to the stable insoluble ions of ferric iron (Fe^{+++}) and manganic manganese (Mn^{++++}), giving water a rust

color. Supplies drawn from anaerobic bottom water of reservoirs, or rivers that have contacted natural formations of iron- and manganese-bearing rock, can contain both reduced and oxidized forms, the latter being complexed frequently with organic matter.

The most popular methods of determining iron and manganese in water use colorimetric procedures. Color development as a technique in testing has the major advantage of being highly specific for the ion involved and generally requires minimum pretreatment of the water sample. Concentration of the substance being evaluated is directly related to intensity of the color developed. The simplest method of color comparison is to make up a set of standards by using known concentrations in a series of glass tubes, commonly referred to as Nessler tubes. After the sample has been prepared, it is placed in an identical glass tube and its color is compared visually against the standards. Or a small comparator (viewing device) can be used to measure color development in a sample against a disk wheel that contains small colored glasses shaded to serve as color standards.

Photoelectric colorimeters of the type shown in Figure 2-10 reduce the human error involved in making color distinctions by eye. In this apparatus, light from an ordinary bulb is transmitted through a lens system and colored filter to obtain a monochromatic light beam. Selected colored filters are used in different chemical tests. Percentage of transmission of the light beam through the sample is measured by using a photoelectric cell connected to a galvanometer. Scales developed for each particular chemical analysis can be positioned behind the galvanometer needle such that the result can be read directly in terms of concentration of the ion that is being tested.

The most accurate laboratory apparatus used in colorimetric measurements is the spectrophotometer. It operates in a fashion similar to the photoelectric colorimeter, but the monochromatic light is developed by a precise prism or grating system that allows the selection of any wavelength in the visible spectrum. Figure 2-11 shows a picture and schematic diagram of a spectrophotometer.

The phenanthroline method is preferred for measuring iron in water. The first step is to ensure that all iron in

Figure 2-10 Photoelectric colorimeter (filter photometer) with built-in color filter wheel for selecting a desired wavelength of a monochromatic light beam. Colorimeter is provided with a pH electrode, a thermometer, reagent and sample bottles, and a variety of direct-reading meter scales for different chemical tests. (Courtesy of Hach Company, Loveland, Colorado.)

solution is in the ferrous state by treating the sample with hydrochloric acid and hydroxylamine as the reducing agent as follows:

$$4Fe^{+++} + 2NH_2OH$$

$$\xrightarrow{pH = 3.2} 4Fe^{++} + N_2O + H_2O + 4H^+ \quad (2\text{-}37)$$

Three molecules of 1,10-phenanthroline sequester each atom of ferrous iron to form an orange-red complex.

$$3 \quad \text{[1,10-phenanthroline]} + Fe^{++} \longrightarrow \left[\begin{array}{c} N-N \\ Fe \\ N-N \quad N-N \end{array} \right]^{++}$$

1,10-phenanthroline   Orange-red complex

A spectrophotometer set for a monochromatic light beam of 510 nanometers is used to measure the light absorption of the colored solution. Concentration of iron in the sample is determined from a percentage transmission versus iron concentration calibration curve prepared from a series of standard iron solutions previously tested.

Manganese is also determined by a colorimetric procedure—the persulfate method. Chloride interference is overcome by adding mercuric sulfate to form relatively insoluble $HgCl_2$. All manganese in the sample is then converted to permanganate by persulfate in the presence of silver as a catalyst.

$$Mn^{++} + S_2O_8^= \xrightarrow{Ag^+} MnO_4^- + SO_4^= \quad (2\text{-}38)$$

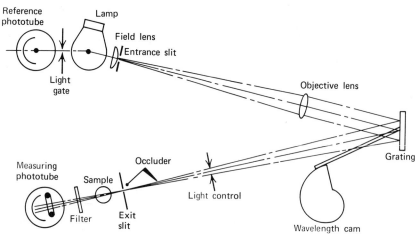

Figure 2-11 Spectrophotometer and optical system schematic. (Courtesy of Bausch and Lomb Inc., Analytical Systems Division.)

The concentration of permanganate ion produced and its resultant color represent the amount of manganese in the sample.

Trace Metals

The toxic trace metals specified in drinking water standards and several other common elements, such as calcium and sodium, are detected by atomic absorption spectrophotometry. The concentration of an element in solution is determined by measuring the quantity of light of a specific wavelength absorbed by atoms of the element released in a flame. As represented graphically in Figure 2-12, an atomic absorption spectrophotometer consists of an atomizer-burner to convert the element in solution to free atoms in an air-acetylene flame, a monochromator (prism and slit) to disperse and iso-late the lightwaves emitted, and a photomultiplier to

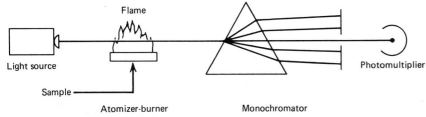

Figure 2-12 Simplified schematic diagram of an atomic absorption spectrophotometer.

detect and amplify the light passing through the monochromator. The light source is a lamp with a cathode formed of the same element being determined, since each element has characteristic wavelengths that are readily absorbed. The light passing through the sample is separated in the monochromator into its component wavelength. The photomultiplier then receives only the isolated resonance wavelength and any absorption of its light by the sample atoms. After the proper lamp for the test element has been inserted, the intensity of the light is measured passing through the unrestricted flame. Then, the sample is introduced into the flame and the concentration of the element in the sample determined by the decrease in light intensity.

Color

Hues in water may result from natural minerals, such as iron and manganese, vegetable origins—humus material and tannins—or colored wastes discharged from a variety of industries including mining, refining, pulp and paper, chemicals, and food processing. True color of water is considered to be only that attributable to substances in solution after removal of suspended materials by centrifuging or filtration. In domestic water, color is undesirable aesthetically and may dull clothes or stain fixtures. Stringent color limits are required for water use in many industries—beverage production, dairy and food processing, paper manufacturing, and textiles.

Standard color solutions are composed of potassium chloroplatinate (K_2PtCl_6) tinted with small amounts of cobalt chloride. The color produced by 1 mg/l of platinum in combination with $\frac{1}{2}$ mg/l of metallic cobalt is

taken as 1 standard color unit. The yellow-brownish hue produced by these metals in solution is similar to that found in natural waters. Comparison tubes containing standard platinum-cobalt solutions ranging from 0 to 70 color units are used for visual measurements; however, laboratories often employ a colorimeter for readings.

Turbidity

Insoluble particles of soil, organics, microorganisms, and other materials impede the passage of light through water by scattering and absorbing the rays. This interference of light passage through water is referred to as turbidity. In excess of 5 units it is noticeable to the average water consumer and accordingly represents an unsatisfactory condition for drinking water. Turbidity in a typical clear lake water is about 25 units, and muddy water exceeds 100 units.

The standard method for determination of turbidity has been based on the Jackson candle turbidimeter consisting of a calibrated glass tube, holder, and candle as shown in Figure 2-13. Units of turbidity using this apparatus are expressed as JTU, Jackson Turbidity Units. With the tube in place over a lighted candle, the water sample is gradually added until the image of the candle flame just disappears from view through the water column in the tube. The longer the light path, the lower the turbidity. For example, a light path of 10.8 cm is 200 JTU, 21.5 cm is 100 JTU, and 72.9 cm is 25 JTU. Since the lowest value that can be measured directly by this technique is 25 units, the Jackson candle turbidimeter is limited in application to turbid natural waters. Measurement of turbidity in treated drinking water, commonly

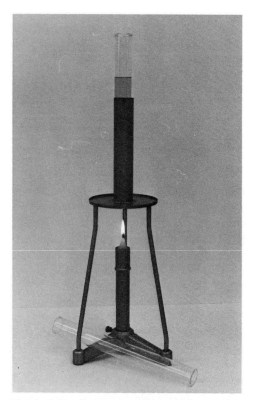

Figure 2-13 Jackson candle turbidimeter for measuring turbidities in excess of 25 JTU in natural waters.

less than one unit, is measured using a precalibrated commercial turbidimeter.

Commercial turbidimeters (nephelometers) for determining low turbidities measure the intensity of light scattered at right angles to the incident light. Because no direct relationship exists between the intensity of light scattered at 90° and Jackson candle turbidity, calibration of a nephelometer in terms of candle units is not valid. To distinguish between the turbidities derived from instrumental and visual methods, nephelometric turbidity units are abbreviated NTU, Nephelometric Turbidity Units. Occasionally, the nephelometric units are referred to as FTU, since Formazin polymer is used as the reference turbidity standard suspension.

The commercial turbidimeter illustrated in Figure 2-14 is a nephelometer that can compensate for color in the

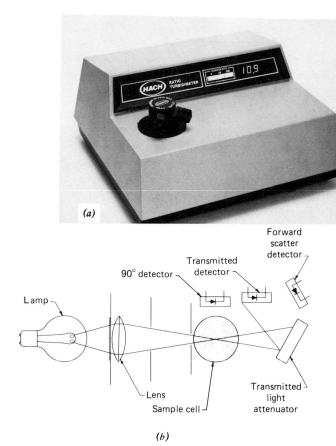

Figure 2-14 Ratio turbidimeter (nephelometer) and light path diagram for measuring turbidities in both cloudy natural waters and treated clear waters with turbidities less than 1 NTU. (Courtesy of Hach Company, Loveland, Colorado.)

water sample. In treated water, the effect of color is generally not significant. Light is focused by a lens before passing horizontally through the sample. In an ordinary turbidimeter, this measurement alone would be used to determine turbidity. In addition, this ratio turbidimeter detects the transmitted light. At low or moderate turbidity levels, the forward scatter signal is negligible compared to the transmitted signal, and the output is simply a ratio of

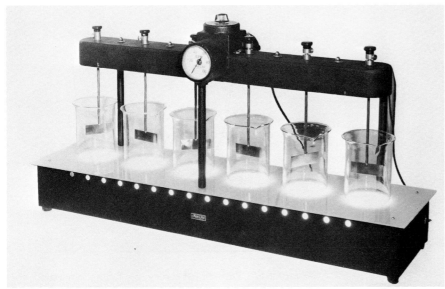

Figure 2-15 Stirring device used in jar tests for chemical coagulation. (Courtesy of Phipps & Bird, Inc.)

the 90° scattered light to transmitted light. This ratioing stabilizes the instrument and negates the affect of color. At high turbidity levels, the instrument combines signals from the forward scatter and transmitted light detectors in the denominator of the ratio, allowing linearity over the entire range.

Jar Tests

Effectiveness of chemical coagulation of water or wastewater can be experimentally evaluated in the laboratory by using a stirring device as illustrated in Figure 2-15. The stirrer consists of six paddles capable of variable-speed operation between 0 and 100 rpm. In making tests, 1 liter or more of water is placed in each of the jars or beakers, and is dosed with different amounts of coagulant. After rapid mixing to disperse the chemicals, the samples are stirred slowly for floc formation and then are allowed to settle under quiescent conditions. The jars are mixed at a speed of 60 to 80 rpm for 1 min after adding the coagulant solution and then are stirred at a speed of 30 rpm for 15 min. After the stirrer is stopped, the nature and settling characteristics of the floc are observed and recorded in qualitative terms, as poor, fair, good, or excellent. A hazy sample indicates poor coagulation, while proper coagulated water contains floc that are well formed with the liquid clear between particles. The lowest dosage that provides good turbidity removal during a jar test is considered the first trial dosage in plant operation. Ordinarily a full-scale treatment plant gives better results than a jar test at the same dosage.

For research or special studies, the beakers used in jar testing may be modified to replicate more closely actual mixing units constructed in treatment plants. Figure 2-16 illustrates a container provided with stators mounted on the inside, above and below the paddle, to provide for more thorough mixing. Special provision is made for applying the coagulant dose through a glass tube extending into the jar near the hub of the impeller. The system also includes siphons that can be used to draw off samples from the same depth in each beaker for chemical testing, for example, turbidity measurements during the settling

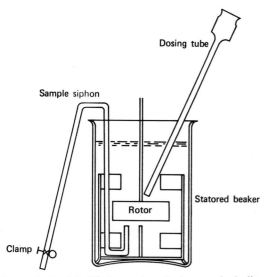

Figure 2-16   Modified container for jar tests including stators, dosing tube, and sampling siphon.

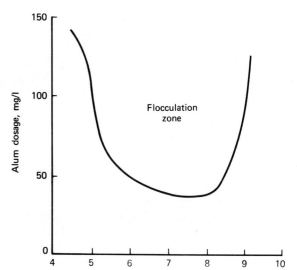

Figure 2-17   Effect of pH on dosage of alum required for good coagulation of a river water.

period. Actual treatment plant operations may dictate a change in mixing and settling times to match those being used in water processing. Results from a jar test for coagulation of a turbid river water using alum are given in Table 2-10. The lowest recommended dosage in treating this water is 40 mg/l of coagulant. Since such other factors as temperature, alkalinity, and pH influence coagulation, jar tests can also be run to evaluate these parameters and to determine optimum dosages under differing conditions. For example, Figure 2-17 shows the results of a series of jar tests, similar to the one illustrated in Table 2-10, that have been run at different pH values. The plot

illustrates the pH of water samples tested versus minimum alum dosage required for good floc formation. In addition to the effect of variation in water quality on chemical dosage, special studies can be conducted to measure optimum application of coagulant aids, such as polymers or activated silica with the primary coagulant.

Fluoride

Excessive fluoride ions in drinking water cause dental fluorosis or mottling of teeth. On the other hand, communities whose drinking water contains no fluoride have a high prevalance of dental caries. Optimum concentrations of fluoride provided in public water supplies, generally in the range of 0.8 to 1.2 mg/l, reduce dental caries to a minimum without causing noticeable dental fluorosis. Several fluoride compounds (Table 2-3) are used in treating municipal water; all of these dissociate readily, yielding the fluoride ion (F^-).

Electrode and colorimetric methods are currently most satisfactory for determining fluoride ion concentration. Both methods are susceptible to interfering substances, for example, chlorine in the colorimetric method and polyvalent cations, such as Al^{+++}, through which complex fluoride ions hinder electrode response. Fluoride can

Table 2-10.   Results of Jar Test for Coagulation

Jar Number	Aluminum (ml/jar)	Sulfate (mg/l)	Dosage (gpg)	Floc Formation
1	1	10	0.58	None
2	2	20	1.17	Smoky
3	3	30	1.85	Fair
4	4	40	2.34	Good
5	5	50	2.92	Good
6	6	60	3.50	Heavy

be separated from other constituents in water by distillation of hydrofluoric acid (HF) after acidifying the sample with sulfuric acid. Pretreatment separation is performed using a distillation apparatus. Samples not containing significant amounts of interfering ions can be tested directly.

The fluoride ion-activity electrode is a specific ion sensor designed for use with a calomel reference electrode and a pH meter having a millivolt scale. The key element in the fluoride electrode is the laser-type doped single lanthanum fluoride crystal across which a potential is established by the presence of fluoride ions. The crystal contacts the sample solution at one face and an internal reference solution at the other. An appropriate calibration curve must be developed that relates meter reading in millivolts to concentration of fluoride. Fluoride activities dependent on total ionic strength of the sample and the electrode do not respond to fluoride that is bound or complexed. These difficulties are overcome by the addition of CDTA (cyclohexylenediaminetetraacetic acid) solution, which is a buffer of high total ionic strength to swamp out variations in sample ionic strength.

The colorimetric method is based on the reaction between fluoride and a zirconium-dye lake. (The term "lake" refers to the color produced when zirconium ion is added to SPADNS dye.) Fluoride reacts with the reddish color-dye lake, dissociating a portion of it into a colorless complex anion ($ZrF_6^=$) and the yellow-color dye. As the amount of fluoride is increased, the color becomes progressively lighter and of a different hue. This bleaching action is directly proportional to fluoride ion concentration. Either a spectrophotometer (Figure 2-11) or a filter photometer (Figure 2-10) may be used to measure sample absorbance for comparison against a standard curve.

Chlorine

The most common method for disinfecting water supplies in the United States is by chlorination. As a strong oxidizing agent, it is also used effectively in taste and odor control, and removal of iron and manganese from water supplies.

Chlorine combines with water to form hypochlorous and hydrochloric acids, Eq. 2-27. Hypochlorous acid reacts with ammonia in water to form monochloramine (NH_2Cl), dichloramine ($NHCl_2$), and trichloramine (NCl_3), depending on the relative amounts of acid and ammonia. Chlorine, hypochlorous acid, and hypochlorite ion are free chlorine residual, while the chloramines are designated as combined chlorine residual.

The N,N-diethyl-p-phenylenediamine (DPD) test differentiates and measures both free and combined chlorine residuals in samples of clear water. In the absence of iodide ion, free available chlorine reacts instantaneously with DPD indicator to produce a red color. Subsequent addition of a small amount of iodide ion acts catalytically with monochloramine to produce color. Addition of excess iodide causes a rapid response from dichloramine. The intensity of the color development, which is proportional to the amount of chlorine residual, can be measured either by titration with ferrous ammonium sulfate (FAS) or by use of a spectrophotometer at a wavelength of 515 nanometers or a photoelectric colorimeter with a filter having maximum transmission in the range of 490–530 nanometers.

The procedure for the colorimetric method is as follows: (1) The sample is diluted with chlorine-demand-free water when the available chlorine in the sample exceeds 4 mg/l. (2) For free chlorine residual, a phosphate buffer solution and DPD indicator solution are mixed with the water, and color development is read immediately and recorded as reading *A*. (3) For determining monochloramine, one small crystal of KI is added and the sample mixed. The intensity of color is read again and recorded as reading *B*. (4) For dichloramine, a few more crystals of KI are dissolved in the sample. After about 2 min, the final color intensity is measured and recorded as reading *C*. The color readings are interpreted as follows: *A* is free chlorine residual, $B - A$ is monochloramine (NH_2Cl), $C - B$ is dichloramine ($NHCl_2$), $C - A$ is the combined residual, and *C* is the total residual. The commercial chlorine test kit illustrated in Figure 2-18 can be used to measure free and total chlorine residuals in samples of clear water.

Residual chlorine in wastewater samples cannot be reliably determined by the DPD technique because of interference from organic matter. The indirect procedure of the iodometric method is much more precise. A wastewater sample is prepared by adding a measured

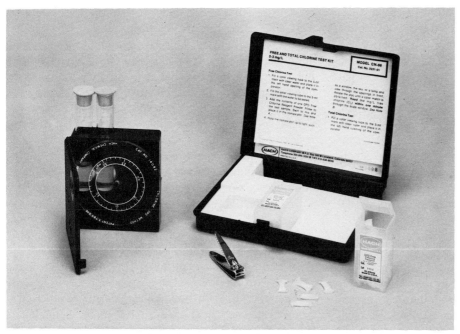

Figure 2-18 Chlorine test kit to measure free and total chlorine residuals in samples of clear water by the DPD colorimetric method. Included are chemical reagents, sample tubes, and a comparator with a color disk for measuring color intensity. (Courtesy of Hach Company, Loveland, Colorado.)

volume of standardized phenylarsine oxide, or thiosulfate, solution; excess potassium iodide; and acetate buffer solution to reduce the pH to between 3.5 and 4.2. Chlorine residual oxidizes an equivalent portion of the excess iodide ion to free iodine that, in turn, is immediately converted back to iodide by a portion of the reducing agent. The amount of phenylarsine oxide remaining is then quantitatively measured by titration with a standard iodine solution. In other words, this analysis reverses the end point by determining the unreacted standard phenylarsine oxide in the treated sample using standard iodine, rather than directly titrating the iodine released by the chlorine. Thus, contact between the full concentration of liberated iodine and the wastewater is avoided. The titration end point may be indicated by using starch, which produces a blue color in the presence of free iodine. An alternative, and more accurate, method for observing the end point involves the use of an amperometric titrator

(Figure 2-19). The meter needle responds to electrical conductivity of the sample solution being titrated. A deflection of the pointer upscale registers the presence of free iodine and the titration end point. This apparatus can be employed to measure free residual, or to differentiate between monochloramine and dichloramine fractions of combined residual.

Nitrogen

Common forms of nitrogen are organic, ammonia, nitrite, nitrate, and gaseous nitrogen. Nitrogenous organic matter, such as protein, is essential to living systems. Industrial wastes are often tested for nitrogen and phosphorus content to ensure that there are sufficient nutrients for biological treatment. Inorganic nitrogen, principally in the ammonia and nitrate forms, is used by green plants in photosynthesis. Since nitrogen in natural waters is pH and to shift the equilibrium ammonia to the right in Eq. 2-39.

$$NH_4^+ \underset{acidic}{\overset{basic}{\rightleftharpoons}} NH_3\uparrow + H^+ \qquad (2\text{-}39)$$

A condenser is then attached to the flask and the mixture is distilled, driving off steam and free ammonia. The steam containing ammonia is condensed and is collected in a boric acid solution. The reaction with boric acid forms ammonium ions, Eq. 2-39 to the left, while producing borate ions from the acid. The amount of ammonia in the sample is determined by the quantity of boric acid consumption in the collecting beaker. This may be measured by back titration of the solution with a standard acid to determine the amount of borate ion produced.

Organic nitrogen in wastewater is determined by digestion of the organic matter, thus releasing ammonia, and then proceeding with the ammonia nitrogen test. Since the bulk of organic nitrogen is in the form of proteins and amino acids, it can be converted to

Figure 2-19 Amperometric titrator for determination of chlorine residual in wastewater by the iodometric method. (Courtesy of Wallace & Tiernan Division, Pennwalt Corp.)

limited, pollution from nitrogen wastes can promote the growth of algae, causing green-colored water. Ammonia is also considered a serious water pollutant because of its toxic effect on fishes.

The test for ammonia nitrogen is a distillation process using the apparatus illustrated in Figure 2-20. The analysis procedure is based on shifting the equilibrium between the ammonium ion and free ammonia, and the release of gaseous ammonia along with the steam that is produced when the water is boiled. After the sample is placed in a flask, a buffer solution is added to increase the

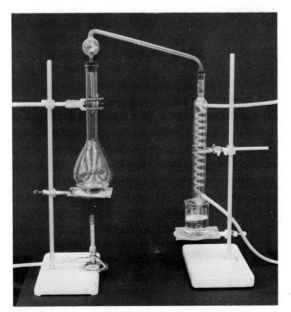

Figure 2-20 Distillation apparatus for ammonia and organic nitrogen tests.

ammonium ion by boiling in an acidified solution with chemical catalysts, Eq. 2-40. When acid is first added to

$$\text{Organic-nitrogen} \xrightarrow[\text{catalysts}]{\text{sulfuric acid}} NH_4^+ \qquad (2\text{-}40)$$

the wastewater sample, the organics turn black. Complete destruction and conversion after 1 to 2 hr of boiling is indicated by a clearing of the liquid. On cooling, a sodium hydroxide reagent is added to raise the pH and to convert the ammonium ion to ammonia. The flask is attached to the connecting bulb of the condenser, and the test is completed by distillation for measurement of the ammonia nitrogen. If both ammonia and organic nitrogen determinations are desired, the sample may be tested first for ammonia and then can be digested and run for organic nitrogen. The sum of these two results is often referred to as total Kjeldahl nitrogen. If separation of organic and ammonia nitrogens is not needed, the total Kjeldahl nitrogen may be determined directly by digesting the entire sample and distilling off both the ammonia nitrogen that originally existed in the sample, as well as that released from digesting the organic nitrogen.

Nitrite (NO_2^-) and nitrate (NO_3^-) in natural and treated waters are routinely determined by colorimetric techniques. Nitrite is determined through formation of a reddish-purple azo dye produced at pH 2.0 to 2.5 by coupling diazotized sulfanilic acid with NED dihydrochloride. If the sample is turbid, suspended solids are removed by filtering through a membrane with a pore diameter of 0.45 μm. The intensity of color development can be determined by a photometric measurement or visual comparison with color standards prepared in Nessler tubes.

Nitrate concentration is determined by the cadmium reduction method. NO_3^- is reduced almost quantitatively to NO_2^- in the presence of cadmium. After turbidity removal and pH adjustment, the reduction is performed by pouring the sample through a 30-cm column packed with commercially available Cd granules treated with $CuSO_4$ to form a copper coating. The nitrite produced by reduction, plus any nitrite in the original sample, is determined by diazotizing with sulfanilamide and coupling with N-(1 naphthyl)-ethylenediamine to form a highly colored azo dye. Color intensity is measured using either a spectrophotometer or a filter photometer.

The specific nitrogen tests conducted on a natural water, treated water, or wastewater depend on the objectives of the study. A drinking water test may require only analysis for nitrate, while a stream pollution survey may be primarily interested in ammonium nitrogen. Total nitrogen in a water is equal to the sum of the organic, ammonia, nitrite, and nitrate nitrogen concentrations. All nitrogen test results are expressed in units of milligrams of N/l.

Phosphorus

The common compounds are orthophosphates ($H_2PO_4^-$, $HPO_4^=$, $PO_4^\equiv$), polyphosphates, such as $Na_3(PO_3)_6$ used in synthetic detergent formulations, and organic phosphorus. All polyphosphates gradually hydrolyze in water to the stable ortho form, while decaying organic matter decomposes biologically to release phosphate. Orthophosphates are, in turn, synthesized back into living animal or plant tissue.

The prime concern in biological waste treatment is ensuring sufficient phosphorus to support microbial growth. Although sanitary wastes normally have a surplus of phosphates, some industrial wastes may be nutrient deficient because of high carbohydrate or hydrocarbon chemical content. In phosphorus pollution control, the primary concern is overfertilization of surface waters, resulting in nuisance growths of algae and aquatic weeds.

A popular procedure for determining orthophosphate is the stannous chloride colorimetric method. Ammonium molybdate and stannous chloride reagents combine with phosphate to produce a blue-colored colloidal suspension. A second technique for orthophosphate is the vanadomolybdophosphoric acid method. Vanadium is used in combination with ammonium molybdate to produce a yellow color with phosphate ion. Color development is quantitatively measured using a spectrophotometer.

Acid hydrolysis of a water sample at boiling-water temperature converts condensed phosphates (pyro, tri, poly, and higher-molecular-weight species, such as hexa-metaphosphate) to orthophosphate. This mild hydrolysis unavoidably releases some phosphate from organic compounds, but this is reduced to a minimum by selection of the acid strength and hydrolysis time and temperature. The condensed phosphate fraction is equal to the acid-hydrolyzable phosphorus determined by this test minus the direct orthophosphate measurement.

Digestion using a strong acid converts all of the phosphorus in a sample, including organic phosphorus, to orthophosphate. Therefore, in a total phosphorus test, the sample is boiled in either a concentrated perchloric acid or sulfuric acid-nitric acid solution to digest the organic matter releasing bound phosphorus. The total phosphorus is measured by testing the digested sample for orthophosphate content.

The classification of phosphorus fractions in a sample includes the physical state of filterable (dissolved) and particulate, as well as chemical types. Separating dissolved from particulate forms is accomplished by filtration through a 0.45 μm membrane filter. A complete phosphorus analysis consists of conducting ortho, acid-hydrolyzable, and total phosphate tests on measured portions of unfiltered and filtered samples. The particulate contents for each of the three phosphorus fractions are calculated by subtracting the filterable measurements from the respective test determinations on the whole sample.

The variety of techniques for pretreatment of samples, measurement of phosphorus concentrations, and expression of results can lead to confusion. The particular procedure used in analysis should be recorded with test results. The most common shortcoming in presenting phosphorus data in printed literature is failure to document collection technique, storage, and filtration or other pretreatment. The thirteenth and subsequent editions of *Standard Methods* have determined total phosphorus by acid digestion, whereas earlier editions considered that portion available after acid hydrolysis as being the total amount, when in fact this includes only the ortho and poly forms. Another modification involved changing the expression of results. *Standard Methods* now calculates results of all phosphate tests in terms of milligrams per liter as phosphorus, while prior to 1971 milligrams per liter as phosphate were used. To convert milligrams per liter as phosphate to milligrams per liter as phosphorus, the value given should be divided by 95 (molecular weight of PO_4) and multiplied by 31 (atomic weight of phosphorus). In other words, 1.00 mg/l PO_4 equals 0.34 mg/l P, or 1.00 mg/l P equals 3.06 mg/l PO_4.

Dissolved Oxygen

Biological decomposition of organic matter uses dissolved oxygen. Levels significantly below saturation values, given in Table 2-5, often occur in polluted surface waters. Since fishes and most aquatic life are stifled by a lack of oxygen, dissolved oxygen determination is a principal measurement in pollution surveys. The rate of air supply to aerobic treatment processes is monitored by dissolved oxygen testing to maintain aerobic conditions, and to prevent waste of power by excessive aeration. DO tests are used in the determination of biochemical oxygen demand of a wastewater. Small samples of wastewater are mixed with dilution water and placed in BOD bottles for dissolved oxygen testing at various intervals of time. Oxygen is a significant factor in corrosion of piping systems. Removal of oxygen from boiler feed waters is common practice, and the DO test is the means of control.

The azide modification of the iodometric method is the most common chemical technique for measuring dissolved oxygen. The standard test uses a 300-ml BOD bottle for containing the water sample (Figure 2-21). The chemical reagents used in the test are manganese sulfate solution, alkali-iodide-azide reagent, concentrated sulfuric acid, starch indicator, and standardized sodium thiosulfate titrant. The first step is to add 1.0 ml of each of the first two reagents to the BOD bottle, restopper with care to exclude air bubbles, and mix by repeatedly inverting the bottle. If no oxygen is present, the manganous ion (Mn^{++}) reacts only with the hydroxide ion to form a pure white precipitate of $Mn(OH)_2$, Eq. 2-41. If oxygen is present, some of the Mn^{++} is oxidized to a higher valence (Mn^{++++}) and precipitates as a brown-colored oxide (MnO_2), Eq. 2-42.

$$Mn^{++} + 2OH^- \rightarrow Mn(OH)_2\downarrow \qquad (2\text{-}41)$$

$$Mn^{++} + 2OH^- + \tfrac{1}{2}O_2 \rightarrow MnO_2\downarrow + H_2O \quad (2\text{-}42)$$

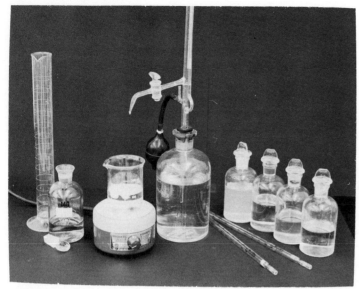

Figure 2-21 Apparatus for chemical determination of dissolved oxygen.

After the bottle has been shaken and sufficient time has been allowed for all the oxygen to react, the chemical precipitates are allowed to settle leaving clear liquid in the upper portion. Then, 1.0 ml of concentrated sulfuric acid is added. The bottle is restoppered and mixed by inverting until the suspension is completely dissolved and the yellow color is uniform throughout the bottle. The reaction that takes place with the addition of acid is shown in Eq. 2-43; the manganic oxide is reduced to manganous manganese while an equivalent amount of iodide ion is converted to free iodine. The quantity of I_2° is equivalent to the dissolved oxygen in the original sample.

$$MnO_2 + 2I^- + 4H^+ \rightarrow Mn^{++} + I_2^\circ + 2H_2O \quad (2\text{-}43)$$

A volume of 201 ml, corresponding to 200 ml of the original sample after correction for the loss of sample displaced by reagent additions, is poured from the BOD bottle into a container for titration with 0.0250 N thiosulfate solution. Thiosulfate in the titrant is oxidized to tetrathionate while the free iodine is converted back to iodide ion, Eq. 2-44.

$$2S_2O_3^= + I_2^\circ \xrightarrow[\text{indicator}]{\text{starch}} S_4O_6^= + 2I^- \quad (2\text{-}44)$$

Since it is impossible to titrate accurately the yellow-colored iodine solution to a colorless liquid, an end point indicator is needed. Soluble starch in the presence of free iodine produces a blue color. Therefore, after titration to a pale straw color, a few drops of starch solution are added and titration is continued to the first disappearance of blue color. If 0.0250 N sodium thiosulfate is used to measure the dissolved oxygen in a volume equal to 200 ml of original sample, 1.0 ml of titrant is equivalent to 1.0 mg/l DO.

Since high concentrations of suspended solids and biological activity of activated-sludge flocs may have high oxygen utilization rates, microbial activity must be stopped at the time of sample collection, and suspended solids must be settled out of solution prior to the iodometric dissolved oxygen test. A common procedure is to use copper sulfate-sulfamic acid inhibitor solution to stop biological activity and to flocculate suspended solids. Recommended sampling procedure is to add 10 ml of inhibitor to a 1-qt, or 1-liter, wide-mouth bottle. For immersion in an aeration tank, the bottle is often placed in a special sampler designed so that the bottle fills from a tube near the bottom and overflows only 25 to 50 percent of the bottle capacity. After it has been removed from the sampler, the sample is stoppered and allowed to settle until a clear supernatant can be siphoned into a BOD bottle. DO concentration is then measured by the azide-iodometric method.

Membrane electrodes are available for measurement of dissolved oxygen without chemical treatment of a sample. A dissolved oxygen probe is composed of two solid metal electrodes in contact with a salt solution that is separated from the water sample by a selective membrane (Figure 2-22). The recessed end of the probe containing the metal electrodes is filled with potassium chloride solution and is covered with a polyethylene or teflon membrane held in place by a rubber O-ring. The probe also has a sensor for measuring temperature. The unit inserted in the bottle in Figure 2-22 is designed specifically for measuring dissolved oxygen in nondestructive BOD testing; the same bottle can be measured for oxygen depletion at various time intervals, restoppering between

readings. The field probe is submersible and can be lowered into water for DO and temperature measurements in lakes and streams. The same unit attached to a long rod can record dissolved oxygen in aeration tanks. Meters used with the probes have both temperature and dissolved oxygen scales, and can be operated by line power in the laboratory or battery operated in the field. Membrane electrodes may be calibrated by reading against air or, more frequently, a water sample of known DO concentration determined by the iodometric method.

Solids

Suspended and dissolved solids, both organic and inert, are common tests of polluted waters. In drinking water, the maximum recommended total dissolved solids concentration is 500 mg/l. Suspended and volatile solids are common parameters used in defining a municipal or industrial wastewater. Operational efficiency of various treatment units is defined by solids removal, for example, suspended solids removal in a settling basin, and volatile solids reduction in sludge digestion.

Total solids, total residue on evaporation, is the term

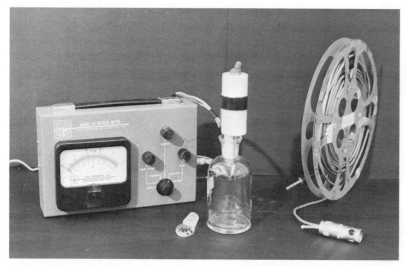

Figure 2-22 Dissolved oxygen meter with the laboratory probe inserted in a BOD bottle. At the right is a field probe with the extension wire wound on a reel.

applied to material left in a dish after evaporation of a sample of water or wastewater and subsequent drying in an oven. A measured volume of sample is placed in an evaporating dish, usually porcelain, as shown in Figure 2-23. After evaporation of the water from the dish on a steam bath, the dish is transferred to an oven maintained at 103° C and is dried to a constant weight. The milligrams of total residue are equal to the difference between the cooled weight of the dish and the original weight of the empty dish. Concentration of total residue is calculated using Eq. 2-45.

$$\text{mg/l total residue} = \frac{\text{mg total residue} \times 1000}{\text{ml sample}} \quad (2\text{-}45)$$

The terms suspended solids and dissolved solids refer to matter that is retained and passed through a filter, respectively. In water analyses they are referred to as nonfiltrable and filtrable residue, whereas in wastewater they are called suspended and dissolved solids. In either case, a measured portion of sample is drawn through a glass-fiber filter, retained in a funnel, by applying a vacuum to the suction flask pictured in Figure 2-23. Rather than a membrane filter apparatus, a Gooch crucible (a small porcelain filtering dish with pinholes in the bottom) may be used to hold the glass-fiber filter. After filtration, the disk is removed from the funnel, dried, and weighed to determine the increase as a result of the residue retained. Calculation of total suspended matter (filtrable residue) uses the same type of formula as Eq. 2-45. Dissolved matter (nonfiltrable residue) is not determined directly but is calculated by subtracting suspended solids concentration from the total residue on evaporation.

Volatile solids are determined by igniting the residue on evaporation, or the filtrable solids, at 550° C in an electric muffle furnace. Residues are burned for 15 to 20 min. Loss of weight on ignition is reported as milligrams per liter of volatile solids, and residue after burning is referred to as fixed solids. The evaporating dish used in volatile solids analysis and the glass-fiber filter disk used for volatile suspended solids must be pretreated by burning in the muffle furnace to determine an accurate initial (empty) weight. Volatile solids in a wastewater are often interpreted as being a measure of the organic matter. However, this is not precisely true, since combustion of many pure organic compounds leaves an ash and many inorganic salts volatilize during ignition.

Following are typical data and calculations in the analysis of a wastewater sludge to determine total solids and total volatile solids:

Figure 2-23 Apparatus for solids determinations including a suction flask with attached funnel, filter pads, and a drying dish.

Weight of empty dish on analytical balance		64.532 g
Weight of dish on simple balance	64.5 g	
Weight of dish plus sludge sample	144.7	
Weight of sample	80.2	
Weight of dish plus dry sludge solids		68.950
Weight of sludge solids		4.418
Percentage of total solids	$\dfrac{4.42}{80.2} \times 100 = 5.51$ percent	
Weight of dish plus ignited solids		65.735
Weight of volatile solids		3.215

Percentage of volatile solids $\dfrac{3.22}{80.2} \times 100 = 4.02$ percent

Percentage of total solids that are volatile $\dfrac{3.22}{4.42} \times 100 = 72.8$ percent

Settleable solids are determined by filling a 1-liter conical-shaped container (Imhoff cone) and allowing 1 hr for sedimentation. The tip at the bottom of the cone is scribed with volume markings so that the quantity of settled matter can be read directly in milliliters per liter. If it is desired to express the settleable matter on a weight basis, milligrams per liter, a sample of the wastewater is poured into a 1-liter graduated cylinder and is allowed to stand quiescent for 1 hr. Solids analysis run on the initial well-mixed sample, and suspended matter remaining at the midpoint of the graduated cylinder after settling are used in the following calculation:

mg/l settleable matter

$$= \text{mg/l suspended} - \text{mg/l nonsettleable} \quad (2\text{-}46)$$

Chemical Oxygen Demand

COD is widely used to characterize the organic strength of wastewaters and pollution of natural waters. The test measures the amount of oxygen required for chemical oxidation of organic matter in the sample to carbon dioxide and water. The apparatus used in the dichromate reflux method pictured in Figure 2-24 consists of an Erlenmeyer flask fitted with a condenser and a hot plate. The test procedure is to add a known quantity of standard potassium dichromate solution, sulfuric acid reagent containing silver sulfate, and a measured volume of sample to the flask. This mixture is refluxed (vaporized and condensed) for 2 hr. Most types of organic matter are destroyed in this boiling mixture of chromic and sulfuric acid, Eq. 2-47. After the mixture has been cooled

Organics $+ Cr_2O_7^= + H^+$

$$\xrightarrow[\text{Ag}^+]{\text{heat}} CO_2 + H_2O + 2Cr^{+++} \quad (2\text{-}47)$$

and diluted with distilled water, and the condenser has been washed down, the dichromate remaining in the

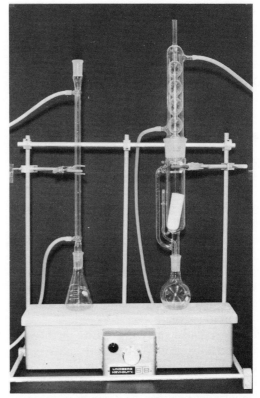

Figure 2-24 Reflux unit for chemical oxygen demand (*left*) and Soxhlet grease extraction apparatus (*right*).

specimen is titrated with standard ferrous ammonium sulfate using ferroin indicator. Ferrous ion reacts with dichromate ion as in Eq. 2-48 with an end point color

$6Fe^{++} + Cr_2O_7^= + 14H^+$

$$\rightarrow 6Fe^{+++} + 2Cr^{+++} + 7H_2O \quad (2\text{-}48)$$

change from blue-green to reddish brown. A blank sample of distilled water is carried through the same COD testing procedure as the wastewater sample. The purpose of running a blank is to compensate for any error that may result because of the presence of extraneous organic matter in the reagents. COD is calculated using Eq. 2-49; the difference between amounts of titrant used for the blank and the sample is divided by volume of sample and multiplied by normality of the titrant. The 8000 multiplier is to express the results in units of milligrams per liter of oxygen, since 1 liter contains 1000 ml and the equivalent weight of oxygen is 8.

$$\frac{COD}{mg/l} = \frac{\begin{array}{c}(\text{ml blank} - \text{ml sample titrant}) \times \\ (\text{normality Fe(NH}_4)_2(\text{SO}_4)_2)8000\end{array}}{\text{ml sample}} \quad (2\text{-}49)$$

Total Organic Carbon

The TOC test measures the organically bound carbon in a water or wastewater sample. Unlike COD, it is independent of the oxidation state of the organic matter. Distinction is made, and must be reported, between purgeable and nonpurgeable TOC. For instance, in the combustion infrared method, the removal of inorganic carbon (CO_2) by purging causes loss of purgeable organics, and the result of the determination is reported as nonpurgeable TOC. Instrumentation is available for determining both purgeable and nonpurgeable organic carbon, and the sum can be reported as true TOC.

Commercially available analyzers oxidize the organic carbon to carbon dioxide by either heat and oxygen, ultraviolet radiation, chemical oxidants, or various combinations of these. The CO_2 can be measured directly by a nondispersive infrared analyzer, or it can be reduced to methane and measured by a flame ionization detector in a gas chromatograph. Free CO_2 in the sample may be titrated chemically. Inorganic carbon must be eliminated or compensated for, since it is usually a very large portion of the total carbon in a water or wastewater sample. The determination of total carbon and total inorganic carbon with the estimation of total organic carbon by difference is a common procedure.

Trace Organics

Minute quantities of the many different organic substances that appear in surface waters have been detected in drinking water. Naturally occurring trace organics include humic and fulvic acids derived from decaying vegetation. Upon chlorination of water containing these compounds, trihalomethanes such as chloroform (trichloromethane) are formed. Trihalomethanes are suspected carcinogens and limited by drinking water standards.

Other organic compounds of health significance are synthetic chemicals from industrial wastewaters and pesticides. The analytical technique for detection of these trace organics is gas chromatography. The methods of sample preparation are complex and involve initial separation of the trace organics from water solution. Chlorinated-hydrocarbon pesticides are extracted by a solvent and concentrated by evaporation prior to injection into a gas chromatograph. Organohalides (trihalomethanes) are purged by bubbling an inert gas through the water sample followed by trapping in a short column containing a suitable sorbant. The compounds are desorbed thermally from the trap and backflushed into a gas chromatographic column.

The main components of a gas chromatograph are a supply of carrier gas with flow control, injection port for introducing the sample, temperature-controlled column containing a specific packing, detector to identify compounds that elute from the column, and a recorder to graph the signal from the detector (Figure 2-25). The carrier gas (nitrogen, argon-methane, helium, or hydrogen) flows continuously through the column at constant temperature and flow rate. A small sample for analysis is injected with a microsyringe through the septum into the sample port where flash vaporization converts the volatile components into gases. Conveyed by the carrier gas, the gaseous organic compounds travel through the column at different rates depending on differences in partition coefficients between the moving carrier gas phase and the stationary phase. The stationary phase is a liquid that has been coated on an inert granular solid, called the column packing, held in a glass or metal tube. Components of the sample that are relatively soluble in the liquid phase move through the column slower than relatively insoluble constituents. As each component of the sample exits the column and passes through the detector, a quantitatively proportional charge in electrical signal is measured on a strip-chart recorder. The retention time in the column permits identification of a particular substance, and the peak height or peak area graphed by the recorder is proportional to its quantity. Methods of detection include thermal conductivity, flame ionization, electron capture, and mass spectrometry. The stationary-phase material and concentration, column

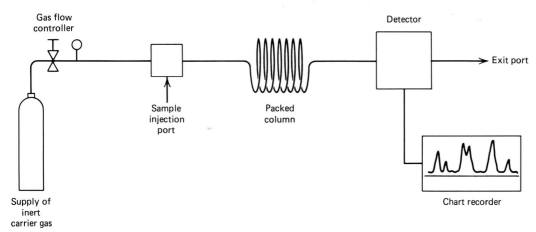

Figure 2-25   Schematic diagram of a gas chromatograph.

length and diameter, operating temperature, carrier-gas flow, and detector type are the controlled variables that allow analysis for different trace organics.

Oil and Grease

A variety of organic substances including hydrocarbons, fats, oils, waxes, and high-molecular-weight fatty acids are collectively referred to as oil and grease. Their importance in municipal and industrial wastes is related to their difficulty in handling and treatment. Because of low solubility, grease separates from water adhering to the interior of pipes and tank walls, reduces biological treatability of a wastewater, and produces greasy sludge solids difficult to process.

Apparatus for grease determination by the Soxhlet extraction method is shown in Figure 2-24. A sample is prepared for grease extraction by filtering 1 liter of acidified wastewater through a filter consisting of muslin cloth overlaid with filter paper supporting diatomaceous-silica filter aid. After the sample has been drawn through, the filter is removed from the funnel, rolled and placed in an extraction thimble, and dried in a hot-air oven at 103° C. The boiling flask is carefully weighed and partially filled with the solvent trichlorotrifluoroethane, which is nonflammable. After the thimble containing the sample is placed into the Soxhlet extension tube, the flask with

solvent is fitted on the bottom, and the condenser is placed on top. Heating the solvent drives vapor up through a side arm in the Soxhlet extension tube to the base of the condenser. Condensed solvent drips down into the thimble containing the grease sample, passes through the filter paper to a space outside the thimble in the extension tube. When filled to the siphon outlet of the tube, the solvent is discharged back down into the flask. This sequence of vaporizing and flushing the solvent through the sample and back to the flask is referred to as a cycle. Extraction in the standard test consists of 20 cycles per hour for 4 hr. After cooling, the flask is removed and placed on a water bath at 70° C for slow distillation. Driving off the solvent leaves the extracted grease in the boiling flask. Grease content is calculated by dividing the increase in weight of the flask by the volume of wastewater sample filtered.

Organic Acids

Measurement of organic acid concentration in anaerobically digesting sludge is used to monitor the digestion process. The acids, principally acetic, propionic, and butyric, are separated from water solution by column chromatography. A small clarified sample is drawn into a Gooch or fritted-glass crucible packed with granular silicic acid to adsorb the organic acids. Chloroform-butanol reagent is then added to the column and drawn through to elute selectively the organic acids from the column packing. After the extract has been purged with nitrogen or carbon dioxide free air, it is titrated with a standard base to measure acid content. Total organic acids are expressed in milligrams per liter as acetic acid.

Gas Analysis

Anaerobic decomposition of sludges produces methane, carbon dioxide, and traces of hydrogen sulfide. Digester operation can be monitored by observing the relative amounts of carbon dioxide and methane in the head gases; the ratio is approximately one third and two thirds, respectively. A common laboratory apparatus consists of a water-jacketed buret, catalytic combustion chambers, and a series of glass units containing selective gas absorption reagents for carbon dioxide, oxygen, and carbon monoxide. The catalytic burners are for combustion of methane. A gas sample is drawn into the buret and initial volume recorded. The confined gas can then be passed through a manifold into any one of the absorbing units or through the combustion chambers, and drawn back to the buret again. After several passages through one of the absorbing units, the quantity of gas remaining is measured, and the percentage of that gas constituent in the sample can be calculated.

REFERENCES

1. *National Interim Primary Drinking Water Regulations*, U.S. Environmental Protection Agency, Office of Water Supply, EPA 570/9-76-003, 1976.
2. Sawyer, C. N., and P. L. McCarty. *Chemistry for Sanitary Engineers*, Third Edition, New York: McGraw-Hill, 1978.

FURTHER READING

Standard Methods for the Examination of Water and Wastewater, 15th Edition, American Public Health Association, American Water Works Association, and Water Pollution Control Federation, 1980.

PROBLEMS

2-1 Using atomic weights given in Table 2-1, calculate the molecular and equivalent weights of (a) $Al_2(SO_4)_3$, (b) lime, (c) ferrous sulfate ($FeSO_4 \cdot 7H_2O$), (d) fluosilicic acid, and (e) soda ash. [*Answers* (a) 342 and 57.0, others are given in Table 2-3]

2-2 What ions are formed when the following compounds dissolve in water: (a) ammonium nitrate, (b) calcium hypochlorite, (c) sodium fluoride, (d) sulfuric acid, and (e) soda ash? [*Answers* (a), (b), (c), and (d) ions and radicals are listed in Tables 2-1 and 2-2, and (e) Ca^{++} and $CO_3^=$ or HCO_3^- depending on pH, Eq. 2-7]

2-3 If 1.5 mg/l of pure sodium fluosilicate is added to water, what is the corresponding increase in fluoride ion concentration?

2-4 If a water contains 40 mg/l of Ca^{++} and 10 mg/l of Mg^{++}, what is the hardness expressed in milligrams per liter as $CaCO_3$? (*Answer* 141 mg/l)

2-5 If a water contains 56 mg/l of Ca^{++} and 13 mg/l of Mg^{++}, what is the hardness expressed in milligrams per liter as $CaCO_3$?

2-6 The alkalinity of a water consists of 16 mg/l of $CO_3^=$ and 120 mg/l of HCO_3^-. Compute the alkalinity in milligrams per liter as $CaCO_3$. (*Answer* 125 mg/l)

2-7 The alkalinity of a water consists of 15 mg/l carbonate ion and 150 mg/l bicarbonate ion. Compute the alkalinity in milligrams per liter as $CaCO_3$.

2-8 Draw a milliequivalents-per-liter bar graph and list the hypothetical combinations of compounds for the following water analysis:

Ca^{++} = 2.6 meq/l	HCO_3^- = 2.7 meq/l
Mg^{++} = 0.6 meq/l	$SO_4^=$ = 0.6 meq/l
Na^+ = 0.3 meq/l	Cl^- = 0.3 meq/l
K^+ = 0.1 meq/l	

2-9 Draw a milliequivalents-per-liter bar graph for the following water analysis:

Ca^{++} = 60 mg/l	
Mg^{++} = 10 mg/l	
Na^+ = 7 mg/l	
K^+ = 20 mg/l	

$HCO_3^- = 115$ mg/l as $CaCO_3$
$SO_4 = 96$ mg/1
$Cl^- = 11$ mg/1

(*Answer* sum of the cations = 4.6 meq/l)

2-10 Draw a milliequivalents-per-liter graph and list the hypothetical combinations of chemicals in solution for the following:

Calcium hardness	= 150 mg/l
Magnesium hardness	= 65 mg/l
Sodium ion	= 8 mg/l
Potassium ion	= 4 mg/l
Alkalinity	= 190 mg/l
Sulfate ion	= 29 mg/l
Chloride ion	= 10 mg/l
pH	= 7.7

2-11 A sulfuric acid solution is added to a scale-forming water to convert calcium carbonate to calcium bicarbonate. Write the chemical equation for this reaction, and calculate the amount of sulfuric acid in milligrams per liter to neutralize 20 mg/l of calcium carbonate. (*Answer* 9.8 mg/l)

2-12 Sodium carbonate can be formed by passing carbon dioxide gas through a solution of sodium hydroxide. Write a balanced equation with water as the second product formed in the reaction. How many pounds of sodium hydroxide are neutralized per pound of carbon dioxide reacted?

2-13 Calcium hydroxide (lime slurry) reacts with calcium bicarbonate in solution to precipitate calcium carbonate. Write a balanced equation, and calculate the amount of calcium oxide required in the slurry to react with 185 mg/l of calcium hardness. (*Answer* 104 mg/l CaO)

2-14 Metal-plating wastes containing sodium cyanide (NaCN) are chemically treated with chlorine and sodium hydroxide to oxidize the cyanide (CN^-) to nitrogen gas according to the following reaction:

$2NaCN + 5Cl_2 + 12NaOH$

$$= N_2\uparrow + 2Na_2CO_3 + 10NaCl + 6H_2O$$

How many pounds of chlorine are needed to destroy one pound of cyanide?

2-15 After desalination of seawater, lime and carbon dioxide are added to the distillate to form $Ca(HCO_3)_2$ in solution to make a stable, noncorrosive water. Write a balanced equation with CaO, water, and carbon dioxide combining to form $Ca(HCO_3)_2$. If the quantity of calcium ion in the treated water is to be 45 mg/l, what concentrations of carbon dioxide and lime as CaO are to be added?

2-16 What is the saturation value of dissolved oxygen in pure water at 20° C at sea level? At an altitude of 5000 ft? (*Answers* 9.2 mg/l, 7.7 mg/l)

2-17 What is the saturation value of dissolved oxygen in a water containing 1000 mg/l of chloride ion with a temperature of 22°C at sea level? At an altitude of 4000 ft?

2-18 Refer to the titration curves in Figure 2-5a for pure water and 2-5c for water containing carbonic acid. Why does the curve in 2-5c have a sloping section in the central portion while the curve in 2-5a is vertical in this same pH range?

2-19 A 100-ml sample of water with an initial pH of 7.7 is titrated to pH 4.5 using 18.5 ml of 0.02 N sulfuric acid. Calculate the alkalinity. (*Answer* 185 mg/l)

2-20 In titration for alkalinity, 0.6 ml of 0.02 N sulfuric acid is used to reach the phenolphthalein end point and an additional 10.8 ml to reach the mixed bromocresol green-methyl red end point for a 100-ml water sample. Calculate the alkalinity values for the two ionic forms present.

2-21 Name the following compounds: $CH_3CH_2CH_3$, CH_3CH_2OH, CH_3COOH, $CH_3CH_2COO^-Na^+$, CH_3—NH_2, and CH_3CH_2—O—CH_2CH_3. Write the formulas for the following chemicals: methane, methyl alcohol, phenol, isopropyl alcohol, acetone, propionic acid, urea, and sodium acetate.

2-22 Describe the three major categories of biodegradable organic compounds.

2-23 Explain why not all of the organic matter in wastewater is converted to carbon dioxide and water in biological treatment.

2-24 Convert the following units: (a) 50 mg/l to pounds per million gallons and pounds per million cubic feet;

(b) 1500 lb/mil gal to milligrams per liter; and (c) 7.5 gpg to milligrams per liter. [*Answers* (a) 417 lb/mil gal, 3120 lb/mil cu ft, (b) 180 mg/l, (c) 128 mg/l]

2-25   A dosage of 24 mg/l of alum is applied in treating a water supply of 5.0 mgd. How many pounds are applied per day?

2-26   A chlorine dosage of 2.4 mg/l is used in treating an average daily water supply of 3.0 mgd. How many 100-lb chlorine containers are needed per month? (*Answer* 18 containers per month)

2-27   Ten chlorine cylinders, each containing 2000 lb, are emptied in one year at a water treatment plant that processes an average of 6.0 mgd. Compute the mean dosage of chlorine in pounds per million gallons and milligrams per liter.

chapter 3
Biology

An understanding of the key biological organisms—bacteria, viruses, algae, protozoa, crustaceans, and fishes—is essential in sanitary technology. Bacteria and protozoa comprise the major groups of microorganisms in the "living" system that is used in secondary treatment of wastewaters. A mixed culture of these microorganisms also performs the reaction in the biochemical oxygen demand test to determine wastewater strength. The aquatic food chain, involving organisms in natural waters, can be distressed by water pollution. Several waterborne diseases of humans are caused by bacteria, viruses, and protozoa. Indicator organisms, particularly coliforms, are used to evaluate the sanitary quality of water for drinking and recreation.

3-1 BACTERIA AND FUNGI

Bacteria (*sing.* bacterium) are simple, colorless, one-celled plants that use soluble food and are capable of self-reproduction without sunlight. As decomposers they fill an indispensable ecological role of decaying organic matter both in nature and in stabilizing organic wastes in treatment plants. Bacteria range in size from approximately 0.5 to 5 μm and are therefore only visible through a microscope. Figures 3-1*a* and *b* are photomicrographs of bacteria magnified 400 times. Individual cells may be spheres, rods, or spiral-shaped and may appear singly, in pairs, packets, or chains.

Bacterial reproduction is by binary fission, that is, a cell divides into two new cells each of which matures and again divides. Fission occurs every 15 to 30 min in ideal surroundings of abundant food, oxygen, and other nutrients. Some species form spores with tough coatings that are resistant to heat, lack of moisture, and loss of food supply as a means of survival.

A binomial system is used to name bacteria. The first word is the genus; the second is the species name. In activated sludge growing in domestic wastewater, a wide variety of bacteria are found, the majority of which appear to be of the genera *Alcaligenes*, *Flavobacterium*, *Bacillus*, and *Pseudomonas*. Identification of particular types in biological floc is performed only in research studies, since it is extremely difficult and of limited value in treatment operations to isolate them. One of the most common problems in aerobic biological treatment is poor settleability of the activated-sludge floc related to filamentous growths that produce more buoyant floc (Figure 3-2*a* and *b*). Often this is caused by the bacterium *Sphaerotilus natans* (Figure 3-1*b*) where the cells grow protected in a long sheath; on maturity individual motile cells swarm out of the protective tube seeking new sites for growth. Perhaps the most frequently referred to bacterium in sanitary work is *Escherichia coli*, a common coliform used as an indicator of the bacteriological quality of water. *E. coli* cells under microscopic examination at 1000 magnification appear as individual short rods.

Bacteria are classified into two major groups as heterotrophic or autotrophic depending on their source of

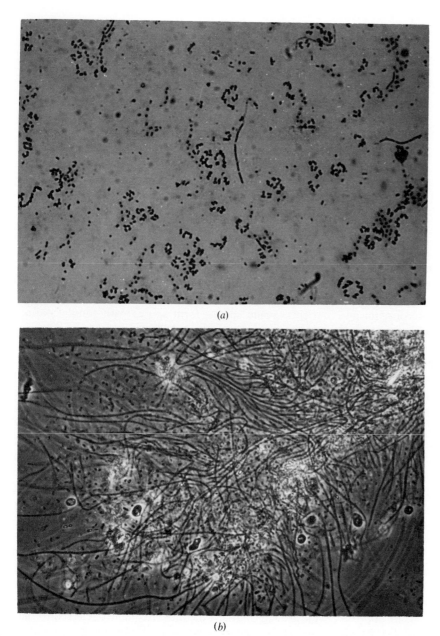

(a)

(b)

Figure 3-1 Photomicrographs of bacterial growths in wastewater. (*a*) Dispersed growth of bacteria (400X). (*b*) Strands of *Sphaerotilus* with swarming cells (400X).

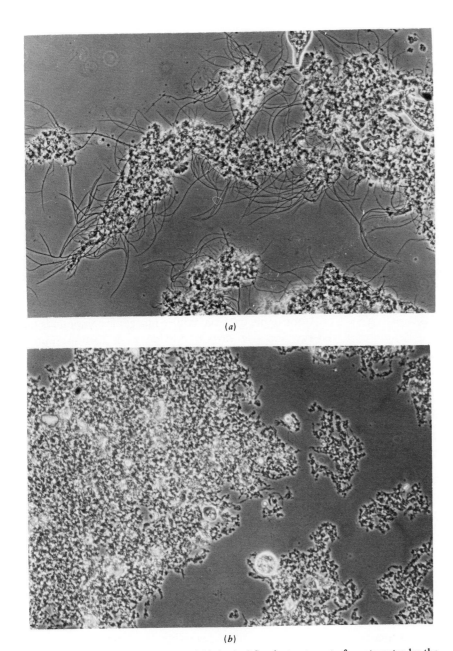

(a)

(b)

Figure 3-2 Photomicrographs of biological floc in treatment of wastewater by the activated-sludge process. (a) Filamentous floc (100X). (b) Normal floc (100X).

nutrients. Heterotrophs, sometimes referred to as saprophytes, use organic matter as both an energy and a carbon source for synthesis. These bacteria are further subdivided into three groups depending on their action toward free oxygen. Aerobes require free dissolved oxygen in decomposing organic matter to gain energy for growth and multiplication, Eq. 3-1. Anaerobes oxidize organics in the complete absence of dissolved oxygen by using oxygen bound in other compounds, such as nitrate and sulfate. Facultative bacteria comprise a group that uses free dissolved oxygen when available but that can also live in its absence by gaining energy from anaerobic reaction. In wastewater treatment, aerobic microorganisms are found in activated sludge and trickling filters, but anaerobes predominate in sludge digestion. Facultative bacteria are active in both aerobic and anaerobic treatment units.

Aerobic:

$$Organics + oxygen \rightarrow CO_2 + H_2O + energy \qquad (3\text{-}1)$$

Anaerobic:

$$Organics + NO_3^- \rightarrow CO_2 + N_2 + energy \qquad (3\text{-}2)$$

$$Organics + SO_4^= \rightarrow CO_2 + H_2S + energy \qquad (3\text{-}3)$$

$$Organics \rightarrow organic\ acids + CO_2$$
$$+ H_2O + energy \quad (3\text{-}4)$$
$$\rightarrow CH_4 + CO_2 + energy \quad (3\text{-}5)$$

The primary reason heterotrophic bacteria decompose organics is gain of energy for synthesis of new cells, and for respiration and motility. A small fraction of the energy is lost in the form of heat, Eq. 3-6. Amount of energy

$$Synthesis: \quad Energy + organics \rightarrow new\ cell\ growth$$
$$\rightarrow respiration\ and\ motility$$
$$\rightarrow lost\ heat \qquad (3\text{-}6)$$

biologically available from a given quantity of matter depends on the oxygen source used in metabolism. The greatest amount is available when dissolved oxygen is used in oxidation, Eq. 3-1, and the least energy yield is derived from strict anaerobic metabolism, Eq. 3-4. Microorganisms growing in wastewater seek the greatest

energy yield in order to have maximum synthesis. For example, consider the reactions that would occur in a sample of fresh aerated wastewater placed in a closed container. Aerobic and facultative bacteria immediately start to decompose waste organics, depleting the dissolved oxygen. Although the strict aerobic organisms could not continue to function, facultative bacteria operating anaerobically can use the bound oxygen in nitrate releasing nitrogen gas, Eq. 3-2. The next most accessible oxygen is available in sulfate (Eq. 3-3) by conversion to hydrogen sulfide (rotten egg odor). Simultaneously, other facultative and anaerobic bacteria partially decompose material to organic acids, alcohols, and other compounds, Eq. 3-4, producing the least amount of energy. If methane-forming bacteria are present, the digestion process is completed by converting the organic acids to gaseous end products of methane and carbon dioxide, Eq. 3-5.

Autotrophic bacteria oxidize inorganic compounds for energy and use carbon dioxide as a carbon source. Nitrifying, sulfur, and iron bacteria are of greatest significance. Nitrifying bacteria oxidize ammonium nitrogen to nitrate in a two-step reaction as follows:

$$NH_3 + oxygen \xrightarrow{\textit{Nitrosomonas}} NO_2^- + energy \quad (3\text{-}7)$$
$$NO_2^- + oxygen \xrightarrow{\textit{Nitrobacter}} NO_3^- + energy \quad (3\text{-}8)$$

Nitrification can occur in biological secondary treatment under the conditions of low organic loading and warm temperature. Although providing a more stable effluent, nitrification is often avoided to reduce oxygen consumption in treatment and to prevent floating sludge on the final clarifier. The latter is caused when sludge solids are buoyed up by nitrogen gas bubbles that are formed as a result of nitrate reduction, Eq. 3-2.

A common sulfur bacterium performs the reaction given in Eq. 3-9, which leads to crown corrosion in sewers.

$$H_2S + oxygen \rightarrow H_2SO_4 + energy \qquad (3\text{-}9)$$

Wastewater flowing through sewers often turns septic and releases hydrogen sulfide gas, Eq. 3-3. This occurs most frequently in sanitary sewers constructed on flat grades in warm climates. The hydrogen sulfide is absorbed in the condensation moisture on the side walls and crown of the pipe. Here sulfur bacteria, able to tolerate pH levels of less than 1.0, oxidize the weak acid H_2S to strong sulfuric acid using oxygen from air in the sewer. The sulfuric acid reacts with concrete, reducing its structural strength. If sufficiently weakened, corrosion can lead to collapse under heavy overburden loads. Using corrosion-resistant pipe material, such as vitrified clay or PVC plastic, is the best protection from corrosion in sanitary sewers. In large collectors where size and economics dictate concrete pipe, crown corrosion can be reduced by ventilation to expel the hydrogen sulfide and to reduce the moisture of condensation, or by chemical treatment of the wastewater as it flows through the sewer to control generation of hydrogen sulfide. Protection of the concrete pipe interior by coatings and linings should be considered.

Iron bacteria are autotrophs that oxidize soluble inorganic ferrous iron to insoluble ferric, Eq. 3-10. The

$$Fe^{++}(ferrous) + oxygen \rightarrow Fe^{+++}(ferric) + energy$$

$$(3-10)$$

filamentous bacteria *Leptothrix* and *Crenothrix* deposit oxidized iron, $Fe(OH)_3$, in their sheath, forming yellow or reddish-colored slimes. Iron bacteria thrive in water pipes where dissolved iron is available as an energy source and bicarbonates are available as a carbon source. With age the growths die and decompose, releasing foul tastes and odors. There is no easy or inexpensive way of controlling bacteria in water distribution systems. The most certain procedure is to eliminate the ferrous iron from solution by water treatment and by controlling internal pipe corrosion. An alternative to removing iron from the water is a continuous ongoing maintenance program of treating and flushing water mains. A section of main to be treated is isolated, and the water is then dosed with an excessive concentration of chlorine or other chemical to kill the bacteria. After several hours of contact the pipe section is flushed and returned to service.

Fungi (*sing.* fungus) refer to microscopic nonphotosynthetic plants including yeasts and molds. Yeasts are used for industrial fermentations in baking, distilling, and brewing. Under anaerobic conditions, yeast metabolizes sugar, producing alcohol with minimum synthesis of new yeast cells. Under aerobic conditions, alcohol is not produced and the yield of new cells is much greater. Therefore, if the objective is to grow fodder yeast on waste sugar or molasses, aerobic fermentation is used.

Molds are filamentous fungi that resemble higher plants in structure with branched, threadlike growths. They are nonphotosynthetic, multicellular, heterotrophic, aerobic, and grow best in acid solutions high in sugar content. Growths are frequently observed on the exterior of decaying fruits. Because of their filamentous nature, molds in activated-sludge systems can lead to a poor settling floc, preventing gravity separation from the wastewater effluent in the final clarifier. Undesirable fungi growths are most frequently created by low pH conditions often associated with treatment of industrial wastes high in sugar content. Addition of an alkali to increase the pH, and sometimes ammonium nitrogen in high-carbohydrate wastewater, is necessary to suppress molds and to enhance the growth of bacteria.

3-2 VIRUSES

Viruses are obligate, intracellular parasites that replicate only in living hosts' cells. Composed largely of nucleic acid and protein, they lack the metabolic systems for self-reproduction. Their small size requires viewing under an electron microscope; most viruses of interest in sanitary technology are 20 to 100 nanometers (millimicrons) in size, which is about one fiftieth the size of bacteria. As illustrated in Figure 3-3, virus particles are generally helical, polyhedral, or combination T-even. Viruses that infect only bacteria are called bacteriophages, or simply phages.

The life-reproductive cycle of a virus of a T-even phage infecting a bacterium consists of six stages as diagramed in Figure 3-4. In the dormant stage, viruses have no interaction with their surroundings. Adsorption of a virus onto the cell wall takes place by means of a complementary association between the attachment site of the virus particle and a receptor site on the cell. Following attachment, the bacteriophage injects genetic material into the bacterium; the protein coat does not enter the cell. In

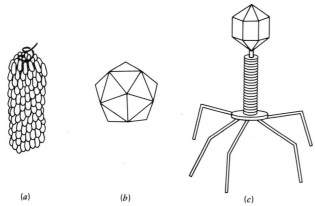

Figure 3-3 Typical structural shapes of viruses. (*a*) Helical. (*b*) Polyhedral. (*c*) Combination T-even.

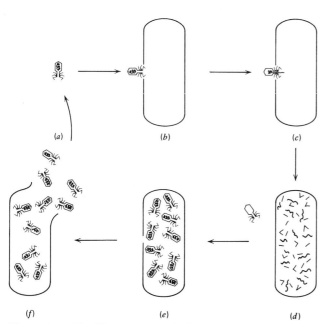

Figure 3-4 The life-reproductive cycle of a virus as illustrated by a T-even bacteriophage infecting a bacterial cell: (*a*) dormancy, (*b*) adsorption, (*c*) penetration, (*d*) replication of new proteins and nucleic acids, (*e*) maturation, and (*f*) release.

contrast, animal viruses break the host cell's membrane and drift in to uncoat the genetic material. Once inside, the virus takes over the cell's metabolism to synthesize new virus proteins and nucleic acids. The subsequent maturation process assembles the protein coat and nucleic acids to form complete virus particles, and finally the bacterium lyses (breaks open) to release the new viruses. For phages infecting bacteria, the whole cycle takes about 30 to 40 min.

Viruses of concern in water pollution are those found in the intestinal tracts of humans. Since viruses do not normally reside in humans, they are excreted with the feces from infected persons, who are mostly infants and children. The frequent means of transmission is from person to person by the fecal-oral route. Nevertheless, viruses are present in wastewater treatment plant effluents discharged to surface waters. The viruses found in relatively large numbers in human feces are adenoviruses associated with respiratory and eye infections; coxsackieviruses, causing aseptic meningitis, fever, hepangina, and myocarditis; echoviruses, causing aseptic meningitis, diarrhea, rash, and respiratory infections; polioviruses, causing aseptic meningitis and poliomyelitis; and infectious hepatitis, causing liver inflammation. Infectious hepatitis is the only viral infection proven and documented to have been transmitted by drinking water, causing outbreaks of disease.

3-3   ALGAE

Algae (*sing.* alga) are microscopic photosynthetic plants of the simplest forms, having neither roots, stems, nor leaves. They range in size from tiny single cells, giving water a green color, to branched forms of visible length that often appear as attached green slime. The term diatom is sometimes used in referring to single-celled algae encased in intricately etched silica shells. Figure 3-5 pictures algae filtered from water of a eutrophic lake magnified 100 times. *Anacystis*, *Anabaena*, and *Aphanizomenon* are blue-green algae associated with polluted water. Long strands of the latter clumped together appear to the naked eye as short grass clippings when suspended in water. *Oocystis* and *Pediastrum* are green algae. There are hundreds of algal species in a wide variety of cell

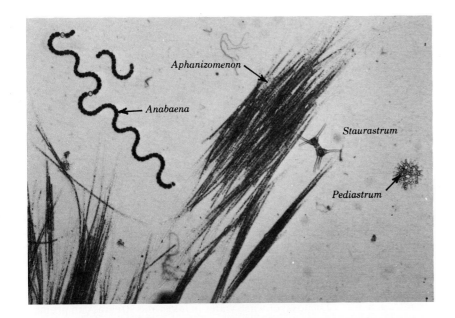

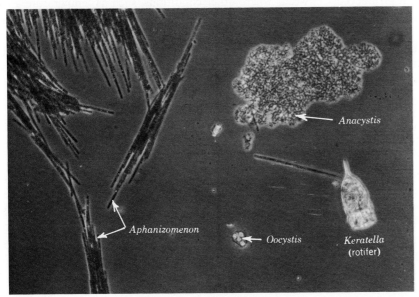

Figure 3-5 Photomicrographs of algae filtered from water of a eutrophic lake (100X).

structures in various shades of green and, less commonly, brown and red. An alga is identified by microscopic observation of its essential characteristics. *Standard Methods*[1] contains colored illustrations of common algae associated with taste and odor, filter clogging, polluted water, clean water, and others.

The process of photosynthesis is illustrated by the equation

$$CO_2 + PO_4 + NH_3 \xrightarrow[\text{sunlight}]{\substack{\text{energy} \\ \text{from}}} \frac{\text{new cell}}{\text{growth}} + O_2 \quad (3\text{-}11)$$

Algae are autotrophic, using carbon dioxide or bicarbonates as a carbon source, and inorganic nutrients of phosphate and nitrogen as ammonia or nitrate. In addition, certain trace nutrients are required, such as magnesium, boron, cobalt, and calcium. Certain species of blue-green algae are able to fix gaseous nitrogen if inorganic nitrogen salts are not available. The products of photosynthesis are new plant growth and oxygen. Energy for photosynthesis is derived from light. Pigments, the most common being chlorophyll, which is green, biochemically convert sunlight energy into useful energy for plant growth and reproduction. In the absence of sunlight, plants use some of the previously formed photosynthate, which when combined with oxygen, provides the necessary energy for continued survival. The rate of this survival reaction is significantly slower than photosynthesis.

The purpose of photosynthesis is to produce new plant life, thereby increasing the number of algae. Given a suitable environment and proper nutrients, algae grow and multiply in abundance. In natural waters the growth of algae may be limited by turbidity blocking sunlight, low temperatures during the winter, or depletion of a key nutrient. Clear, cold, mountain lakes tend to support few algae, while warm-water lakes enriched with nitrogen and phosphorus from land runoff, or wastewaters, exhibit heavy algal growths creating green-colored, turbid water during the plant growing season. Wastewater stabilization ponds support luxurious blooms of algae to the point where the suspension becomes self-shading, that is, surplus nitrogen, phosphorus, and carbon nutrients cannot be synthesized because turbidity caused by the algae limits penetration of sunlight.

Macrophytes are aquatic photosynthetic plants, excluding algae, that occur as floating, submersed, and immersed types. Floating plants are not rooted and ride on the water surface; duckweed, a small three-leafed floating plant about 5 mm in diameter is common, and another is water hyacinth with its flowering tops. All or most of the foliage of submersed plants—pond weed, water weed, and coontail—grow beneath the water surface. These may root at depths greater than 10 ft below the surface, depending on the clarity of the water. Immersed plants, such as pickerelweed and cattail, attached by roots to the bottom mud, extend their principal foliage into the air above the surface. Rock- and gravel-bottom lakes and nutrient-poor bodies support limited growths of aquatic plants, while eutrophic lakes have extensive weed beds in shallow bays and along shore lines. Disposal of wastewater in lakes and streams can promote weeds where other conditions, such as temperature and sunlight, are favorable. Unwanted weeds in wastewater stabilization ponds are controlled by constructing rather steep-side slopes and maintaining a minimum depth of liquid of not less than 3 ft to block sunlight penetration to the bottom.

3-4   PROTOZOA AND MULTICELLULAR ANIMALS

Protozoa (or protozoans, *sing.* protozoan) are single-celled aquatic animals that multiply by binary fission. They have complex digestive systems and use solid organic matter as food. Protozoa are aerobic organisms found in activated sludge, trickling filters, and oxidation ponds treating wastewater, as well as in natural waters. By ingesting bacteria and algae, they provide a vital link in the aquatic food chain.

Flagellated protozoa (Figure 3-6a) are the smallest type ranging in size from 10 to 50 μm. Long hairlike strands (flagella) provide motility by a whiplike action. While many ingest solid food, some flagellated species take in soluble organics. Amoeba (Figure 3-6b) move and take in food through the action of a mobile protoplasm. Although not as common as other forms of protozoa,

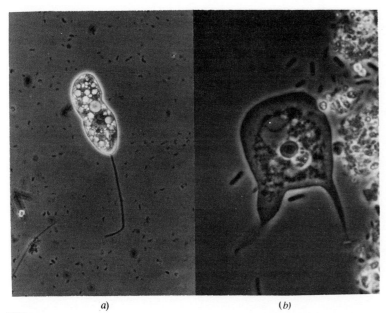

a)   *(b)*

Figure 3-6 Photomicrographs of protozoa found in wastewater aeration basins. (*a*) Flagellated protozoan (1000X). (*b*) Amoeba, mobile protoplasm (1000X).

amoeba are often found in the slime coating on trickling filter media and aeration basin walls. Free-swimming protozoa (Figure 3-7*a*) have cilia, small hairlike processes, used for propulsion and gathering in organic matter. These are easily observed in a wet preparation under a microscope because of their rapid movement and relatively large size, 50 to 300 μm. Stalked forms (Figure 3-7*b*) attach by a stem to suspended solids and use cilia to propel their heads about and bring in food.

Rotifers are simple, multicelled, aerobic animals that metabolize solid food. The rotifer illustrated in Figure 3-8 uses two circular rows of head cilia for catching food. Its head and foot are telescopic and it moves with a leechlike action, attaching the foot to a surface. Figure 3-5 pictures an entirely different type of rotifer with neither appendages nor telescopic head or foot. *Keratella* has a hard protective shell, an anus and, on the front, six spines along with cilia. Rotifers are found in natural waters, stabilization ponds, and extended aeration basins under low organic loading.

Crustaceans are multicellular animals, typically 2 mm in size, easily visible with the naked eye. They have branched swimming feet or a shell-like covering with a variety of appendages. In the aquatic food chain they serve as herbivores, ingesting algae and, in turn, being eaten by fishes. Most protozoa and higher animals can be identified by microscopic examination using a recognition key as an aid.[2]

3-5 FISHES

Fishes are cold-blooded animals typically with backbones, gills, fins, and usually scales. The internal structure is similar to that of other animals with flexible backbone and ribs, gullet, stomach, intestine, liver, spleen, and kidney. A two-chambered heart pumps blood to the gills to pick up oxygen before going to other parts of the body. Fishes also have a brain protected by the skull, spinal cord through the backbone, and nerves extending to internal organs and muscles. An air bladder, below the backbone,

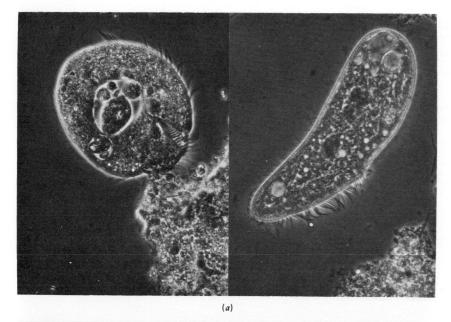

(a)

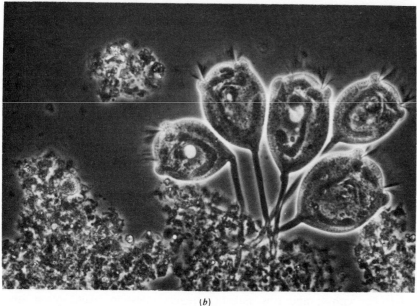

(b)

Figure 3-7   Photomicrographs of protozoa found in wastewater aeration basins. (a) Free-swimming protozoa (400X). *Euplotes* (*left*) and *Blepharisma* (*right*). (b) Stalked protozoa, *Vorticella* (400X).

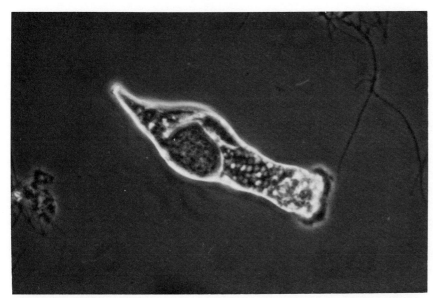

Figure 3-8 Photomicrograph of a rotifer, *Rotaria*, found in aerobic wastewater treatment (200X).

is used in maintaining buoyancy in water. Like all animals, fishes have a full complement of diseases that arise either internally or as a result of external agents. Illnesses include common organic and degenerative disorders of the internal organs. From outside come viruses, fungi, bacteria, parasitic protozoa, and worms. If they are not killed by disease or disorder, fishes must still survive predators, fishermen, and adverse environment. Often, the latter is caused by water pollution, which not only affects adults directly but also their reproductive success.

Fish gills act as lungs. Fine red filaments attached to the gill arch bond contain numerous blood cells that absorb dissolved oxygen from water and give off carbon dioxide. Insufficient oxygen causes death by suffocation. This may occur naturally during the winter in shallow lakes and ponds covered with ice and snow for extended periods of time. However, more frequently it results from bacterial decomposition of waste organics depleting the dissolved oxygen. Municipal wastewater disposed of in a stream without adequate treatment, or polluted land drainage from storm runoff, can reduce dissolved oxygen levels and drive fishes away or trap them, resulting in a fish kill. The deep unmixed bottom water in a eutrophic lake may have

too little oxygen in late summer, after several months of thermal stratification, thereby limiting fishes to the warmer surface water. Excessive blooms of algae in this upper layer can, under certain conditions, result in oxygen depletion during the night, causing fish deaths.

The body of a fish is covered with scales and/or skin. Scales, formed of bonelike material, are contained in pockets in the skin. A protective slimy substance is produced to help prevent bacteria and other disease-producing organisms from penetrating the skin. Chemical pollutants may attack this protective slime, making the fish more vulnerable to diseases. Fins are used in swimming and to maintain balance. Fishes use an internal ear to detect vibrations and lidless, movable eyes for vision. Predator species, often considered by fisherman as the best game, are seriously impaired by turbid water, which limits their ability to catch prey. Silt from erosion of cultivated fields, gravel washing, quarrying, and other industrial operations frequently contributes to unwanted turbidity.

The common reproductive process of fishes is deposition of eggs by the female and fertilization by the male. Some species construct saucer-shaped depressions in which they deposit their eggs and guard their young; others spawn at random and leave the fate of their brood to nature. Sunfishes, crappies, and largemouth bass nest in a circular depression in mud or sand, often among roots of aquatic plants, while others, such as smallmouth bass, prefer a gravel bottom for nesting. Northern pike, walleyes, and carp scatter their eggs over bottom areas in shallow water. Some fishes, for example, trout and salmon, ascend streams to spawn. Success in reproduction can be severely affected by pollution. Siltation covers fish eggs, certain heavy metals and organic chemicals reduce hatch, and stream contamination may inhibit migration to spawning grounds.

Generally, fish can be identified by their external appearance. Size and shape of scales, fins, teeth, head, and body are common criteria. In ichthyology (a branch of zoology dealing with fishes) an elaborate classification has been developed that groups fishes that look alike or have similar characteristics. The most meaningful grouping for a sanitary technologist is by tolerance level. Intolerant species—grayling, white fishes, brook and rainbow trout—require cold water, high dissolved oxygen concentrations, and low turbidity. These are the choice food fishes and, unfortunately, also the most susceptible to adverse environmental changes. Tolerant fishes can withstand wide fluctuations in surroundings, warm water, low dissolved oxygen, high carbon dioxide, and turbidity. Carp, black bullhead, and bowfin are examples. Many moderately tolerant fishes, such as walleye and northern pike, are both good angling and eating. Their location in warm-water lakes near densely populated areas has promoted them as prized game fishes by sportsmen. However, the waterborne wastes from surrounding agriculture, business, and housing have polluted these warm-water streams and lakes, causing a shift in fish populations to more tolerant species such as bluegill and yellow perch.

3-6 AQUATIC FOOD CHAIN

Aquatic productivity is limited by the supply of raw materials and the biological efficiency of converting them into various life forms. Rivers and lakes well supplied with oxygen, carbon dioxide, nitrogen, phosphorus, and sufficient sunlight are rich in plant and animal life. If any of these nutrients is scarce, or if the water is polluted or does not receive enough sunlight, the production of life is low. Basic forms of life, algae and other green plants, are called primary producers, since they use the energy of sunlight to synthesize inorganic substances into living tissues. Rooted plants, although usually the most conspicuous water vegetation, play a relatively minor role in river or lake productivity. The most numerous plants are algae. Animals, being unable to manufacture their own food, obtain energy and nutrients secondhand by eating either plants or smaller animals. Flow of energy and material from one form of life to another is known as a food chain. The aquatic food chain in Figure 3-9 shows algae as the primary producers. Then come the first-order consumers—animals, such as fly nymphs, copepods, and water fleas. These small herbivores (plant eaters) are in turn eaten by second-order consumers—carnivores (flesh eaters), such as sunfishes. A third-order consumer, such as a bass, may then eat the sunfishes. If the bass dies, it is consumed by scavengers like crayfishes; or the nutrients stored in the dead fish are released by bacterial decomposers and recycled into algae by photosynthesis. Examining a single food chain reveals only part of nature's complex process since hundreds of them exist. Most of the chains are interconnected such that the organization of a community is more accurately described as a food web. For example, rather than dying the bass might be carried away by a heron or captured underwater by a mink. The fly nymph can leave the water environment by emerging as an adult and flying away to be consumed by a swallow.

The complex system of aquatic food relationships can be upset by pollution. For example, primary production is modified by increasing turbidity or overfertilization. Murky water blocks out sunlight, slowing plant growth; by chain reaction it reduces the supply of food to all higher forms of life. Overfertilization of water with plant nutrients can be equally disastrous, particularly in lakes. Uninhibited photosynthesis produces more algae than the herbivores can consume. Dense blooms of algae create turbidity, floating scum, and reduction in dissolved oxygen when the mass of algae decompose. More

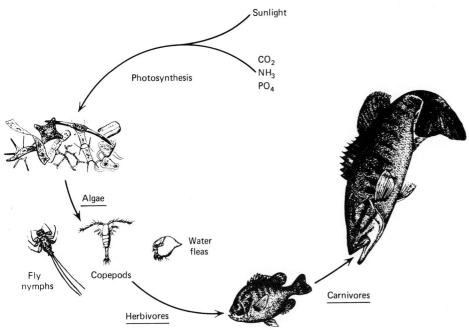

Figure 3-9   Aquatic food chain. Energy and material pass from green plants formed by photosynthesis through small animals and immature insects that graze on algae to second- and third-order consumers, which are fishes.

available nutrients also increase aquatic weed growths, which in combination with increased turbidity from algae, upset the predator-prey relationship between carnivores. Prey species, such as bluegills, can effectively hide from predatory fishes. Thus, the total number of smaller fishes increases dramatically compared to larger game species.

Industrial wastes—acids, oils, heavy metals, or other chemicals—have a direct influence on aquatic life. Many wastes can directly kill fishes if they shift the pH of the water sufficiently, cover the surface causing a reduction in dissolved oxygen, or build up to toxic levels. A more subtle effect results if the contaminant dosage is sublethal to the fishes but eliminates their food supply by poisoning the algae or herbivores. Some chemicals are magnified biologically as they pass through the food chain. If traces of mercury, lead, or chlorinated hydrocarbons are in water, their concentration found in algae is much greater. Herbivores consuming the algae accumulate these contaminants in their bodies, which, in turn, are concentrated

to even higher levels in fishes. The seriousness of biological magnification varies with different chemicals. Most frequently the influence is indirect, often resulting in decreased population, either by decreasing reproductive success or by influencing body functions, making the fish a less successful predator. In other instances, the fish flesh may be tainted or unedible.

Biological life in clean water is exemplified by a large number of different species of plants and animals with generally balanced populations. Polluted water supports large populations with little variety. Small polluted lakes may have strong dominance of one type of fish; a pond turbid with silt may support a large population of black bullheads, while a heavily eutrophic lake may dominate in bluegills. A short distance downstream from the introduction of a large quantity of organic matter the dissolved oxygen decreases sharply. Here the most abundant form of life is bacteria, such as *Sphaerotilus*, so-called sewage fungus, rat-tailed maggots, mosquito larvae, and sludge worms. Further downstream protozoa, rotifers, certain fly larvae, polluted-water algae, and tolerant fish species appear. Finally, after the stream has returned to a high dissolved oxygen content, a broad spectrum of life reappears on each level of the food web.

3-7 WATERBORNE DISEASES

Several infectious, enteric (intestinal) diseases of humans are transmitted through fecal wastes. Pathogens, disease-producing agents excreted in the feces of infected persons, include all major categories: bacteria, viruses, protozoa, and parasitic worms. The resultant diseases are most prevalent in those areas where sanitary disposal of human feces is not practiced. Transmission can be by direct contact of infected individuals, via insect vectors such as flies, or through contaminated food and water. Complete control involves instituting an environmental health program that incorporates personal and household hygiene, control of fly species and other insects, monitoring of food processing, proper waste disposal, immunization of the populace where possible, and isolation and treatment of diseased persons.

Three of the most common waterborne bacterial diseases are typhoid fever caused by *Salmonella typhi*, cholera (*Vibrio cholera*), and bacillary dysentery (*Shigella dysenteriae*). Typhoid is an acute infectious disease characterized by a continued high fever and infection of the spleen, gastrointestinal tract, and blood. Cholera symptoms include diarrhea, vomiting, and dehydration. Dysentery causes diarrhea, bloody stools, and sometimes fever. All of these diseases are debilitating and often cause death. Transmission is by direct contact, food, milk, shellfish, and water. Although these diseases are still prevalent in underdeveloped countries and have been responsible for millions of deaths historically, they have been virtually eliminated in the United States by environmental control through pasteurization of milk and chlorination of water supplies.

Viruses of poliomyelitis and infectious hepatitis are excreted in the stools of infected persons, but the modes of transmission are not completely defined. Transfer of polio virus appears to be by direct contact with some evidence of transmission by food and water. Waterborne outbreaks of infectious hepatitis have occurred, although spreading by person-to-person contact appears to be the chief means of transmission. Most authorities believe that outbreaks from municipally treated water supplies are unlikely. The symptoms of hepatitis are loss of appetite, fatigue, nausea, and pain. The most characteristic feature of the disease is a yellow color that appears in the white of the eyes and skin, hence the common name yellow jaundice. Although more severe in older persons, it is seldom fatal.

Two waterborne parasitic protozoa are *Entamoeba histolytica*, causing the severe disease of amoebic dysentery, and *Giardia lamblia*, causing a less severe gastrointestinal infection. Amoebic dysentery is transmitted by direct human contact and by contaminated food and water in tropical climates, but it is nontransmittable in temperate climates. Water carriage can be prevented by filtration treatment with heavy chlorination, or by individuals boiling their drinking water. Giardiasis is an intestinal disease prevalent in the United States, causing diarrhea, nausea, vomiting, and fatigue lasting from a few days to several weeks. The incubation period in the host before symptoms appear is usually one to two weeks. *Giardia* enter a water supply through the feces of an infected host. Although humans and many other animals can be infected, beavers are considered the common hosts. Consequently, streams flowing in mountainous regions are most likely to be contaminated. The life cycle of *Giardia lamblia* has an inactive cyst stage and a disease-causing stage as microscopic trophozoites. Cysts can exist for long periods of time in water with longer survival times at cooler temperatures. After cysts are ingested, trophozoites come out of the cysts and infect the host's intestines. Those trophozoites not remaining in the intestinal tract are shed in cyst form and excreted in the host's feces. Granular-media filtration following chemical coagulation is the most effective way to remove *Giardia* cysts from water. If their presence is suspected, a high degree of water treatment is recommended. Preferably, the turbidity after filtration should be less than 0.1 NTU. Furthermore, disinfection at the usual levels of chlorine residual may be inadequate since the cysts are resistant to destruction by chlorination.

Schistosomiasis, often called bilharziasis, is a parasitic disease caused by a small, flat worm that infests human internal organs—heart, lungs, veins, and liver. Eggs of the schistosome worms, living in human abdominal organs, are passed into water with fecal discharges. They hatch into miracidia, enter snails, and develop into sporocysts that produce fork-tailed cercariae. These leave the snail, attach themselves to humans, bore through the skin, and enter blood vessels migrating to the internal organs where they grow into adulthood. This debilitating disease, for which there is no immunization, is the world's worst health problem in agricultural communities in Asia, Africa, and South America. The disease does not occur in the continental United States, since the intermediate snail host must be one of several specific species that are not found in this country. Duck schistosomiasis, transmitted in a fashion similar to the human disease, is found in United States lakes frequented by wild ducks. If infected snails inhabit bathing beaches on these lakes, the cercariae released may come in contact with and penetrate the skin of bathers. The resultant infection, neither a contagious nor a fatal disease, is referred to as swimmer's itch. Red spots appear where the cercariae have penetrated, accompanied by severe irritation that lasts several hours. During the following two days, extremely itchy small lumps may appear, sometimes swelling and filling with pus.

Safety surveys of sanitation workers have shown that incidence of waterborne disease is no greater in this group than in the population as a whole. Although wastewater must be considered potentially pathogenic, the reduced incidence of waterborne disease in the United States has diminished the possibility of treatment plant employees becoming infected. Most safety manuals stress that the best defense against infection is the practice of good personal hygiene and prompt medical care for any injury that breaks the skin. The latter appears to be the most critical problem, since infected wounds are far more common than enteric diseases. The most common source of *Clostridium tetani* is human feces; therefore, it is recommended that sanitation workers receive artificial immunity by tetanus toxoid injections.

3-8   INDICATOR ORGANISMS FOR WATER QUALITY

Testing a water for pathogenic bacteria might at first glance be considered a feasible method for determining bacteriological quality. However, on closer examination, this technique has a number of shortcomings that preclude its application. Laboratory analyses for pathogenic bacteria are difficult to perform and generally are not quantitatively reproducible. Furthermore, the demonstrated absence of *Salmonella*, for example, does not exclude the possible presence of *Shigella*, *Vibrio*, or disease-producing viruses. Finally, since few pathogens are present in polluted waters, the high frequency of negative results would lead to questioning the validity of the test procedures. It is far more reassuring to an analyst if a few tests yield positive results. For these reasons, the bacteriological quality of water is based on testing for nonpathogenic indicator organisms, principally the coliform group.

Coliform bacteria, as typified by *Escherichia coli*, and fecal streptococci (enterococci) residing in the intestinal tract of humans, are excreted in large numbers in feces of humans and other warm-blooded animals, averaging about 50 million coliforms per gram. Untreated domestic wastewater generally contains more than 3 million coliforms per 100 ml. Pathogenic bacteria and viruses causing enteric diseases in humans originate from the same source, namely, fecal discharges of diseased persons. Consequently, water contaminated by fecal pollution is identified as being potentially dangerous by the presence of coliform bacteria.

Drinking water standards generally specify that a water is safe provided that testing in a specified manner does not reveal more than an average of 1 coliform organism per 100 ml. This criterion is supported by the following arguments. The number of pathogenic bacteria, such as *Salmonella typhi*, in domestic wastewater in the United States is generally less than 1 per million coliforms, and measured virus densities are in a wide range of approximately 10 to 200 enteric viruses per million coliforms. The die-off rate of pathogenic bacteria is greater than the death rate of coliforms outside of the intestinal tract of animals; thus, exposure to treatment and residence in surface waters reduce the number of pathogens relative to coliforms. Based on these premises, water that meets a standard of less than 1 coliform per 100 ml is, statistically speaking, safe for human consumption because of the improbability of ingesting any pathogens. Strictly speaking, this coliform standard as established by the Environmental Protection Agency[3] applies only to processed water where treatment includes chlorination.

Extension of coliform criteria to water quality for purposes other than drinking is poorly defined. Since these bacteria can originate from warm-blooded animals, soil, and cold-blooded animals in addition to feces of humans, presence of coliforms in surface waters indicates any one or a combination of three sources: wastes of humans, farm animals, or soil erosion. Although a special test can be run to separate fecal coliforms from soil types, there is no way of distinguishing between the human bacteria and those of animals. The significance of coliform testing in pollution surveys then depends on a knowledge of the river basin and the most probable source of the observed coliforms.

Numerical coliform criteria for body-contact water use and general recreation have been established by most states at the upper limits of 200 fecal coliforms per 100 ml and 1000 total coliforms per 100 ml. These values are only considered to be guidelines, since there is no positive epidemiological evidence that bathing beaches with higher coliform counts are associated with transmission of enteric diseases. In fact, these standards appear to be far too conservative from a standpoint of realistic public health risk. A survey of illnesses developed by persons as a result of bathing on public beaches showed that the majority were infections of skin, eye, ear, nose, or throat with only about one tenth related to gastrointestinal disorders. In recent years in the United States, no cases of enteric disease have been linked directly to recreational water use, while several thousand persons have drowned and even greater numbers have been disabled by automobile accidents in traveling to recreational areas. Although substantial risk is involved in recreational water use, the greatest danger is apparently not related to disease transmission. Hence, coliform standards applied to water used for swimming are linked to water-associated diseases of the skin and respiratory passages rather than enteric diseases. This is, of course, completely different than the purpose of the coliform standard for drinking water, which is related to enteric disease transmission. Here tighter restrictions are imperative, since a water distribution system has the potential of mass transmission of pathogens in epidemic proportions.

Streptococci could be used as indicator organisms rather than coliforms. But the use of coliforms was established first, and there does not appear to be any distinct advantages to warrant shifting to a fecal streptococcus system.

3-9   TESTS FOR THE COLIFORM GROUP

Extensive laboratory apparatus is needed to conduct bacteriological tests, including a hot-air sterilizing oven, autoclave, incubators, sample bottles, fermentation tubes with inverted vials, dilution bottles, petri dishes and dish containers, graduated pipettes and pipette containers, a wire inoculating loop, culture media, and preparation utensils. The hot-air oven is used for sterilizing glassware, empty sample bottles, pipettes, and petri dishes. Heating to a temperature of 160 to 180° C for a period of $1\frac{1}{2}$ hr is adequate to kill microbial cells and spores. An autoclave is used to sterilize culture media and dilution water under steam pressure. Recommended operation is for 15 min at 15 psi steam pressure, which corresponds to a temperature of 121.6° C. Sample bottles with a volume of about 120 ml are usually soft glass with screw-top closures suitable to withstand sterilization temperatures. Fermentation tubes are round-bottomed glass tubular containers designed to accept either a screw cap, a slip-on stainless steel closure, or an inserted cotton plug. The inverted vial is a small tube placed upside down in the culture medium in the fermentation tube to collect gas generated by bacterial growth. Screw-cap dilution bottles are to autoclave measured volumes of distilled water for use in diluting water samples prior to testing. Petri dishes are small shallow circular plates with glass covers for holding solid culture media. The dishes are placed in a stainless steel container for sterilization in a hot-air oven. Pipettes used to transfer samples from one container to another are also hot-air treated to prevent sample contamination. A small, thin, wire loop of Chromel or platinum is used to aseptically transfer small quantities of culture. The wire is sterilized by heating red-hot in the flame of a bunsen burner.

Culture media are special formulations of organic and inorganic nutrients to support the growth of microorganisms. Dehydrated culture media (powders) are available commercially. Preparation involves placing a measured amount in distilled water and heating in a double boiler to dissolve without scorching. Lactose broth containing beef extract, peptone (protein derivatives), and lactose (milk sugar) is the primary medium used in testing for coliform bacteria. Some laboratories use lauryl tryptose broth, which also contains the critical lactose ingredient. Since occasionally noncoliform aerobic spore-forming bacteria, or groups of bacteria working together, can produce gas in lactose broth, a fermentation tube showing positive results is tested further to confirm the presence of coliforms. The medium used is brilliant green lactose bile broth. The green dye inhibits noncoliform growth while the presence of lactose stimulates the coliforms to overcome the toxic effect of the dye. There are two types of solid inhibitory media—eosin methylene blue (EMB) and Endo agar—that are used in coliform testing. Both contain inhibitory dyes and agar for solidification. These solid media are prepared by dissolving in distilled water, dispensing into bottles for autoclaving, and pouring in the bottom plate of sterile petri dishes to harden. The surface is inoculated by streaking with a wire loop that has been dipped into a culture growing in a fermentation tube. EC medium is a nutrient broth used to test for fecal coliforms.

Examination of Potable Water

Collection of water samples for bacteriological examination must be done carefully to prevent contamination. The sampling fixture is generally a faucet, petcock, or small valve; fire hydrants are poor sampling stations because of the impossibility of sterilizing the long barrel. After the tap has been drained for several minutes to flush out the service line to the water main, a sterilized bottle is used to catch a sample from the stream of water. Care must be taken not to accidentally contaminate the cap or mouth of the bottle with the fingers. All samples of water collected from chlorinated sources must be dechlorinated at the time of collection. This is achieved by adding a measured amount of prepared sodium thiosulfate solution to the empty sample bottle before sterilization. Its presence in the bottle at the time of collection neutralizes any residual chlorine and prevents continuation of the disinfection action during the time the sample is in transit to the laboratory.

The minimum number of water samples collected from a distribution system and examined each month for coliforms depends on the population that is served by the system. For example, the minimum number required for populations of 1000, 10,000, and 100,000 are 2, 12, and 100, respectively. Compliance with the microbiological requirements of drinking water standards is determined by the number of positive tests. When 10-ml standard portions are examined, not more than 10 percent in any month should be positive, in other words, the upper limit of coliform density is an average of one per 100 ml.

The coliform group is defined as all aerobic and facultative anaerobic, nonspore forming, Gram-stain negative rods that ferment lactose with gas production within 48 hr of incubation at $35°C$. The initial coliform analysis is the presumptive test based on gas production from lactose. Ten-milliliter portions of a water sample are transferred using sterile pipettes into prepared fermentation tubes. The tubes contain lactose, or lauryl tryptose, broth, and inverted vials. Inoculated tubes are placed in a warm-air incubator at $35°C \pm 0.5°C$. Growth with the production of gas, identified by the presence of bubbles in the inverted vial, is a positive test indicating coliform bacteria may be present. A negative reaction, either no growth or growth without gas, excludes the coliform group.

The confirmed test is used to substantiate, or deny, the presence of coliforms in a positive presumptive test. One of two procedures as outlined on the schematic diagram in Figure 3-10 is used. A wire loop is used to transfer growth from a positive lactose tube, either after 24 or 48 hr incubation, to a fermentation tube containing brilliant green lactose bile broth. If growth with gas occurs within 48 hr at $35°C$, the presence of coliforms is confirmed. If no gas is produced, the test is negative. The alternate confirmed test is to streak the culture from a positive presumptive test across the surface of Endo or EMB agar in prepared petri dishes by using a sterile wire loop. The inoculated plates are incubated for 24 hr at $35°C$. Bacterial colonies growing on the surface of the solid media that exhibit a green sheen, referred to as typical coliform colonies, confirm the presence of coliform bacteria. Atypical growths, light-colored colonies without dark centers or a green sheen, neither confirm nor deny the presence of the coliform group. Absence of bacterial growth in the dishes is a negative test. Figure 3-11 shows fermentation tubes prior to inoculation, positive growth with gas, and a negative test of growth without gas; one of the Endo agar plates pictured exhibits typical coliform colonies.

Normally, in examining potable water, coliform testing is terminated with the confirmed test using brilliant green

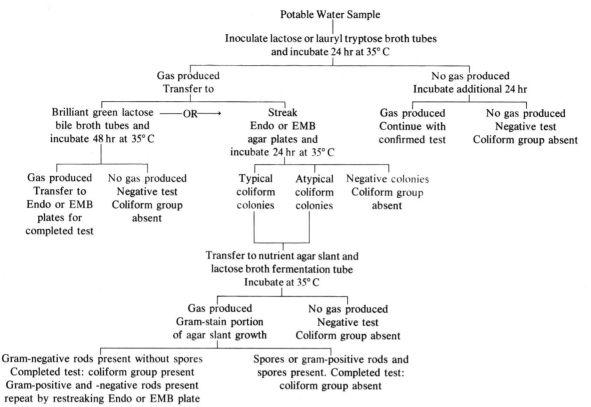

Figure 3-10 Schematic diagram for detection of the coliform group in potable water.

bile broth. Occasionally one may desire to run a completed test as is outlined in Figure 3-10. This involves transferring a colony from an Endo or EMB plate to nutrient agar and into lactose broth. If gas is not produced in the lactose fermentation tube, the colony transferred did not contain coliforms and the test is negative. If gas is produced in the tube, a portion of growth on the nutrient agar is smeared onto a glass slide and prepared for observation under a microscope by using the Gram-stain technique. If the bacteria are short rods, with no spores present, and the Gram-strain is negative (stained cells are pink in color), the coliform group is present and the test is completed. If the culture Gram-stains positive (purple colored), the completed test is proved negative.

Examination of Surface Water

An elevated temperature coliform test separates microorganisms of the coliform group into those of fecal origin and those of nonfecal sources. This technique is applicable to investigations of stream pollution, water supply sources, wastewater treatment systems, bathing waters, and general water quality monitoring. The procedure is not recommended as a substitute for the coliform tests used in examination of potable waters. The test procedure, diagramed schematically in Figure 3-12, cannot be used for direct isolation of coliforms from water but is used as a confirmatory procedure following presumptive testing. The first step is to transfer a measured volume of sample into fermentation tubes containing

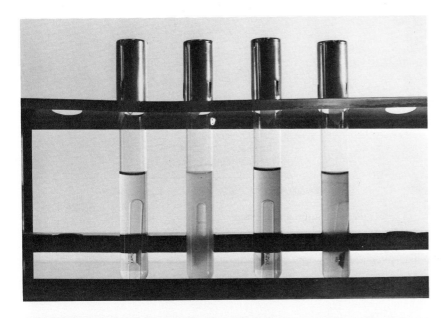

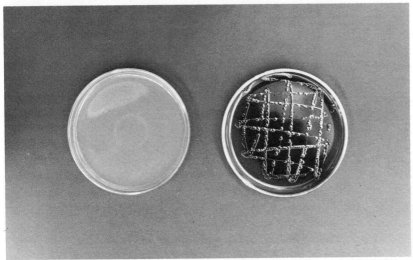

Figure 3-11 Presumptive and confirmed tests for the coliform group. Fermentation tubes shown are, from left to right, sterile lactose broth, growth with gas produced (positive test), sterile brilliant green lactose bile broth, and growth without gas produced (negative test). The petri dish on the left is sterile Endo agar while the plate on the right exhibits typical coliform colonies.

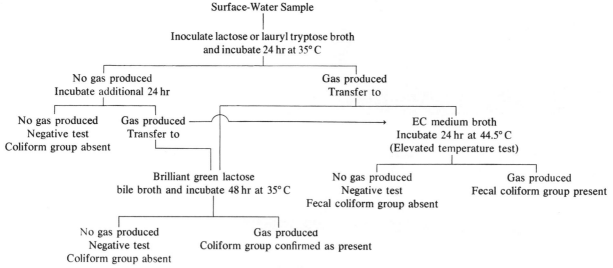

Figure 3-12 Schematic diagram for identification of the coliform group and the fecal-coliform group in water-quality monitoring.

lactose or lauryl tryptose broth. Those tubes showing growth with gas after 24 or 48 hr are used to inoculate culture tubes containing EC medium broth, which are incubated in a water bath at $44.5 \pm 0.2°$ C by immersion to the upper level of the medium. Because of the close temperature tolerance required in the fecal coliform test, the standard warm-air incubators normally used in bacteriological testing are not satisfactory. Growth with gas is positive for the fecal coliform group; no gas demonstrates absence.

In water-quality surveys, evaluation of the number of coliform bacteria present is necessary to determine if a water meets the appropriate standard. The multiple-tube fermentation technique can be used to enumerate positive presumptive, confirmed, and fecal coliform tests. Results of the test are expressed as the most probable number (MPN), since the count is based on statistical analysis of sets of tubes in a series of serial dilutions. MPN is by definition related to a sample volume of 100 ml; hence, an MPN of 10 means 10 coliforms per 100 ml of water.

The common procedure for MPN determination involves the use of sterile pipettes calibrated in 0.1 ml increments, sterile screw-top dilution bottles containing 99 ml of water, and a rack containing six sets of five

lactose broth fermentation tubes. Figure 3-13 is a sketch showing test preparation. A sterile pipette is used to transfer 1.0 ml portions of the sample into each of five fermentation tubes, followed by dispensing of 0.1 ml to a second set of five. For the next higher dilution, the third, only 0.01 ml of sample water is required, which it is impossible to pipette accurately. Therefore, 1.0 ml of sample is placed in a dilution bottle containing 99 ml of sterile water and mixed. Now, 1.0 ml portions containing 0.01 ml of the surface-water sample can be pipetted into the third set of five tubes. The fourth set receives 0.1 ml from this same dilution bottle. The process is then carried one more step by transferring 1.0 ml from the first dilution bottle into 99 ml of water in the second for another 100-fold dilution. Portions from this dilution bottle are pipetted into the fifth and sixth tube sets. After incubation for 48 hr at 35° C, the tubes are examined for gas production, and the number of positive reactions for each of the serial dilutions is recorded.

The MPN index and confidence limits for various combinations of positive and negative multiple-tube fermentation results can be determined from Table 3-1. The first three columns in the table refer to the number of positive tubes out of five containing 10-ml, 1-ml, and

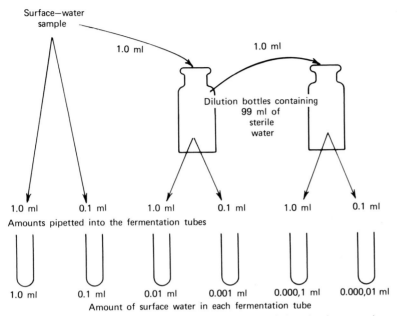

Figure 3-13   Procedure for preparation of the multiple-tube fermentation technique for determining the most probable number (MPN) of coliform bacteria in a surface-water sample.

0.1-ml sample portions. If smaller amounts of sample were used in the tubes, which is usually the case, the tabulated answers must be adjusted. If 1.0, 0.1, and 0.01 ml are used, the resultant MPN is 10 times the value given in the table; if the sample quantities are 0.1, 0.01, and 0.001 ml, record 100 times the tabulated value; and so on for other combinations. When more than three dilutions are employed in a decimal series of dilutions, the results from only three of these are used in computing the MPN. The three dilutions selected are the highest dilution giving positive results in all five portions tested (no lower dilution giving any negative results) and the two next succeeding higher dilutions.

The presumptive MPN test can be extended to confirmed testing and fecal coliform determination as is outlined on Figure 3-12. Growth from each of the positive lactose tubes is transferred aseptically using a wire loop to brilliant green bile lactose broth and EC medium broth. These are incubated to confirm the presence of coliforms and to determine whether they are of fecal origin.

EXAMPLE 3-1

The following are results from multiple-tube fermentation analyses of a polluted river water sample for the presumptive, confirmed, and fecal coliform groups. The serial dilutions were set up as is illustrated in Figure 3-12. Determine the MPN for each of the three tests.

		Number of Positive Reactions out of Five Tubes		
Serial Dilution	Sample Portion (ml)	Lactose Broth	Brilliant Green Bile	EC Medium
0	1.0	5	5	5
1	0.1	5	5	4
2	0.01	5	3	1
3	0.001	1	1	1
4	0.000,1	0	0	0
5	0.000,01	0	0	0

Table 3-1.   MPN Index and 95 Percent Confidence Limits for Coliform Counts by the Multiple-Tube Fermentation Technique for Various Combinations of Positive and Negative Results When Five 10-ml, Five 1-ml, and Five 0.1-ml Portions Are Used

No. of Tubes Giving Positive Reaction out of			MPN Index per 100 ml	95 Percent Confidence Limits		No. of Tubes Giving Positive Reaction out of			MPN Index per 100 ml	95 Percent Confidence Limits	
5 of 10 ml Each	5 of 1 ml Each	5 of 0.1 ml Each		Lower	Upper	5 of 10 ml Each	5 of 1 ml Each	5 of 0.1 ml Each		Lower	Upper
0	0	0	0			4	2	1	26	9	78
0	0	1	2	<0.5	7	4	3	0	27	9	80
0	1	0	2	<0.5	7	4	3	1	33	11	93
0	2	0	4	<0.5	11	4	4	0	34	12	93
1	0	0	2	<0.5	7	5	0	0	23	7	70
1	0	1	4	<0.5	11	5	0	1	31	11	89
1	1	0	4	<0.5	11	5	0	2	43	15	110
1	1	1	6	<0.5	15	5	1	0	33	11	93
1	2	0	6	<0.5	15	5	1	1	46	16	120
						5	1	2	63	21	150
2	0	0	5	<0.5	13						
2	0	1	7	1	17	5	2	0	49	17	130
2	1	0	7	1	17	5	2	1	70	23	170
2	1	1	9	2	21	5	2	2	94	28	220
2	2	0	9	2	21	5	3	0	79	25	190
2	3	0	12	3	28	5	3	1	110	31	250
						5	3	2	140	37	340
3	0	0	8	1	19						
3	0	1	11	2	25	5	3	3	180	44	500
3	1	0	11	2	25	5	4	0	130	35	300
3	1	1	14	4	34	5	4	1	170	43	490
3	2	0	14	4	34	5	4	2	220	57	700
3	2	1	17	5	46	5	4	3	280	90	850
3	3	0	17	5	46	5	4	4	350	120	1000
4	0	0	13	3	31	5	5	0	240	68	750
4	0	1	17	5	46	5	5	1	350	120	1000
4	1	0	17	5	46	5	5	2	540	180	1400
4	1	1	21	7	63	5	5	3	920	300	3200
4	1	2	26	9	78	5	5	4	1600	640	5800
4	2	0	22	7	67	5	5	5	≦2400		

Source: Standard Methods for the Examination of Water and Wastewater.

Solution

(a) **Presumptive Coliform MPN.** The three dilutions selected for MPN determination show 5, 1, and 0 positive tubes. The index value from Table 3-1, without regard to size of sample portion, is 33. However, since the sample portions corresponding to the 5, 1, and 0 serial dilutions are 0.01, 0.001, and 0.000,1 ml, the MPN value must be multiplied by 1000. Therefore, the MPN index for the presumptive test series is 33,000, with 95 percent confidence limits of 11,000 and 93,000. (MPN is the most probable number in 100 ml of water, and the confidence limits mean that 95 percent of all MPN analyses run on the same sample should, statistically speaking, fall within this range of limiting values.)

(b) **Confirmed Coliform MPN.** Selected dilutions are 0.1, 0.01, and 0.001 ml showing 5, 3, and 1 positive tubes, respectively. The multiplication factor is 100. From Table 3-1, MPN equals 11,000, with confidence limits of 3100 to 25,000.

(c) **Fecal Coliform MPN.** Selected dilutions are 5, 4, and 1, with a correction factor of 10. MPN equals 1700, with confidence limits of 430 to 4900.

Membrane Filter Technique for Coliform Testing

This method consists of drawing a measured volume of water through a filter membrane fine enough to take out bacteria, and then placing the filter on a growth medium in a petri dish. Each bacterium retained by the filter grows and forms a small colony. The number of coliforms present in a filtered sample is determined by counting the number of colonies and expressing this value in terms of number per 100 ml of water. The membrane filter technique has been widely adopted for use in water-quality monitoring studies, since it requires much less laboratory apparatus than the standard multiple-tube technique; and, portable membrane filter apparatus have been developed for conducting coliform tests in the field.

Special apparatus required to conduct membrane filter coliform tests include filtration units, filter membranes, absorbent pads, forceps, and culture dishes. The common laboratory filtration unit (Figure 3-14) consists of a funnel that fastens to a receptacle bearing a porous plate to support the filter membrane. The filter-holding assembly, constructed of glass, porcelain, or stainless steel, is sterilized by boiling, autoclaving, or ultraviolet radiation. For filtration, the assembly is mounted on a side-arm filtering flask, which is evacuated to draw the sample through the filter. For field use a small hand-sized plunger pump (syringe) is used to draw a sample of water through the small assembly holding the filter membrane.

Filter membranes, available from several commercial manufacturers, are 2-in.-diameter disks with pore openings of 0.45 ± 0.02 μm, small enough to retain microbial cells. Filters used in determining bacterial counts have a grid printed on the surface for ease in counting colonies. Filter membranes are sterilized prior to use, either in a glass petri dish or wrapped in heavy paper. Absorbent pads for nutrients are normally furnished with the filters by the manufacturer. After sterilization these pads are placed in culture dishes to absorb the nutrient media on which the membrane filter is placed. During the testing procedure, the filters are handled on the outer edges with forceps that are sterilized before use by dipping in alcohol and then ignition.

Either glass or disposable plastic culture dishes may be used. If glass petri dishes are employed, a humid environment must be maintained during incubation to prevent loss of media by evaporation, since these dishes have loose-fitting covers. More popular are disposable plastic dishes with tight-fitting lids, reducing the problem of dehydration. Suitable sterile plastic dishes are available commercially; if reuse is necessary, the dishes should be treated by immersion in 70 percent ethanol for 30 min, air dried, and reassembled.

The standard volume filtered for drinking water samples is 100 ml. For surface-water analyses, the size of the sample to be filtered is governed by the anticipated bacterial density. An ideal quantity results in growth of about 50 coliform colonies and not more than 200 colonies of all types. Since it may be difficult to anticipate the number of bacteria in a sample, two or three volumes of the same sample should be tested. When the portion being filtered is less than 20 ml, a small amount of sterile dilution water is added to the funnel before filtration to

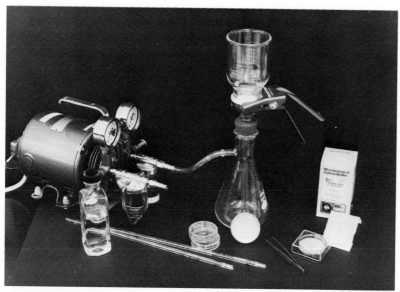

Figure 3-14 Laboratory apparatus for coliform testing by the membrane filter technique.

uniformly disperse the bacterial suspension over the entire surface of the filter. Filtration units should be sterilized at the beginning of each filtration series. Rapid decontamination between successive filtrations is accomplished by use of an ultraviolet sterilizer, flowing steam, or boiling water.

The main steps of the test procedure are filtration, incubation, and enumeration. The filter-holding assembly is placed on the suction flask, a sterile filter is placed grid side up over the porous plate of the apparatus using sterile forceps, and the funnel is locked in place holding the membrane. Filtration is accomplished by passing the sample through the filter under partial vacuum. A culture dish is prepared by placing a sterile absorbent pad in the upper half of the dish and pipetting enough enrichment media on top to saturate the pad. M-Endo medium is used for the coliform group and M-FC for fecal coliforms. The prepared filter is removed from the filtration apparatus using a forceps and is placed directly on the pad in the dish. The cover is replaced, and the culture is incubated for 24 hr at 35° C. Incubation for fecal coliforms is performed by placing the culture dishes in watertight plastic bags and submerging them in a water bath for incubation at 44.5° C. Typical coliform colonies are pink to dark red with a metallic surface sheen that may vary in size from a small pinhead to complete coverage of the colony surface. Plates with counts within the 20 to 80 coliform colony range are considered to be most valid. Coliform density is calculated in terms of coliforms per 100 ml by multiplying the colonies counted by 100 and dividing this by the milliliters of the sample filtered.

Turbidity and noncoliform bacteria may interfere with the performance of the membrane filter technique on surface waters. Turbid waters result in suspended solids interfering with filtration and colony development, and water containing a large number of noncoliform bacteria, which develop on the culture medium, can overgrow the coliform colonies. The membrane filter method is suitable for testing potable water supplies. When the technique is introduced for the first time in a laboratory, parallel testing is desirable to demonstrate applicability and to familiarize technicians with the procedure.

3-10 TESTING FOR ENTERIC VIRUSES

Viruses excreted with human feces include several groups (Section 3-2) consisting of different serological types so that more than 100 human enteric viruses are recognized. Because of the complexity in isolation and identification, examination of water and wastewater for viruses is restricted to special circumstances, such as disease outbreaks and research studies. Virus assay should be performed by a trained virologist working in a specially equipped virology laboratory.

Testing for viruses requires extraction, concentration, and identification. Problems associated with virus detection relate to the small size of viruses, their low concentration in water and the variability in amounts and types that may be present, their instability, presence of dissolved and suspended solids interfering with concentration, and the limitations of identification methods. Since viruses are obligate intracellular parasites, the basis for virus detection and identification is replication in living hosts' cells leading to cell destruction. The difficulty is the specificity of the host for different viruses. No single, universal host system can be used in testing for all enteric viruses. The two major host-cell systems are whole animals (usually mice) and mammalian cell cultures of primate origin.

Virus separation and concentration from relatively unpolluted waters is likely to require greater than 100 gal of sample; for a wastewater sample, 1 gal may suffice. A common concentration technique is adsorption to and elution from microporous filters. After chemicals have been added to maximize virus adsorption, the water is forced under pressure through a microporous filter of cellulose nitrate or fiberglass. The filtered matter containing the viruses is eluted by pumping a small volume of eluent liquid, such as an alkaline buffer containing glycine, through the filter. Further concentration is performed by a second adsorption-elution step or by chemical precipitation and centrifugation. During the process of concentration, only a portion of the viruses present in the original sample is captured. The efficiency of separation varies widely depending on water quality. Therefore, the procedure must be conducted on samples to which known suspensions of one or more test virus types have been added to establish recovery efficiency.

Sample concentrates can be stored at room temperature for up to 2 hr or at refrigerator temperature up to 48 hr. If assay of the concentration is to be delayed beyond 48 hr, the sample should be frozen and stored at $-70°C$ or lower. Sample concentrates from wastewater contain bacteria and fungi that can overgrow cell cultures and interfere with detection and assay. The common technique for decontamination is the addition of penicillin and streptomycin antibiotics to destroy the bacteria and amphotericin B for fungi.

Cell culture systems, rather than mice or other animals, are preferred for virus detection and enumeration. African green monkey kidney cells are sensitive hosts for several enteric viruses. In cell cultures, monolayers of these cells are grown attached to the interior glass surface of culture flasks. A portion of the virus sample concentrate is diluted and spread on the surface of the cell tissue. Following incubation, the virus concentration in the applied sample is determined by counting the plaques and correcting for dilution. (A plaque is a clear area in the cell monolayer produced by viral destruction of the cells.) Virus concentration in a water or wastewater sample is expressed as the number of plaque-forming units (PFU) per liter. Further examination is required to identify the virus type creating a plaque. Preliminary identification can be made from microscopic examination of the visible effects of the virus on the infected animal cells. Precise identification involves testing with blood serums, additional cell cultures, and inoculation of test animals.

3-11 BIOCHEMICAL OXYGEN DEMAND

Biochemical oxygen demand (BOD) is the most commonly used parameter to define the strength of a municipal or organic industrial wastewater. Widest application is in measuring waste loadings to treatment plants and in evaluating the efficiency of such treatment systems. In addition, the BOD test is used to determine the relative oxygen requirements of treated effluents and polluted waters. However, it is of limited value in measuring the actual oxygen demand of surface waters, and extrapolation of test results to actual stream oxygen demands is highly questionable, since the laboratory environment cannot reproduce the physical, chemical, and biological stream conditions.

BOD is by definition the quantity of oxygen utilized by a mixed population of microorganisms in the aerobic oxidation (of the organic matter in a sample of wastewater) at a temperature of 20° C. Measured amounts of a wastewater, diluted with prepared water, are placed in 300-ml BOD bottles (Figure 3-15). The dilution water, containing phosphate buffer (pH 7.2), magnesium sulfate, calcium chloride, and ferric chloride, is saturated with dissolved oxygen. Seed microorganisms are supplied to oxidize the waste organics if sufficient microorganisms are not already present in the wastewater sample. The general biological reaction that takes place is Eq. 3-12. The wastewater supplies the organic matter (biological food), and the dilution water furnishes the dissolved oxygen. The primary reaction is metabolism of the organic matter and uptake of dissolved oxygen by bacteria, releasing carbon dioxide and producing a substantial increase in bacterial population. The secondary reaction results from the oxygen used by the protozoa-consuming bacteria, a predator-prey reaction. Depletion of dissolved oxygen in the test bottle is directly related to the amounts of degradable organic matter. The BOD of a wastewater where microorganisms are already present in the sample, requiring no outside seed, is calculated using Eq. 3-13. (The standard test has an incubation period of five days at 20° C.)

$$\underset{\text{matter}}{\text{Organic}} \xrightarrow[\text{bacteria}]{\overset{\text{dissolved}}{\overset{\text{oxygen}}{\searrow}}} CO_2 + \underset{\text{cells}}{\text{Bacterial}} \xrightarrow[\text{protozoa}]{\overset{\text{dissolved}}{\overset{\text{oxygen}}{\searrow}}} CO_2$$

$$+ \underset{\text{cells}}{\text{Protozoal}} \qquad (3\text{-}12)$$

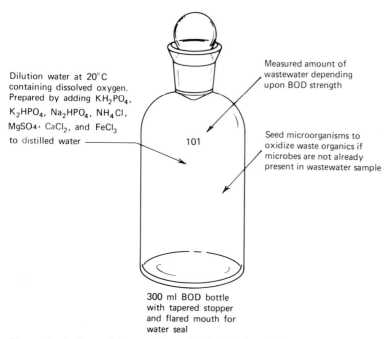

Dilution water at 20°C containing dissolved oxygen. Prepared by adding KH_2PO_4, K_2HPO_4, Na_2HPO_4, NH_4Cl, $MgSO_4$, $CaCl_2$, and $FeCl_3$ to distilled water

101

Measured amount of wastewater depending upon BOD strength

Seed microorganisms to oxidize waste organics if microbes are not already present in wastewater sample

300 ml BOD bottle with tapered stopper and flared mouth for water seal

Figure 3-15   Essential constituents in the biochemical oxygen demand (BOD) test are a measured amount of the wastewater sample being analyzed, prepared dilution water containing dissolved oxygen, and seed microorganisms if not present in the wastewater.

$$BOD = \frac{D_1 - D_2}{P} \qquad (3\text{-}13)$$

where BOD = biochemical oxygen demand, milligrams per liter

D_1 = initial DO of the diluted wastewater sample about 15 min after preparation, milligrams per liter

D_2 = final DO of the diluted wastewater sample after incubation for five days, milligrams per liter

P = decimal fraction of the wastewater sample used

$$= \frac{\text{milliliters of wastewater sample}}{\text{milliliter volume of the BOD bottle}}$$

The biochemical oxygen demand of a wastewater is in reality not a single point value but time dependent. The curve in Figure 3-16 shows BOD exerted, dissolved oxygen depleted, as the biological reactions progress with time. Carbonaceous oxygen demand, Eq. 3-12, progresses at a decreasing rate with time, since the rate of biological activity decreases as the available food supply diminishes. The shape of the hypothetical curve is best expressed mathematically by first-order kinetics in the form of Eq. 3-14, where t is time in days and k is a rate constant. Based on this equation for a k equal to 0.1 per day, a common value for domestic waste, 68 percent of the ultimate carbonaceous BOD is exerted after five days.

$$\begin{array}{c}\text{BOD at any}\\\text{time, } t\end{array} = \begin{array}{c}\text{ultimate}\\\text{BOD}\end{array} (1 - 10^{-kt}) \qquad (3\text{-}14)$$

Nitrifying bacteria can exert an oxygen demand in the BOD test as in Eqs. 3-7 and 3-8. Fortunately, growth of nitrifying bacteria lags behind that of the microorganisms performing the carbonaceous reaction. Nitrification generally does not occur until several days after the standard five-day incubation period for BOD tests on untreated wastewaters. Treatment plant effluents and stream waters may show early nitrification where the sample has a relatively high population of nitrifying bacteria. The

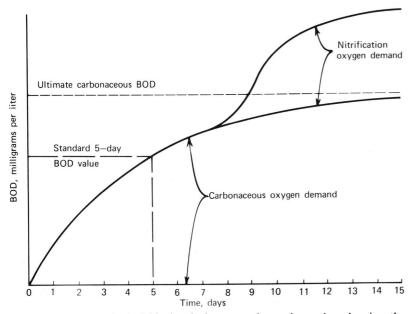

Figure 3-16 Hypothetical biochemical oxygen demand reaction showing the carbonaceous and nitrification demand curves.

standard method recommended for preventing nitrification is the addition of 10 mg/l of 2-chloro-6(trichloro methyl)pyridine to the dilution water to stop nitrate formation.

The degree of reproducibility of the BOD test cannot be defined precisely because of variations that occur in bacterial decomposition of various organic substances. Laboratory technicians can periodically evaluate their procedures by conducting BOD analyses on a standard solution of glucose and glutamic acid containing 150 mg/l of each component. The mean BOD value should fall in the range of 200 ± 37 mg/l, and the k-rate in the bounds of 0.16 to 0.19. This represents a deviation of about ± 18 percent from the average value. Tests on real wastewaters normally show observations varying from 10 to 20 percent on either side of the mean. The larger variations occur in testing industrial wastewaters that require seeding, contain substances that inhibit biological activity, and treatment plant effluent samples that are affected by nitrification.

Contaminated dilution water and dirty incubation bottles can distort BOD test results. *Standard Methods*[1] emphasizes proper cleaning of the bottles and analysis of dilution water blanks during BOD testing as a check on the quality of unseeded dilution water. The DO uptake for unseeded dilution water with phosphate buffer and inorganic nutrient solutions added should not be more than 0.2 mg/l and preferably not more than 0.1 mg/l after five days incubation at $20°$ C. If the quality of the dilution water is questionable, it can be checked for acceptability before using in BOD analyses. The procedure is to add sufficient seed material to the dilution water being tested to produce a DO uptake of 0.05 to 0.1 mg/l in five days. The actual oxygen utilization should not exceed 0.1 to 0.2 mg/l.

No uniform relationship exists between the COD and BOD of wastewaters, except that the COD value must be greater than the BOD. This is because chemical oxidation decomposes nonbiodegradable organic matter, and the standard BOD test measures only the oxygen used in metabolizing the organic matter for five days. The correlation of COD to BOD for a particular wastewater can be determined by statistical comparison of several laboratory analyses. Unfortunately, such a relationship may be invalidated by the simple day-to-day variations in quality of a municipal wastewater. Occasionally, out of necessity for converting oxygen demand data, even if the results are of questionable accuracy, the COD of a soluble wastewater is assumed to be numerically equal to the ultimate BOD.

BOD of Municipal Wastewater

Following is the description of a typical BOD test on a municipal wastewater. The first step is to determine the portion of sample to be placed in each BOD bottle. The information in either Table 3-2 or Eq. 3-13 may be used to select appropriate amounts. For example, for a wastewater with an estimated BOD of 350 mg/l, Table 3-2 indicates sample volumes in the range of 2 to 5 ml per 300-ml bottle would be appropriate; or substituting into Eq. 3-13 a BOD equal to 350 mg/l, bottle volume of 300 ml, and a desired dissolved oxygen decrease of 5 mg/l, the calculated wastewater portion is 4.3 ml. For a valid BOD test, at least 2 mg/l of dissolved oxygen should be used, but the final DO should not be less than 1 mg/l. Since the initial dissolved oxygen in dilution water is about 9 mg/l, an average amount of 5 mg/l is available for biological uptake between the minimum desired 2 mg/l and the maximum of 8 mg/l.

The distilled water for dilution is stabilized at $20°$ C by placing in the BOD incubator. It is then removed and prepared by aeration and addition of the four buffer and nutrient solutions (Figure 3-15). Outside seed microorganisms are not required for municipal wastewater, since they are already present in the sample. To ensure that some of the bottles prepared provide valid test data, three different dilutions of two or three bottles each are set up, plus three bottles for initial dissolved oxygen measurement and three blanks of dilution water only. Measured portions of the wastewater are pipetted directly into empty BOD bottles, which are then filled with prepared dilution water by siphoning through a hose. The sample preparation and dissolved oxygen measurements for this illustration are summarized in Table 3-3. Wastewater volumes of 2.0 ml, 4.0 ml, and 6.0 ml were used. The first three bottles were titrated immediately for dissolved oxygen, using the azide modification of the iodometric

Table 3-2. Suggested Wastewater Portions and Dilutions in Preparing BOD Tests

By Direct Measurement of Wastewater into a 300-ml BOD Bottle		By Mixing Wastewater into Dilution Water $\left[\dfrac{\text{Wastewater Volume}}{\text{Total Volume of Mixture}}\right]$	
Wastewater (ml)	Range of BOD (mg/l)	Percentage of Mixture	Range of BOD (mg/l)
0.20	3000 to 10,500	0.10	2000 to 7000
0.50	1200 to 4200	0.20	1000 to 3500
1.0	600 to 2100	0.50	400 to 1400
2.0	300 to 1050	1.0	200 to 700
5.0	120 to 420	2.0	100 to 350
10.0	60 to 210	5.0	40 to 140
20.0	30 to 105	10.0	20 to 70
50.0	12 to 42	20.0	10 to 35
100	6 to 21	50.0	4 to 14

Table 3-3. Typical BOD Data from the Analysis of a Municipal Wastewater to Determine the Five-Day BOD Value and Estimate the k-Rate of the Biological Reaction.[a]

Bottle Number	Wastewater Portion (ml)	Initial DO (mg/l)	Incubation Period (days)	Final DO (mg/l)	DO Drop (mg/l)	Calculated BOD (mg/l)
BOD Tests to Determine Average Five-Day Value						
1	2.0	8.3	0			
2	4.0	8.4	0			
3	6.0	8.4	0			
		Average = 8.4				
4	2.0	8.4	5.0	5.9	2.5	375
5	2.0	8.4	5.0	6.0	2.4	360
6	4.0	8.4	5.0	3.8	4.6	345
7	4.0	8.4	5.0	3.5	4.9	365
8	6.0	8.4	5.0	0	(Invalid test)	
9	6.0	8.4	5.0	0	(Invalid test)	
					Average = 360	
BOD Analyses to Permit Calculation of k-Rate						
10	4.0	8.4	0.5	7.2	1.2	90
11	4.0	8.4	0.5	7.4	1.0	75
12	4.0	8.4	1.0	6.2	2.2	165
13	4.0	8.4	1.0	5.9	2.5	190
14	4.0	8.4	2.0	5.2	3.2	240
15	4.0	8.4	2.0	5.2	3.2	240
16	4.0	8.4	3.0	4.4	4.0	300
17	4.0	8.4	3.0	4.6	3.8	285

[a] Results are plotted in Figure 3-17. All BOD bottles had a volume of 300 ml.

method. The other nine test bottles and three blank bottles of dilution water were incubated for five days at 20° C and then were titrated for remaining dissolved oxygen. The BOD value for each bottle was calculated using Eq. 3-13. The average BOD for four of the six tests was 360 mg/l; two bottles were considered invalid, since the dissolved oxygen was depleted prior to the end of the five-day incubation period. The three blanks showed less than 0.2 mg/l of DO uptake.

Determination of BOD *k*-Rate

The rate constant *k* in Eq. 3-14 can be computed from BOD values measured at various times. Bottles numbered 10 to 17 of the analyses in Table 3-3 were analyzed after incubation periods ranging from 0.5 to 3.0 days. These data along with the average five-day value are plotted in Figure 3-17. The shape of the BOD-time curve is typical of a municipal wastewater derived primarily from domestic sources. The upper portion of Figure 3-17 illustrates a graphical method for estimating *k*-rate. Values of the cube root of time in days over BOD in milligrams per liter are calculated from the laboratory data and are plotted as ordinates against the corresponding times. The best fit line drawn through these points is used to calculate the *k*-rate by the following relationship:

$$k = 2.61 \frac{B}{A} \qquad (3\text{-}15)$$

where k = rate constant, per day
   A = intercept of the line on the ordinate axis
   B = slope of the line

BOD Analysis of Industrial Wastewaters

Most organic wastes from food-processing industries, and other sources, that are susceptible to biological decomposition, can be tested for BOD. However, particular care must be taken in properly neutralizing the wastewaters, seeding the test bottles or dilution water with microorganisms from aged wastewater or waste-polluted stream water, and using sufficient dilution so that the effect of any

toxicity is reduced and the maximum BOD value is obtained. In addition to careful collection and compositing of samples according to wastewater flow variation, special data should be recorded for industrial wastes. The industry's production, namely, the type and quantity of product manufactured, and specific operational conditions existing during the sampling period should be recorded for correlation with the quantity and strength of the wastewater produced. Several composite samples over an extended period of time are often required for industries with variable production schedules to key the amount of the wastes produced to the quantity of product manufactured.

Initial pretreatment is to neutralize, if necessary, the sample to pH 7.0 with sulfuric acid or sodium hydroxide to remove caustic alkalinity or acidity. The pH of the dilution water should not be changed by addition of the wastewater in preparing a BOD test bottle of lowest dilution. Samples containing residual chlorine must be dechlorinated prior to setup. Often the residual dissipates if samples are allowed to stand for 1 or 2 hr, but higher chlorine residuals should be destroyed by adding sodium sulfite solution. Industrial wastewaters containing other toxic substances require special study and pretreatment. In extreme cases where a technique for neutralizing the toxin cannot be developed, BOD testing is abandoned and replaced with COD analyses.

The presence of unknown substances that inhibit biological growth are often detected by carefully conducted BOD tests. Table 3-4 and Figure 3-18 summarize the results from analysis on a food-processing wastewater containing an interfering compound. The source of the toxicity was a bactericide used in disinfecting production pipelines and tanks. The first evidence of the potential problem is revealed by comparing the rather wide range of BOD values observed in test bottles at the same dilution—at 0.67 percent the numbers range from 220 mg/l to 485 mg/l. More importantly, the tests show increasing BOD values with increasing dilutions. The concentration of toxin, and consequently the inhibition of biological activity, is greater in lower dilutions (3.0 ml wastewater in a 300-ml bottle) than at higher dilutions (1.0 ml/300 ml). The reported BOD should be the highest value obtained in valid tests, this being the aver-

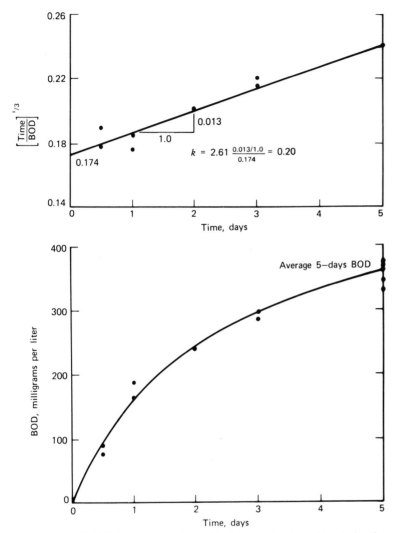

Figure 3-17 BOD-time curve and graphical determination of k-rate for the
wastewater data given in Table 3-3.

age of the highest dilutions providing a minimum uptake
of 2.0 mg/l dissolved oxygen.

Very few industrial wastewaters have sufficient biological populations to perform BOD testing without providing an acclimated seed. The ideal seed is a mixed culture
of bacteria and protozoa adapted to decomposing the
specific industrial wastewater organics, with a low
number of nitrifying bacteria. Microorganisms for food-
processing wastes, and similar organics, can be obtained
from aged untreated domestic wastewater. The seed
material is the supernatant liquid from a sample of
domestic wastewater that has been allowed to age
and settle in an open container for about 24 hr at
room temperature. Biota for industrial wastewaters, with

Table 3-4. BOD Data from the Analysis of a Food-Processing Wastewater Containing a Bactericidal Agent Causing an Increase in BOD Value with Increasing Dilution.[a]

Wastewater Portion (ml)	Percentage Dilution of Wastewater	Measured Five-day BOD (mg/l)
3.0	1.00	240
3.0	1.00	205
3.0	1.00	250
3.0	1.00	145
		Average = $\overline{210}$
2.0	0.67	365
2.0	0.67	485
2.0	0.67	315
2.0	0.67	220
		Average = $\overline{350}$
1.0	0.33	520
1.0	0.33	440
1.0	0.33	565
1.0	0.33	520
		Average = $\overline{510}$

[a] Figure 3-18 is a graphical presentation of the results.

biodegradable organic compounds not abundant in municipal wastewater, should be obtained from a source having microorganisms acclimated to these wastewaters. If the industrial discharge is being treated by a biological system, activated sludge from an aeration basin, slime growth from a trickling filter, or water from a stabilization pond contains acclimated microorganisms. When the wastewater is disposed of in a watercourse, the stream water several miles below the point of discharge may be a good source for seed. An adapted seed culture can be developed by using a small laboratory draw-and-fill activated-sludge unit on a mixed feed of industrial and domestic wastewater if a natural source of microbial seed is not readily available, or proves to be unsatisfactory.

The amount of seed material added to the dilution water, or each individual BOD bottle, must be sufficient to provide substantial biological activity without adding too much organic matter. When aged domestic or polluted water is used, the rule of thumb is to add an amount of seed wastewater such that 5 to 10 percent of the total BOD exerted by the sample results from oxygen demand of the seed alone. For example, assume that the water being used for seed has an estimated BOD of 150 mg/l. Based on Table 3-2, a 10-ml volume per bottle would be

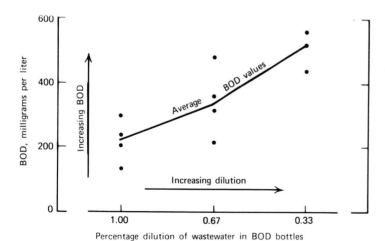

Figure 3-18 Plot of BOD values, given in Table 3-4, showing increasing BOD for the same industrial sample set up at increasing dilutions. The sample was a food-processing wastewater containing a biologically inhibiting substance.

used to run a BOD test on this wastewater; however, for seeding purposes, only 5 to 10 percent of this amount is desired. Hence, 0.5 to 1.0 ml should be added in setting up each industrial wastewater test bottle. If the culture is to be placed in the dilution water, rather than directly into each BOD bottle, the percentage-of-mixture data from Table 3-2 is helpful. To conduct a BOD test on a 150-mg/l waste, a 3.5 percent mixture is recommended. Therefore, seeded dilution water should be between 0.17 and 0.35 percent seed wastewater, prepared by adding 1.7 to 3.5 ml to each liter of dilution water.

Oxygen demand of the seed is compensated for in computing BOD of a seeded industrial wastewater test by using Eq. 3-16. The terms D_1 and D_2 are the initial and final dissolved oxygen concentrations in the seeded wastewater bottles containing an industrial wastewater fraction of P. The terms B_1 and B_2 are initial and final dissolved oxygen values from a separate BOD test on the seed material, and f is the ratio of seed volume used in the industrial wastewater test to the amount used in the test on the seed. Hence, $(B_1 - B_2)f$ is the oxygen demand of the seed.

$$\text{BOD} = \frac{(D_1 - D_2) - (B_1 - B_2)f}{P} \qquad (3\text{-}16)$$

where D_1 = DO of diluted seeded wastewater sample about 15 min after preparation

D_2 = DO of wastewater sample after incubation

B_1 = DO of diluted seed sample about 15 min after preparation

B_2 = DO of seed sample after incubation

f = ratio of seed volume in seeded wastewater test to seed volume in BOD test on seed

$= \dfrac{\text{percentage or milliliters of seed in } D_1}{\text{percentage or milliliters of seed in } B_1}$

P = decimal fraction of wastewater sample used

$= \dfrac{\text{volume of wastewater}}{\text{volume of dilution water plus wastewater}}$

A time lag, resulting from insufficient seed, unacclimated microorganisms, or presence of inhibiting substances, often occurs in the oxygen-demand reaction on an industrial wastewater. To determine BOD accurately, existence of a time lag must be identified during the first few days of incubation. This can be accomplished by either preparing extra bottles for dissolved oxygen testing after $\frac{1}{2}$, 1, 2, and 3 days, or if a dissolved oxygen probe is available, by monitoring the rate of oxygen depletion in two or three bottles each day during incubation. Figure 3-19 illustrates dissolved oxygen testing on a chemical-manufacturing wastewater that exhibits a two-day lag period. The most probable BOD can be approximated in a test of this nature by reestablishing time zero at the end of the lag period and by measuring the five-day value based on this new origin.

Data throughout the incubation period may also reveal other irregularities influencing selection of a five-day value. Figure 3-20a is an effluent from an extended aeration plant. Because of the high population of nitrifying bacteria, nitrification started on the third day of incubation and exerted approximately one half of the total oxygen demand by the fifth day of the test. The diphasic curve in Figure 3-20b is a result of rapid oxygen uptake during the first two days caused by a large amount of discharge from a pudding factory into the municipal sewer. Since soluble pudding waste is more easily decomposed by bacteria, it is metabolized first at a very high rate ($k_1 = 0.65$) while the remaining domestic waste organics continue their oxygen demand at a slower rate throughout the duration of the test.

Industrial wastes frequently have high strengths that make it difficult, or even impossible, to pipette accurately the small quantity desired for a single test bottle. In this case, the wastewater can be diluted by serial dilution to volumes that can be accurately measured. For example, only 0.5 ml of a 3000-mg/l BOD waste is required per BOD bottle. However, if 100 ml of the 3000-mg/l waste are diluted to 1000 ml with distilled water, 5.0 ml of the mixture can be pipetted accurately into each bottle. Wastewaters high in suspended solids may be difficult to mix with water; one alternative is to homogenize the sample in a blender to aid dispersion in the dilution water.

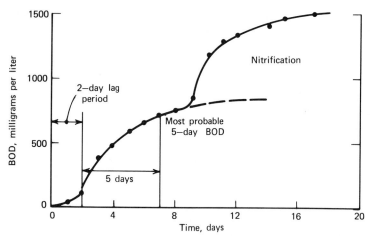

Figure 3-19   Reaction-rate curve from a seeded BOD test on a chemical-manufacturing wastewater exhibiting a time lag in the biological reaction.

The purpose of this discussion is to present some of the problems and pitfalls in BOD testing of industrial wastewaters. Although difficult to perform, these tests are common in evaluating manufacturing wastes to assess sewer use fees or treatment plant loadings. Improper seeding and unrecognized influence of inhibiting substances cause erroneous results that lead to confusion and consternation. Although occasionally industrial wastewaters defy biological testing, more often they can be tested with reasonable accuracy by following very careful analytical procedures.

Manometric Measurement of BOD

Manometric apparatus to measure oxygen uptake for determination of BOD, as shown in Figure 3-21, is marketed for general laboratory use; however, it does not replace the *Standard Methods* dilution procedure. Measured samples of wastewater are placed in each of the brown glass bottles, the amount depending on the anticipated BOD concentration. Buffer and nutrient chemicals are added to the wastewater portions to ensure sufficient nutrients for biological growth and buffer capacity to prevent a change in pH. A small magnetic stirring bar is placed in each bottle, and a seal cup containing a few crystals of lithium hydroxide is placed

under the bottle cap. The prepared bottles are connected to mercury manometers and are set in position on the stirring apparatus. The samples are mixed continuously by turning the stirring bars driven by a rotating magnet under each bottle. The electric stirring motor is enclosed at one end of the unit. After the initial stirring to reach equilibrium, the bottle closures are tightened, and caps are placed on the manometers to prevent the affect of barometric pressure changes. As the microorganisms take up dissolved oxygen in the liquid, gaseous oxygen is absorbed from air in the closed sample bottle. Carbon dioxide given off by the microorganisms is absorbed and is converted to carbonate ion by the lithium hydroxide crystals in the seal cup underneath the bottle cap. Absorption reduces the volume of carbon dioxide gas to zero. Volumetric reduction of the air in a bottle, representing the oxygen demand, is indicated by the manometer sight glass, which is calibrated to read directly in milligrams per liter BOD. To maintain 20° C, which is required for the standard test, the entire apparatus is placed in an incubator.

Studies have been conducted to compare manometric BOD measurements to those performed by the standard dilution method. For domestic wastewater and synthetic glucose-glutamic acid, the two techniques have provided comparable results at 20° C incubation temperature.

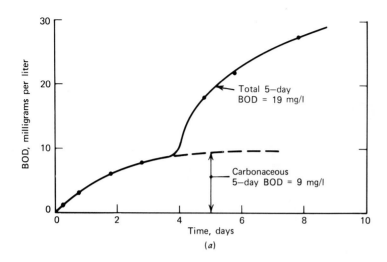

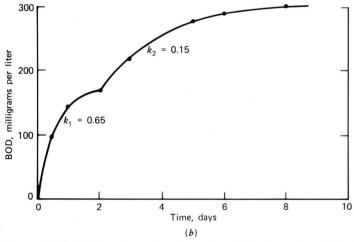

Figure 3-20 BOD analyses showing early nitrification and a diphasic carbonaceous stage. (*a*) BOD curve of a treated wastewater from an extended aeration plant showing nitrification starting about the fourth day of incubation. (*b*) BOD curve of a raw municipal wastewater exhibiting a diphasic carbonaceous stage resulting from a large amount of soluble pudding-manufacturing waste combined with the domestic wastewater.

However, tests on industrial wastewaters often show significant variance, primarily because the sample being tested in the manometric bottle is undiluted, leading to greater influence of toxic or inhibiting substances. If the manometric method is applied on a regular basis in testing municipal discharges that contain a substantial fraction of industrial wastewaters, validity of the results should be verified by comparison with the standard dilution technique. The main advantages given in manufacturers' literature are ease of operation, reduced testing

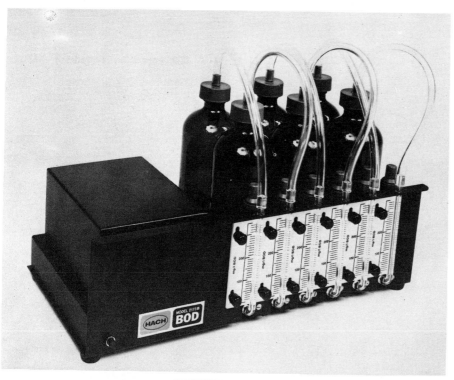

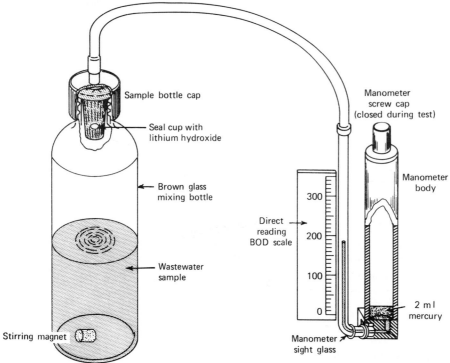

Figure 3-21 Diagram and picture of a manometric BOD apparatus. (Courtesy of Hach Company, Loveland, Colorado.)

costs, direct reading of BOD without calculations, and continuous readings for monitoring the rate of reaction. Also undiluted tests purportedly are more representative of actual treatment conditions.

EXAMPLE 3-2

Data from an unseeded domestic wastewater BOD test are 5.0 ml of wastewater in a 300-ml bottle, initial DO of 7.8 mg/l, and five-day DO equal to 4.3 mg/l. Compute (a) the BOD, and (b) the ultimate BOD, assuming a k-rate of 0.10 per day.

Solution

(a) From Eq. 3-13,

$$\text{BOD} = \frac{7.8 - 4.3}{5.0/300} = 210 \text{ mg/l}$$

(b) Using Eq. 3-14,

$$\text{Ultimate BOD} = \frac{\text{BOD}}{1 - 10^{-kt}} = \frac{210}{1 - 10^{-0.1 \times 5.0}}$$

$$= \frac{210}{1 - 0.32} = 310 \text{ mg/l}$$

EXAMPLE 3-3

A seeded BOD test is to be conducted on meat-processing wastewater with an estimated strength of 800 mg/l. The seed is supernatant from aged, settled, domestic wastewater with a BOD of about 150 mg/l. (a) What sample portions should be used for setting up the middle dilutions of the wastewater and seed tests? (b) Calculate the BOD value for the industrial wastewater if the initial DO in both seed and sample bottles is 8.5 mg/l, and the five-day DO's are 4.5 mg/l and 3.5 mg/l for the seed test bottle and seeded wastewater sample, respectively.

Solution

(a) Based on data in Table 3-2,
Volume required for BOD test on 150-mg/l seed = 10.0 ml pipetted directly into a 300-ml bottle.
Volume of meat-processing wastewater required, assuming 800 mg/l = 2.0 ml, pipetted into a BOD bottle.

Amount of aged wastewater for seeding wastewater sample = 0.10 × 10.0 = 1.0 ml pipetted into the BOD bottle, or the addition of 3.33 ml/l of dilution water for a mixture of 0.33 percent.
(b) Substituting into Eq. 3-16,

$$\text{BOD} = \frac{(8.5 - 3.5) - (8.5 - 4.5)1.0/10.0}{2.0/300} = 690 \text{ mg/l}$$

3-12 BIOLOGICAL TREATMENT SYSTEMS

Biological processing is the most efficient way of removing organic matter from municipal wastewaters. These living systems rely on mixed microbial cultures to decompose, and to remove colloidal and dissolved organic substances from solution. The treatment chamber holding the microorganisms provides a controlled environment; for example, activated sludge is supplied with sufficient oxygen to maintain an aerobic condition. Wastewater contains the biological food, growth nutrients, and inoculum of microorganisms. Persons who are not familiar with wastewater operations often ask where the "special" biological cultures are obtained. The answer is that the wide variety of bacteria and protozoa present in domestic wastes seed the treatment units. Then by careful control of wastewater flows, recirculation of settled microorganisms, oxygen supply, and other factors, the desirable biological cultures are generated and retained to process the pollutants. The slime layer on the surface of the media in a trickling filter is developed by spreading wastewater over the bed. Within a few weeks the filter is operational, removing organic matter from the liquid trickling through the bed. Activated sludge in a mechanical, or diffused-air, system is started by turning on the aerators and feeding the wastewater. Initially a high rate of recirculation from the bottom of the final clarifier is necessary to retain sufficient biological culture. However, within a short period of time a settleable biological floc matures that efficiently flocculates the waste organics. An anaerobic digester is the most difficult treatment unit to start up, since the methane-forming bacteria, essential to digestion, are not abundant in raw wastewater. Furthermore, these anaerobes grow very slowly and require optimum environmental conditions. Start-up of an anaerobic digester can be hastened considerably by filling the tank with wastewater and seeding with a substantial quantity of digesting sludge from a nearby treatment plant. Raw sludge is then fed at a reduced initial rate, and lime is supplied as necessary to hold pH. Even under these conditions, several months may be required to get the process fully operational.

Enzymes are organic catalysts that perform biochemical reactions at temperatures and chemical conditions compatible with biological life. Chemical decomposition of potato or meat requires boiling in a strong acid solution, as performed in the COD test. However, these same foods can be readily digested by microorganisms, or in the stomach of an animal, at a much reduced temperature without strong mineral acids through the action of enzymes. Most enzymes cannot be isolated from living organisms without impairing their functioning capability. Although a discussion of enzymes is beyond the scope of this book, technicians in the field of sanitary science must be aware that enzyme additives sold to enhance biological treatment processes are ineffective. The label on the container generally uses highly scientific terms to convince the purchaser of product worthiness, for example, enzymes for wastewater (or for anaerobic digestion, stabilization ponds, septic tanks, etc.), minimum of 10 billion colonies per gram, excellent diastic, proteolytic, amylolytic, and lipolytic activity, a special formulation of enzymes, aerobic and anaerobic bacteria, and the like. In reality, domestic wastewater contains an abundant supply of all these enzymes, and to pour in more at a cost of $5 to $10 per pound can be figuratively described as pouring money down the drain.

Factors Affecting Growth

The most important factors affecting biological growth are temperature, availability of nutrients, oxygen supply, pH, presence of toxins and, in the case of photosynthetic plants, sunlight. Bacteria are classified according to their optimum temperature range for growth. Mesophilic bacteria grow in a temperature range of 10 to 40° C, with an optimum of 37° C. Aeration tanks and trickling filters generally operate in the lower half of this range with wastewater temperatures of 20 to 25° C in warm climates

and 8 to 10° C during the winter in northern regions. If cold well water serves as a water supply, wastewater temperatures can be lower than 20° C during the summer, and winter operation during extremely cold weather may result in ice formation on the surface of final clarifiers and freezing of stabilization ponds. Anaerobic digestion tanks are normally heated to near the optimum level of 35° C (98° F).

The rate of biological activity generally doubles or halves for every 10 to 15° C temperature rise or decrease within the range of 5 to 35° C. The mathematical relationship for the change in the reaction-rate constant with temperature (adopted from chemical kinetics, Section 2-5, Eq. 2-26) is expressed as

$$k = k_{20} \Theta^{T-20} \qquad (3-17)$$

where k = reaction-rate constant at temperature T,
        per day
   k_{20} = reaction-rate constant at 20° C, per day
   Θ = temperature coefficient, dimensionless
   T = temperature of biological reaction,
        degrees Celsius

The value of Θ is 1.072 if the rate of biological activity doubles or halves with a 10° C temperature change and is 1.047 if it doubles or halves with a 15° C change. The rate of a biological reaction does not remain constant over the entire 5 to 35° C range; in fact, considerable variation may occur. The general reaction-rate–temperature curve sketched in Figure 3-22 shows the reaction rate increasing with rising temperature; the slope of the curve becomes steeper with increasing temperature. Therefore, for this reaction-rate–temperature curve, the value of Θ becomes greater with higher reaction temperatures.

Above 40° C, mesophilic activity drops off sharply and thermophilic growth starts. Thermophilic bacteria have a range of approximately 45 to 75° C, with an optimum near 55° C. This higher temperature range is rarely used in waste treatment, since it is difficult to maintain that high an operating temperature, and because thermophilic bacteria are more sensitive to small temperature changes. Of particular importance is the dip between mesophilic and thermophilic ranges; operation in this region should be avoided. A technician seeking increased activity in his

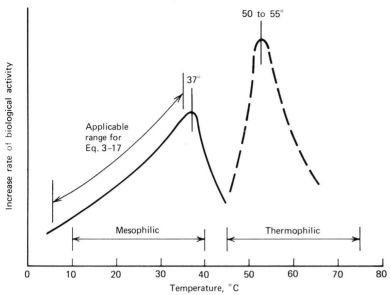

Figure 3-22   Effect of temperature on the rate of biological activity and relative positioning of the mesophilic and thermophilic zones.

or her anaerobic digester may gradually increase the temperature of the digesting sludge and thus realize improved gas production and efficiency. However, if the sludge is heated in excess of the mesophilic optimum, the rate of biological activity will decrease sharply, adversely affecting operation.

Municipal wastewaters commonly contain sufficient concentrations of carbon, nitrogen, phosphorus, and trace nutrients to support the growth of a microbial culture. Theoretically, a BOD to nitrogen to phosphorus ratio of 100/5/1 is adequate for aerobic treatment, with small variations depending on the type of system and mode of operation. Average domestic wastewater exhibits a surplus of nitrogen and phosphorus with a BOD/N/P ratio of about 100/17/5. If a municipal wastewater contains a large volume of nutrient-deficient industrial waste, supplemental nitrogen is generally supplied by the addition of anhydrous ammonia (NH_3) or ammonium nitrate (NH_4NO_3) and phosphate as phosphoric acid (H_3PO_4).

Diffused and mechanical aeration basins must supply sufficient air to maintain dissolved oxygen for the biota to use in metabolizing the waste organics. Rate of microbial activity is independent of dissolved oxygen concentration above a minimum critical value, below which the rate is reduced by the limitation of oxygen required for respiration. The exact minimum depends on the type of activated sludge process and the characteristics of the wastewater being treated. The most common design criterion for critical dissolved oxygen is 2.0 mg/l, but in actual operation values as low as 0.5 mg/l have proved satisfactory. Anaerobic systems must, of course, operate in the complete absence of dissolved oxygen; consequently, digesters are sealed with floating or fixed covers to exclude air.

Hydrogen ion concentration has a direct influence on biological treatment systems that operate best in a neutral environment. The general range of operation of aeration systems is between pH 6.5 and 8.5. Above this range microbial activity is inhibited, and below pH 6.5 fungi are favored over bacteria in the competition for metabolizing the waste organics. Normally the bicarbonate buffer capacity of a wastewater is sufficient to prevent acidity and reduced pH, while carbon dioxide production by the microorganisms tends to control the alkalinity of high pH wastewaters. Where industrial discharges force the pH of a municipal wastewater outside the optimum range, addition of a chemical may be required for neutralization. In that case, it is more desirable to have the industry pretreat its waste by equalization and neutralization prior to disposal in the sewer rather than contend with the problem of pH control at the city's disposal plant.

Anaerobic digestion has a small pH tolerance range of 6.7 to 7.4 with optimum operation at pH 7.0 to 7.1. Domestic waste sludge permits operation in this narrow range except during start-up or periods of organic overloads. Limited success in digester pH control has been achieved by careful addition of lime with the raw sludge feed. Unfortunately, the buildup of acidity and reduction of pH may be a symptom of other digestion problems, for example, accumulation of toxic heavy metals that the addition of lime cannot cure.

Biological treatment systems are inhibited by toxic substances. Industrial wastes from metal finishing industries often contain toxic ions, such as nickel and chromium; chemical manufacturing produces a wide variety of organic compounds that can adversely affect microorganisms. Since little can be done to remove or neutralize toxic compounds in municipal treatment, pretreatment should be provided by industries prior to discharging wastes to the city sewer.

Population Dynamics

In biological processing of wastes, the naturally occurring biota are a variety of bacteria growing in mutual association with other microscopic plants and animals. Three of the major factors in population dynamics are competition for the same food, predator-prey relationship, and symbiotic association. When organic matter is fed to a mixed population of microorganisms, competition arises for this food, and the primary feeders that are most competitive become dominant. Under normal operating conditions, bacteria are the primary feeders in both aerobic and anaerobic operations. Protozoa consuming bacteria is the common predator-prey relationship in activated sludge and trickling filters. In stabilization ponds, protozoa and rotifiers graze on both algae and bacteria. Symbiosis is the living together of organisms for mutual benefit such that the association produces more vigorous growth of both species. An excellent example of this is the relationship between bacteria and algae in a stabilization pond.

In an activated-sludge process, waste organics serve as food for the bacteria, and the small population of fungi that might be present. Some of the bacteria die and lyse, releasing their contents, which are resynthesized by other bacteria. The secondary feeders (protozoa) consume several thousand bacteria for a single reproduction. The benefit of this predator-prey action is twofold: (1) removal of bacteria stimulates further bacterial growth, accelerating metabolism of the organic matter; and (2) settling characteristics of the biological floc are improved by reducing the number of free bacteria in solution. The effluent from the process consists of nonsettleable organic matter and dissolved inorganic salts (Figure 3-23).

Control of the microbial populations is essential for efficient aerobic treatment. If wastewater were simply aerated, the liquid detention times would be intolerably long, requiring a time period of about five days at 20° C for 70 percent reduction. However, extraction of organic matter is possible within a few hours of aeration provided that a large number of microorganisms is mixed with the wastewater. In practice this is achieved by settling the microorganisms out of solution in a final clarifier and returning them to the aeration tank to metabolize additional waste organics (Figure 3-23). Good settling characteristics occur when an activated sludge is held in the endogenous (starvation) phase. Furthermore, a large population of underfed biota remove BOD very rapidly from solution. Excess microorganisms are wasted from the process to maintain proper balance between food supply and biological mass in the aeration tank. This balance is referred to as the food-to-microorganism ratio (F/M) which is normally expressed in units of pounds of BOD applied per day per pound of MLSS in the aeration basin (MLSS is the mixed liquor suspended solids). Operation at a high F/M ratio results in incomplete

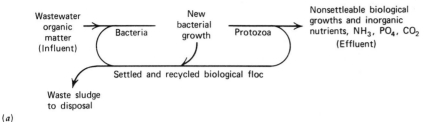

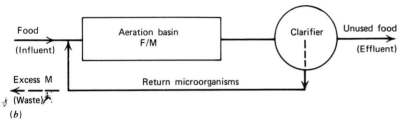

Figure 3-23 Generalized biological population dynamics in the activated-sludge wastewater treatment process. (*a*) Predator-prey relationship between protozoa and bacteria in the activated-sludge process. (*b*) Sedimentation and recirculation maintains the desired food-to-microorganism (F/M) ratio in the aeration basin.

metabolism of the organic matter, poor settling characteristics of the biological floc and, consequently, poor BOD removal efficiency. At a low F/M ratio, the mass of microorganisms are in a near starvation condition that results in a high degree of organic matter removal, good settleability of the activated sludge, and efficient BOD removal.

The relationship that exists between bacteria and algae in a small pond is illustrated in Figure 3-24. Bacteria decompose the organic matter, yielding inorganic nitro-

gen, phosphates, and carbon dioxide. Algae use these compounds, along with energy from sunlight, in photosynthesis, releasing oxygen into solution. This oxygen is, in turn, taken up by the bacteria, thus closing the cycle. The effluent from a stabilization pond contains suspended algae and excess bacterial decomposition end products. In the summer, BOD reduction in lagoons is very high, commonly in excess of 95 percent. However, during cold temperature operation, microbial activity is reduced, and BOD removal relies to a considerable extent on dilution

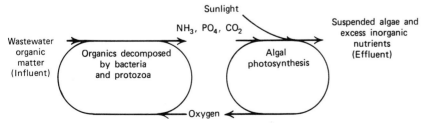

Figure 3-24 Schematic diagram of the mutually beneficial association (symbiotic relationship) between bacteria and algae in a stabilization pond.

of the inflowing raw wastewater into the large volume of impounded water. Liquid detention in a stabilization pond is rarely less than 90 days and generally is considerably greater. Following a cold winter, particularly if the pond was covered with ice and snow, odorous conditions can be anticipated during the spring thaw. With increased temperature, the organic matter accumulated during the winter is rapidly decomposed by the bacteria using dissolved oxygen at a faster rate than can be absorbed from the air or supplied by the algae. After a few days to several weeks, depending on climatic conditions and waste load on the lagoon, the algae become reestablished and again supply oxygen to the bacterial cycle. Once this symbiotic relationship is again operational, aerobic conditions are firmly established and odorous emissions cease.

3-13   BIOLOGICAL KINETICS

The characteristic growth pattern for a single bacterial species in a batch culture is sketched in Figure 3-25. This growth with time occurs when a sterile liquid substrate is inoculated with a small number of bacteria. After a short lag period, the bacteria reproduce exponentially by binary fission, rapidly increasing the number of viable cells and biomass in the medium. The presence of excess substrate promotes the maximum rate of growth possible, limited only by the ability of the bacteria to reproduce. In this *exponential growth phase*, the increase in both the number of viable cells and the accumulation of biomass is represented by Eq. 3-18, in which μ is a proportionality constant.

$$r_g = \mu X \qquad (3\text{-}18)$$

where r_g = biomass growth rate, milligrams per liter per day
μ = specific growth rate (rate of growth per unit of biomass), per day
X = concentration of biomass, milligrams per liter

The *declining growth phase* is the result of diminishing substrate limiting bacterial growth. The rate of reproduction decreases and some cells die so that the total biomass exceeds the mass of viable cells. When the

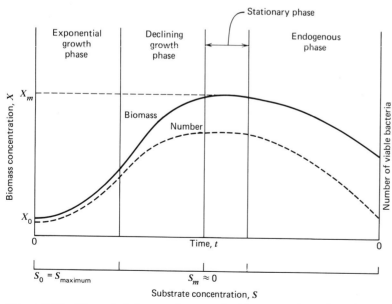

Figure 3-25   Characteristic growth curves for biomass and number of viable bacteria in pure, batch culture.

substrate is depleted at the end of declining growth, the number of viable bacteria and biomass remain relatively constant, resulting in a *stationary phase*. Rate of growth in the declining growth phase is described by the Monod equation

$$\mu = \mu_m\left(\frac{S}{K_S + S}\right) \qquad (3\text{-}19)$$

where μ_m = maximum specific growth rate, per day
   S = concentration of growth-limiting substrate, milligrams per liter
   K_S = saturation constant (equal to the limiting substrate concentration at one half the maximum growth rate), milligrams per liter

This mathematical relationship between the concentration of the growth-limiting substrate and the specific growth rate of biomass is a hyperbolic function as plotted in Figure 3-26. The constant K_S is equal to the concentration of substrate S when the specific growth rate μ equals one half of the growth rate μ_m. If the specific growth rate relationship in Eq. 3-19 is substituted into Eq. 3-18, the rate of biomass growth in a substrate-limiting solution is

$$r_g = \frac{\mu_m X S}{K_S + S} \qquad (3\text{-}20)$$

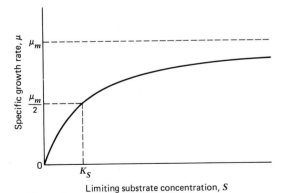

Figure 3-26 Curve of specific growth rate versus concentration of the growth-limiting substrate for a pure bacterial culture during the declining growth phase, defined mathematically by the Monod relationship in Eq. 3-19.

The *endogenous growth phase* (Figure 3-25) is a period of decreasing metabolism with a resulting decrease in both the biomass and the number of viable bacteria. The bacteria remaining compete for the small amount of substrate still in solution. Aged cells die and lyse, releasing nutrients back into solution. The action of lysis decreases both the number of cells and total biomass. The rate of biomass decrease is proportional to the biomass present, so that

$$r_d = -k_d X \qquad (3\text{-}21)$$

where r_d = biomass decay rate, milligrams per liter per day
   k_d = biomass decay coefficient, per day
   X = concentration of biomass, milligrams per liter

The net growth rate of biomass during the endogenous growth phase is the sum of Eqs. 3-20 and 3-21

$$r'_g = \frac{\mu_m X S}{K_S + S} - k_d X \qquad (3\text{-}22)$$

where r'_g = net biomass growth rate, milligrams per liter per day.

The Monod relationship (Eq. 3-19) modified for net specific growth rate is

$$\mu' = \mu_m\left(\frac{S}{K_S + S}\right) - k_d \qquad (3\text{-}23)$$

where μ' = net maximum specific growth rate, per day.

Growth yield is the incremental increase in biomass resulting from metabolism of an incremental amount of substrate. In a batch culture (Figure 3-25), the maximum yield is the biomass increase during the exponential and declining growth phases $(X_m - X_0)$ relative to the substrate used during this same period $(S_0 - S_m)$. The substrate concentration S_m approaches zero; therefore,

$$Y = \frac{X_m - X_0}{S_0} = \frac{r_g}{r_{su}} \qquad (3\text{-}24)$$

where Y = growth yield, milligrams per liter of biomass increase per milligram per liter of substrate metabolized
   r_{su} = rate of substrate utilization, milligrams per liter per day

If the growth yield is incorporated into Eq. 3-22, the net growth rate in the endogenous phase is

$$r'_g = Yr_{su} - k_dX \qquad (3-25)$$

The following equation is for the observed growth yield that accounts for the affect of the endogenous decay in biomass.

$$Y_{obs} = \frac{r'_g}{r_{su}} \qquad (3-26)$$

where Y_{obs} = observed growth yield in a laboratory test.

REFERENCES

1. *Standard Methods for the Examination of Water and Wastewater*, 15th Edition, American Public Health Association, American Water Works Association, and Water Pollution Control Federation, 1980.
2. Needham, J. G., and P. R. Needham. *A Guide to the Study of Fresh-water Biology.* San Francisco: Holden-Day, 1962.
3. *National Interim Primary Drinking Water Regulations*, U.S. Environmental Protection Agency, Office of Water Supply, EPA 570/9-76-003, 1976.

PROBLEMS

3-1 A fresh wastewater containing nitrate ions, sulfate ions, and dissolved oxygen is placed in a sealed jar without air. In what sequence are these oxidized compounds reduced by the bacteria? When do obnoxious odors appear?

3-2 State why some bacteria convert ammonia to nitrate (Eqs. 3-7 and 3-8) while others reduce nitrate to nitrogen gas (Eq. 3-2).

3-3 Write the equation for acid production in the moisture of condensation inside a sanitary sewer. What preventative measures can be taken to reduce or prevent damage to the interior of the pipes?

3-4 What are the major nutrients required for growth of iron bacteria inside water pipes? How do these bacterial growths contribute foul tastes and odors to the water?

3-5 What is a bacteriophage?

3-6 Compare the life-reproductive cycle of viruses and heterotrophic bacteria.

3-7 List the major differences and similarities between algae and nitrifying bacteria.

3-8 What is the principal role of protozoa in biological wastewater treatment and the aquatic food chain?

3-9 List several ways that fishes can be adversely affected by water pollution.

3-10 In the aquatic food chain, describe the organisms that are the first-order consumers.

3-11 Name the common waterborne bacterial and viral diseases.

3-12 Explain the reason mountain streams are the flowing waters most likely to be contaminated by *Giardia lamblia*.

3-13 Why can coliform bacteria be used as indicators of quality for drinking water?

3-14 Coliform bacteria in surface waters can originate from feces of humans, wastes of farm animals, or soil erosion. Can the coliforms from these three different sources be distinguished from one another?

3-15 A multiple-tube fermentation analysis of a river water yielded the following results. What are the MPN and confidence limits? (*Answer* 7000, 2300–17,000)

Serial Dilution	Sample Portion (ml)	Positive Tubes
0	1.0	5 of 5
1	0.1	5 of 5
2	0.01	2 of 5
3	0.001	1 of 5
4	0.000,1	0 of 5

3-16 Multiple-tube fermentation analyses of a river water yielded the following results. What are the MPN and confidence limits for the presumptive and confirmed coliform tests and fecal coliform count?

Serial Dilution	Sample Portion (ml)	Number of Positive Tubes Out of Five		
		Lactose	BGB	EC
0	1.0	5	5	5
1	0.1	5	5	2
2	0.01	2	2	2
3	0.001	2	0	0
4	0.000,1	0	0	0

3-17 The membrane filter technique is used to test a polluted water for the coliform group. Three different water sample volumes (5 ml, 50 ml, and 500 ml) were filtered through five filter membranes. The colonies' counts were as follows: 5-ml portions—4, 8, 5, 2, 3; 50-ml portions—24, 38, 25, 32, 30; and 500-ml portions—250, 320, 360, 210, 280. Calculate the coliform count per 100 ml using the most valid data.

3-18 List some of the problems associated with detection and identification of viruses.

3-19 In cell-culture testing, how are viruses in a sample concentrate detected and enumerated? Why is the viral count expressed in terms of PFU per liter?

3-20 Calculate the five-day BOD of a domestic wastewater based on the following data: volume of wastewater added to 300-ml bottle = 6.0 ml, initial DO = 8.1 mg/l, five-day DO = 4.2 mg/l. What is the ultimate BOD assuming a k-rate of 0.1 per day? (*Answer* 195 mg/l and 290 mg/l).

3-21 A seeded BOD test is to be conducted on a poultry-processing waste with an estimated five-day value of 600 mg/l. The seed taken from an existing preaeration tank at the industrial site has an estimated BOD of 200 mg/l. (a) What sample portions should be used for setting up the middle dilutions of the wastewater and seed tests? (b) Compute the BOD value for the poultry waste if the initial DO in both the seed and sample bottles is 8.2 mg/l and the five-day values are 3.5 mg/l and 4.0 mg/l for the seed test bottle and seeded wastewater sample, respectively. Use the volumetric additions from part (a) and assume that the volume of seed used in

the waste BOD bottle is 10 percent of that used in the seed test. [*Answers* (a) 2.0 ml poultry waste plus 0.7 ml of seed, 7.0 ml of seed sample (b) 560 mg/l].

3-22 A seeded BOD analysis was conducted on a high strength, food-processing wastewater. Twenty-milliliter portions were used in setting up the 300-ml bottles of aged, settled, wastewater seed. The seeded sample BOD bottles contained 1.0-ml of industrial waste and 2.0 ml of seed material. The results for this series of test bottles are listed below. Calculate the BOD values. (B_1 = 8.2 mg/l, D_1 = 8.2 mg/l, f = 0.10, P = 1/300). Plot a BOD-time curve; what is the five-day value? Determine the k-rate. (*Answers* five-day BOD = 900 mg/l and k = 0.09 per day)

Time (days)	Seed Tests B_2 (mg/l)	Sample Tests D_2 (mg/l)	Time (days)	Seed Tests B_2 (mg/l)	Sample Tests D_2 (mg/l)
0	8.2	8.2	7.9	4.5	4.1
1.0	7.3	7.2	9.8	4.1	3.8
1.9	6.5	6.4	11.8	3.9	3.5
2.9	5.9	5.8	14.0	3.4	2.9
3.8	5.5	5.3	15.6	3.6	2.7
4.7	5.1	5.0	19.0	3.8	2.1
6.0	4.8	4.6			

3-23 A BOD analysis was conducted on the unchlorinated effluent from a municipal treatment plant. Several identical tests with 30 ml of wastewater added to each 300-ml bottle were prepared and chemically tested for dissolved oxygen after various intervals of time. After three samples were titrated immediately, the initial dissolved oxygen was determined as 8.7 mg/l. The time-residual DO data for the incubated bottles were t = 1.2 days, DO = 7.6 mg/l; 2.1 days, 6.7 mg/l; 3.6 days, 5.9 mg/l; 5.8 days, 4.9 mg/l; and 9.2 days, 2.5 mg/l. Plot a BOD-time curve. Determine the carbonaceous k-rate and five-day BOD value.

3-24 Determine the BOD of a wastewater from candy manufacturing based on the following data. The seed was settled, aged, domestic wastewater.

Wastewater Portion (ml)	Seed Portion (ml)	Initial DO (mg/l)	Five-day DO (mg/l)
Bottles tested for initial dissolved oxygen			
5.0	1.5	8.8	
6.0	1.5	8.7	
7.0	1.5	9.0	
Tests for determining the BOD of the seed			
15	0		5.4
15	0		5.5
15	0		5.7
Tests for determining the BOD of the wastewater			
8.0	1.5		0.5
8.0	1.5		0.8
8.0	1.5		0.4
6.0	1.5		3.1
6.0	1.5		2.9
6.0	1.5		2.9
4.0	1.5		4.6
4.0	1.5		4.5
4.0	1.5		4.8

3-25 A seeded BOD analysis was conducted on a meat-processing wastewater. Dissolved oxygen data for the seed test and seeded wastewater are listed below. The BOD bottles for the seed (aged, settled, domestic wastewater) were set up by adding 15 ml per 300-ml bottle. The test bottles for the meat-processing wastewater were set up by adding 2.0 ml of wastewater and 1.5 ml of seed.

	DO Measurements in Bottles	
Time (days)	Seed (mg/l)	Seeded Wastewater (mg/l)
0	7.9	8.1
1.0	6.5	5.5
2.2	5.4	4.2
3.0	4.9	3.7
4.0	4.0	2.5
5.0	3.8	2.1
6.0	3.8	2.1
7.0	3.6	2.0

Based on the averages of three tests, the initial DO in the seed bottles was 7.9 mg/l and the initial DO in the wastewater bottles was 8.1 mg/l. Sketch BOD-time curves for both the seed wastewater and meat-processing wastewater. Determine the five-day BOD values and k-rates using the graphical procedure.

3-26 Solids analyses were conducted on the meat-processing wastewater that was tested for BOD in Problem 3-25. The tests were conducted in triplicate. Based on the test data below, determine the concentrations of total solids, total volatile solids, suspended solids, and volatile suspended solids.

Total Solids Test Data			
Dish number	1	2	3
Weight of empty dish, g	51.494	51.999	50.326
Weight of dish + sample, g	109.5	103.4	118.4
Weight of dish + dry solids, g	51.587	52.081	50.437
Weight of dish + ignited solids, g	51.541	52.042	50.383
Suspended Solids Test Data			
Test number	1	2	3
Weight of filter, g	0.1154	0.1170	0.1170
Sample volume, ml	25	25	30
Weight of filter + dry solids, g	0.1241	0.1259	0.1278
Weight of filter + ignited solids, g	0.1160	0.1173	0.1178

3-27 A seeded BOD analysis was conducted on a soybean-processing wastewater. Dissolved oxygen data for the seed test and seeded wastewater test are listed below. The BOD bottles for the seed (aged, settled, domestic wastewater) were set up by adding 15 ml per 300-ml bottle. The test bottles for the soybean-processing wastewater were set up by adding 4.0 ml of wastewater and 1.5 ml of seed. Based on the average of three tests, the initial DO in the seed bottles was 7.9 mg/l, and the initial DO in the wastewater bottles was 8.1 mg/l. Sketch BOD-time curves for both the seed wastewater and soybean-processing wastewater. Determine the five-day BOD values and k-rates using the graphical procedure.

	DO Measurements in Bottles	
Time (days)	Seed (mg/l)	Seeded Wastewater (mg/l)
0	7.9	8.1
1.0	6.3	6.3
2.2	5.3	5.2
2.9	5.1	4.9
3.8	4.4	4.2
4.9	3.8	3.9
5.8	4.0	4.1
6.8	3.6	3.7

3-28 Solids analyses were conducted on the soybean-processing wastewater that was tested for BOD in Problem 3-27. The tests were conducted in triplicate. Based on the test data below, determine the concentrations of total solids, total volatile solids, suspended solids, and volatile suspended solids.

Total Solids Test Data			
Dish number	1	2	3
Weight of empty dish, g	65.842	53.252	53.073
Weight of dish + sample, g	123.0	104.0	102.8
Weight of dish + dry solids, g	65.993	53.375	53.220
Weight of dish + ignited solids, g	65.965	53.355	53.173

Suspended Solids Test Data			
Test number	1	2	3
Weight of filter, g	0.1160	0.1165	0.1152
Sample volume, ml	48	72	72
Weight of filter + dry solids, g	0.1215	0.1259	0.1257
Weight of filter + ignited solids, g	0.1180	0.1205	0.1189

3-29 Are special microorganisms added to initiate microbial activity in biological wastewater treatment systems?

3-30 List the major factors influencing the rate of biological growth.

3-31 How does temperature affect biological processes in wastewater treatment?

3-32 The rate of BOD reduction in aeration of a synthetic wastewater decreased by 20 percent when the temperature of the laboratory fermentation tank was lowered from 20 to 16°C. Using Eq. 3-17, calculate the temperature coefficient.

3-33 State two reasons why protozoa are significant in activated sludge.

3-34 Why must an activated-sludge process be operated in the starvation (low F/M) stage?

3-35 What's wrong with this statement: Algae consume waste organics applied to a stabilization pond; therefore, if the lagoon water is not green colored, odors result from undecomposed wastes.

3-36 The following loading-temperature-efficiency data were collected during full-scale studies of an extended aeration plant treating domestic wastewater. Why is the effluent BOD (treatment efficiency) apparently independent of temperature?

Season of Year	BOD Loading (lb/1000 cu ft)	Effluent BOD (mg/l)	BOD Removal (percent)	Temperature of Mixed Liquor (°C)
Winter	11	17	91	11
Summer	12	16	92	19

3-37 Why do wastewater stabilization ponds support dense populations of algae?

3-38 A series of fermentation tubes containing varying concentrations of glucose in a nutrient broth were inoculated with a pure bacterial culture. The concentrations of cells in the broth media were determined after 16 hr of incubation at 37°C. Rates of growth and initial glucose concentrations are listed below. Plot growth rate expressed as cell divisions per hour versus initial glucose concentration. Estimate the maximum growth rate and the saturation constant (glucose concentration at one half the maximum growth rate). Write an equation in the form of Eq. 3-19 and draw the curve for this equation on the graph with the plotted data. Does the growth of this pure culture appear to be a hyperbolic function as defined by the Monod relationship?

Glucose (moles × 10^{-4})	Cell Divisions (per hr)	Glucose (moles × 10^{-4})	Cell Divisions (per hr)
0.1	0.23	0.8	0.94
0.1	0.28	1.6	1.06
0.2	0.32	3.2	1.15
0.4	0.71		

chapter 4

Hydraulics and Hydrology

The following sections present the basic principles of hydraulics applicable to water distribution and wastewater collection systems. Low-flow characteristics of streams and stratification of lakes provide background for the study of water pollution. The basic concepts of groundwater hydrology are essential to understanding water well construction and operation.

4-1 WATER PRESSURE

Mass per unit volume is referred to as the density of a fluid. The density of water at a temperature of 60° F (15° C) and pressure 1 atmosphere is 1.94 slugs/cu ft (999 kg/m^3). The force exerted by gravity on 1.0 cu ft (1.0 m^3) of water is 62.4 lb (9.80 kN/m^3). This is equal

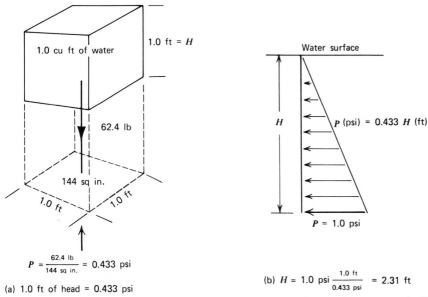

(a) 1.0 ft of head = 0.433 psi

(b) $H = 1.0 \text{ psi} \dfrac{1.0 \text{ ft}}{0.433 \text{ psi}} = 2.31 \text{ ft}$

Figure 4-1 Basic relationships between water pressure in psi (pounds per square inch) and feet of water head.

to the density multiplied by the acceleration of gravity, which is 32.2 ft/sec² (9.81 m/s²).

Pressure is the force exerted per unit area. Figure 4-1a shows that the pressure exerted on the bottom of a 1.0-cu ft container filled with water is equal to 0.433 psi. In engineering hydraulics, water pressure is frequently expressed in terms of feet of head, as well as psi. The relationship between these units is visually shown in Figure 4-1a where a height of 1.0 ft of water head exerts a pressure of 0.433 psi. Water pressure increases with depth below the surface linearly such that the pressure in psi is equal to 0.433 times the depth in feet. The sketch in Figure 4-1b shows the pressure acting only horizontally for ease of illustration. In fact, water pressure is exerted equally in all directions. The computation given under the diagram states that a head of 2.31 ft produces a water pressure of 1.0 psi. Thus, 1.0 ft of head is equivalent to 0.433 psi or, alternately, a pressure of 1.0 psi is equivalent to 2.31 ft of head.

The relationships for water pressure and water head in SI metric units are illustrated in Figure 4-2. The pressure exerted on the bottom of a 1.0-m³ container filled with water at 15° C equals 9.80 kPa. Therefore, 1.0 m of water head exerts a pressure of 9.80 kPa, or 1.0 kPa of pressure is equal to 0.102 m of water head.

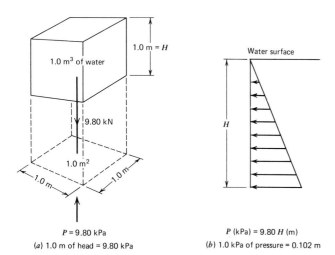

$P = 9.80$ kPa

(a) 1.0 m of head = 9.80 kPa

P (kPa) = 9.80 H (m)

(b) 1.0 kPa of pressure = 0.102 m

Figure 4-2 Basic relationships between water pressure in kilopascals and meters of water head.

A simple piezometer, illustrated in Figure 4-3a, consists of a small tube rising from a container of water under pressure. The height of the water in the piezometer tube denotes the pressure of the confined water. Water piezometers are rarely practical for pressure measurements, since values generally exceed the height of water columns that are convenient to read. A mercury column can be used to measure relatively high pressure values within a limited head range, since mercury is approximately 13.6 times heavier than water, that is, a 1.0-ft head of mercury is equivalent to a pressure of 5.9 psi. Figure 4-3b is a sketch of a mercury manometer. The pressure reading is equal to the difference of the fluid pressures in the two legs of the U-shaped tube. The use of manometers is generally restricted to indoor applications where the unit is in a fixed location.

Water pressure is commonly measured by a Bourdon gauge (Figure 4-3c). A hollow metal tube of elliptical cross section is bent in the form of a circle with a pointer attached to the end through a suitable linkage. As pressure inside the tube increases, the elliptical cross section tends to become circular, and the free end of the Bourdon tube moves outward. The dial of the instrument can be calibrated to read gauge pressure in psi. Bourdon tubes and manometers actually measure pressure relative to the atmosphere pressure. [Atmospheric or barometric pressure is caused by the weight of the thick mass of air above the earth's surface. Under standard atmospheric conditions, the barometric pressure at sea level is 14.7 psi (101 kPa).] If the pressure measured is greater than atmospheric, this value is sometimes called gauge pressure. If the pressure measured is less than atmospheric, it is referred to as a vacuum. Absolute pressure is the term used for a pressure reading that includes atmospheric pressure, that is, the pressure relative to absolute zero.

4-2 PRESSURE-VELOCITY-HEAD RELATIONSHIPS

The association between quantity of water flow, average velocity, and cross-sectional area of flow is given by the equation

$$Q = VA \tag{4-1}$$

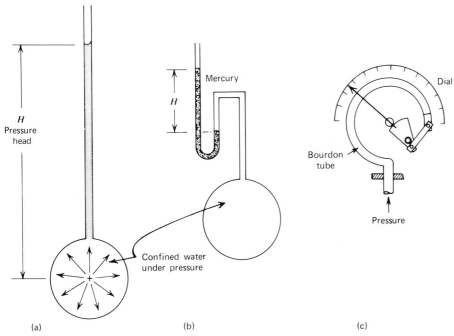

(a)   (b)   (c)

Figure 4-3   Water pressure measuring devices. (*a*) Piezometer. (*b*) Manometer. (*c*) Pressure gauge.

where Q = quantity, cubic feet per second
                (cubic meters per second)
        V = velocity, feet per second
                (meters per second)
        A = cross-sectional area of flow, square feet
                (square meters)

This formula is known as the continuity equation. For an incompressible fluid such as water, if the cross-sectional area decreases (the water main becomes smaller), the velocity of flow must increase; conversely, if the area increases, the velocity decreases (Figure 4-4).

The total energy at any point in a hydraulics system is equal to the sum of the elevation head, pressure head, and velocity head.

$$\frac{\text{Total}}{\text{Energy}} = \frac{\text{Elevation}}{\text{Head}} + \frac{\text{Pressure}}{\text{Head}} + \frac{\text{Velocity}}{\text{Head}} \quad (4\text{-}2)$$

$$E = Z + \frac{P}{w} + \frac{V^2}{2g}$$

where E = total energy head
        Z = elevation above datum
        P = pressure
        V = velocity of flow
        w = unit weight of liquid
        g = acceleration of gravity

The vertical line at point 1 on the left-hand side of Figure 4-5 graphically illustrates the total energy equation. The pressure head is equal to the height to which the water would rise in a piezometer tube inserted in the pipe. The

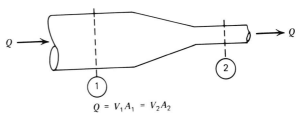

$$Q = V_1 A_1 = V_2 A_2$$

Figure 4-4   Flow equation for incompressible liquids.

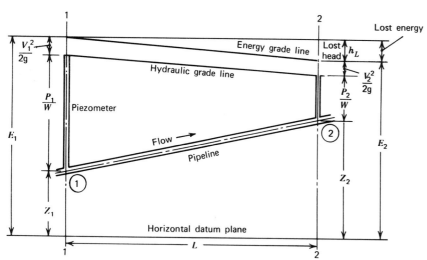

Figure 4-5 Energy equation parameters as related to water flow in a pipe.

elevation head is equal to the vertical distance Z from an assumed datum plane to the pipe. The sum of these two heads is the hydraulic head. The velocity head is equal to the kinetic energy in the water flow, and when added to the hydraulic head yields total energy. If ft-lb-sec units are used in Eq. 4-2, each of the separate head calculations will be in feet.

If the energy is compared in English units at two different points in a piping system (Figure 4-5), Eq. 4-3 results. The term h_L represents the energy losses that occur in any real system. The major loss of energy is due to friction between the moving water and pipe wall; however, energy losses also occur from flow disturbance caused by valves, bends in the pipeline, and changes in diameter. The imaginary line connecting all points of total energy is called the energy gradient or grade line. This line must always slope in the direction of flow showing a decrease in energy, unless external energy is added to the system, for example, by a pump. The hydraulic gradient is defined as the line connecting the level of elevation plus pressure energies, this being defined by the water surfaces in imaginary piezometer tubes inserted in the piping.

$$Z_1 + \frac{P_1}{62.4} + \frac{V_1^2}{64.4} = Z_2 + \frac{P_2}{62.4} + \frac{V_2^2}{64.4} + h_L \quad (4\text{-}3)$$

where Z = elevation, feet
$\quad P$ = pressure, pounds per square foot
$\quad V$ = velocity, feet per second
$\quad h_L$ = head loss, feet
$\quad 62.4$ = unit weight of water, pounds per cubic foot
$\quad 64.4$ = 2g, feet per second squared

The expression for Eq. 4-2 in SI metric units is

$$Z_1 + \frac{P_1}{9.80} + \frac{V_1^2}{19.6} = Z_2 + \frac{P_2}{9.80} + \frac{V_2^2}{19.6} + h_L \quad \text{(SI units)}$$
$$(4\text{-}4)$$

where Z = elevation, meters
$\quad P$ = pressure, kilopascals (kilonewtons per square meter)
$\quad V$ = velocity, meters per second
$\quad h_L$ = head loss, meters
$\quad 9.80$ = specific weight of water, kilonewtons per cubic meter
$\quad 19.6$ = 2g, meters per second squared

Head loss as a result of pipe friction can be computed by using the Darcy Weisbach equation:

$$h_L = f\frac{LV^2}{D2g} \quad (4\text{-}5)$$

where h_L = head loss, feet
  f = friction factor, Figure 4-6
  L = length of pipe, feet
  V = velocity of flow, feet per second
  D = diameter of pipe, feet

The friction factor f is related to the relative roughness of the pipe material and the fluid flow characteristics. For turbulent water flow, the f value for Eq. 4-5 can be taken from Figure 4-6 based on pipe diameter and material roughness.

Valves, fittings, and other appurtenances disturb the flow of water, causing losses of head in addition to the friction loss in the pipe. These units in a typical distribution system are at infrequent intervals; consequently, losses due to appurtenances are relatively insignificant in comparison with pipe friction losses. In the case of pumping stations and treatment plant piping, the minor losses in valves and fittings are significant and constitute a major part of the total losses. Unit head losses may be expressed as being equivalent to the loss through a certain length of pipe or by the formula

$$h_L = \frac{kV^2}{2g} \qquad (4\text{-}6)$$

where h_L = head loss, feet (meters)
  V = velocity, feet per second (meters per second)
  k = loss coefficient

The equivalent length of pipe that results in the same head loss is expressed in terms of the number of pipe diameters. Table 4-1 gives values for k and equivalent lengths of pipe for various fittings and valves.

EXAMPLE 4-1

Calculate the head loss in a 24-in.-diameter, 5000-ft-long, smooth-walled concrete ($\varepsilon = 0.001$) pipeline carrying a water flow of 10 cu ft/sec.

Solution

Using Eq. 4-1,

$$V = \frac{Q}{A} = \frac{10 \text{ cu ft/sec}}{\pi \text{ sq ft}} = 3.2 \text{ ft/sec}$$

From Fig. 4-6 entering with $d = 24$ in. and $\varepsilon = 0.001$, $f = 0.017$. Then substituting into Eq. 4-5,

$$h_L = 0.017 \frac{5000}{2} \frac{(3.2)^2}{2 \times 32.2} = 6.8 \text{ ft}$$

EXAMPLE 4-2

A pump discharge line consists of 200 ft of 12-in. new cast-iron pipe, three 90° medium-radius bends, two gate valves, and one swing check valve. Compute the head loss through the line at a velocity of 3.0 ft/sec.

Solution

If equivalent lengths for the fittings and valves as given in Table 4-1 are used, the total equivalent pipe length is

$$200 + 3 \times 27 + 2 \times 17 + 1 \times 135 = 450$$

From Figure 4.6, $f = 0.019$; then using Eq. 4-5,

$$h_L = 0.019 \frac{450}{1.0} \frac{(3.0)^2}{2 \times 32.2} = 1.2 \text{ ft}$$

EXAMPLE 4-3

Calculate the head loss in the pipeline illustrated in Figure 4-5 based on the following: $Z_1 = 4.5$ m, $P_1 = 280$ kPa, $V_1 = 1.2$ m/s, $Z_2 = 9.3$ m, $P_2 = 200$ kPa, $V_2 = 1.2$ m/s.

Solution

Substituting into Eq. 4-4,

$$4.5 + \frac{280}{9.80} + \frac{(1.2)^2}{19.6} = 9.3 + \frac{200}{9.80} + \frac{(1.2)^2}{19.6} + h_L$$

$$4.5 + 28.6 + 0.07 = 9.3 + 20.4 + 0.07 + h_L$$

$$h_L = 3.4 \text{ m} = 33 \text{ kPa}$$

4-3   FLOW IN PIPES UNDER PRESSURE

The Darcy Weisbach equation for computing head loss is cumbersome and not widely used in waterworks design and evaluation. A trial-and-error solution is required to determine pipe size for a given flow and head loss, since the friction factor is based on the relative roughness,

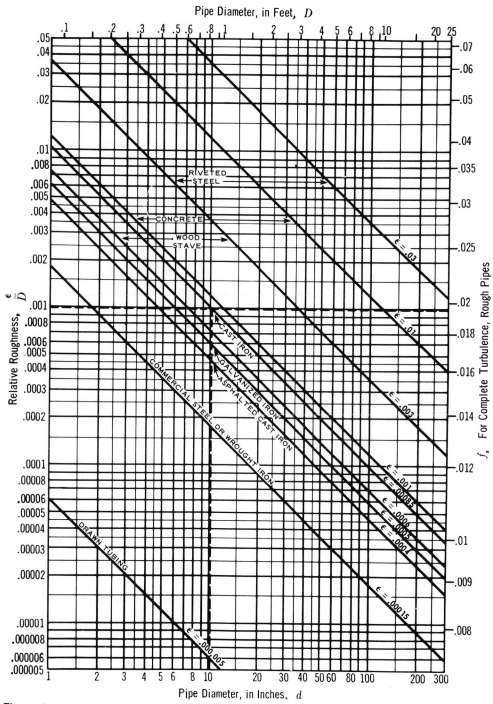

Figure 4-6 Relative roughness of pipe materials and friction factors for turbulent pipe flow. (From *Flow of Fluids*, Crane Co.)

Table 4-1. Approximate Minor Head Losses in Fittings and Valves

Fitting or Valve	Loss Coefficient k	Equivalent Length (Diameters of Pipe)
Tee (run)	0.60	20
Tee (branch)	1.80	60
90° bend-	—	—
Short radius	0.90	32
Medium radius	0.75	27
Long radius	0.60	20
45° bend	0.42	15
Gate valve (open)	0.48	17
Swing check valve (open)	3.7	135
Butterfly valve (open)	1.2	40

which involves the pipe diameter. Because of this practical shortcoming, exponential equations are commonly used for flow calculations.

The most common pipe flow formula used in the design and evaluation of a water distribution system is the Hazen Williams, Eq. 4-7. This equation relates the velocity of turbulent water flow with hydraulic radius, slope of the hydraulic gradient, and a coefficient of friction depending on roughness of the pipe. Coefficient values for different pipe materials are given in Table 4-2.

Table 4-2. Values of Coefficient C for the Hazen Williams Formula, Equations 4-7, 4-8, and 4-9

Pipe Material	C
Asbestos cement	140
Ductile iron	
Cement lined	130 to 150
New, unlined	130
5-year-old, unlined	120
20-year-old, unlined	100
Concrete	130
Copper	130 to 140
Plastic	140 to 150
New welded steel	120
New riveted steel	110

$$V = 1.318CR^{0.63}S^{0.54} \qquad (4\text{-}7)$$

where V = velocity, feet per second
C = coefficient that depends on the material and age of the conduit, Table 4-2
R = hydraulic radius, feet (cross-sectional area divided by the wetted perimeter)
S = slope of the hydraulic gradient, feet per foot

By combining the Hazen Williams formula with Eq. 4-1 and substituting the value of $D/4$ for R, the formula for quantity of flow in circular pipes flowing full is

$$Q = 0.281CD^{2.63}S^{0.54} \qquad (4\text{-}8)$$

where Q = quantity of flow, gallons per minute
C = coefficient, Table 4-2
D = diameter of pipe, inches
S = hydraulic gradient, feet per foot

The nomograph shown in Figure 4-7 solves the above equation for a coefficient equal to 100, represents 15- to 20-year-old ductile-iron pipe. Given any two of the parameters (discharge, diameter of pipe, loss of head, or velocity), the remaining two can be determined from the intersections along a straight line drawn across the nomograph. For example, the flow of 500 gpm in an 8-in.-diameter pipe has a velocity of 3.2 ft/sec with a head loss of 8.5 ft/1000 ft (0.0085 ft/ft). Head losses in pipes with coefficient values other than 100 can be determined by using the correction factors in Table 4-3. For example, if

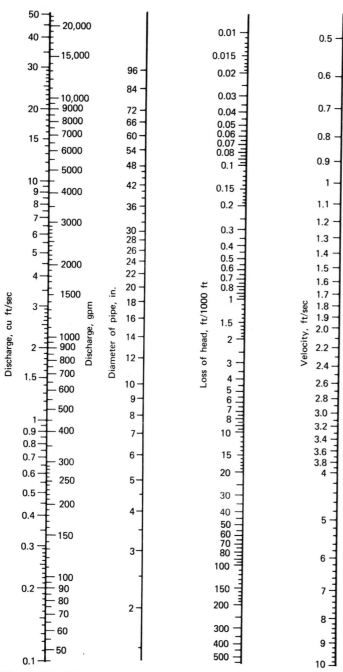

Figure 4-7 Nomograph for Hazen Williams formula, Eq. 4-8, based on $C = 100$.

Table 4-3. Correction Factors to Determine Head Losses from Figure 4-7 at Values of C Other Than $C = 100$

Corrected $h_L = K \times h_L$ at $C = 100$			
C	K	C	K
80	1.51	120	0.71
100	1.00	130	0.62
110	0.84	140	0.54

the head loss at $C = 100$ is $8.5/1000$ ft, the head loss at $C = 130$ would equal $0.62 \times 8.5 = 5.3$ ft/1000 ft.

The Hazen Williams formula, Eq. 4-8, expressed in SI metric units is

$$Q = 0.278CD^{2.63}S^{0.54} \quad \text{(SI units)} \quad (4\text{-}9)$$

where Q = quantity of flow, cubic meters per second
C = coefficient, Table 4-2
D = diameter of pipe, meters
S = hydraulic gradient, meters per meter

The nomograph in Figure 4-7 can be used to solve a pipe flow problem given in metric data by applying the appropriate conversion factors given in the Appendix.

EXAMPLE 4-4

If a 10-in. water main is carrying a flow of 950 gpm, what is the velocity of flow and head loss for (a) $C = 100$, and (b) $C = 140$?

Solution

(a) From Figure 4-7, $V = 3.9$ ft/sec and $h_L = 9.0$ ft/1000 ft
(b) $V = 3.9$ ft/sec (same as part a)
Using K from Table 4-3, $h_L = 0.54 \times 9.0 = 4.9$ ft/1000 ft

EXAMPLE 4-5

If a 200-mm water main ($C = 100$) is carrying a flow of 30 l/s, what is the velocity of flow and head loss?

Solution

$Q = 30$ l/s $= 0.030$ m³/s and $D = 200$ mm $= 0.20$ m

The cross-sectional area of flow $= \pi(0.20/2)^2 = 0.0314$ m².
Using Eq. 4-1,

$$0.030 \text{ m}^3/\text{s} = V \times 0.0314 \text{ m}^2$$
$$V = 0.030/0.0314 = 0.955 \; m/s$$

Substituting into Eq. 4-9,

$$0.030 \text{ m}^3/\text{s} = 0.278 \times 100(0.20 \text{ m})^{2.63} S^{0.54}$$

$$S^{0.54} = \frac{0.030}{0.278 \times 100(0.20)^{2.63}} = \frac{0.030}{0.403} = 0.0744$$

$$S = 0.0744^{1/0.54} = 0.0744^{1.85} = 0.00817 \text{ m/m}$$
$$= 8.17 \text{ m/1000 m}$$

Using the nomograph in Figure 4-7, enter with

$$Q = 30 \text{ l/s} \times 15.85 = 480 \text{ gpm and}$$
$$D = 200 \text{ mm} \times 0.03937 = 7.9 \text{ in.}$$

Read velocity and slope, and convert answers to metric units.

$$V = 3.2 \text{ ft/sec} \times 0.3048 = 0.97 \text{ m/s and}$$
$$S = 8.2 \text{ ft/1000 ft} = 8.2 \text{ m/1000 m}$$

EXAMPLE 4-6

An extremely simplified water supply system consisting of a reservoir with lift pumps, elevated storage, piping, and load center (withdrawal point) is shown in Figure 4-8. (a) Based on the following data, sketch the hydraulic gradient for the system:

$$Z_A = 0 \text{ ft}, P_A = 80 \text{ psi}, Z_B = 30 \text{ ft}, P_B = 30 \text{ psi},$$
$$Z_C = 40 \text{ ft}, P_C = 100 \text{ ft (water level in tank)}$$

(b) For these conditions compute the flow available at point B from both the supply pumps and elevated storage. Use $C = 100$ and pipe sizes as shown in the diagram.

Solution

(a) Hydraulic head at $A = 0$ ft $+ 80$ psi $\times 2.31$ ft/psi $= 185$ ft; at $B = 30 + 30 \times 2.31 = 99$ ft; at $C = 40 + 100 = 140$ ft.
The hydraulic gradient is shown as straight lines connecting these hydraulic heads drawn vertically.

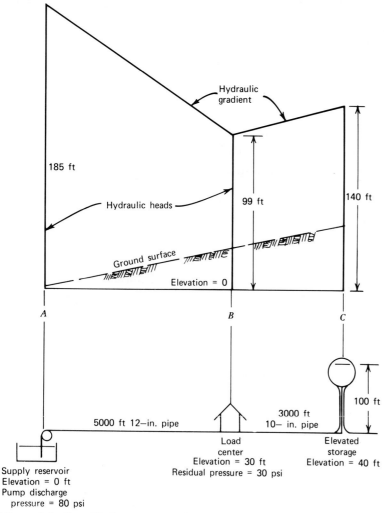

Hydraulic gradient

185 ft

Hydraulic heads

99 ft

140 ft

Ground surface

Elevation = 0

A

B

C

5000 ft 12–in. pipe

Load center
Elevation = 30 ft
Residual pressure = 30 psi

3000 ft
10– in. pipe

100 ft

Elevated storage
Elevation = 40 ft

Supply reservoir
Elevation = 0 ft
Pump discharge
  pressure = 80 psi

Figure 4-8   Simplified water system for Example 4-6.

(b) h_L between A and $B = 185 - 99 = 86$ ft
h_L per 1000 ft $= 86 \div 5 = 17.2$ ft
Using Figure 4-7, align $h_L = 17.2$ ft/1000 ft and 12-in. diameter, read $Q = 2160$ gpm

$$h_L \text{ per 1000 ft } B \text{ to } C = \frac{140 - 99}{3} = 13.7 \text{ ft}$$

For $h_L = 13.7$ ft/1000 ft and 10 in., $Q = 1180$ gpm. Hence, total Q available at $B = 2160 + 1180 = 3340$ gpm.

4-4   CENTRIFUGAL PUMP CHARACTERISTICS

Pumps are used for a variety of functions in water and wastewater systems. Low-lift pumps are used to elevate water from a source, or wastewater from a sewer, to the treatment plant; high-service pumps are used to discharge water under pressure to a distribution system or wastewater through a force main; booster pumps are used to increase pressure in a water distribution system; recirculation and transfer pumps are used to move water being processed within a treatment plant; well pumps are used to lift water from shallow or deep wells for water supply; still other types are used for chemical feeding, sampling, and fire fighting. Centrifugal pumps are used commonly for low and high service to lift and transport water, reciprocating positive-displacement and progressing cavity pumps are used to move sludges, vertical turbine pumps are used for well pumping, and pneumatic ejectors are used for small wastewater lift stations. Air-lift, peristaltic, rotary displacement, and turbine pumps are used in special applications. The discussion in this section, however, is restricted to the characteristics of centrifugal pumps; specific types are illustrated and discussed in relationship to their applications in other sections of the book.

Centrifugal pumps are popular because of their simplicity, compactness, low cost, and ability to operate under a wide variety of conditions. The two essential parts are a rotating member with vanes, the impeller, and a surrounding case (Figure 4-9). The impeller driven at a high speed throws water into the volute, which channels it through the nozzle to the discharge piping. This action depends partly on centrifugal force, hence, the particular name given to the pump. The function of the pump passages is to develop water pressure by efficient conversion of kinetic energy. A closed impeller is generally used in pumping water for higher efficiency, while an open unit is used for wastewater containing solids. The casing may be in the form of a volute (spiral) or may be equipped with diffuser vanes. Several other variations in design involve casing and impeller modifications to provide a wide range of pumps with special operating features.

Pump Head-Discharge Curve

The head developed by a particular pump at various rates of discharge at a constant impeller speed is established by pump tests conducted by the manufacturer. The head given is the discharge pressure with the inlet static water level at the elevation of the pump centerline and excluding losses in suction and discharge piping. Consider the test arrangement illustrated schematically in Figure 4-10 where the discharge is controlled by a throttling valve, the discharge pressure measured by a gauge, and the rate of discharge recorded by a flow meter. Simultaneously, the power input is measured and efficiency determined. With the valve in the discharge pipe closed, the rotating impeller simply churns in the water causing the pressure at the outlet of the pump to rise to a value referred to as the shutoff head. As the valve is gradually opened allowing increasing flow of water, the pump head decreases as drawn in Figure 4-11. The pump efficiency rises

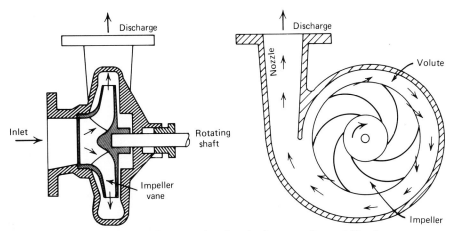

Figure 4-9   Cross-sectional diagrams showing the features of a centrifugal pump.

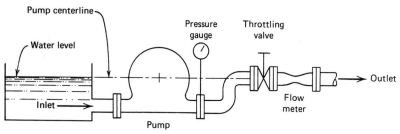

Figure 4-10   Schematic of a pump head-discharge test.

with increasing rate of discharge to an optimum value and then decreases. The flow rate at peak efficiency is determined by pump design and the rotational speed of the impeller.

A centrifugal pump is designed to operate near the point of best operating efficiency. Radial loads on the bearings are at a minimum when the impeller is operating at a balanced load at this point. As the pump discharge increases beyond optimum operation, the radial loads increase and cavitation is a potential problem. Cavitation occurs when vapor bubbles form at the pump inlet and are carried into a zone of higher pressure where they collapse abruptly. The force of the surrounding water rushing in to fill the voids creates a hammering action and high localized stresses that can pit the pump impeller. When

the rate of pump discharge decreases toward the shutoff head, recirculation of water within the casing can cause vibration and hydraulic losses in the pump. For these reasons, operating a pump at a rate of discharge within a range of between 60 and 120 percent of the best efficiency point (bep) is good practice.

Pump Characteristics

Normally, a pump can be furnished with impellers of different diameters, within a specified range, for each size casing. Since discharge of a pump changes with impeller diameter and operating speed, manufacturers publish data showing the characteristic curves for impellers of several diameters operating at two or more speeds. For a given impeller diameter operated at different speeds, the discharge is directly proportional to the speed (Eq. 4-10), head is proportional to the square of the speed (Eq. 4-11), and power input varies with the cube of the speed (Eq. 4-12).

$$\frac{Q_1}{Q_2} = \frac{N_1}{N_2} \tag{4-10}$$

$$\frac{H_1}{H_2} = \frac{N_1^{\,2}}{N_2^{\,2}} \tag{4-11}$$

$$\frac{P_{i_1}}{P_{i_2}} = \frac{N_1^{\,3}}{N_2^{\,3}} \tag{4-12}$$

where Q = discharge, gallons per minute (liters per second)

H = head, feet (meters)

P_i = power input, horsepower (kilowatts)

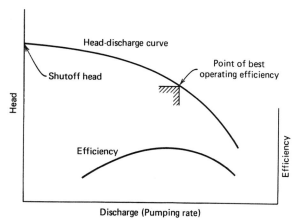

Figure 4-11   Characteristic curves for a centrifugal pump operating at a constant speed.

For a pump operating at the same speed, a change in impeller diameter affects discharge, head, and power input approximately as follows:

$$\frac{Q_1}{Q_2} = \frac{D_1}{D_2} \tag{4-13}$$

$$\frac{H_1}{H_2} = \frac{D_1^2}{D_2^2} \tag{4-14}$$

$$\frac{P_{i_1}}{P_{i_2}} = \frac{D_1^3}{D_2^3} \tag{4-15}$$

where D = impeller diameter, inches (centimeters).

Power and Efficiency

The power output of a pump is the work done per unit of time in lifting the water to a higher elevation. The equation relating discharge and head to power delivered is

$$P_o = \frac{wQH}{550 \times 60} = \frac{QH}{3960} \tag{4-16}$$

where P_o = power output, horsepower
Q = discharge, gallons per minute
H = head, feet
w = unit weight of water = 8.34 lb/gal
550 = foot pounds/second per horsepower
60 = seconds per minute
3960 = 550 × 60/8.34

In SI metric units, power output of a pump is

$$P_o = wQH = 0.0098QH \quad \text{(SI units)} \tag{4-17}$$

where P_o = power output, kilowatts
(kilonewton · meters per second)
Q = discharge, liters per second
H = head, meters
w = unit weight of water = 0.0098 kN/1
The efficiency of a pump is the ratio of the power output to measured power input. (Power input is sometimes called brake horsepower, bhp.)

$$E_p = \frac{P_o}{P_i} \tag{4-18}$$

where E_p = pump efficiency, dimensionless

P_i = power input (brake horsepower), horsepower (kilowatts)
P_o = power output, horsepower (kilowatts)
Centrifugal pump efficiency usually ranges from 60 to 85 percent.

EXAMPLE 4-7

The characteristics of a centrifugal pump operating at two different speeds are listed below. Graph these curves and connect the best efficiency points (bep) with a dashed line. Calculate head-discharge values for an operating speed of 1450 revolutions per minute (rpm) and plot the curve. Finally, sketch the pump operating envelope between 60 and 120 percent of the best efficiency points.

Speed = 1750 rpm			Speed = 1150 rpm		
Discharge (gpm)	Head (ft)	Efficiency (%)	Discharge (gpm)	Head (ft)	Efficiency (%)
0	220	—	0	96	—
1500	216	63	1000	93	65
2500	203	81	1500	89	77
3000	192	85	2000	82	83
3300	182	86[a]	2200	77	84[a]
3500	176	85	2500	70	83
4500	120	72	3000	49	71

[a] best efficiency point

Solution

The characteristic curves for 1750 rpm and 1150 rpm are drawn in Figure 4-12, with efficiency values listed at the plotting points. The best efficiency points are connected with a dashed line.

The head-discharge values for an operating speed of 1450 rpm are calculated using Eqs. 4-10 and 4-11. For example, using the data of 1500 gpm and 216 ft at 1750 rpm, one plotting point is calculated as

$$Q_2 = Q_1 \frac{N_2}{N_1} = 1500 \frac{1450}{1750} = 1240 \text{ gpm}$$

$$H_2 = H_1 \left(\frac{N_2}{N_1}\right)^2 = 216\left(\frac{1450}{1750}\right)^2 = 148 \text{ ft}$$

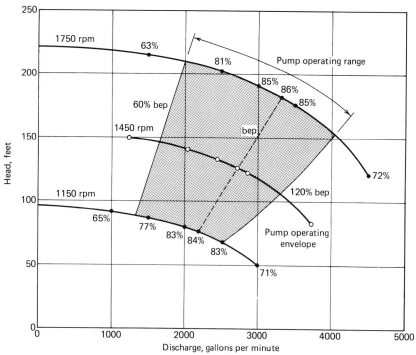

Figure 4-12 Characteristic pump curves for Example 4-7 showing the pump operating envelope.

All of the values calculated from the data at 1750 rpm are 1240 gpm at 148 ft, 2040 at 140, 2450 at 132, 2730 at 125 (bep), 2860 at 121, and 3730 gpm at 82.

The boundaries of the operating envelope at 1750 rpm are calculated as 60 and 120 percent of the discharge at the best operating point of 3300 gpm.

$$0.60 \times 3300 = 2000 \text{ gpm}$$

$$1.20 \times 3300 = 4000 \text{ gpm}$$

At 1150 rpm for a bep of 2200 gpm, the limits are 1300 gpm and 2600 gpm. The pump operating envelope is shown shaded in Figure 4-12.

4-5   SYSTEM CHARACTERISTICS

When a centrifugal pump lifts water from a reservoir into a piping system, the resistance to flow at various rates of discharge is described by a system head curve. The two components of discharge resistance are the static head, which is the elevation difference between the water levels in the suction reservoir and the discharge tank or point of discharge, and friction head loss that increases with pumping rate.

Consider the simplified water system illustrated in Figure 4-13. Water is being pumped through a pipe to discharge either into elevated storage (outlet 1) or at the load center (outlet 2) or both. Hydraulic grade line 1 is for the entire pump flow Q discharging into elevated storage at outlet 1. Hydraulic grade line 2 is for the entire pump flow discharging from outlet 2. Under the condition of grade line 2, water is also flowing out of elevated storage and discharging from outlet 2. The system head-discharge curves for these hydraulic grade lines are shown in Figure 4-14. Curve 1 is for the pump flow entering elevated storage, and curve 2 is for the pump flow discharging at outlet 2, the load center. The head-discharge operating

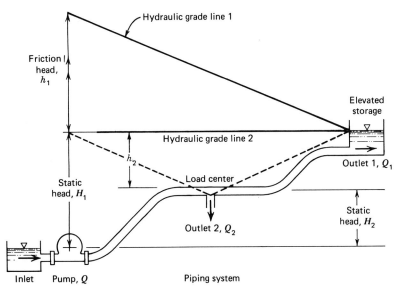

Figure 4-13 Schematic of a simplified water-supply system consisting of a pipeline, a lift pump, elevated storage, and a withdrawal outlet at the load center. The hydraulic grade lines are 1 for discharge at outlet 1 only and 2 for discharge at outlet 2 only.

point for this piping and elevated storage system can lie anywhere between these two curves depending on the relative amounts of water discharging at outlets 1 and 2. For a real water distribution network, the system head curve is actually a curved band to envelope a series of individual curves for different conditions of pumping, storage, and withdrawal.

Constant-Speed Pumps

Continuity of flow rate and water pressure must exist at the common boundary between a pump and piping system. Hence, a constant-speed pump operates at the head-discharge point defined by the intersection of the pump head-discharge curve and the system head-discharge curve. Figure 4-15 graphs the system curves 1 and 2 from Figure 4-14 with a constant-speed pump head-discharge curve. When the entire discharge flows into

elevated storage, pump operation is at the point labeled A. If free discharge is allowed only from the piping system at outlet 2, the pump operates at point B on the head-discharge curves. The quantity of flow Q_2 is greater than Q_1, since the operating head (static head plus friction head loss) at B under condition 2 is less than the operating head at A under condition 1. Starting from the condition of hydraulic grade line 2, consider throttling the discharge at outlet 2. The result is raising of both the hydraulic grade line and the system head-discharge curve so that the new operating point along the pump head-discharge curve is at some point *C* between *A* and *B*. A constant-speed centrifugal pump always operates at some point along its pump head-discharge curve, and the system head-discharge curve must intersect at this operating point.

Pumping stations with two constant-speed pumps of the same capacity may be used in small water systems. For example, the water can be pumped directly into elevated

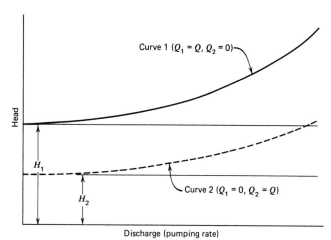

Figure 4-14   System head-discharge curves for the simplified water system illustrated in Figure 4-13. Curve 1 is for discharge at outlet 1 only and curve 2 is for discharge at outlet 2 only.

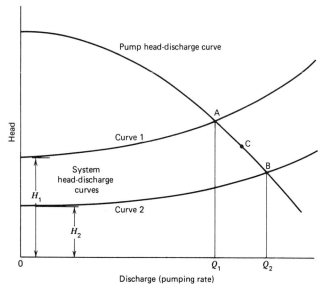

Figure 4-15   The head-discharge curve for a constant-speed pump drawn with system head-discharge curves to illustrate pump operating points. The related hydraulic scheme is given in Figure 4-13, and the system head-discharge curves are redrawn from Figure 4-14.

storage that, in turn, supplies the distribution network. The pump operates intermittently, controlled by the fluctuation of water level in the tank, and the elevated storage maintains pressure in the distribution system independent of pump operation. In larger systems, the installation of at least three pumps is desirable to cover the extremes of water demand and to provide a standby pump in case one unit is out of service. Since a larger system has continuous water demand, the pumps can discharge directly into the distribution piping, and elevated storage tanks are connected to the pipe network.

Parallel operation of constant-speed pumps is illustrated graphically in Figure 4-16. Pumps 1 and 2 are identical units, and pump 3 has a greater capacity and higher shutoff head, and head-discharge curves include friction losses in suction and discharge piping. The pumps are operated individually or in combination to meet the water demand by discharging into a common header and outlet pipe. When two or more pumps operate simultaneously, the combined head-discharge curve is determined by adding the rates of discharge at the same head of the individual curves. The point of intersection of the combined head-discharge curve and the system curve gives the combined rate of discharge of the pumps and the operating head of the pumps.

Variable-Speed Pumps

The best way to maintain a constant pump discharge pressure over a wide range of flow rates is by varying the rotational speed of the pump impeller. Although the discharge of a pump running at constant speed can be controlled by use of a throttling valve installed in the pump outlet, restricting discharge causes the impeller to recirculate water in the casing, reducing efficiency and possibly damaging the pump bearings and impeller. Speed control of a centrifugal pump is accomplished by using an electric motor designed for stepless-speed drive.

The head-discharge curves for a pump operating at two impeller speeds are given in Figure 4-17a, with the lines of equal efficiency plotted between the two curves. Actually, the pump can operate on an infinite number of curves between the minimum and maximum speeds. The

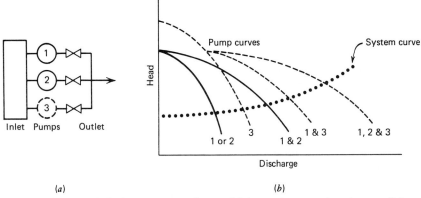

Figure 4-16 Head-discharge curves for multiple-pump operation in parallel. (*a*) Parallel installation arrangement. Pump numbers 1 and 2 are the same size; number 3 is larger. (*b*) Head-discharge curves for the pumps operating individually and in various combinations, with the system head-discharge curve shown as a dotted line.

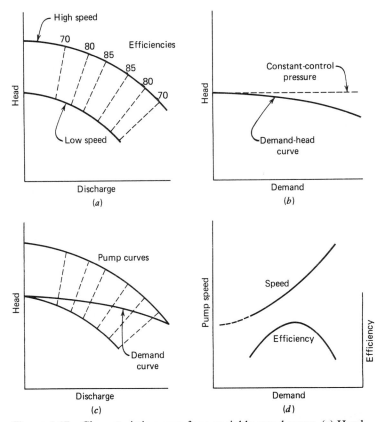

Figure 4-17 Characteristic curves for a variable-speed pump. (*a*) Head-discharge curves with efficiency values for a pump operating at two impeller speeds. (*b*) System demand-head curve for a constant pressure discharge corrected for transducer operation. (*c*) Superimposed pump head-discharge and demand curves. (*d*) Curves of speed and efficiency versus demand.

demand-head curve, which is the required head boost by the pump versus flow demand of the system, turns downward from the theoretical constant-control pressure as sketched in Figure 4-17b. Pump speed increases when the pump discharge pressure reduces as a result of increasing demand, and decreases with increasing discharge pressure. The transducer of a variable-speed drive transfers the signal from the pressure switches to the drive motor. To prevent cycling between pressure sensors, the decrease pressure switch is usually adjusted to be actuated when the pressure is slightly higher than the pressure that actuates the increase pressure switch. This transducer span results in an outlet pressure at maximum pump discharge (maximum demand), approximately 10 percent less than the outlet pressure at zero discharge. Transducer span and downward curvature of the demand-head curve can be reduced to zero with a more complex pump control system that includes error detection and a correction loop; however, this feature is usually unnecessary in water pumping applications. The pump head-discharge and demand-head curves are graphed together in Figure 4-17c. Values from the intersections of the demand-head curve with the pump speed and efficiency lines can be used to plot speed and efficiency versus demand (Figure 4-17d).

A variable-speed drive must be prevented from operating a pump at extremely low speeds. When the demand is less than the minimum required discharge, the pump is protected from damage by recirculating water through the pump. The recommended minimum discharge rate is generally 25 to 35 percent of the pumping rate at the best operating efficiency. Two control techniques are shown schematically in Figure 4-18. The first system uses a flow meter to operate a modulating valve in a by-pass to maintain a nearly constant flow through the pump when the demand is less than the minimum recommended rate of discharge. A second system employing a back-pressure regulating valve in the by-pass can be used if the discharge curve turns downward. The regulating valve is adjusted to maintain a pump discharge pressure approximately 10 percent above the desired output head at a pump discharge equal to the minimum rate of discharge. Therefore, at high output pressure corresponding to lowering demand, the valve in the by-pass begins to open, allowing recirculation flow.

Variable-speed pumps can be operated in parallel in multiple-pump installations. Based on the choice of design, the pumps may function by either load sharing or staggered operation. In load sharing, all pumps run at the same speed and discharge at equal rates. In staggered operation, one or more of the pumps runs at optimum efficiency (constant speed) while the speed of only one pump is varied to meet changing demand. The power input for alternative systems should be calculated to determine the design for most economical operation. Variable-speed pumps can also be used in combination with constant-speed pumps. Often the variable-speed unit meets the low flow demands in staggered operation. The flow head-discharge curves and demand-head curves can be added for multiple-pump operation in the manner described for multiple constant-speed pump installations (Figure 4-16).

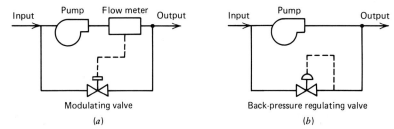

Figure 4-18   Arrangements for protecting variable-speed pumps at low discharges. (a) Flow meter-modulating valve system. (b) Back-pressure regulating valve arrangement.

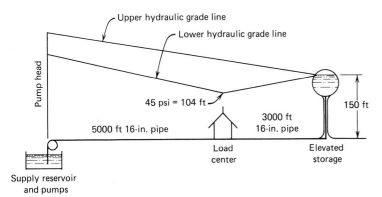

Figure 4-19 Simplified water system for Example 4-8.

EXAMPLE 4-8

Draw system head-discharge curves for the two operating conditions for the simplified water system diagramed in Figure 4-19. The highest system head-discharge curve (upper boundary) occurs when pump discharge enters the elevated storage tank with no system withdrawal. The lowest anticipated system head-discharge curve occurs when the pressure at the load center is 45 psi and flow is entering the system from both pump discharge and elevated storage. On the same head-discharge diagram, draw the characteristic pump curve from Figure 4-12 at an operating speed of 1750 rpm. Label the range of pump operation.

Solution

Under the condition of no system withdrawal, the static head transmitted to the pumps by elevated storage is 150 ft. The friction head losses in the 8000 ft of 16-in. pipe is determined for different pump discharges. The total pump head is the static head plus the friction head loss. For example, at 2000 gpm the head loss in 8000 ft of 16-in. pipe equals 8000 ft × 0.0037 ft/ft or 30 ft, and the total pump head is 150 ft plus 30 ft, for 180 ft. Therefore, one plotting point on the upper system head-discharge curve is 2000 gpm and 180 ft. Other calculated points are shown along the curve plotted in Figure 4-20.

For discharge at the load center, the static head is 45 psi or 104 ft, and the friction head loss is determined for

5000 ft of 16-in. pipe. For example, at 2000 gpm the total head is 104 ft plus 5000 ft × 0.0037 ft/ft, equaling 123 ft. This is one of the plotting points on the lower system head-discharge curve.

The pump head-discharge curve is the same as the pump characteristic curve at 1750 rpm from Figure 4-12. During operation along the upper system head-discharge, the pump head is 200 ft and the discharge is 2620 gpm. The intersection of the lower system head-discharge curve and the pump curve is at 165 ft and 3770 gpm. The range of pump operation straddles the best efficiency point and is within the normal pump operating range of 60 to 120 percent of the bep.

4-6 FLOW IN PIPE NETWORKS

Analysis of a water distribution system includes determining quantities of flow and head losses in the various pipelines and resulting residual pressures. Computations for a large network can often be made easier if a series of pipes of varying diameter are replaced with equivalent pipes. An equivalent pipe is an imaginary conduit that replaces a section of a real system such that the head losses in the two systems are identical for the quantity of flow. For example, pipes of differing diameters connected in series can be replaced by an equivalent pipe of one diameter as follows: Assume a quantity of flow and determine the head loss in each section of the line for this flow; then, using the sum of the sectional head losses and

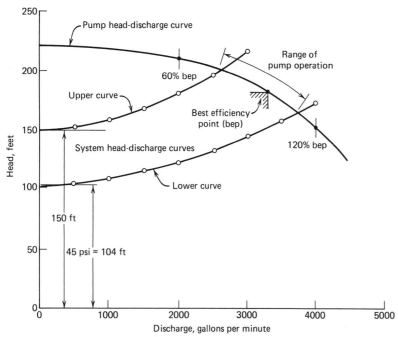

Figure 4-20 Solution for Example 4-8. System head-discharge curves for simplified water system diagramed in Figure 4-19, and characteristic pump curve at 1750 rpm from Figure 4-12.

the assumed flow, enter the nomogram to find the equivalent pipe diameter. For parallel pipe systems, a head loss is assumed, and the quantity of flow through each of the pipes is calculated for that head loss. Then the sum of the flows and the assumed head loss are used to determine the equivalent pipe size. Example 4-9 illustrates equivalent pipe calculations for lines in series and parallel.

EXAMPLE 4-9

Determine an equivalent pipeline 2000 ft in length to replace the pipe system illustrated in Figure 4-21.

Solution

First replace the parallel pipes between *B* and *C* with one equivalent line 1000 ft in length. Assume a head loss of 10 ft between *B* and *C*. Based on the nomograph, the flow in 1000 ft of 8-in. pipe at a head loss of 10 ft/1000 ft =

550 gpm, and the flow in the 6-in. pipe at a loss of 10 ft/800 ft (12.5 ft/1000 ft) = 290 gpm. The equivalent pipe *B* to *C*, 1000 ft in length, is then that size pipe that has a 10 ft/1000 ft head loss at a discharge of 550 + 290 = 840 gpm. From the nomograph, diameter = 9.4 in.

Next consider the three pipes in series: 400 ft of 8 in., 1000 ft of 9.4 in., and 600 ft of 10 in. At an assumed flow of 500 gpm the head losses in pipes are 0.4 × 8.3 = 3.3 ft, 3.8 ft, and 0.6 × 2.7 = 1.6 ft, respectively, for a total of 8.7-ft head loss in 2000 ft. The equivalent pipe *A* to *D* is then that diameter that exhibits a loss of 4.4 ft/1000 ft at a flow = 500 gpm. The answer is 9.2 in.

The method of equivalent pipes cannot be applied to complex systems, since crossovers result in pipes being included in more than one loop and because a number of withdrawal points are normally throughout the system. Network analysis can be performed using the Hardy Cross technique. Assumed flows are assigned to each of

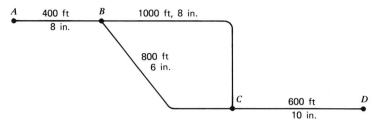

Figure 4-21 Pipe system for Example 4-9.

the pipelines and then corrections are applied, based on hydraulic calculations, until the computed flows and resultant head losses are in acceptable balance.

The derivation of the expression for flow corrections follows. The Hazen Williams equation (4-8) can be rewritten as

$$Q = KS^{0.54} \qquad (4\text{-}19)$$

where K is a constant when dealing with a single pipe of specific size and roughness. Rearranging and substituting head loss, h_L, for the hydraulic gradient, S, results in an equation relating lead loss as a function of flow to the 1.85 power for the Hazen-Williams formula.

$$h_L = KQ^{1.85} \qquad (4\text{-}20)$$

In balancing a given loop within a network the actual quantity of flow, Q, is determined by applying a correction ΔQ to the assumed flow, Q_1.

$$Q = Q_1 + \Delta Q \qquad (4\text{-}21)$$

When this term is inserted in Eq. 4-20 and the binomial is expanded, the equation becomes

$$KQ^{1.85} = K(Q_1 + \Delta Q)^{1.85}$$
$$= K(Q_1^{1.85} + 1.85Q_1^{0.85}\,\Delta Q + \cdots) \quad (4\text{-}22)$$

Then, the sum of the head losses must equal zero for balanced flow,

$$\sum h_L = \sum KQ^{1.85}$$
$$= \sum KQ_1^{1.85} + \sum 1.85KQ_1^{0.85}\,\Delta Q = 0 \quad (4\text{-}23)$$

And finally, solving for ΔQ and substituting h_L for the $KQ^{1.85}$ from Eq. 4-20, the term for flow correction is Eq. 4-24.

$$\Delta Q = -\frac{\sum h_L}{1.85 \sum (h_L/Q)} \qquad (4\text{-}24)$$

The procedure for solving a pipe network using the Hardy Cross method is illustrated in Example 4-10 and involves the following steps:

1. Assume a direction and quantity of flow for each pipe in the system such that the amount entering each pipe junction, or discharge point, equals the quantity flowing out.
2. Select one pipe loop in the system and compute the net head loss for that circuit based on the assumed flows. (The nomograph in Figure 4-7 is normally used for these calculations.)
3. Without regard to sign, calculate the sum of the h_L/Q values.
4. Compute the flow correction using Eq. 4-24, and correct each of the flows in the loop by this amount.
5. Apply this procedure to each pipe loop in the system, repeating earlier circuits as necessary, to arrive at answers within the desired accuracy. (Computational accuracy depends on the precision in reading the nomograph. In general, an error of 5 to 10 percent in calculating the head losses around a loop is satisfactory.)

Use of the special nomograph for head losses, shown in Figure 4-22, permits a simplified method for determining flows in pipe networks. The lines on the chart are Q, h_L, $1.85h_L/Q$, and L for 8-in.-diameter pipe with a $C = 100$. Table 4-4 gives multiplication factors to convert lengths of various size pipes to equivalent lengths of 8-in. pipe for the same head losses. To use this nomograph in solving for flows in a pipe network, all lines in the system are

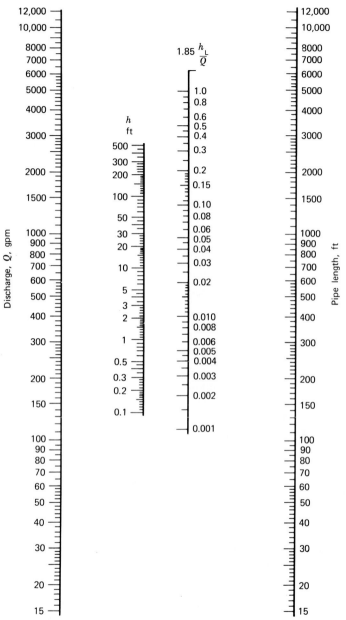

Figure 4-22   Special nomograph for head losses and 1.85 h_L/Q values based on 8-in. pipe with $C = 100$.

Table 4-4.   Conversion Factors for Various Sized Pipes[a]

Diameter (in.)	Factor	Diameter (in.)	Factor
1	24,000	14	0.066
2	840	15	0.047
$2\frac{1}{2}$	286	16	0.034
3	120	18	0.020
4	29	20	0.012
5	9.7	21	0.0093
6	4.06	22	0.0074
8	1.00	24	0.0048
10	0.34	30	0.0016
12	0.14	36	0.00066

[a] Multiply length of pipe by factor given to obtain the equivalent length of 8-in. pipe.

converted to equivalent lengths of 8-in. pipe with $C = 100$. For example, 1000 ft of 10-in. pipe is equivalent to $1000 \times 0.34 = 340$ ft 8-in. pipe (Table 4-4). If the C value also changes, the equivalent length must be adjusted accordingly, using values given in Table 4-3. Convert 1000 ft 10-in. pipe with $C = 120$ to 8-in. pipe with $C = 100$: $1000 \times 0.34 \times 0.71 = 240$ ft. Figure 4-22 can then be used to find values of h_L and $1.85 h_L/Q$ for a given Q and L of equivalent 8-in. pipe by means of a straight edge aligned across the nomograph. For example, 500 gpm flowing through 1000 ft of 8-in. pipe produces a head loss of 8.0 ft and the $h_L/Q = 0.03$. When the equivalent pipe is very short, a length 10 times as great can be used and the resulting h_L and h_L/Q values can be divided by 10.

EXAMPLE 4-10

Using the Hardy Cross method, compute the quantities of flow for the pipe network shown in Figure 4-23. Water enters the distribution system from the well supply, 1350 gpm at point A, and storage reservoir, 800 gpm at point J.

Solution

The assumed direction and quantities of flow are shown for each line on Figure 4-23. These estimates are not entirely arbitrary but are based on careful assessment of the network, thus providing the most probable flow

pattern. For example, at junction A a significantly greater flow is assumed for the 12-in. diameter pipe AD than the 6-in. line AB, and the quantity entering at J is directed primarily into pipe JI heading to the heavy discharge at point H.

The trial computations are given in Table 4-5. After the information on pipe diameter and length for each line in the loop are listed, head losses are determined from the nomograph (Figure 4-7), using the assumed flows. All clockwise flows are positive, and all counterclockwise flows are negative. Head loss resulting from positive flow is considered positive, and negative flows yield negative losses. The head loss is determined with respect to sign for each line, and then is summed to provide total h_L. Values of h_L/Q are computed by dividing the head loss in each line by the assumed flow. The sum of h_L/Q is performed without regard to sign. The flow correction is then calculated using Eq. 4-24. The correction values in this problem are rounded to the nearest 10 gpm, which is consistent with the accuracy of reading the nomograph.

The first trial—loop I—resulted in a negligible value for ΔQ, indicating a balance of the head losses for the assumed flows. As is shown in trial four, this does not mean that these flows are correct, since two of the lines (DE and EF) are common with other loops. The trial number was placed in parentheses after the assumed flows in loop I, for example, line AB 250 (1).

The second trial—loop II—reversed the assumed flow in line GC and reduced the flow in line DE, common with loop I, from 400 to 300 gpm. Trials three to six were sufficient to provide the required balancing of flows in the network. The error in head loss calculations for the sixth trial—loop III—was 7 percent. Final quantities for flow for each pipe are underlined.

Alternate Method

The monograph in Figure 4-22 can be used to solve Hardy Cross problems by tabulating values as are shown in Table 4-6. The same trial calculations as in Table 4-5 could be performed in the same sequence by using this format. The final results would be the same. Rather than repeat the entire analysis of the network in Figure 4-23, Table 4-6 loops the outside circuit $ABFJIHGCDA$, using the underlined values as the assumed flows.

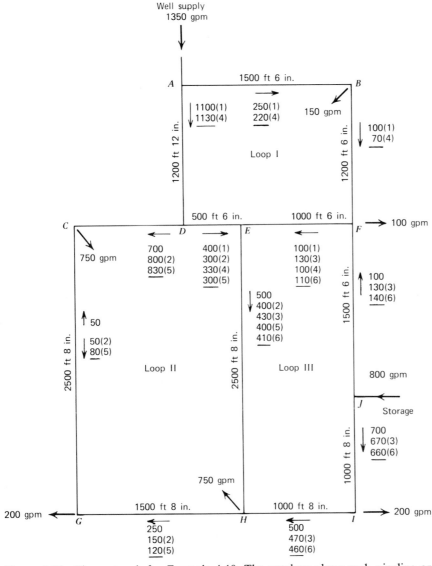

Figure 4-23 Pipe network for Example 4-10. The numbers along each pipeline are quantities of flow from several trial calculations given in Table 4-5. For example, 220(4) on line *AB* means 220 gpm computed during trial 4.

4-7 COMPUTER ANALYSIS OF PIPE NETWORKS

Several mathematical programs have been developed for computer analyses of water distribution systems. The common choice is a steady-state simulation model that represents the system for any prescribed set of flow and pressure conditions. An unsteady-state model simulates the behavior of a system during a series of successive steps over time. The numerical solution techniques used in programs may be the Hardy Cross method, adaptable to

Table 4-5. Computations for the Hardy Cross Analysis of the Pipe Network in Figure 4-23, Example 4-10

Line	Diameter (in.)	Length (1000 ft)	Flow (gpm)	Head Loss (ft/1000 ft)	Total h_L (ft)	$\dfrac{h_L}{Q}$	Corrected Flow
			First Trial, Loop I				
AB	6	1.5	+250	+9.2	+13.8	0.055	+250
BF	6	1.2	+100	+1.6	+1.9	0.019	+100
FE	6	1.0	+100	+1.6	+1.6	0.016	+100
ED	6	0.5	−400	−23	−11.5	0.029	−400
DA	12	1.2	−1100	−5.0	−6.0	0.005	−1100
					−0.2	0.124	

$$\Delta Q = -\frac{\sum h_L}{1.85 \sum h_L/Q} = -\frac{-0.2}{1.85 \times 0.124} = <1 \text{ gpm (negligible)}$$

Line	Diameter (in.)	Length (1000 ft)	Flow (gpm)	Head Loss (ft/1000 ft)	Total h_L (ft)	$\dfrac{h_L}{Q}$	Corrected Flow
			Second Trial, Loop II				
CD	8	1.0	−700	−16	−16.0	0.024	−800
DE	6	0.5	+400	+23	+11.5	0.029	+300
EH	8	2.5	+500	+8.3	+20.8	0.042	+400
HG	8	1.5	+250	+2.3	+3.4	0.014	+150
GC	8	2.5	+50	+0.1	+0.2	0.004	−50
					+19.9	0.113	

$$\Delta Q = -\frac{+19.9}{1.85 \times 0.113} = -97 \text{ gpm} (-100 \text{ gpm})$$

Line	Diameter (in.)	Length (1000 ft)	Flow (gpm)	Head Loss (ft/1000 ft)	Total h_L (ft)	$\dfrac{h_L}{Q}$	Corrected Flow
			Third Trial, Loop III				
EF	6	1.0	−100	−1.6	−1.6	0.016	−130
FJ	6	1.5	−100	−1.6	−2.4	0.024	−130
JI	8	1.0	+700	+16	+16.0	0.023	+670
IH	8	1.0	+500	+8.3	+8.3	0.017	+470
HE	8	2.5	−400	−5.5	−13.7	0.034	−430
					+6.6	0.114	

$$\Delta Q = -\frac{+6.6}{1.85 \times 0.114} = -30 \text{ gpm}$$

Line	Diameter (in.)	Length (1000 ft)	Flow (gpm)	Head Loss (ft/1000 ft)	Total h_L (ft)	$\dfrac{h_L}{Q}$	Corrected Flow
			Fourth Trial, Loop I				
AB	6	1.5	+250	+9.2	+13.8	0.055	+220
BF	6	1.2	+100	+1.6	+1.9	0.019	+70
FE	6	1.0	+130	+2.7	+2.7	0.021	+100
ED	6	0.5	−300	−13.0	−6.5	0.022	−330
DA	12	1.2	−1100	−5.0	−6.0	0.005	−1130
					+5.9	0.122	

$$\Delta Q = -\frac{+5.9}{1.85 \times 0.122} = -30 \text{ gpm}$$

Table 4-5 (*Continued.*)

Line	Diameter (in.)	Length (1000 ft)	Flow (gpm)	Head Loss (ft/1000 ft)	Total h_L (ft)	$\dfrac{h_L}{Q}$	Corrected Flow
			Fifth Trial, Loop II				
CD	8	1.0	−800	−20	−20.0	0.025	−830
DE	6	0.5	+330	+16	+8.0	0.024	+300
EH	8	2.5	+430	+6.5	+16.2	0.038	+400
HG	8	1.5	+150	+0.9	+1.4	0.009	+120
GC	8	2.5	−50	−0.1	−0.2	0.004	−80
					+5.4	0.100	

$$\Delta Q = -\frac{+5.4}{1.85 \times 0.100} = -30 \text{ gpm}$$

Line	Diameter (in.)	Length (1000 ft)	Flow (gpm)	Head Loss (ft/1000 ft)	Total h_L (ft)	$\dfrac{h_L}{Q}$	Corrected Flow
			Sixth Trial, Loop III				
EF	6	1.0	−100	−1.6	−1.6	0.016	−110
FJ	6	1.5	−130	−2.7	−4.0	0.031	−140
JI	8	1.0	+670	+14.8	+14.8	0.022	+660
IH	8	1.0	+470	+7.4	+7.4	0.016	+460
HE	8	2.5	−400	−5.5	−13.7	0.034	−410
					+2.9	0.119	

$$\Delta Q = -\frac{+2.9}{1.85 \times 0.119} = -10 \text{ gpm}$$

$$\text{Error} = \frac{\text{net } h_L}{\sum h_L} = \frac{2.9}{1.6 + 4.0 + 14.8 + 7.4 + 13.7} \times 100 = 7 \text{ percent}$$

Table 4-6.   Computations for the Hardy Cross Analysis of Circuit *ABFJKHGCDA*, Figure 4-23, Example 4-10, Using Special Nomograph Figure 4-22 and Table 4-4

Line	Length (ft)	Diameter (in.)	Equivalent 8 in. (ft)	Q (gpm)	h_L (ft)	$\dfrac{1.85 h_L}{Q}$	Correct Q
AB	1500	6	6100	+220	+11.0	0.090	
BF	1200	6	4900	+70	+1.0	0.028	Same as
FJ	1500	6	6100	−140	−4.6	0.062	the
JI	1000	8	1000	+660	+13.8	0.038	assumed
IH	1000	8	1000	+460	+7.2	0.028	values
HG	1500	8	1500	+120	+0.8	0.013	of Q
GC	2500	8	2500	−80	−0.6	0.016	
CD	1000	8	1000	−830	−20.5	0.046	
DA	1200	12	170	−1130	−6.5	0.012	
					+1.6	0.333	

$$\Delta Q = -\frac{+1.6}{0.333} = -4.7 \text{ gpm (negligible)}$$

small computers, or matrix routines that require greater computer capacity but arrive at a solution much more rapidly. Head losses are normally calculated in the program using the Hazen Williams equation. After an appropriate computer program has been selected, the data on the water system are entered and the model calibrated based on field measurements. The computer printout sheets usually organize the results in a columnar format listing flows in pipes, pressures at pipe junctions, hydraulic gradients, pump discharge pressures, and all of the input data. The physical components describing a water distribution system are the pipe network, storage reservoirs, pumps, valves, and other appurtenances.

A computer program stores the pipe network as a mathematically defined map by incorporating data on pipe lengths and diameters, roughness values (Hazen Williams C values), pipe junctions (nodes) including elevations and connecting pipes by number, check valves, and pressure regulators. Pipe roughness factors are generally assumed based on the age of the pipe and then adjusted during calibration of the model. In some cases, actual C values for larger mains are determined by field testing. Some computer programs allow the exclusion of selected pipes when running an analysis. With this option, the model can incorporate anticipated main extensions that can then be either included to evaluate proposed expansions or deleted for analysis of the existing network. In order to minimize the size of the network model, small-diameter pipes can be either ignored or combined with adjacent lines by substituting hydraulically equivalent pipes. A carefully selected skeletal pipe network generally simplifies the model with no loss of accuracy in analysis. Storage reservoirs are defined by location in the pipe network and their operating water level.

Pumping stations are described by elevation and station head-discharge curves, which are defined mathematically in the normal operating range of the pumps. The pump head-discharge curves from manufacturers cannot be incorporated directly into the program, since they do not account for suction water level and head losses in station piping, column head losses in wells, wear of impellers, or other factors. Pump characteristics for modeling are best determined by field measurements at the pump station or well under a range of discharge

pressures. Simultaneous measurements of water level in the suction reservoir or well casing are necessary to adjust the pumping curves for different conditions.

With the physical system modeled, the next step is to incorporate water use data by assigning water withdrawals at the nodes (junctions in the pipe network) for a known condition of water consumption. One approach is to assign water withdrawals for commercial and industrial users to nearby nodes and distribute nodal domestic consumptions based on residential population density and type of dwellings. If the sum of the withdrawals does not equal the average daily water production, the unaccounted-for water may be distributed among the nodes. An alternate approach is to start by assigning the average daily water withdrawal to large commercial and industrial users and assign the remaining consumption to nodes on the basis of land area or length of piping. Obviously, actual water consumption records for residential, commercial, and industrial areas from meter readings provide the best data on water use.

The mathematical model of the system in the computer program must be calibrated to be sure that it represents the real system as closely as possible. Normally, the procedure involves recording of flows, pressures, and operational conditions for selected test days. These data are then entered for computer analysis and the results checked against field measurements. Preliminary calibration is done by measuring pressures at various locations throughout the pipe network when the distribution of withdrawals is typical of the average consumption, which is usually 8 A.M. to 4 P.M. on weekdays. In addition to static pressure recordings at hydrants, essential data include elevations of water in reservoirs, discharge pressures and rates of flow from pumping stations and wells, and flow measurements of meters in the distribution system and at major water customers. If simulation is not satisfactory because of a localized deviation, the possibility of a closed valve or other irregularity in the distribution system should be investigated. A model is generally brought into calibration by changing C values in the piping and, if necessary, adjusting the distribution of water withdrawals. If the model is calibrated to reproduce the pressure observed during average water consumption, accurate prediction is not assured under extreme conditions, such as fire demand or maximum daily use. A major problem is that compensating errors can be introduced into the model during preliminary calibration. For example, higher estimates than actual water use offset the selection of incorrectly high C values, yet the model will appear to be calibrated.

Predictive capability of a model necessitates collection and analysis of independent sets of data for both calibration and verification to eliminate compensating errors. Verification can be accomplished by measurements of peak flows during periods of high seasonal water consumption and by conducting fire flow tests at important locations in the distribution system. Accuracy of calibration is checked by comparing the computer output values of hydraulic parameters with field measurements during verification studies. Predicted pressures should be within ± 5 psi. Accuracy in reading a hydrant pressure gauge is approximately 2 psi, and a 10-ft error in elevation of a test hydrant results in a 4-psi error. Since pressure readings are subject to the influence of inaccurate elevation data, an alternate method to determine the degree of calibration is to compare flows predicted by field test and computer simulation at a common arbitrary residual pressure, such as the 20-psi value used in calculating available fire flows. A third criterion is the ratio of observed to predicted head loss, which also indicates whether to modify the estimated C values or water withdrawals to achieve calibration.

A reliable computer model of a distribution system has several advantages. The hydraulics of the system can be evaluated for optimum energy efficiency. For periods of both low and high water demand, pumps and wells can be analyzed for best operation. Emergency situations, such as a major fire demand or main break, can be studied, and the effect of a potential major industrial customer can be analyzed. In the planning of new facilities, a calibrated model can be used to test alternative schemes and establish priority for improvements. Finally, the model is essential for design in determining the size of new mains, location of distribution storage, and selection of pumps.

The process of writing and calibrating a computer model requires both complete understanding of the water distribution system being modeled and expertise in network analysis. Although writing the mathematical model can be accomplished in a short period of time, calibration and verification require collection of field data intermittently throughout at least one entire year. The activity involving model calibration and optimization of operations provides an educational opportunity for system employees, whereas the program analyses evaluating improvements are often of particular interest in management of the system.

4-8 GRAVITY FLOW IN CIRCULAR PIPES

Sanitary and storm sewers are designed to flow as open channels, not under pressure, although storm sewers may occasionally be overloaded when water rises in the manholes surcharging the sewer. The wastewater flows downstream in the pipe moved by the force of gravity. Velocity of flow depends on steepness of the pipe slope and frictional resistance. The Manning formula, Eq. 4-25, is used for uniform, steady, open channel flow. The coefficient of roughness n depends on the condition of the pipe surface, alignment of pipe sections, and method of jointing. The common sewer pipe materials of asbestos cement, vitrified clay, and smooth concrete have n values in the range of 0.011 to 0.015. The lower value is applicable to clear water and smooth joints, whereas the greater roughness is for wastewater and poor joint construction. Corrugated steel pipe used for culverts has considerably greater roughness in the range of 0.021 to 0.026. The common value of n adopted for sewer design is 0.013. Even though certain pipe materials in a new condition may exhibit a lower n, such as plastic or asbestos cement, after pipes are placed in use they accumulate grease and other solids that modify the interior and disturb wastewater flow.

$$Q = \frac{1.49}{n} A R^{2/3} S^{1/2} \qquad (4\text{-}25)$$

where Q = quantity of flow, cubic feet per second
n = coefficient of roughness depending on material
A = cross-sectional area of flow, square feet

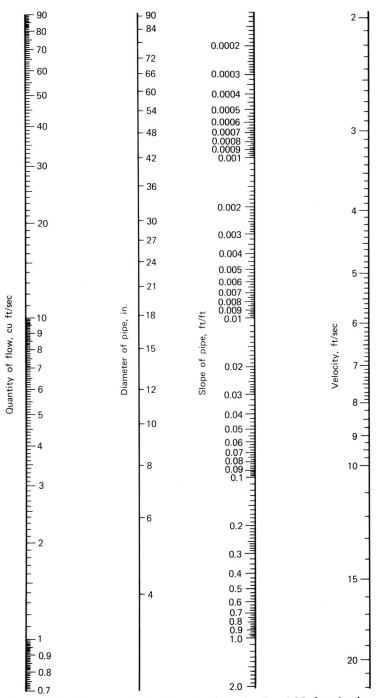

Figure 4-24 Nomograph for Manning formula, Eq. 4-25, for circular pipes flowing full based on $n = 0.013$.

R = hydraulic radius, feet (cross-sectional area divided by the wetted perimeter)

S = slope of the hydraulic gradient, feet per foot

The Manning formula, Eq. 4-25, expressed in SI metric units is

$$Q = \frac{1.00}{n} AR^{2/3}S^{1/2} \quad \text{(SI units)} \quad (4\text{-}26)$$

where Q = quantity of flow, cubic meters per second

n = coefficient of roughness depending on material

A = cross-sectional area of flow, square meters

R = hydraulic radius, meters

S = slope of hydraulic gradient, meters per meter

The nomograph shown in Figure 4-24 solves the Manning equation (Eq. 4-25) for circular pipes flowing full based on a coefficient of roughness of 0.013. Given any two of the parameters (quantity of flow, diameter of pipe, slope of pipe, or velocity), the remaining two can be determined from the intersections along a straight line drawn across the nomograph. For example, an 8-in.-diameter pipe set on a slope of 0.02 ft/ft (2.0%) conveys a quantity of flow of 1.7 cu ft/sec flowing full with an average velocity of 4.9 ft/sec. Quantity of flows at n values other than 0.013 can be calculated by multiplying the nomograph value by 0.013 and dividing the resultant by the desired n factor. The Manning formula can also be solved by using the diagram in Figure 4-25 without using a straightedge. The horizontal and vertical grid lines are quantity of flow and slope of pipe (ft/100 ft), respectively; the slanting grid includes pipe diameter and velocity. For the same example, enter the bottom at 2 ft/100 ft and move vertically to the intersection with the diagonal line representing 8-in. pipe diameter. From this point read 1.7 cu ft/sec along the vertical scale and 4.9 ft/sec on the other diagonal grid line.

Hydraulic problems of circular pipes flowing partly full are solved by using Figure 4-26. To plot these curves, ratios of hydraulic elements were calculated at various depths of flow in a circular pipe using the Manning equation. The symbols q, v, and a relate to the partial flow condition while Q, V, and A are at full flow. Consider flow in a pipe at a depth of 30 percent of the pipe diameter, represented by the horizontal line labeled 0.3. Where this line intersects the curved lines for quantity of flow, area of flow, and velocity, project downward to the horizontal scale and read, respectively, the values 0.2, 0.25, and 0.78; these values mean that partial flow at this depth conveys a quantity of flow equal to 20 percent of the flowing full quantity, the cross-sectional area of flow is 25 percent of the total open area of the pipe, and average velocity of flow is 78 percent of the flowing full velocity. This diagram shows that a pipe flowing at one-half depth conveys one half the flowing full quantity of flow at a velocity equal to the flowing full velocity. Also, the greatest quantity of flow occurs when the pipe is flowing at about 0.93 of the depth, and maximum velocity is at about 0.8 depth. The reason for this is that as the section approaches full flow, the additional frictional resistance caused by the crown of the pipe has a greater effect than the added cross-sectional area. For practical application, Figure 4-26 yields only approximate results, since the theoretical curves are based on steady, uniform flow, which does not precisely characterize the nature of actual flow in sewers. Examples 4-11 through 4-14 illustrate the use of the Manning equation, nomographs, and partial-flow diagram.

EXAMPLE 4-11

If a 10-in. sewer is placed on a slope of 0.010, what is the flowing full quantity and velocity for (a) $n = 0.013$, and (b) $n = 0.015$?

Solution

(a) Project across Figure 4-24 through slope = 0.010 and diameter = 10 in., read $Q = 2.2$ cu ft/sec and $V = 4.0$ ft/sec.

(b) From Eq. 4-25, Q and V are both inversely proportional to n; therefore,

$$Q \text{ at } n = 0.015 = \frac{0.013 \times 2.2}{0.015} = 1.9 \text{ cu ft/sec}$$

$$V \text{ at } n = 0.015 = \frac{0.013 \times 4.0}{0.015} = 3.5 \text{ ft/sec}$$

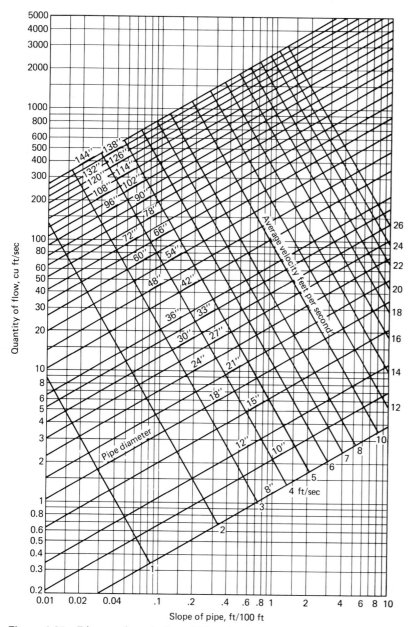

Figure 4-25 Diagram for solution of Manning formula (Eq. 4-25) for circular pipes flowing full based on $n = 0.013$. (From *Concrete Pipe Design Manual*, American Concrete Pipe Association.)

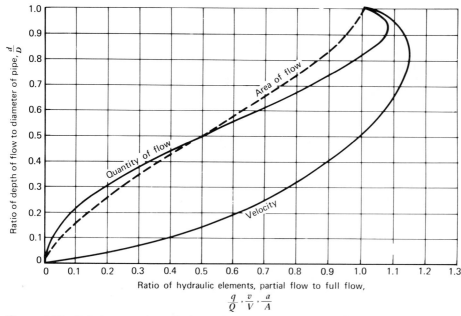

Figure 4-26   Relative quantity, velocity, and cross-sectional area of flow in a circular pipe for any depth of flow.

Alternate solution

(a) On Figure 4-25 locate the intersection of the vertical line for 1.0 ft/100 ft and the diagonal for a 10-in. size, read $V = 4.0$ ft/sec on the opposing diagonal line and project horizontally to $Q = 2.2$ cu ft/sec.

EXAMPLE 4-12

The measured depth of flow in a 48-in. concrete storm sewer on a grade of 0.0015 ft/ft is 30 in. What is the calculated quantity and velocity of flow?

Solution

From either Figure 4-24 or Figure 4-25, the flowing full Q and V are 56 cu ft/sec and 4.4 ft/sec, respectively. The depth of flow to diameter of pipe ratio is

$$\frac{d}{D} = \frac{30}{48} = 0.62$$

Entering Figure 4-26 with a horizontal line at this ratio, read q/Q and v/V by projecting downward at the intersections with the flow and velocity curves. Read $q/Q = 0.72$ and $v/V = 1.08$.
Then, flow at a depth of 30 in. $= 0.72 \times 56 = 40$ cu ft/sec and $v = 1.08 \times 4.4 = 4.7$ ft/sec.

EXAMPLE 4-13

An 18-in. sewer pipe, $n = 0.013$, is placed on a slope of 0.0025. At what depth of flow does the velocity of flow equal 2.0 ft/sec?

Solution

Velocity flowing full $= 3.0$ ft/sec (Figure 4-25).
Calculate $v/V = 2.0/3.0 = 0.67$.
Enter Figure 4-26 vertically at 0.67, intersect the velocity line, and project horizontally to read $d/D = 0.23$.
Depth at a velocity of 2.0 ft/sec $= 0.23 \times 18 = 4.1$ in.

EXAMPLE 4-14

What is the flowing full quantity and velocity of flow for a 450-mm-diameter sewer, $n = 0.013$, on a slope of 0.00412 m/m? Determine the quantity of flow for a depth of flow equal to 300 mm.

Solution

Cross-sectional area of pipe

$$= \pi \left(\frac{0.450}{2} \right)^2 = 0.159 \text{ m}^2$$

Hydraulic radius flowing full

$$= \frac{\pi d^2/4}{\pi d} = \frac{d}{4} = \frac{0.45}{4} = 0.112 \text{ m}$$

Using Eq. 4-26,

$$Q_{\text{full}} = \frac{1.00}{0.013} (0.159)(0.112)^{2/3}(0.00412)^{1/2}$$

$$= 0.182 \text{ m}^3/\text{s} = 180 \text{ l/s}$$

$$V_{\text{full}} = \frac{Q}{A} = \frac{0.182}{0.159} = 1.14 \text{ m/s}$$

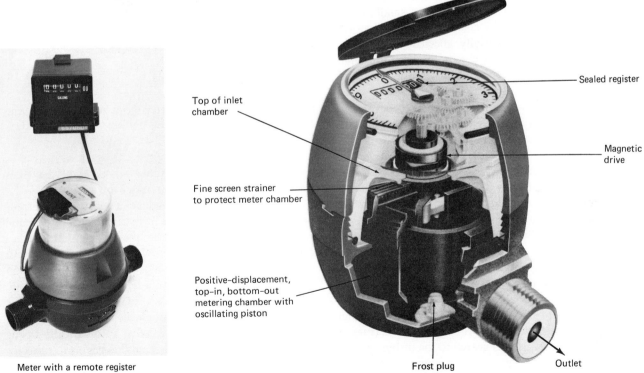

Meter with a remote register

Figure 4-27   Cutaway section of a household water meter with an oscillating disk that transmits a metering signal to the register by a magnetic drive. The insert shows the meter connected to a remote register that can be mounted on the outside of the house. (Courtesy of Kent Meter Sales, Inc.)

Enter Figure 4-26 with

$$\frac{d}{D} = \frac{300 \text{ mm}}{450 \text{ mm}} = 0.667$$

and read $q/Q = 0.78$.

Hence Q at d of 300 mm is $0.78 \times 180 = 140$ l/s.

4-9 FLOW MEASUREMENT IN PIPES

The common positive-displacement water meter has a measuring chamber of known volume containing a disk that goes through a cyclic motion as water passes through (Figure 4-27). The rotation resulting from filling and emptying of the chamber is transmitted to a recording register. The advantages of this meter are simplicity of construction, high sensitivity and accuracy, small loss of head, and low maintenance costs. Also, the accuracy of registration is not materially affected by position. This oscillating disk meter is commonly used for small customer services, such as individual households and apartments.

Several meter manufacturing firms have developed remote registration systems. The simplest is a remotely mounted digital register that is placed so that the person reading the register can do so from outside the house (Figure 4-27). A more sophisticated system consists of an outdoor receptacle read by plugging in a small portable recording device. The recorder punch card, or magnetic tape, is later used in data processing equipment at the central office for billing and recording in the memory system. Automatic remote meter reading and billing by computer processing reduces costs for reading of meters and preparation of water bills.

The measuring device in a current (velocity) meter is a bladed wheel that rotates at a speed in proportion to the quantity of water that passes through the blades (Figure 4-28). A recording register is geared to the turbine wheel. The shortcoming of a current meter is poor accuracy at low flow rates when the water is not moving at sufficient velocity to rotate the blades. Consequently, current meters are of limited value in water systems and are used only in restricted applications, for instance, on sprinkler carts that draw water from hydrants or on water towers.

Figure 4-28 Turbine-type water meter used for measuring continuous high rates of flow. (Courtesy of Hersey Products Inc.)

Current meters are unsuitable for customer services because of their inability to register low flow rates.

A general-service compound meter consists of a current meter, a positive-displacement meter, and an automatic valve arrangement that directs the water to the current meter during high rates of flow and to the displacement meter at low rates (Figure 4-29). The advantages are high accuracy at all flow rates and a relatively wide operating range. The switchover from low- to high-flow operation is controlled by head loss through the disk meter; when it exceeds a certain amount, a valve automatically opens and the current meter in the main line begins to measure the higher flow. Compound meters are used on larger services where widely varying flow rates are typical, such as motels, office buildings, factories, and commercial properties.

Proportional meters are sometimes referred to as "compound meters." The name is derived from the fact

Figure 4-29   Compound water meter designed for services requiring accurate measurements of both large and small flows. (Courtesy of Rockwell International Corporation.)

that only a portion of the total flow through the meter passes through a by-pass measuring chamber, which is the actual measuring device. The by-pass meter, either positive displacement or current type, is adjusted to register the total flow through the meter. Different methods of by-passing the water are utilized, such as an orifice in the main throat or a diverging tube. The proportional meter is capable of measuring large volumes of flow at comparatively low head loss and unrestricted flow, as well as accuracy of measurement.

Differential pressure meters, including venturi tubes, orifices, and nozzles, placed in a pipe section are the primary flow measuring devices in water systems. The meter shown in Figure 4-30 consists of a venturi tube with a converging portion, throat, and diverging portion and differential pressure gauge. As water flows through the tube, velocity is increased in the constricted portion and temporarily lowers the static pressure, in accordance with the energy Eq. 4-2. The pressure difference between inlet and throat is measured and correlated to the rate of flow. A flow recorder with a moving chart and ink pen plots variation of flow with respect to time, and a digital totalizer registers total flow. Venturi meters are shaped to maintain streamline flow for minimum head loss.

The equation for calculating flow through a venturi meter is

$$Q = CA_2[2g(H_1 - H_2)]^{1/2} \qquad (4\text{-}27)$$

where Q = quantity of flow, cubic feet per second (cubic meters per second)

C = discharge coefficient, commonly in the range of 0.98 to 1.02

A_2 = cross-sectional area of the throat, square feet (square meters)

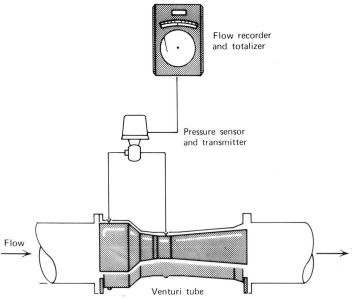

Figure 4-30 Venturi (differential pressure) meter inserted in a water pipeline, with sensor and recorder.

$\qquad$ g = acceleration of gravity,
$\qquad\qquad$ 32.2 ft/sec^2 (9.81 m/s^2)
$H_1 - H_2$ = difference of pressure heads between the
$\qquad\qquad$ inlet and throat, feet (meters)

4-10   FLOW MEASUREMENT IN OPEN CHANNELS

Wastewater contains suspended and floating solids that prohibit the use of enclosed meters. Furthermore, wastewater is commonly conveyed by open channel flow rather than in pressure conduits. Therefore, the Parshall flume is the most common device used to measure wastewater flows. A typical flume (Figure 4-31) consists of a converging and dropping open channel section. Flow moving freely through the unit can be calculated by measuring the upstream water level. A stilling well is normally provided to hold a float, bubble tube, or other depth-measuring device, which is connected to a transmitter and flow recorder similar to that shown in Figure 4-30. Flumes are commercially available to set in cut-open sections of

sewer lines to measure flow from portions of a collection system, or individual industrial wastewater discharges. The advantages of an open channel flume are low head loss and self-cleansing capacity.

For a Parshall flume under free-flowing (unsubmerged) conditions, the formula for calculating discharge for a flume with a throat width between 1 and 8 ft is

$$Q = 4BH^{1.522B^{0.026}} \qquad (4\text{-}28)$$

where Q = quantity of flow, cubic feet per second
$\qquad$ B = throat width, feet
$\qquad$ H = upper head, feet

Weirs can also be used for measuring flow of water in open channels. Water flowing over the sharp edge crest must discharge to the atmosphere, that is, air must be allowed to pass freely under the jet. If these conditions are met, the rate of flow can be directly related to the height of water measured behind the weir. Many weirs have been developed and calibrated with the majority having V-notch or rectangular openings.

The most common weir used for measuring wastewater flows is the 90° V-notch weir, illustrated in Figure 4-32. It

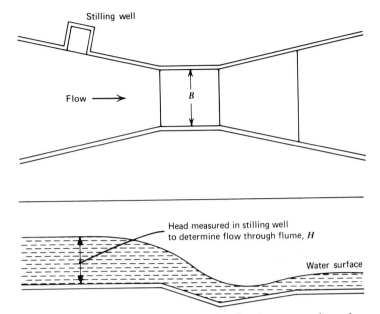

Figure 4-31 Parshall flume for measuring flow in an open channel.

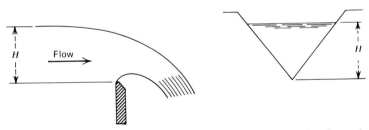

Figure 4-32 Ninety-degree V-notch weir used for measuring flow of water in a channel; refer to Eq. 4-29 and Figure 4-33.

is particularly well adapted to recording wide variations in flow, and may be used in treatment plants too small to warrant continuous flow recording and the more expensive Parshall flume. V-notch weirs are commonly installed on a temporary basis to make flow measurements associated with industrial wastewater surveys. Discharge over a 90° V-notch weir can be calculated using the following equation:

$$Q = 2.5H^{5/2} \qquad (4\text{-}29)$$

where Q = free discharge over 90° V-notch weir, cubic feet per second
      H = vertical distance (head) from crest of weir to the free water surface, feet

In SI metric units, Eq. 4-29 for a 90° V-notch weir is

$$Q = 1.4H^{5/2} \qquad \text{(SI units)} \qquad (4\text{-}30)$$

where Q = free discharge, cubic meters per second
      H = vertical head, meters.

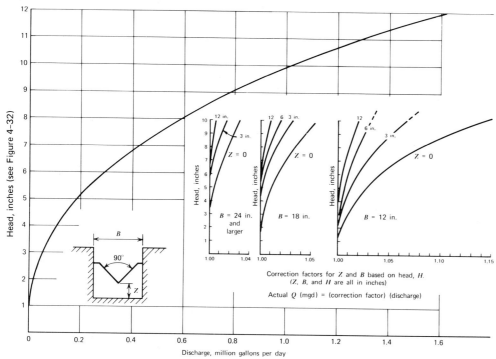

Figure 4-33 Discharge curve for 90° V-notch weir and correction curves for height of weir crest (Z) and various channel widths.

Figure 4-33 is a plot of Eq. 4-29 giving the head in inches and discharge in million gallons per day (mgd). Since a weir is commonly placed in a relatively shallow narrow channel for flow measurement, the calculated flow must be adjusted by using correction factors for width and depth as is shown in Figure 4-33. Example 4-15 illustrates the use of this diagram. The height of flow, vertical distance from the weir crest to the free water surface, is measured a short distance behind the weir in a stilling well. In wastewater survey measurements, it is common to place the float of the flow recorder inside a stovepipe held partly submerged below the water surface in the channel some distance behind the weir.

EXAMPLE 4-15

A 90° V-notch weir is placed in an 18-in.-wide channel with the base of the notch 6 in. from the channel bottom.

If the height of flow over the weir is 7 in., what is the quantity of flow?

Solution

Using Figure 4-33, the flow from the curve is 0.41 mgd. Using the correction factor curve for $B = 18$ in., locate the intersection of $H = 7$ in. and $Z = 6$ in. Dropping vertically, read a correction factor of 1.01, which is negligible.

Therefore, actual $Q = 0.41$ mgd.

4-11   AMOUNT OF STORM RUNOFF

The rational method for calculating quantity of runoff for storm sewer design is defined by the relationship

$$Q = CIA \qquad (4-31)$$

Table 4-7 Coefficients of Runoff for the Rational Method Eqs. 4-31 and 4-32 for Various Areas and Types of Surfaces

Description	Coefficient
Business areas depending on density	0.70 to 0.95
Apartment-dwelling areas	0.50 to 0.70
Single-family areas	0.30 to 0.50
Parks, cemeteries, playgrounds	0.10 to 0.25
Paved streets	0.80 to 0.90
Watertight roofs	0.70 to 0.95
Lawns depending on surface slope and character of subsoil	0.10 to 0.25

where Q = maximum rate of runoff, cubic feet per second

C = coefficient of runoff based on type and character of surface, Table 4-7

I = average rainfall intensity, for the period of maximum rainfall of a given frequency of occurrence having a duration equal to the time required for the the entire drainage area to contribute flow, inches per hour

A = drainage area, acres

Equation 4-31 in SI metric units is

$$Q = 0.278CIA \quad \text{(SI units)} \quad (4\text{-}32)$$

where Q = maximum rate of runoff, cubic meters per second

C = coefficient of runoff, Table 4-7

I = rainfall intensity, millimeters per hour

A = drainage area, square kilometers

Rainfall may be intercepted by vegetation, retained in surface depressions and evaporate, infiltrate into the soil, or drain away over the surface. The coefficient of runoff is that fraction of rainfall that contributes to surface runoff from a particular drainage area. The coefficients given in Table 4-7 show that the majority of rain falling on paved and built-up areas runs off, while open spaces with grassed surfaces retain the bulk of rain water. The size of the drainage area is determined by field survey, or measurement from a map.

The most complex parameter in the rational formula is the rainfall intensity. The U.S. Weather Bureau maintains recording gauges that automatically chart rainfall rates with respect to time. These data can be compiled and organized statistically into intensity-duration curves like those shown in Figure 4-34. The following illustrates

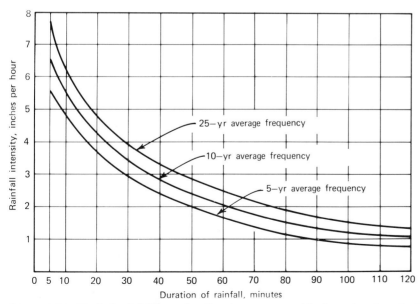

Figure 4-34 Typical rainfall intensity-duration curves used in the rational method for calculating quantity of storm water runoff.

how to read this diagram: For storms with a duration of 30 min, the maximum average rainfall anticipated once every 5 years is 2.9 in./hr, and the 25-year frequency is 3.9 in./hr. Although curves are a popular method of presenting rainfall information, precipitation formulas have been developed for various parts of the United States. In design, 5-year storm frequency is used for residential areas, 10-year frequency for business sections, and 15-year frequency for high-value districts where flooding would result in considerable property damage. The duration of rainfall used to enter Figure 4-34 depends on the time of concentration of the watershed. It is the time required for the maximum runoff rate to develop during a continuous uniform rain or, in other words, duration of rainfall required for the entire watershed to be contributing runoff. If a portion of the area being considered drains into an inlet and through a storm sewer, the time of concentration equals the inlet time plus the time of flow through the pipe. Inlet times generally range from 5 to 20 min based on characteristics of the drainage area, including amount of lawn area, slope of street gutters, and spacing of street inlets. Example 4-16 illustrates application of the rational formula.

EXAMPLE 4-16

Compute the diameter of the outfall sewer required to drain storm water from the watershed described in Figure 4-35, which gives the lengths of lines, drainage areas, and inlet times. Assume the following: a rainfall coefficient of 0.30 for the entire area, the five-year frequency curve from Figure 4-34, and a flowing full velocity of 2.5 ft/sec in the sewers.

Solution

Flow time in sewer from manhole 1 to manhole 2

$$= \frac{400 \text{ ft}}{2.5 \text{ ft/sec} \times 60 \text{ sec/min}} = 2.7 \text{ min}$$

From manhole 2 to manhole 3

$$= \frac{600}{2.5 \times 60} = 4.0 \text{ min}$$

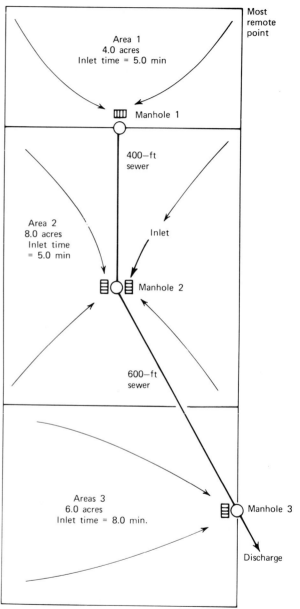

Figure 4-35 Watershed for Example 4-16, illustrating storm-runoff calculations using the rational method.

Times of concentration from remote points of the three separate areas to manhole 3 are $5.0 + 2.7 + 4.0 = 11.7$ min for Area 1; $5.0 + 4.0 = 9.0$ min for Area 2; and 8.0 min (inlet time only) for Area 3. Entering Figure 4-34 with the maximum time of concentration (duration of rainfall) for the watershed of 11.7 min, the rainfall intensity I is 4.6 in./hr for a five-year frequency.

The sum of the CA values

$$= 0.3 \times 4.0 + 0.3 \times 8.0 + 0.3 \times 6.0 = 5.4$$

Substituting into Eq. 4-31,

$$Q = 4.6 \times 5.4 = 25 \text{ cu ft/sec (acre-inches per hour)}$$

For $Q = 25$ cu ft/sec and $V = 2.5$ ft/sec from Figure 4-25, the required sewer diameter is 42 in. set on a slope of 0.055 ft/100 ft.

4-12 FLOW IN STREAMS AND RIVERS

Permanent gauging stations have been located on streams and rivers throughout the United States by several governmental agencies, including the Corps of Engineers, Bureau of Reclamation, Soil Conservation Service, and Geological Survey. A typical recording gauge installation is a stilling well, connected to the river by an intake pipe, with a float-operated water-level recorder housed in a shelter above the well. River stage is continuously scribed with a pen on a roll of paper driven by a clock mechanism. In some instances a control weir is constructed across a stream to provide sensitivity at low flows, but more frequently a rating curve is developed for the natural river channel. Data for a stage-discharge curve are gathered by using current meter measurements for various river discharges. Water levels during the times of survey are plotted against the measured flows. Rating curves must be reexamined periodically to account for shifts that occur as a result of channel filling or scouring. Continuous stage readings recorded on the chart paper are translated into average daily flow by using the rating curve; these are tabulated and published, usually under the title of surface-water records, and are distributed to libraries and interested governmental agencies.

Streamflows are of concern in water pollution, since treated wastewater effluents are often disposed of by dilution in rivers and streams; of greatest interest are low flows when the least dilution capacity is provided. The stream classification system of water-quality standards establishes maximum allowable concentration of pollutants; for example, ammonia nitrogen in warm-water streams is set at a maximum allowable level of 3.5 mg/l. These standards must be keyed to a specific quantity of flow to determine the amount of pollutant that can be discharged to the watercourse without exceeding the specified concentration. The one-in-ten-year seven-consecutive-day low flow is the value that is commonly adopted. Of course, if the stream has intermittent flow, the critical condition is during the dry period.

The following procedure is used to develop a frequency curve of seven-consecutive-day low flows for determining the one-in-ten-year value. All of the daily discharge records for gauging stations on the stream being evaluated are assembled. After the data for each year are scanned, the seven consecutive days of lowest flows are identified and averaged arithmetically. These values are listed by year as shown in Table 4-8. The following statistical method, illustrated in Table 4-9 and Figure 4-36, is used to calculate and plot the frequency curve from the yearly low flow values.

1. Arrange the minimum annual flows from historical records in order of severity, that is, from highest to lowest flow.
2. Assign a serial number m to each of the n values: 1, 2, 3, . . . , n.
3. Compute the probability plotting position for each serial value as m divided by $n + 1$.
4. Plot flow on the vertical logarithmic scale, and the corresponding probability value along the horizontal axis.
5. Draw the best fit line through the plotted data.

The frequency curve in Figure 4-36 is read by entering the diagram from either the top or bottom, the return period or probability, and reading the corresponding low flow on the vertical scale. For a return period of ten years, or probability of 90 percent, the minimum flow is 22.6 cu ft/sec. In other words, during one seven-day period every ten years, the lowest average flow for those seven days is expected to be 22.6 cu ft/sec, no other

Table 4-8. Streamflow Records Listing the Lowest Mean Discharge for Seven Consecutive Days for Each Year from 1961 to 1982[a]

Year	Lowest Mean Flow in Cubic Feet per Second for 7 Consecutive Days
1961	19.6
1962	28.6
1963	18.1
1964	34.3
1965	29.3
1966	35.7
1967	35.0
1968	27.0
1969	35.0
1970	36.9
1971	90.3
1972	50.6
1973	35.3
1974	59.4
1975	26.3
1976	30.1
1977	29.4
1978	29.7
1979	30.4
1980	49.6
1981	36.6
1982	59.1

[a] The average annual discharge for this period was 178 cu ft/sec.

Table 4-9. Streamflow Data from Table 4-8 Organized for Statistical Evaluation as Shown in Figure 4-36

Minimum Flows in Order of Severity (cu ft/sec)	Serial Number, m	$\dfrac{m}{n+1}$
90.3	1	$\dfrac{1}{23} = 0.0435$
59.4	2	$\dfrac{2}{23} = 0.087$
59.1	3	0.130
50.6	4	0.174
49.6	5	0.217
36.9	6	0.261
36.6	7	0.305
35.7	8	0.347
35.3	9	0.390
35.0	10	0.435
35.0	11	0.477
34.3	12	0.520
30.4	13	0.565
30.1	14	0.605
29.7	15	0.650
29.4	16	0.695
29.3	17	0.740
28.6	18	0.782
27.0	19	0.825
26.3	20	0.870
19.6	21	0.912
18.1	$22 = n$	0.955

seven-day period having a lower flow. Ninety percent of the flows during seven-day intervals are expected to be greater than this value. Logarithmic-probability paper is used since the hydrologic extremes, floods and droughts, are skewed and do not follow a normal symmetrical distribution. The break in the line drawn in Figure 4-36 indicates that drought flows did not occur during the five years when the low flows were greater than 40 cu ft/sec. The importance of understanding the concept of low flows for wastewater dilution is pointed up by the fact that the average flow in this stream throughout the period of record was 178 cu ft/sec. If the waste assimilative capacity were to be based on average flow rather than the calculated low flow, excessive pollution of the stream would occur a considerable portion of the time. On the other hand, using the lowest daily flow, in this case 18.1 cu ft/sec, leads to extreme conservation in the opinion of many authorities.

4-13 HYDROLOGY OF LAKES AND RESERVOIRS

The parameters used to define the physical characteristics of a lake are surface area, mean depth, volume, retention time (volume divided by influent flow), color and turbidity of the water, currents, surface waves, thermodynamic

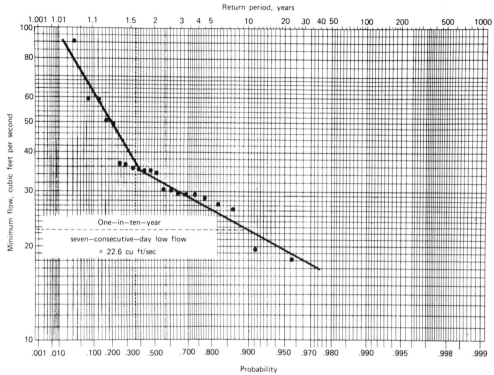

Figure 4-36   Plot of yearly low flows from Table 4-9 on logarithmic probability paper to determine the one-in-ten-year, seven-consecutive-day low flow.

relations, and stratification. All of these influence the chemistry and biology of a lake or reservoir and, hence, water quality. Thermal stratification is the most important phenomenon with regard to water supply and eutrophication. Lakes in the temperate zone, or in higher altitudes in subtropical regions, have two circulations each year, in spring and autumn. Thermal stratification is inverse in winter and direct in summer. Lakes of the warmer latitudes in which the water temperature never falls below 4° C at any depth have one circulation each year in winter, and directly stratify during the summer. For example, a lake may stratify from May through September, and may circulate continuously from October to April.

The seasons in a lake are defined graphically in Figure 4-37. In winter the densest water sinks to the bottom, and

ice near 0° C covers the surface; the maximum density of fresh water is at 4° C. Shading caused by ice and snow cover inhibits photosynthesis and, if the lake is rich in organic matter, the dissolved oxygen near the bottom gradually decreases. In spring after melting of the ice, the surface waters warm to 4° C and begin to sink, while the less dense bottom waters rise. These convection currents, aided by wind, mix the lake thoroughly for several weeks while the water temperature gradually increases. This is called the spring overturn, or spring circulation. With the approach of summer, the surface waters warm more rapidly and brisk spring winds subside, such that a lighter surface layer is formed. As the summer season progresses, resistance to mixing between the top and bottom layers of different density becomes greater and thermal stratification is established. The epilimnion, warm surface layer, is

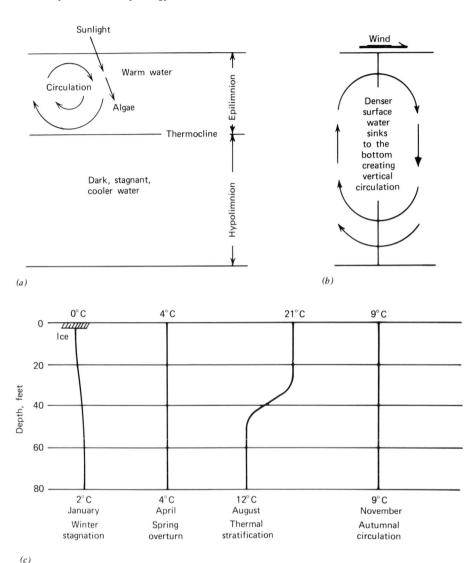

Figure 4-37   Thermal stratification and circulation of a dimictic lake in northern United States. (*a*) Thermal stratification during late summer. (*b*) Spring and autumn overturns. (*c*) Temperature profiles of a lake showing stratification and mixing.

continuously mixed by wind and density currents and supports the growth of algae. The hypolimnion, cooler bottom layer, is dark and stagnant. Although the bulk of fish food is found in the epilimnion, many species find the cooler bottom water a more suitable environment. In nutrient-rich bodies of water, the hypolimnion increases in carbon dioxide content and may become devoid of dissolved oxygen after many weeks of stratification. The thermocline is the thin zone of rapid temperature drop between the water layers. The approach of autumn with shorter, cooler days causes the lake to lose heat faster than it is absorbed. When the surface waters are cooled to a higher density than the water in the hypolimnion, vertical currents lead to autumnal circulation. This mixing is supported by wind action until finally the densest water stays on the bottom and the surface freezes.

Thermal stratification in reservoirs and lakes has a direct influence on quality of water supply. In the summer, water drawn from near the surface is warm and may contain algae that cause filter clogging and taste and odor problems. Stagnant, cooler hypolimnion water may be devoid of dissolved oxygen, high in carbon dioxide, and may contain the products of anaerobic conditions, such as hydrogen sulfide, odorous organic compounds, or reduced iron. Usually the region just below the thermocline provides the most satisfactory water quality during stratification. During winter stagnation, water closer to the surface is likely to be more desirable, since the quality adjacent to the bottom may be poor because of contact with decaying organic matter. The vertical variation that may occur in a body of water illustrates the importance of having a water intake tower with ports at various depths so that the water supply can be drawn from the most advantageous level in the water profile. Spring and autumn circulation mixes the water, spreading any undesirable matter throughout the entire profile. Treatment for taste and odor control may have to be intensified, particularly in the autumn when decaying algae and anaerobic bottom waters are mixed.

4-14 GROUNDWATER HYDROLOGY

Groundwater originates as infiltration from precipitation, stream flow, lakes, and reservoirs. As the water percolates vertically down through the near-surface soils, thin layers of water referred to as soil moisture are left behind, coating soil grains. Eventually, the water enters the zone of saturation where porous soil or fissured rock is filled with water. The surface of the saturated zone is called the water table, and its depth is described by the level of free water in an observation well extending into the saturated zone. A water table may fluctuate up and down with the seasonal supply and demand of groundwater. Horizontal movement is dictated by the hydraulic gradient, flowing downgrade with little vertical mixing.

The porosity of a soil or fissured rock is an expression of the void space defined as

$$n = \frac{V_v}{V} \qquad (4\text{-}33)$$

where n = porosity
V_v = volume of voids
V = total volume

Typical values of porosity are 0.2 to 0.4 for sands and gravels, depending on grain size, size distribution, and degree of compaction; 0.1 to 0.2 for sandstone; and 0.01 to 0.1 for shale and limestone, depending on texture and the size of the fissures. When groundwater drains from an aquifer as a result of lowering the water table, some water is retained in the voids. The quantity draining out is the specific yield or effective porosity; the amount retained is the specific retention. The specific yield for alluvial sand and gravel deposits ranges from 90 to 95 percent. For instance, if a coarse sand has a porosity of 0.40 and specific yield of 90 percent, the volume draining from 1.0 cu ft of aquifer is calculated as follows: 0.40 × 0.90 = 0.36 cu ft of water.

Aquifers are defined as permeable geologic strata that convey groundwater. They act as reservoirs, discharging by gravity flow or well extraction and recharging by infiltration. Common aquifers are valley fills composed of sand, gravel, and silt adjacent to streams; granular deposits along coastal plains; alluvium and loess of high plains; waterborne deposits from glaciation; terrains of volcanic origin; fractured limestone or dolomite rock; and poorly cemented sandstones. As illustrated in Figure 4-38, aquifers may be either unconfined or confined. The upper boundary of an unconfined aquifer is the water table, which is free to move up and down, changing the saturation zone. A well in an unconfined aquifer is referred to as a water table well. A perched water table is a special case of an unconfined aquifer where groundwater lies on a relatively impermeable stratum of small areal extent above the main body of groundwater. Confined aquifers, also known as artesian or pressure aquifers, exist where groundwater is confined by relatively impermeable

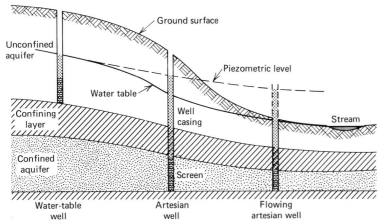

Figure 4-38 A profile of unconfined and confined aquifers illustrating water-table and artesian wells. (From Hammer, M. J., and K. A. Mac Kichan, *Hydrology and Quality of Water Resources,* John Wiley & Sons, New York, 1981.)

strata. The groundwater under pressure in a confined aquifer rises in a well to the piezometric level.

Permeability is the ability of a porous medium to transmit water. The coefficient of permeability K is defined by Darcy's law stating that the velocity of flow is directly proportional to the hydraulic gradient.

$$v = Ki \qquad (4\text{-}34)$$

where v = velocity of flow, feet per second (millimeters per second)
K = coefficient of permeability, feet per second (millimeters per second)
i = hydraulic gradient, feet per foot (meters per meter)

Values for K cover a wide range from less than 10^{-5} ft/sec (0.003 mm/s) for fine-grain deposits to over 1 ft/sec (300 mm/s) for coarse gravels.

A typical water well is constructed by drilling a deep, small-diameter borehole into the ground, which is held open by inserting a casing. The well is screened in the aquifer to allow groundwater to enter the borehole. Pump impellers and a suction pipe are hung down into the casing to lift the water out of the well. The movement of

groundwater near a pumping well depends on the characteristics of the aquifer and construction of the well. When pumping is started, the water table is drawn down forming a cone of depression. Water flows radially toward the well with steadily increasing velocity. Based on Darcy's law, the hydraulic gradient must also increase as groundwater approaches the well; therefore, the drawdown curve has a continuously steeper slope near the well. The size and shape of the cone of depression for a particular well are related to the withdrawal rate, duration of pumping, slope of the water table, sources of recharge, and aquifer characteristics.

Steady radial flow to a well penetrating an ideal unconfined aquifer is drawn in Figure 4-39. Ideal steady-state flow is defined by uniform withdrawal, a stable drawdown curve, uniform horizontal and laminar groundwater flow, a velocity of flow proportional to the tangent of the hydraulic gradient, and a homogeneous aquifer. Assuming these conditions, the well discharge is related to the coefficient of permeability, depth of aquifer, and shape of the drawdown curve as follows:

$$Q = \pi K \frac{h_0^2 - h_w^2}{\log_e(r_0/r_w)} \qquad (4\text{-}35)$$

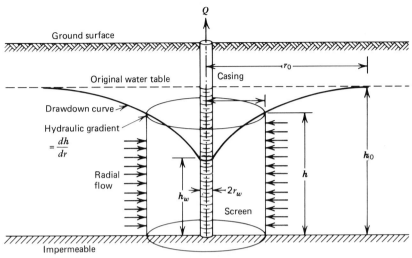

Figure 4-39 Steady radial flow to a well penetrating an ideal unconfined aquifer, Eq. 4-35. (From Hammer, M. J., and K. A. Mac Kichan, *Hydrology and Quality of Water Resources*, John Wiley & Sons, New York, 1981.)

where Q = well discharge, cubic feet per second (liters per second)

K = coefficient of permeability, feet per second (millimeters per second)

h_0 = saturated thickness of aquifer before pumping, feet (meters)

r_0 = radius of the cone of depression, feet (meters)

h_w = depth of water in well while pumping, feet (meters)

r_w = radius of well, feet (meters)

If this equation is used, prediction of well discharge Q based on a known permeability K, or determination of K from a measured Q, is reasonably accurate for natural unconfined aquifers that approach ideal conditions. Application of Eq. 4-35 requires estimating the radius of influence, r_0, or installation of observation wells in the drawdown area to determine the radius of influence.

Steady radial flow to a well penetrating a confined aquifer is shown in Figure 4-40. For ideal conditions, the equation for well discharge is

$$Q = 2\pi Kb \frac{h_0 - h_w}{\log_e(r_0/r_w)} \qquad (4\text{-}36)$$

where b = thickness of aquifer in feet (meters). Values of r_0 and h_0 may be assumed or measured from observation well data.

The permeability of an aquifer surrounding a well can be determined in the field by conducting a pumping test. Observation wells are needed to record drawdown at various distances from the test well as illustrated in Figure 4-41. After establishing steady-state conditions under continuous well discharge, permeability of an unconfined aquifer is calculated as

$$K = \frac{Q}{\pi(h_2^2 - h_1^2)} \log_e\left(\frac{r_2}{r_1}\right) \qquad (4\text{-}37)$$

For a confined aquifer (Figure 4-40), permeability is calculated from observation well data as

$$K = \frac{Q}{2\pi b(h_2 - h_1)} \log_e\left(\frac{r_2}{r_1}\right) \qquad (4\text{-}38)$$

EXAMPLE 4-17

A well with a diameter of 2.0 ft is constructed in a confined aquifer as illustrated in Figure 4-40. The sand aquifer has

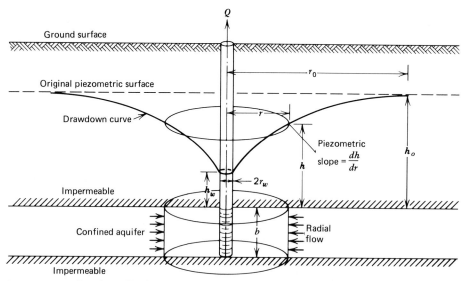

Figure 4-40 Steady radial flow to a well completely penetrating an ideal confined aquifer, Eq. 4-36. (From Hammer, M. J., and K. A. Mac Kichan, *Hydrology and Quality of Water Resources*, John Wiley & Sons, New York, 1981.)

a uniform thickness of 50 ft overlain by an impermeable layer with a depth of 115 ft. A pumping test was conducted to determine the coefficient of permeability of the aquifer. The initial piezometric surface was 49.0 ft below the ground-surface datum of the test well and observation wells. After water was pumped at a rate of 0.46 cu ft/sec for several days, water levels in the wells stabilized with the following drawdowns: 21.0 ft in the test well, 12.1 ft in the observation well 30 ft from the test well, and 7.9 ft in a second observation well at a distance of 100 ft. From these

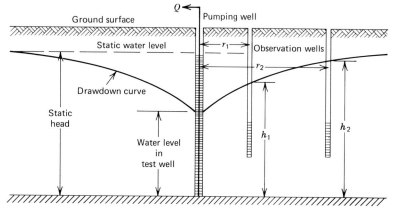

Figure 4-41 Flow of water toward a well during a pumping test in an unconfined coarse-grained aquifer. The dimensions shown apply to Eq. 4-37. (From Hammer, M. J., and K. A. Mac Kichan, *Hydrology and Quality of Water Resources*, John Wiley & Sons, New York, 1981.)

test data, calculate the permeability of the aquifer. Then, using this K value, estimate the well discharge with the drawdown in the well lowered to the top of the confined aquifer.

Solution

Assume a datum at the top of the aquifer as in Figure 4-40.

$$h_0 = 115 - 49.0 = 66.0 \text{ ft}$$

$$h_w = 66.0 - 21.0 = 45.0 \text{ ft}$$

$$h_1 = 66.0 - 12.1 = 53.9 \text{ ft}$$

$$h_2 = 66.0 - 7.9 = 58.1 \text{ ft}$$

Substituting into Eq. 4-38,

$$K = \frac{0.46 \text{ cu ft/sec}}{2\pi(50 \text{ ft})(58.1 \text{ ft} - 53.9 \text{ ft})}$$

$$\times \log_e\left(\frac{100 \text{ ft}}{30 \text{ ft}}\right) = 0.00042 \text{ ft/sec}$$

With the drawdown in the well at the top of the sand aquifer, $h_0 = 66.0$ ft and $h_w = 0.0$ ft. The radius of the well is 1.0 ft, and the distance to the edge of the cone of depression is assumed to be 700 ft. The estimated well discharge at maximum drawdown is calculated using Eq. 4-36 as follows:

$$Q = 2\pi(0.00042 \text{ ft/sec})(50 \text{ ft}) \frac{66.0 \text{ ft} - 0.0 \text{ ft}}{\log_e(700 \text{ ft}/1.0 \text{ ft})}$$

$$= 1.3 \text{ cu ft/sec}$$

FURTHER READING

1. Hammer, M. J., and K. A. Mac Kichan. *Hydrology and Quality of Water Resources.* New York: John Wiley, 1981.

PROBLEMS

4-1 (a) What is the hydraulic head in feet equivalent to 40 psi? (b) What is the static pressure equivalent to 80 ft of head? [*Answers* (a) 92 ft, (b) 35 psi]

4-2 If the pressure in a water main is 50 psi, what is the remaining pressure at a faucet in a building 30 ft above the main, assuming a head loss of 20 psi in the service connection?

4-3 (a) What is the hydraulic head in meters equivalent to 280 kPa? (b) What is the static pressure equivalent to 40 m of head?

4-4 Compute the total energy in a pipeline with an elevation head of 100 ft, water pressure of 50 psi, and velocity of flow equal to 2.0 ft/sec; use Eq. 4-3.

4-5 Compute the total energy in a pipeline with an elevation head of 9.0 m, water pressure of 410 kPa, and velocity of flow equal to 1.2 m/s; use Eq. 4-4. (*Answer* 51 m)

4-6 Calculate the head loss in 2000 ft of 14-in.-diameter pipe for a flow rate of 1500 gpm. The friction factor f for the pipe material is 0.025. (*Answer* 6.5 ft)

4-7 Calculate the head loss in 1000 ft of 8-in.-diameter cast-iron pipe at a flow rate of 500 gpm. Use the Darcy Weisbach equation and select the appropriate f value from Figure 4-6.

4-8 Using the nomograph in Figure 4-7 and Eqs. 4-7 and 4-8 for a C of 100, determine the head loss and velocity of flow in a 10-in. ductile-iron pipe carrying 800 gpm. What are the values of head loss and velocity for a $C = 140$?

4-9 Calculate the head loss and velocity of flow in a 150 mm pipe carrying 22 l/s. (*Answers* 19 m/1000 m, 1.3 m/s)

4-10 What size pipe should be used to supply 1600 gpm so that the velocity does not exceed 4.0 ft/sec assuming a $C = 100$. (*Answer* 14 in.)

4-11 A flow test was conducted on an old existing pipeline to determine the coefficient C for the Hazen Williams formula. The pipeline was a straight, horizontal section with a diameter of 8 in. The pressure loss was measured as 6.8 psi in 2000 ft of pipe at a water flow of 300 gpm. Calculate the C value of the pipe.

4-12 Review Example 4-6. What is the calculated discharge at point B in Figure 4-8 if water is flowing neither to nor from the elevated storage tank? (*Answer* 1500 gpm)

4-13   A simplified water system is illustrated in Figure 4-42. Calculate the rate of discharge from elevated storage, and draw the hydraulic gradient for the system based on a pump inflow of 2000 gpm at A and outflows of 2500 gpm at B and 1500 gpm at D. Assume a value of $C = 100$ for all pipes.

4-14   Draw the hydraulic gradient for the system shown in Figure 4-43. The pump inflow at A is 100 1/s at a discharge pressure of 550 kPa, and the outflow at B is 40 1/s. Determine the outflow at point C in liters per second. Assume a value of $C = 100$ for all pipes.

4-15   Refer to the simplified water system illustrated in Figure 4-8. For the same physical layout, calculate the pump discharge pressure required at A to provide fire flow and maximum daily discharge of 5000 gpm at 20 psi at point B. Assume $C = 100$. (*Answer* 300 ft)

4-16   At night, water is pumped from a treatment plant reservoir through distribution piping to an elevated storage tank. Using the energy equation, calculate the pump discharge pressure required to supply a flow of 1000 gpm to the tank. Assume no other withdrawals from the system. The water surface in the supply reservoir is at

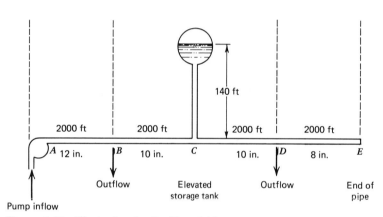

Figure 4-42   Illustration for Problem 4-13.

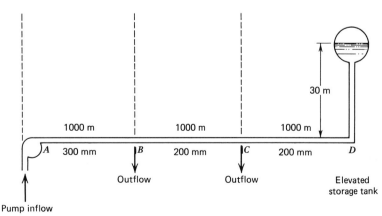

Figure 4-43   Illustration for Problem 4-14.

10-ft elevation; the lift-pump elevation is 20 ft; the water level in the elevated tank is 140 ft. The pipe network may be considered equivalent to 5000 ft of 10-in.-diameter cast-iron pipe, $C = 100$. (*Answer* 74 psi)

4-17 The discharge, head, and efficiency data of a centrifugal pump are tabulated below. The best efficiency point (bep) is at 2500 gpm. The impeller diameter is 15 in. and the operating speed is 2500 rpm. Plot the characteristic curves of head-discharge and efficiency as diagramed in Figure 4-11, and locate the bep and the recommended pump operating range. Calculate the power output and power input at the bep.

Discharge (gpm)	Head (ft)	Efficiency (%)	Discharge (gpm)	Head (ft)	Efficiency (%)
0	105	—	2000	83	85
500	100	41	2500	72	88
1000	96	63	3000	58	84
1500	91	78	3500	42	65

4-18 Draw a head-discharge curve for the pump described in Problem 4-17 operating at 1700 rpm. Locate points along the curve at 60 percent bep, bep, and 120 percent bep, and sketch the pump operating envelope as shown in Figure 4-12. (*Answer* bep: 1700 gpm, 33 ft)

4-19 Draw a head-discharge curve for the pump described in Problem 4-17 operating at 2500 rpm with a 12-in.-diameter impeller. (*Answer* bep: 2000 gpm, 46 ft)

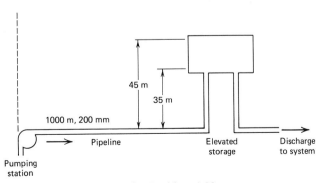

Figure 4-44   Illustration for Problem 4-20.

4-20 Water is pumped into an elevated storage tank through a 1000-m, horizontal pipeline with a $C = 100$, as shown in Figure 4-44. Draw system head-discharge curves at the low and high water levels in the tank for the range of inflows from 0 to 75 l/s.

4-21 Draw system head-discharge curves for the simplified water system diagramed in Figure 4-8. The lowest anticipated system head-discharge curve is when the hydraulic gradient at the load center is 99 ft, as drawn in Figure 4-8. The highest anticipated head-discharge curve is when the discharge at B is zero and the hydraulic head at C is 140 ft.

4-22 How many parallel 4-in.-diameter pipes are needed to be equivalent to one 8-in.-diameter pipe? (*Answer* 6)

4-23 A 6-in. pipe with a $C = 100$ is laid in parallel with an 8-in. pipe with a $C = 140$. What is the diameter of an equivalent pipe with a C-value of 120?

4-24 Find an equivalent, 1200-ft single pipe to replace the parallel pipes shown in Figure 4-45.

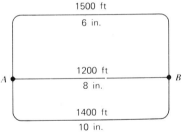

Figure 4-45   Pipes in parallel for Problem 4-24.

4-25 Determine the diameter of an equivalent, 4100-ft pipe between points A and B in Figure 4-46.

Figure 4-46   Pipes in series for Problem 4-25.

4-26 Determine the diameter of an equivalent pipe with a $C = 100$ for a 2000-ft, 8-in. pipe with a $C = 100$ in series with a 1000-ft, 6-in. pipe with a $C = 140$. (*Answer* 7.5 in. at $C = 100$)

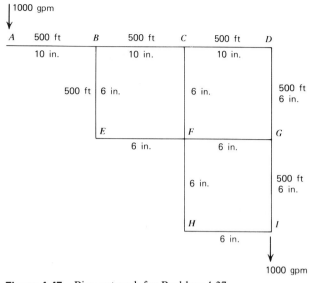

Figure 4-47   Pipe network for Problem 4-27.

4-27   Calculate the water pressure at point I in the pipe network in Figure 4-47, if the pressure at point A is 80 psi. Assume that all points are at the same elevation. Use the Hardy Cross method to estimate quantities of flow in the pipes; use $C = 100$. (*Answers* 63 psi, $AB = 1000$ gpm, $BC = 720$, $CD = DG = 360$, $GI = 580$, $BE = EF = 280$, $FH = HI = 420$ gpm)

4-28   Estimate the flows in the pipe network illustrated in Figure 4-48 using the Hardy Cross method. (*Answers* $AB = 1315$ gpm, $BE = 730$, $AD = DE = 685$, $BE = 585$, $EF = 875$, $FC = 70$, $EG = GH = 395$, $FH = 805$ gpm)

4-29   The pipe network in Figure 4-49 is a small portion of a water distribution system that includes an elevated storage tank. For the conditions shown, calculate the inflow from the elevated tank, and by the Hardy Cross method determine the flows in each pipe. Based on head

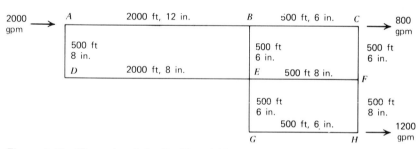

Figure 4-48   Pipe network for Problem 4-28.

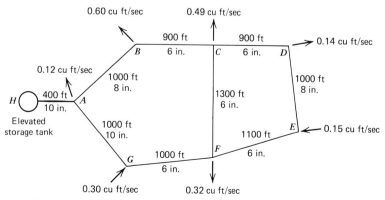

Figure 4-49   Pipe network for Problem 4-29.

losses using a $C = 100$, compute the water pressure at E assuming the water level in the storage tank is 90 ft above the elevation of point E.

4-30   Determine the flows in the pipe network drawn in Figure 4-50 using the Hardy Cross method. If the elevation at E is 20 ft and the water pressure 75 psi, what is the pressure at junction D where the elevation is 50 ft?

4-31   Figure 4-51 is a portion of a water distribution network of a small town. The discharges at the pipe junctions are maximum daily consumption with fire flow from junctions of F, G, J and K. Estimate the quantity of flow in the pipes, and calculate the water pressures at junctions J and G using head loss and elevation data. (*Answer* $P_J = 31$ psi)

4-32   Outline the data on a water distribution system that are necessary to program the system for computer analysis. How is a program calibrated and verified?

4-33   What is the flowing full quantity and velocity of flow for a 12-in. sewer on a 0.0060 ft/ft slope using (a) $n = 0.013$ and (b) $n = 0.011$? [*Answers* (a) 2.7 cu ft/sec, 3.5 ft/sec; (b) 3.2 cu ft/sec, 4.1 ft/sec]

4-34   What is the flowing full quantity and velocity of flow for an 18-in.-diameter sewer on a slope of 0.00412 ft/ft using (a) $n = 0.013$ and (b) $n = 0.011$? (c) For a roughness of 0.013, compute the quantity of flow for a depth of flow equal to 12 in.

4-35   What is the minimum slope required for a 24-in.-diameter sewer to maintain an average velocity of flow equal to 3.0 ft/sec when the quantity of flow is 20 percent of the flowing full capacity? (*Answer* 0.0028 ft/ft)

4-36   What size sewer pipe do you recommend to convey 5.4 cu ft/sec based on the following limitations: maximum allowable velocity of 10 ft/sec and a maximum allowable pipe slope of 3.0 ft/100 ft? (*Answer* 12 in.)

4-37   What size sewer pipe do you recommend to convey 15 cu ft/sec based on the following limitations: minimum allowable flowing full velocity of 2.0 ft/sec, maximum allowable velocity of 15 ft/sec, and a maximum pipe slope of 3.0 ft/100 ft, which is specified to reduce the quantity of trench cut requiring rock excavation?

4-38   Compute the design population that can be served by an 8-in. sanitary sewer laid on a slope of 0.40 percent. Assume that the design flow per person is 400 gpd. (*Answer* 1200 persons)

4-39   A 33-in. sewer pipe ($n = 0.013$) is placed on a slope of 0.40 ft/100 ft. (a) At what depth of flow is the velocity equal to 2.0 ft/sec? (b) If the depth of flow is 18 in., what is the discharge?

4-40   What slope is required to produce a velocity of flow equal to 1.0 m/s when the depth of flow is 90 mm in a 450-mm-diameter vitrified clay sewer?

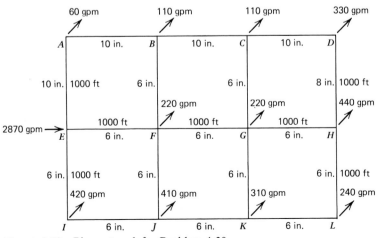

Figure 4-50   Pipe network for Problem 4-30.

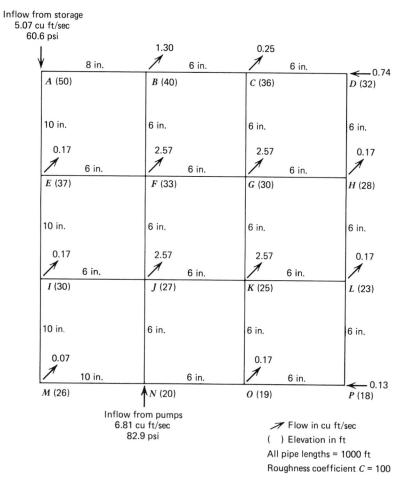

Inflow from storage
5.07 cu ft/sec
60.6 psi

Inflow from pumps
6.81 cu ft/sec
82.9 psi

↗ Flow in cu ft/sec
( ) Elevation in ft
All pipe lengths = 1000 ft
Roughness coefficient C = 100

Figure 4-51 Pipe network for Problem 4-31.

4-41 A 60-in.-diameter storm sewer conveys a flow of 55 cu ft/sec at a depth of flow equal to 40 in. What is the unused capacity of the pipe in cubic feet per second? (*Answer* 14 cu ft/sec)

4-42 Explain why a compound meter is suitable for customer services while a current meter is not.

4-43 Referring to the energy equation, Eq. 4-2, explain the operating principle of a venturi meter (Figure 4-30).

4-44 A 90° V-notch weir is placed in a 12-in.-wide channel with the base of the notch 3 in. from the channel bottom. Using Figure 4-33, determine the correction factor and discharge for a height of flow equal to (a) 4.5 in.,

and (b) 8.0 in. measured in a stilling well behind the weir. [*Answers* (a) 1.01 and 0.14 mgd, (b) 1.05 and 0.62 mgd]

4-45 Given the drainage area in Figure 4-52, calculate the discharge at the outfall using the rational method. Use the

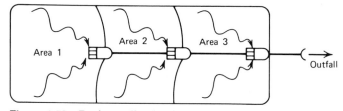

Figure 4-52 Drainage diagram for Problem 4-45.

five-year rainfall intensity-duration curve in Figure 4-34. Other data are for Area 1, $C = 0.50$, area $= 1.3$ acres, and inlet time $= 7$ min; for Area 2, $C = 0.40$, area $= 2.5$ acres, and inlet time $= 5$ min; for Area 3, $C = 0.70$, area $= 3.9$ acres, and inlet time $= 5$ min; sewer lines in Areas 2 and 3 are each 500 ft in length; and the average velocity of flow in the sewers may be assumed to be 3.0 ft/sec. (*Answer* 20 cu ft/sec)

4-46 Determine the size outfall sewer below MH 1 needed to serve the 12-acre drainage area in Figure 4-53. Each area has an inlet time of 10 min. The coefficient of runoff for the two housing areas is 0.45, and the park is 0.15. The distance between manholes is 600 ft, and all pipes are set on a slope of 0.0020. The rainfall intensity-duration relationship is $i = 131$ divided by $t + 19$, where $i =$ inches per hour and $t =$ minutes.

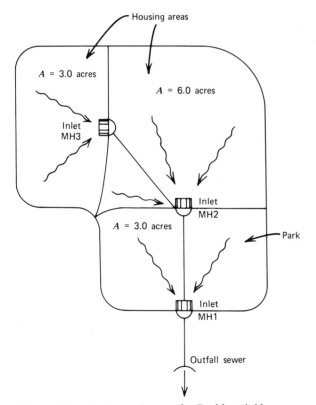

Figure 4-53   Drainage diagram for Problem 4-46.

4-47 Refer to Figure 4-35. Size the storm sewers from MH1 to MH2 to MH3, and the outfall from MH3 based on the following: (a) areas, inlets times, and sewer lengths as given on Figure 4-35; (b) a runoff coefficient of 0.40 for all areas; (c) the ten-year frequency curve from Figure 4-34; (d) circular pipe with an n of 0.013; and (e) pipe slope limitations: a maximum flowing full velocity of 10 ft/sec, minimum velocity of 3.0 ft/sec, and maximum allowable slope of 0.030 ft/ft.

4-48 Determine the one-in-ten-year seven-consecutive-day low flow given the following flow data. (Two-cycle logarithmic probability paper is provided in the appendix.) (*Answer* 85 cu ft/sec)

Year	cu ft/sec	Year	cu ft/sec
1965	120	1973	170
1966	162	1974	82
1967	142	1975	74
1968	137	1976	110
1969	254	1977	184
1970	367	1978	121
1971	145	1979	208
1972	153	1980	145
		1981	198

4-49 Determine the one-in-ten-year, seven-consecutive-day low flow given the following flow data.

Year	cu ft/sec	Year	cu ft/sec	Year	cu ft/sec
1953	62.7	1963	89.6	1973	77.4
1954	15.1	1964	149.0	1974	88.4
1955	54.1	1965	121.0	1975	36.1
1956	27.4	1966	108.0	1976	23.0
1957	48.4	1967	81.4	1977	123.0
1958	52.9	1968	98.0	1978	136.0
1959	39.0	1969	191.0	1979	159.0
1960	42.4	1970	188.0	1980	186.0
1961	123.0	1971	271.0	1981	135.0
1962	145.0	1972	191.0	1982	117.0

4-50   Explain why lakes thermally stratify, and why the impounded water in a dimictic lake circulates twice a year.

4-51   A pumping test was conducted in an unconfined aquifer using a test well penetrating to the underlying impervious stratum at a depth of 60 ft. Two observation wells were located at distances of 60 ft and 360 ft from the main well. Before the test was started, the static water level in all three wells was at a depth of 15.0 ft below the ground surface datum. Upon reaching equilibrium conditions after pumping the test well at a rate of 550 gpm for 2 days, the water level drawdowns were measured as 9.1 ft and 2.4 ft in the observation wells at distances of 60 ft and

360 ft, respectively. Calculate the field coefficient of permeability. (*Answer* $K = 0.0013$ ft/sec)

4-52   A well is installed in a 50-ft-thick, sandstone aquifer confined under an impervious overburden with a depth of 100 ft. The diameter of the screen and gravel pack in the aquifer is 2.0 ft. Based on test well data, the piezometric surface is at a depth of 30 ft below the ground surface, the radius of the cone of depression is 1000 ft, and the permeability of the sandstone is 2.5×10^{-4} ft/sec. Calculate the estimated pumping rate that will lower the water level in the well to the top of the sandstone aquifer.

chapter 5

Water Quality and Pollution

The many ways in which water promotes the economic and general well-being of society are known as beneficial uses. The major ones, shown in Figure 5-1, are water supply, recreation, and aquatic life. The relative importance of beneficial uses for any particular stream, lake, or estuary depends on the economy of the area and the desires of its people. Many applications are restricted within narrow ranges of water quality, such as public and industrial water supply. Unregulated wastewater disposal conflicts with use of the water as a municipal source. Therefore, control of quality is required to ensure that the best employment of the water is not prevented by indiscriminate use of watercourses for disposition of wastes.

Historically, water quantity has been considered in water law. Riparian rights, derived from British common law, state that the user of water is not entitled to diminish it in quantity. Recently, laws have been developed to protect water quality for downstream users. Past preoccupation with the quantity of water, without regard to quality, has been a handicap in the development of water resources planning.

The Federal Water Pollution Control Act, as amended by the Water Quality Act of 1965, authorizes states to establish water-quality standards for interstate waters. Its primary purpose is to protect water character for present and future beneficial uses. After water uses for each drainage basin were enumerated, standards with a reason-able margin of safety were set to protect quality for these uses. The standards are chemical, physical, and biological degradation limits beyond which a water may not be polluted for the defined beneficial use.

5-1 TYPES AND SOURCES OF POLLUTION

All domestic, industrial, and agricultural wastes affect in some way the normal life of a river or lake. When the influence is sufficient to render the water unacceptable for its best usage, it is said to be polluted. The following is a discussion of the common types of pollutants. The principal sources of water pollution are shown in Figure 5-2.

Oxygen-Consuming Matter

Deoxygenation can be caused by reducing agents that have an immediate oxygen demand, or by biological decomposition of waste organic matter. The latter is a relatively slow reaction, gradually depleting the dissolved oxygen in a river as the water flows downstream. Oxygen is replaced by aeration at the surface and photosynthetic activity of green plants. The maximum oxygen deficit depends on the interrelationship of biological oxygen utilization and reaeration. Fishes and most aquatic life are stifled by a lack of oxygen, and unpleasant tastes

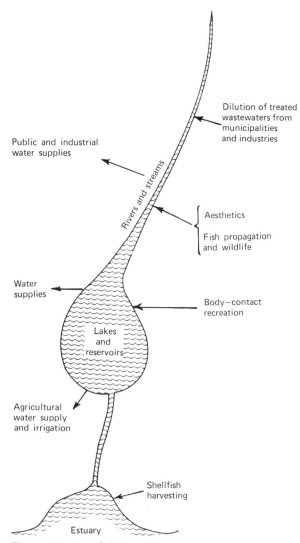

Figure 5-1   Beneficial uses of surface waters.

Inorganic Solids

Inert suspended solids, such as silt and mine slurries, produce turbidity that reduces light penetration and, therefore, interferes with photosynthesis. Solids that settle out of solution blanket the streambed, smothering bottom organisms and hindering the reproduction cycle of fishes.

Poisons

Acids, alkalis, and toxic chemicals adversely affect aquatic life and impair recreational uses of water. Sharp deviations of pH at the point of discharge into a river or lake eliminate less tolerant animal and plant species, and have considerable influence on the toxicity of some poisons. For example, ammonia is much more toxic in alkaline water than acid because free ammonia (NH_3) is more inhibiting than the ionized form (NH_4^+). Certain chemicals are poisonous to humans as well as to aquatic life, rendering the water unsuitable for domestic supplies. Heavy metals, such as mercury, are serious pollutants, since they form stable compounds that persist in nature and are concentrated in the aquatic food chain. The fishing industry has sustained economic losses in recent years because unacceptable levels of mercury and other heavy metals were discovered in fishes from contaminated waters, resulting in government condemnation of the affected catches.

Nontoxic Salts

Buildup of salts from domestic wastewaters and waste brines can interfere with water reuse by municipalities, industries manufacturing textiles, paper, and food products, and agriculture for irrigation water. Salts like sodium chloride and potassium sulfate pass through conventional water and wastewater treatment plants unaffected.

Inorganic phosphorus and nitrogen salts induce the growth of algae and aquatic weeds in surface waters. The majority of phosphates originate from fertilizer washed from agricultural land and phosphate builders used in

and odors are produced if the content is sufficiently reduced. Settleable organic solids can create sludge deposits that decompose, causing regions of high oxygen demand and intensified odors. Floating solids are unsightly and obstruct passage of light vital to plant growth Thin films of oil can also reduce the rate of reoxygenation.

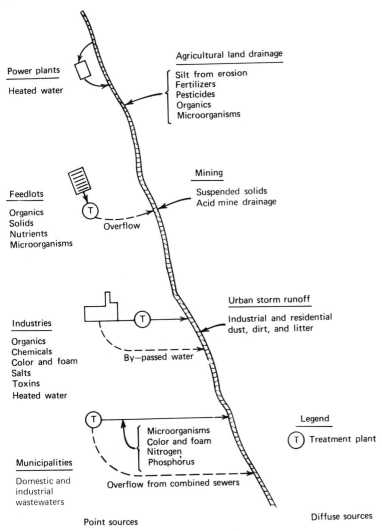

Figure 5-2 Principal sources of water pollution.

synthetic detergents. The latter source contributes approximately 60 percent of the phosphorus in domestic wastewater, and often the majority found in industrial wastewaters. Ammonia nitrogen is extremely soluble and is readily transported by surface runoff from cultivated farmland. In wastewater treatment, the nitrogen in organic compounds is released as soluble inorganic nitrogen. Removal of nitrogen and phosphorus in con-

ventional biological wastewater treatment is generally only 30 to 50 percent.

Unesthetic Wastes

Foam-producing matter and color, although often not harmful, lend an undesirable appearance to receiving streams; they are considered indicators of contamination.

Taste- and odor-producing compounds interfere with the palatability of the water for drinking purposes. Their source may be an industrial wastewater discharge, or may result from blooms of algae encouraged by nutrient enrichment from waste disposal. An increase in water temperature often magnifies the offensiveness of polluted water. Discharging heated water, such as cooling water from power plants, accelerates dissolved oxygen depletion, promotes the growth of blue-green algae, intensifies tastes and odors, and can stress fishes and other aquatic life.

5-2   WATER-QUALITY CHANGES

The water-quality changes resulting from municipal use are shown schematically in Figure 5-3. Groundwaters and surface waters drawn for community supplies have a wide range of quality. Although a few well waters may be adequate for public use without treatment, the majority of sources must be processed to improve water quality to the level required for public use. The primary function of municipal water is to carry away unwanted wastes from domestic and industrial operations. Contaminants in-

clude human excreta, food-processing scraps, and a wide variety of organic and inorganic trade wastes. Wastewaters are treated to improve quality prior to disposal to surface waters. Conventional treatment, including biological secondary, is the lowest level of approved wastewater processing. Considerable dilution in a natural watercourse is necessary to permit indirect reuse for water supplies. Advanced wastewater-treatment techniques may be applied in certain cases to remove additional organic matter, phosphates, nitrogen compounds, or other pollutants that degrade surface waters where sufficient dilution is not available. Water reclamation is a combination of conventional and advanced treatment processes employed to return wastewater to nearly original quality, thus reclaiming the water for reuse. Although not used for public water supply, renovated water can be beneficially used for a number of agricultural and industrial applications.

A quality comparison of drinking water with a typical municipal wastewater after conventional treatment reveals the following approximate increases of pollutants: 300 mg/l total solids, 200 mg/l volatile solids, 30 mg/l suspended solids, 30 mg/l BOD, 20 mg/l nitrogen, and 7

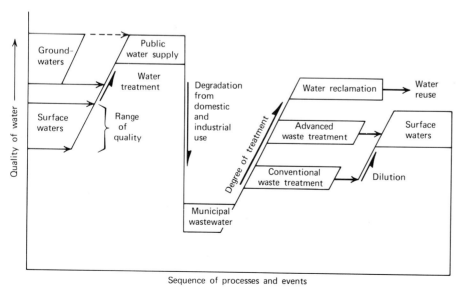

Sequence of processes and events

Figure 5-3   Water-quality changes resulting from municipal use.

mg/l phosphorus. The following industrial contaminants are often present in wastewaters from industrialized cities: phenol, cadmium, chromium, lead, and manganese. Most of these can be removed by additional physical and chemical processes applied in advanced wastewater treatment and reclamation. Methods available for removing inorganic nitrogen and phosphorus are both difficult and expensive, and removal of dissolved salts is rarely performed. The latter involves demineralization processes, such as electrodialysis, reverse osmosis, ion exchange, or distillation to separate the salts. Their cost is high and it is not anticipated that the reduction of dissolved solids by any of these methods will find wide application in wastewater treatment. Because of these factors, the scheme in Figure 5-3 does not show any wastewater treatment process that returns the water used in a public water supply to drinking water quality.

5-3 WATER-QUALITY STANDARDS

The first water-quality standards were established in 1914 for drinking water. Surface-water standards to control

wastewater treatment and disposal practices were not introduced until decades later. With population growth and dramatic industrial expansion, untreated wastewater discharges exceeded the self-purification capacity of many surface waters, and new emphasis on pollution control was needed. Deterioration of surface water sources for public water supplies was of particular concern to the waterworks industry. Certain substances are not significantly affected by conventional chemical water processing. These refractory compounds were often identified with municipal and industrial wastewaters discharged to watercourses. A public awareness of water pollution forced new legislation in the early 1960s. Surface-water standards were established first, including standards for public water supply sources. Next, effluent standards and sewer use ordinances were instituted to limit discharges, particularly those substances dangerous to human health and aquatic life. Currently, comprehensive standards are used to manage an integrated water and wastewater system.

Figure 5-4 shows how various water-quality standards relate to water use and treatment. Surface-water

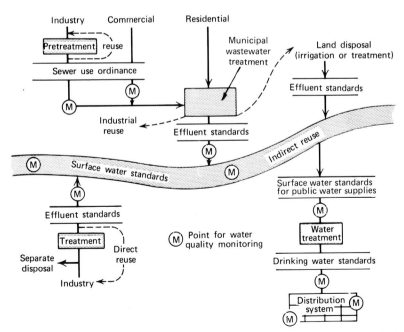

Figure 5-4 This illustration diagrams the various water-quality standards used to manage an integrated water and wastewater system.

standards define the quality acceptable for beneficial uses, for example, for public water supplies. Wastewater effluent standards regulate discharges of industries and municipalities to ensure their compliance with surface-water standards; industries located in cities are controlled by sewer use ordinances. Finally, drinking water standards apply to potable public supplies.

5-4   DRINKING WATER STANDARDS

Drinking water regulations published by the U.S. Environmental Protection Agency (EPA)[1] apply to all public water systems that have at least 15 service connections or regularly serve an average of 25 or more persons daily at least 60 days out of the year. The primary standards that specify approval limits for health give maximum contaminant levels for microbiological contaminants, inorganic chemicals, organic chemicals, radionuclides, and turbidity. Secondary standards that recommend contaminant levels for esthetics include inorganic chemicals, total dissolved solids, corrosivity, color, and odor. A summary of the drinking water standards is given in Table 5-1.

Microbiological Contaminants

Monitoring of microbiological quality requires testing of treated water at various points in the distribution system for the coliform group. (Sections 3-8 and 3-9 discuss in detail coliforms as indicator organisms and the laboratory testing procedures for these bacteria.) Microbial samples should be taken in known problem areas, for instance, reservoirs, dead-end mains, and remote areas of the distribution system. The number of water samples tested for coliforms in a small public water system serving under 1000 persons is one per month. In larger systems, the minimum number required increases with size of population at the rate of approximately one additional test per 1000 residents. With the approval of the state surveillance agency, chlorine residual tests may be substituted for coliform sampling in areas of a distribution system, provided that an approved minimum chlorine residual is maintained. Chlorine residual testing is better suited to operational control, since it is fast, easy, and provides a performance record warning of abnormal operation.

Inorganic Chemicals

Inorganic chemicals are derived from both natural processes of chemical weathering and soil leaching and human activities, for example, wastes from mining and manufacturing. Corrosion in a distribution system can also add inorganic chemicals to treated water before it reaches customers. To arrive at the maximum contaminant levels, water consumption was assumed to be two liters per day and the total environmental exposure to a specific toxic substance was considered. The lowest practical level was selected to minimize the amount of toxicant contributed by water, particularly when other sources like milk, food, and air are known to represent the major exposure to humans. Analyses for inorganic chemicals detrimental to human health are conducted yearly for community water systems using surface waters, and at three-year intervals for systems using groundwater. Their presence in excess of the maximum contaminant levels for human health constitutes grounds for rejection of the water supply.

Arsenic, barium, cadmium, chromium, lead, mercury, selenium, and silver are poisons that affect the internal organs of the human body. Because of the wide industrial use of cadmium, it can be introduced into surface waters in amounts significant from a health standpoint by improper disposal of industrial wastewaters. Yet, for the general population, the major source of cadmium is in food. When lead is found in drinking water, it is commonly traced to dissolution from the plumbing systems by corrosive water. Selenium is a trace metal naturally occurring in soils derived from some sedimentary rocks. Effects on human health have not been clearly established, and the current maximum contaminant level appears to be well below the no-effect concentration. The trace metals of arsenic, barium, hexavalent chromium (the toxic form), mercury, and silver are rarely found in significant concentrations in public drinking waters.

Table 5-1. Interim Drinking Water Standards of the United States Environmental Protection Agency

Primary Standards for Maximum Contaminant Levels
(Approval Limits for Health)

Microbiological Contaminants

When 10-ml portions of water are tested by the multiple-tube fermentation method, not more than 10 percent in any month shall show the presence of coliform bacteria. No more than 3 portions from 1 sample shall contain coliforms where less than 20 samples are tested per month; in larger systems, no more than 3 portions may be positive in 5 percent of the samples analyzed. (If the portions tested are 100 ml, not more than 60 percent shall show presence of coliforms.) When the membrane filter technique is used with 100-ml portions, the arithmetic mean coliform density shall not exceed 1 per 100 ml. The maximum density in 1 sample is 4 per 100 ml for less than 20 samples per month, or 5 percent of the samples where more than 20 samples are tested per month.

Inorganic Chemicals in Milligrams per Liter

Arsenic	0.05
Barium	1.0
Cadmium	0.010
Chromium	0.05
Lead	0.05
Mercury	0.002
Nitrate (as N)	10.0
Selenium	0.01
Silver	0.05

Recommended and approval limits in milligrams per liter for fluoride are based on the annual average of the maximum daily air temperatures.

Temperature °F	Recommended Optimum	Approval Limit
53.7 and below	1.2	2.4
53.8 to 58.3	1.1	2.2
58.4 to 63.8	1.0	2.0
63.9 to 70.6	0.9	1.8
70.7 to 79.2	0.8	1.6
79.3 to 90.5	0.7	1.4

Table 5-1. (*continued*)

Organic Chemicals in Milligrams per Liter

Chlorinated hydrocarbons	
Endrin	0.0002
Lindane	0.004
Methoxychlor	0.1
Toxaphene	0.005
Chlorophenoxys	
2,4-D	0.1
2,4,5-TP (Silvex)	0.01

Trihalomethanes

Total trihalomethanes	0.10

Radionuclides in Picocuries per Liter

Natural	
Gross alpha activity	15
Radium-226 + radium-228	5
Artificial (human-made)	
Gross beta activity	50
Tritium	20,000
Strontium-90	8

Turbidity

The monthly average shall not exceed 1 turbidity unit (TU). (With state approval 5 TUs may be allowed provided this amount does not interfere with disinfection, maintenance of chlorine residual, or bacteriological testing.) The maximum two-day average is 5 TUs.

Secondary Standards for Recommended Contaminant Levels
(Limits for Esthetics)

Chloride	250 mg/l
Color	15 color units
Copper	1 mg/l
Corrosivity	Noncorrosive
Foaming agents	0.5 mg/l
Iron	0.3 mg/l
Manganese	0.05 mg/l
Odor	3 threshold odor number
pH	6.5–8.5
Sulfate	250 mg/l
Total dissolved solids	500 mg/l
Zinc	5 mg/l

Source: National Interim Primary Drinking Water Regulations, U.S. Environmental Protection Agency, Office of Water Supply, EPA 570/9-76-003, 1976.

Nitrate is the common form of inorganic nitrogen found dissolved in water. In agricultural regions, groundwater can have significant concentrations of nitrate from unused fertilizer leaching into the underlying aquifers. For good crop yields, the inorganic nitrogen content in the soil moisture surrounding the root zone is often much greater than the maximum contaminant level allowed by drinking water standards. Porous soil profiles permit rainfall and irrigation water to transport this high-nitrate porewater to the groundwater table without measurable denitrification loss. Surface waters can be polluted by nitrogen from both discharge of municipal wastewater and drainage from agricultural lands. The health hazard of ingesting excessive nitrate in water is infant methemoglobinemia. In the intestines of an infant, particularly one under three to six months of age, nitrate can be reduced to nitrite that is absorbed into the blood, oxidizing the iron of hemoglobin. This interferes with oxygen transfer in the blood, resulting in cyanosis, which gives the baby a blue color. Incidents of infant methemoglobinemia, however, are extremely rare, since most mothers in regions of known high-nitrate drinking water use either bottled water or a liquid formula requiring little or no dilution. Methemoglobinemia is readily diagnosed by a medical doctor and rapidly reversed by injecting methylene blue into the infant's blood. Healthy adults are able to consume large quantities of nitrate in drinking water without adverse effects. The principal sources of nitrate in the adult diet are saliva and vegetables.

Fluoride is found in groundwaters as a result of dissolution from geologic formations. Surface waters generally contain much smaller concentrations of fluoride. Absence or low concentration of fluoride in drinking water causes the formation of tooth enamel less resistant to decay, resulting in a high incidence of dental caries in children's teeth. Excessive concentration of fluoride in drinking water causes fluorosis, also referred to as mottling. This dental disease in mildest form results in opaque whitish areas on the posterior teeth. With greater severity, the fluorosis is widespread and the color of the teeth is yellow-brown. The approval limit in the drinking water standards is established to prevent unsightly fluorosis. The optimum fluoride concentration in drinking water protects teeth from decay without causing noticeable fluorosis. Cities with water supplies deficient in natural fluoride are successfully providing supplemental fluoridation to optimum levels to reduce the incidence of dental caries in children. Since water consumption is influenced by climate, the recommended optimum concentrations and approval limits listed in Table 5-1 are based on the annual average of the maximum air temperatures obtained for a minimum record of five years.

Organic Chemicals

Organic chemicals in drinking water are found in trace amounts often so low in concentration that predicting any potential effect on human health is difficult. Reliable data on the toxicity to humans of most organic compounds are not readily obtained, since the information available is usually from uncontrolled accidental or occupational exposures. Therefore, long-term animal studies are used to evaluate chronic exposure and carcinogenic risk of organic compounds to humans. In feeding studies of laboratory animals, the dosage of toxin is assumed to be physiologically equivalent to humans on the basis of body weight. After the acceptable daily intake value for a no-observed-effect level in animals is determined, a safety factor is applied to reduce the allowable human intake to account for the uncertainties involved in extrapolating from animals to humans. The safety factors used in establishing drinking water standards are a factor of 10 when chronic human exposure data are available and are supported by chronic oral toxicity data in animals; a factor of 100 when good chronic oral toxicity data are available in some animal species but not in humans; and a factor of 1000 with limited chronic animal toxicity data. The hazard of ingesting a chemical assessed as a confirmed or a suspected carcinogen is evaluated in terms of dose-related risk. The estimate of risk resulting from animal bioassays is made by first converting the laboratory animal dose to the physiologically equivalent human dose on the basis of relative surface areas of test animals and the human body. Then, a mathematical model is used to relate dose to effect, and the dose rate associated with cancer risk is calculated on the basis of daily ingestion over a human lifetime of 70 years.

Chlorinated hydrocarbons can be absorbed into the human body through the lungs, gastrointestinal tract, and skin, and most are stored in fatty tissue. Endrin, lindane, methoxychlor, and toxaphene are insecticides. Endrin has a dangerous effect on the central nervous system of humans and higher animals and presents the greatest hazard of all residual pesticides in water; its maximum contaminant level is 0.0002 mg/l. For lindane, the maximum contaminant level of 0.004 mg/l is based on chronic exposure in dogs and application of a safety factor equal to 500. Lindane toxicity is also related to carcinogenic effects observed in mice. Methoxychlor is an insecticide of very low mammalian toxicity developed for general use. It is neither highly bioconcentrated nor stored in animal fatty tissue. The widely used insecticide toxaphene apparently does not cause environmental damage, since it has been used in agriculture for many years with no measurable adverse effects. Chlorophenoxys are herbicides used to control broadleaf plants and to regulate plant growth. The chemicals 2,4-D and 2,4,5-TP can be applied to control aquatic weeds in impounded waters that may be sources of drinking water. The closely related compound of 2,4,5-T has been banned from major aquatic uses.

Trihalomethanes (THMs) are organohalogen derivatives of methane where three of the four hydrogen atoms have been replaced by three atoms of chlorine, bromine, or iodine. Detectable concentrations of chloroform (trichloromethane), bromodichloromethane, dibromochloromethane, bromoform (tribromomethane), and dichloroiodomethane have been found in drinking waters. Chloroform is the trihalomethane most commonly found in drinking water and is usually present in the highest concentration. In some cases, bromated trichloromethanes dominate as a result of naturally occurring bromide in the water. THMs are produced during treatment of surface waters as a result of the chemical interaction of chlorine applied for disinfection and organic substances naturally present in the raw water. The natural organic substances include humic and fulvic acids produced from decaying vegetation. The general reaction producing trihalomethanes is

Chlorine + (bromide ion + iodide ion)
  + organic substances
    = trihalomethanes and other
      halogenated compounds   (5-1)

Formation of trihalomethanes is not instantaneous but continues for an extended period of time following chlorination. Thus, their concentrations can increase in chlorinated water held in a distribution system. A stringent limit of 0.10 mg/l for total trihalomethanes in drinking water was established because of carcinogenicity in laboratory animals. This maximum contaminant level considers both the risk to a lifetime of exposure and the status of current technology to meet the standard without unreasonable cost. Hence, the standard is a balance of public health considerations and the feasibility of meeting the approval limit in public water systems in the United States. The 0.10 mg/l limit for total THMs is based on a 12-month running average value rather than for a single test. A minimum of four water samples are collected quarterly from the distribution system, with 25 percent from the extremities of the pipe network and 75 percent based on population distribution. The total THM concentration is the sum of the individual concentrations of $CHCl_3$, $CHBrCl_2$, $CHBr_2Cl$, and $CHBr_3$. The value for that quarter is the arithmetical average of the total THM content of all samples, and the maximum contaminant level is the average of this quarterly value and the average concentrations measured during the previous three quarters.

Radionuclides

Radioactivity in drinking water can be from natural or artificial radionuclides. Background radiation results from cosmic ray and terrestrial sources. Radium-226 is found in groundwater from geological formations, and radioactivity from radium is widespread in surface waters because of fallout from testing of nuclear weapons. The majority of human exposure is unavoidable background radiation from these sources. Only a small portion results from drinking water containing small releases from nuclear power plants, hospitals, and industrial users of radioactive materials. The recommended allowable dose from radioisotopes in drinking water supplies is very low, amounting to less than 1 percent of the normal background. Nevertheless, the EPA has adopted maximum contaminant levels for radionuclides, given in Table 5-1, because of the many uncertainties of the environmental effects of radioactivity.

Turbidity

The maximum contaminant levels for turbidity in treated water are a monthly average of one turbidity unit and a maximum two-day average of five turbidity units. The justification for this standard is to reduce interference of particulate matter with disinfection by sheltering microorganisms, maintenance of a chlorine residual, and problems that can arise in microbiological testing resulting from high bacterial populations. Furthermore, turbidity greater than one unit is often related to poor treatment because of inadequate facilities or improper operation. Turbidity measurements to determine compliance with the drinking water standard are made at least once each day on the treated water entering the distribution system.

Secondary Standards

The recommended contaminant levels for aesthetics are based on factors that render a water less desirable for use. Excessive color, foaming, or odor cause customers to question the safety of a drinking water and result in complaints from users. Chloride, sulfate, and dissolved solids have taste and laxative properties, and highly mineralized water affects the quality of coffee and tea. Both sodium sulfate and magnesium sulfate are well-known laxatives with the common names of Glauber salt and Epsom salt, respectively. The laxative effect is commonly noted by travelers or new consumers drinking waters high in sulfates; however, most persons become acclimated in a relatively short time. Copper is an essential nutritional element and does not constitute a health hazard. The recommended limit is established to prevent a copperous taste. Zinc is also an essential

element in the human diet, but excessive amounts act as a gastrointestinal irritant. Iron and manganese are objectionable because of brownish-colored stains imparted to laundry and porcelain, and the bittersweet taste contributed by iron. A noncorrosive water with a neutral pH is desirable to reduce the probability of pipe corrosion contributing trace metals to the water.

5-5 SURFACE-WATER STANDARDS

The Water Pollution Control Act of 1972 established a national program to protect the quality of both interstate and intrastate waters. The general aesthetic requirements are that all surface waters should be capable of supporting aquatic life and should be free of substances attributable to wastewater discharges. When a river or lake is classified in accordance with intended uses, specific physical, chemical, biological, and temperature quality standards are established to ensure that the most beneficial use will not be deterred by pollution. Table 5-2 lists the six most common water use classifications; some states also list navigation and treated waste transportation. In addition to the dissolved oxygen, solids, and coliform criteria listed in Table 5-2, limits are generally specified for pH, temperature, color, tastes and odors, toxic substances, and radioactivity.

The primary purpose for establishing a limit for minimum dissolved oxygen is for protection and propagation of fishes and wildlife, although maintaining adequate oxygen also enhances recreation and reduces the possibility of odor problems that result from the decomposition of waste organics. Cold-water fishes require more stringent limitations, 6 mg/l and 7 mg/l at spawning times, while warm-water species, being more tolerant, require a 4- to 5-mg/l limit. Dissolved solids are restricted since high concentrations interfere with most water uses; furthermore, they are difficult to remove by water treatment. Coliforms, as discussed in Section 3-8, are indicators of domestic pollution. The strictest coliform standard applies to shellfish cultivation and harvesting, since shellfish are often eaten without being cooked. The next most stringent is for water contact recreation, where persons are likely to ingest water while swimming, wading, or water skiing.

Table 5-2. Beneficial Uses and Quality Standards for Surface Waters

| Water Use | Dissolved Oxygen Minimum Allowable (mg/l) | Solids Allowable | | Coliforms Maximum Allowable per 100 ml |
		Dissolved (mg/l)	Other	
Public water supply	4.0	500 to 750	No floating solids or settleable solids that form deposits	2000 fecal 10,000 total
Water contact recreation	4.0 to 5.0	None	Same as above	Mean of 1000 (200 fecal) with not more than 10 percent samples exceeding 2000 (400 fecal)
Fish propagation and wildlife	4.0 to 6.0 depending on warm or cold water fishes, fresh water or saltwater	None	Same as above	Mean of 5000
Industrial water supply	3.0 to 5.0 depending on use	750 to 1500	Same as above	Generally none specified
Agricultural water supply	3.0 to 5.0 based on application	750 to 1500 based on use and climate	Same as above	Generally none specified
Shellfish harvesting	4.0 to 6.0 depending on local conditions	None	Same as above	Mean of 14 with no more than 10 percent of samples exceeding 43

The allowable pH range is normally specified as 6.5 to 9.0 for protection of fish life and to prohibit heavy alkali or acid waste discharges. Temperature standards usually recommend an upper limit of 90° F, with a maximum permissible rise above the naturally existing temperature of 5° F in streams and 3° F in lakes. Trout and salmon waters should not be warmed in order to protect these less tolerant species. In general, discharges of toxic pollutants that cause death, disease, behavioral abnormalities, mutations, or physiological malfunctions in humans or other organisms are tightly controlled. Bioassays are the best method for determining safe concentrations of toxicants for aquatic organisms. Test organisms are exposed to various levels of poison concentrations for a specified time span, 96 hr or less, in a laboratory apparatus. The median lethal concentration (LC50) is the concentration that kills 50 percent of the test organisms. The maximum allowable toxicant concentration in surface waters is usually between 0.1 and 0.01 of the LC50 value. This factor of safety takes into account long-term exposure, and the fact that other materials already present in the receiving water may create additional stresses on aquatic life.

In addition to classifying watercourses and specifying numerical criteria, surface-water standards also provide means for controlling, monitoring, legal enforcement actions, and water resources planning. Effluent standards prescribe the degree of treatment required for particular wastewater discharges. The limit may be a specified amount of pollutant or zero discharge depending on local conditions. Effluent limitations are usually based on the best available control technology, unless more stringent controls are necessary to meet water quality standards. The common interpretation of this rule for municipal wastewater processing is a secondary treatment producing an average BOD of less than 30 mg/l and a suspended solids content of less than 30 mg/l. Advanced wastewater treatment may be deemed necessary to protect public water supplies. Removal of phosphorus and ammonia nitrogen is common where effluents enter lakes subject to eutrophication.

A permit system under the National Pollutant Discharge Elimination System (NPDES) was established to control wastewater discharges. Both publicly owned utilities and industries are required to obtain permits to discharge pollutants to surface waters. The purpose is to ensure that existing treatment plants comply with effluent limitations and performance standards, and that new plants are designed to provide the greatest degree of purification based on the best available technology and operating methods. The program also includes requirements for monitoring and reporting discharges, and adequate funding and staffing with qualified personnel.

Area-wide planning is essential to urban industrial areas with substantial water pollution problems. The Environmental Protection Agency obligates each state to develop regional plans for water-quality management. To receive federal financial support, publicly owned treatment plants must conform to the area-wide plan. This involves preparation of an environmental impact statement to determine if a proposed facility will have adverse effects on public health and welfare, or on environmental quality. Finally, state standards have an antidegradation clause, which recognizes that the existing high quality in certain waters may be better than the assigned classification. This purity is to be maintained unless it can be demonstrated that other uses and different standards are justifiable as a result of necessary economic or social development. Therefore, all proposed new or increased sources of pollution are required to provide the necessary degree of wastewater treatment to maintain high water quality.

5-6   STANDARDS FOR WATER-SUPPLY SOURCES

Drinking water standards dictate the quality of water that should be achieved in municipal water treatment without reference to desirable raw-water quality. Although it is possible to renovate highly polluted surface waters to these standards, the processes required would be both complex and expensive. Even then, some pollutants, such as chloride and nitrate, would not be removed. Raw-water quality standards have been developed to aid in selection of water sources such that the surface supply chosen can be treated to drinking standards by commonly applied treatment processes that have been proved by demonstration, and that are within reasonable economic limits. Two sets of surface-water standards for public supplies are listed in Table 5-3. Those characteristics in concentrations under the permissible column allow production of a safe, potable water meeting the limits of the drinking water standards after treatment. Desirable standards represent high-quality water that allows treatment at less cost and greater factors of safety than is possible with waters meeting permissible standards. All of the compounds to which the table footnote applies are refractory to the typical processes used in treatment of surface waters. Although many of these resistant substances may be altered or extracted by more extensive treatment processes, others would pass unchanged through any or all of the systems in current use.

Several words are used in Table 5-3 in lieu of specific values. "Narrative" means that the subject is discussed in

Table 5-3. Suggested Surface-Water Standards for Water-Supply Sources

Constituent or Characteristic	Permissible Standards	Desirable Standards
Physical:		
Color (color units)	75	<10
Odor	Narrative	Virtually absent
Temperature[a]	do	Narrative
Turbidity	do	Virtually absent
Microbiological:		
Coliform organisms	10,000/100 ml	<100/100 ml
Fecal coliforms	2,000/100 ml	<20/100 ml
Inorganic chemicals:	(mg/l)	(mg/l)
Alkalinity	Narrative	Narrative
Ammonia	0.5 (as N)	<0.01
Arsenic[a]	0.05	Absent
Barium[a]	1.0	do
Boron[a]	1.0	do
Cadmium[a]	0.01	do
Chloride[a]	250	<25
Chromium,[a] hexavalent	0.05	Absent
Copper[a]	1.0	Virtually absent
Dissolved oxygen	≥4 (monthly mean) ≥3 (individual sample)	Near saturation
Fluoride[a]	Narrative	Narrative
Hardness[a]	do	do
Iron (filterable)	0.3	Virtually absent
Lead[a]	0.05	Absent
Manganese[a] (filterable)	0.05	do
Nitrates plus nitrites[a]	10 (as N)	Virtually absent
pH (range)	6.0–8.5	Narrative
Phosphorus[a]	Narrative	do
Selenium[a]	0.01	Absent
Silver[a]	0.05	do
Sulfate[a]	250	<50
Total dissolved solids[a] (filterable residue)	500	<200
Zinc[a]	5	Virtually absent
Organic chemicals:		
Foaming agents[a]	0.5	Virtually absent
Oil and grease[a]	Virtually absent	Absent
Pesticides:		
Endrin[a]	0.0002	do
Lindane[a]	0.004	do
Methoxychlor[a]	0.1	do
Toxaphene[a]	0.005	do

Table 5-3. (*continued*)

Constituent or Characteristic	Permissible Standards	Desirable Standards
Herbicides:		
2,4-D[a]	0.1	do
2,4,5-TP[a]	0.01	do
Radioactivity:	(pc/l)	(pc/l)
Gross alpha[a]	15	do
Gross beta[a]	50	do
Radium-226[a]	5	do
Strontium-90[a]	8	do

[a] Substances that are not significantly affected by the following treatment process: coagulation (less than about 50 mg/l alum, ferric sulfate, or ferrous sulfate with alkali addition as necessary but without coagulant aids or activated carbon), sedimentation (6 hr or less), rapid sand filtration (3 gpm/sq ft or less), and disinfection with chlorine (without consideration to concentration or form of chlorine residual).

the following paragraph. "Absent" means that the constituent cannot be detected by the most sensitive analytical procedure, and "virtually absent" implies that the substance is present in a very low concentration. The word "do" is an abbreviation for ditto, in this sense meaning "same as first value."

The effectiveness of removing odorous materials from water is highly variable depending on the nature of the source and method of treatment. For this reason, it is not feasible to specify an odor criterion. In general, raw water should be free of objectionable odor, and any present should be removable by conventional means. Surface-water temperatures vary with location and climatic conditions; consequently, no fixed criteria are feasible. Turbidity in raw water must be readily removable by chemical treatment. It is desirable not to have frequently changing and varying turbidity characteristics, since such changes cause upset in water treatment plant processing. The standard for a particular water supply should be keyed to the capacity of the treatment works to remove the turbidity adequately and continuously at reasonable cost.

Alkalinity in water should be sufficient to enable chemical coagulation reactions but not so high as to cause physiological stress of consumers. The minimum alkalinity is about 30 mg/l, while the maximum amount should not exceed 400 to 500 mg/l. Fluoride ion is refractory to

common water treatment processes, except perhaps lime softening; therefore, the criteria for raw water are the same as for drinking water given in Table 5-1. A singular limit for maximum hardness is not possible, since hardness is a result of natural geological formations and public acceptance of it varies. Excess phosphorus concentrations in raw water are associated with eutrophication problems and coagulation difficulties attributed to complex phosphates. Because of the complicated nature of these topics, phosphorus standards must be established for each particular water supply.

5-7   STREAM POLLUTION

A stream or river that is used for wastewater dilution depends on natural self-purification to assimilate wastes and to restore its quality. The capacity to recover from a waste discharge is determined by the character of the river, including its climatic setting. A deep meandering river with natural pools has poor reaeration and a slow time of passage. Stream channels that are shallow and steep exhibit good reaeration and high velocities. The most critical condition for waste assimilation normally occurs in the fall of the year, resulting from low flows and high water temperature. For others, the worst circumstances may result in winter under ice cover.

Effect of Dilution and Decay

The mixing zone is an area unavoidably polluted for blending of a wastewater in a receiving stream. Since this zone may constitute a barrier that blocks migration of fishes and other aquatic organisms, it should be kept as short as possible. Where several discharges are located close together along a river, the mixing regions should lie along the same side to allow continuous passage for aquatic organisms on the opposite side.

The concentration of a pollutant below the mixing zone resulting from a wastewater discharge can be calculated using the following relationship:

$$C = \frac{C_1 \times Q_1 + C_2 \times Q_2}{Q_1 + Q_2} \qquad (5\text{-}2)$$

where C = concentration in combined flows
Q_1 = quantity of flow in stream
C_1 = concentration of constituent in Q_1
Q_2 = quantity of wastewater discharge
C_2 = concentration of constituent in Q_2

Many pollutants are reactive and dissipate after the mixing zone by adsorption, reaction, or biological decay. For example, bacteria from a domestic wastewater discharge decline in numbers because of unfavorable environmental conditions including reduced substrate, adverse temperature, and the abundance of predators. Chemicals in wastewaters decrease by reaction with substances in the flowing water and the streambed. Their rate of disappearance may be expressed mathematically by kinetics equations as either zero-order or first-order reactions, discussed in Section 2-5. In contrast, some pollutants are nonreactive and diminish in concentration only with increasing dilution. In passage down a river, greater runoff provided by the increasing drainage area decreases the original concentration below the mixing zone.

Organic Pollution

Many wastewater discharges from municipalities and industries contain organic compounds that decompose using dissolved oxygen. The rate of biological stabilization is a time-temperature function with deoxygenation

increasing as the temperature rises. Oxygen is replenished in the stream water primarily by reaeration from the atmosphere. Thus, quantity of flow, time of passage down the river, water temperature, and reaeration are the four major factors that govern self-purification from organic wastes.

A stream polluted from a substantial point source of organic matter exhibits four fairly well-defined zones (Figure 5-5). The zone of degradation, immediately following the sewer outfall, has a progressive reduction of dissolved oxygen used up in satisfying BOD. The zone of active decomposition exhibits the characteristics of significant pollution. Dissolved oxygen is at a minimum level, and often anaerobic decomposition of bottom muds results in offensive odors. Higher forms of life, particularly fishes, find the environment of these polluted zones undesirable. Bacteria and fungi thrive on the decomposition of organics, decreasing the BOD and increasing ammonia nitrogen. In the zone of recovery, reaeration exceeds the rate of deoxygenation and the level of dissolved oxygen increases slowly. Ammonia nitrogen is converted biologically to nitrate. Rotifers, crustaceans, and tolerant fish species reappear. Algae thrive on the increase in inorganic nutrients that result from the stabilization of the organic matter. The zone of clear water supports a wide variety of aquatic plants and animals, and more sensitive fishes. Dissolved oxygen returns to its original value, and the BOD has been nearly eliminated. The permanent changes in water quality prior to the wastewater discharge and the clear water zone include an increase in inorganic compounds, such as nitrate, phosphates, and dissolved salts. These nutrients produce and support higher algal populations if other environmental conditions of sunlight, pH, and temperature are adequate.

Oxygen Sag Equation

The rate of biochemical oxygen demand of organic matter diluted into a stream is proportional to the remaining concentration of unoxidized material. This first-order kinetics relationship is the same as the one used in developing the theoretical equations for the laboratory BOD test. A hypothetical deoxygenation curve without reaeration is shown as a dashed line in Figure 5-5.

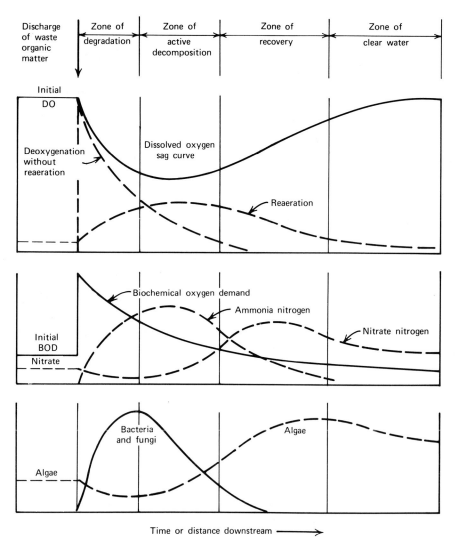

Figure 5-5   Generalized effects of organic pollution on a stream.

Reaeration from the atmosphere is the primary source of oxygen input to a stream. In addition, runoff entering a stream may add oxygen along with dilution water, and algae can produce oxygen by photosynthesis. The latter is neither dependable nor always available, nor can it be expressed in mathematical terms.

The rate of oxygen absorption into water is equal to a coefficient k_2 times the dissolved oxygen deficit (satu-ration concentration minus the existing DO concen-tration). This phenomenon is shown diagrammatically in Figure 5-5 by the dashed reaeration curve; oxygen uptake is least when the DO level is high and greatest at the bottom of the sag curve. The combined effect of deoxygen-ation and reaeration results in a dissolved oxygen sag curve that can be defined mathematically by Eq. 5-3. By rearranging the terms in this formula, Eq. 5-4 can be developed to calculate critical time, that is, time of travel to the minimum DO level of the sag curve.

$$D = \frac{k_1 L_0}{k_2 - k_1} (10^{-k_1 t} - 10^{-k_2 t}) + D_0(10^{-k_2 t}) \quad (5\text{-}3)$$

where D = dissolved oxygen deficit in time t, milligrams per liter
D_0 = initial DO deficit at $t = 0$, milligrams per liter
L_0 = initial ultimate carbonaceous BOD, milligrams per liter
k_1 = deoxygenation rate, per day
k_2 = reaeration rate, per day
t = time of travel, days

$$t_c = \frac{1}{k_2 - k_1} \log \left[\frac{k_2}{k_1} \left(1 - D_0 \frac{k_2 - k_1}{k_1 L_0} \right) \right] \quad (5\text{-}4)$$

where t_c = critical time, days, time of travel to the minimum DO level

Application of these formulas requires prior determination of the deoxygenation and reaeration k-rates. The value of k_1 depends on the rate of oxygen demand of the organic matter in the water. Laboratory BOD test data are generally used to compute the deoxygenation rate, even though this is considered to be a questionable procedure; the laboratory environment does not reproduce stream conditions, particularly as they are related to temperature, sunlight, biological populations, and water movement. Depending on the nature of the wastewater and degree of treatment prior to disposal, k_1 may fall anywhere in the range between 0.05 and 0.2 per day with the most commonly selected value of 0.10 per day at 20° C. A k-rate determined in the laboratory at 20° C can be corrected for river temperature using Eq. 5-5. This equation states that the rate constant increases, or decreases, 4.7 percent for each degree of temperature rise, or fall; and doubles or halves in rate for a 15° C temperature change.

$$k_T = k_{20} \times 1.047^{T-20} \quad (5\text{-}5)$$

where k_T = rate constant at temperature T, per day
k_{20} = rate constant at 20° C, per day
T = temperature, degrees Celsius

Although oxygen deficit regulates the rate of reaeration, the actual amount of oxygen absorbed depends on the physical and hydrological characteristics of the river course, temperature, water depth, surface area, and other factors. Several relationships have been developed for calculating the value of k_2 based on depth and velocity of flow. These formulas usually take on the general form of Eq. 5-6, with specific values for c, n, and m based on characteristics of the river under study. The second part shown in Eq. 5-6 includes some general numerical constants developed from available field and laboratory experimental data. The value of k_2 is also modified by temperature but to a lesser extent than k_1. One degree of temperature rise increases the reaeration coefficient about 2.2 percent; this can be expressed in the form of Eq. 5-5 by substituting 1.022 for 1.047.

$$k_2 = c \frac{V^n}{H^m} = 3.3 \frac{V}{H^{1.33}} \quad (5\text{-}6)$$

where k_2 = reaeration rate at 20° C, per day
V = mean velocity of flow, feet per second
H = mean depth of flow, feet

Distance downstream and time of passage are related by the mean velocity of flow. For accurate analysis, the channel must be cross sectioned at regular intervals along its course to calculate volume, surface area, and effective depth. With these data, the reaeration rate and travel distance can be calculated for a given quantity of flow in the stream.

Many abnormalities in self-purification are not taken into account in the formulation of the sag equation. These include discharge of a wastewater that has an immediate oxygen demand, for example, an inorganic chemical that reacts instantaneously with DO; nitrification that results in biological uptake of oxygen; deposits of sludge in pool areas along the river channel that exert abnormal oxygen demand; and the abnormality of biological extraction where masses of microbial growth attach to the streambed. Further limitations restrict the practical usefulness of the sag equation. The mathematics assumes uniform steady-state conditions all along the river channel below a single point source of organic discharge. Therefore, additional runoff entering from tributaries and

k-rate variations resulting from changes in the river channel are not taken into account. Common misuse of the sag equation results from estimating k_2 from limited measurements of channel characteristics, and applying this single value over too long a reach of river. For complex situations, the stream must be subdivided into reaches that take into account changes in time of passage and reaeration rate.

Along rivers containing large numbers of waste sources, changes in channel characteristics, and tributary streams, the simplified DO sag equation may not be able to provide satisfactory results. In that case, deoxygenation and reaeration should be considered separately so that various k_2 values can be applied over relatively short reaches of the river. This rational method involves isolating one section and accounting for oxygen inputs and withdrawals with the net oxygen asset, or deficit, shown as the increase, or decrease, in DO level in the water.

Computed dissolved oxygen profiles must be verified by field observation, regardless of the method used in analysis. Such studies reveal abnormalities or limitations of the mathematical methods applied. The ideal study entails a comprehensive sampling survey under conditions of known waste loads and stable river hydrology. Monitoring must be sufficiently extensive to describe the entire river profile. Intensive sampling over a short time span of known stable conditions is superior to grab samples taken over long periods in which hydrologic conditions or waste loads may vary. Although they are less than desired, these survey data are valuable if passage curves, channel parameters, and waste loadings can verify calculated values. After the assimilative capacity of a river has been defined by verified self-purification factors, expected stream conditions can be forecast for anticipated waste loadings under various hydrologic conditions.

EXAMPLE 5-1

A wastewater effluent of 20 cu ft/sec with a BOD = 50 mg/l, DO = 3.0 mg/l, and temperature of $23°$ C enters a river where the flow is 100 cu ft/sec with a BOD of 4.0 mg/l, DO of 8.2 mg/l, and temperature of $17°$ C.

From laboratory BOD testing, k_1 of the waste is 0.10 per day at $20°$ C. The river downstream has an average velocity of 0.60 ft/sec and depth of 4.0 ft. Calculate the minimum dissolved oxygen level and its distance downstream by using the oxygen sag equation.

Solution

The combined flow downstream from the discharge is

$$Q_1 + Q_2 = 100 + 20 = 120 \text{ cu ft/sec}$$

Using Eq. 5-2, the BOD, DO, and temperature after mixing of the wastewater with the river are calculated as follows:

BOD

$$= \frac{100 \text{ cu ft/sec} \times 4.0 \text{ mg/l} + 20 \text{ cu ft/sec} \times 50 \text{ mg/l}}{100 \text{ cu ft/sec} + 20 \text{ cu ft/sec}}$$

$$= 11.7 \text{ mg/l}$$

$$\text{DO} = \frac{100 \times 8.2 \text{ mg/l} + 20 \times 3.0 \text{ mg/l}}{120} = 7.3 \text{ mg/l}$$

$$T = \frac{100 \times 17° \text{ C} + 20 \times 23° \text{ C}}{120} = 18° \text{ C}$$

The results of these computations are illustrated in Figure 5-6.

To apply the oxygen sag equations, k_1, k_2, L_0, and D_0 must be determined. Equation 5-5 is used to adjust $k_1 = 0.10$ at $20°$ C to $18°$ C.

$$k_1 \text{ at } 18° \text{ C} = 0.10 \times 1.047^{18-20}$$

$$= \frac{0.10}{(1.047)^2} = 0.09 \text{ per day}$$

From Eq. 5-6,

$$k_2 = 3.3 \frac{0.60 \text{ ft/sec}}{(4.0 \text{ ft})^{1.33}} = 0.31 \text{ per day}$$

Correcting for $18°$ C,

$$k_2 \text{ at } 18° \text{ C} = 0.31 \times 1.022^{18-20}$$

$$= \frac{0.31}{(1.022)^2} = 0.30 \text{ per day}$$

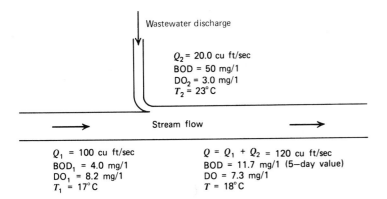

Wastewater discharge

Q_2 = 20.0 cu ft/sec
BOD = 50 mg/l
DO_2 = 3.0 mg/l
T_2 = 23°C

Stream flow

Q_1 = 100 cu ft/sec
BOD_1 = 4.0 mg/l
DO_1 = 8.2 mg/l
T_1 = 17°C

$Q = Q_1 + Q_2$ = 120 cu ft/sec
BOD = 11.7 mg/l (5—day value)
DO = 7.3 mg/l
T = 18°C

Figure 5-6 Results of flow and concentration calculations for Example 5-1.

Rearranging Eq. 3-14,

$$\text{Ultimate carbonaceous BOD} = \frac{\text{5-day BOD}}{(1 - 10^{-5k})}$$

Substituting in the initial five-day BOD value of 11.7 mg/l and $k_1 = 0.10$,

$$L_0 = \frac{11.7}{(1 - 10^{-0.10 \times 5})}$$

$$= \frac{11.7}{(1 - 10^{-0.50})} = \frac{11.7}{0.68} = 17.1 \text{ mg/l}$$

DO saturation at 18°C from Table 2-5 is 9.5; therefore, the initial DO deficit is

$$D_0 = 9.5 - 7.3 = 2.2 \text{ mg/l}$$

Time of travel to minimum DO of sag curve is determined using Eq. 5-4

$$t_c = \frac{1}{0.30 - 0.09} \log \left[\frac{0.30}{0.09} \left(1 - 2.2 \frac{0.30 - 0.09}{0.09 \times 17.1} \right) \right]$$

$$= 1.8 \text{ days}$$

Distance downstream equals time of travel multiplied by velocity

$$\text{Distance} = \frac{1.8 \text{ day} \times 0.60 \text{ ft/sec} \times 86,400 \text{ sec/day}}{5280 \text{ ft/mile}}$$

$$= 17 \text{ miles}$$

The DO deficit at critical time is calculated using Eq. 5-3

$$D = \frac{0.09 \times 17.1}{0.30 - 0.09} (10^{-0.09 \times 1.8} - 10^{-0.30 \times 1.8})$$

$$+ 2.2(10^{-0.30 \times 1.8})$$

$$= 7.3(10^{-0.16} - 10^{-0.54}) + 2.2(10^{-0.54})$$

$$= 3.6 \text{ mg/l}$$

Hence, minimum DO is $9.5 - 3.6 = 5.9$ mg/l.

5-8 EUTROPHICATION

Eutrophication is the process whereby lakes become enriched with nutrients that make the water undesirable for human use, both for water supplies and recreation. Limnologists categorize lakes according to their biological productivity. Oligotrophic lakes are nutrient poor. Typical examples are a cold-water mountain lake and a sand-bottomed, spring-fed lake characterized by transparent water, very limited plant growth, and low fish production. A slight increase in fertility results in a mesotrophic lake with some aquatic plant growth, greenish water, and moderate production of game fish. Eutrophic lakes are nutrient rich. Plant growth, in the forms of microscopic algae and rooted aquatic weeds, produces a water quality undesirable for body-contact recreation.

Effects of Eutrophication

The process of eutrophication is directly related to the aquatic food chain (Figure 5-7). Algae use carbon dioxide, inorganic nitrogen, orthophosphate, and trace nutrients for growth and reproduction. These plants serve as food for microscopic animals (zooplankton). Small fishes feed on zooplankton, and large fishes consume small ones. Productivity of the aquatic food chain is keyed to the availability of nitrogen and phosphorus, often in short supply in natural waters. The amount of plant growth and normal balance of the food chain are controlled by the limitation of plant nutrients. Abundant nutrients un-

balance the normal succession and promote blooms of blue-green algae that are not easily utilized as food by zooplankton. Thus, the water becomes turbid and under extreme conditions takes on the appearance of "pea soup." Floating masses of algae are windblown to the shore, where they decompose producing malodors. Decaying algae also settle to the bottom, reducing dissolved oxygen. Shorelines and shallow bays become weed-choked with the prolific growth of rooted aquatics. Preferred food-fishes cannot survive in these unfavorable conditions and, as eutrophication intensifies, are replaced by a succession of increasingly tolerant fishes. Trout are succeeded by such warm-water fishes as perch, walleye,

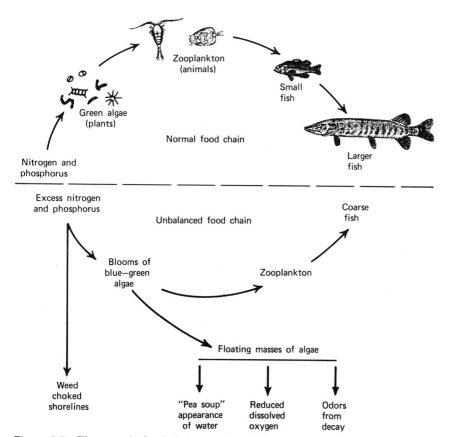

Figure 5-7 The aquatic food chain unbalanced by eutrophication compared with normal succession.

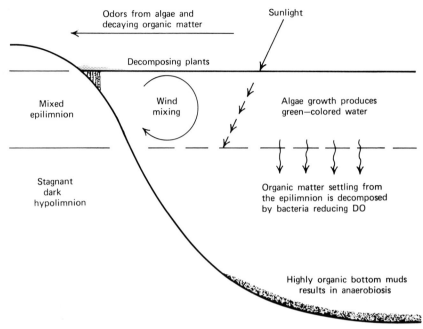

Figure 5-8 Eutrophic lake in the autumn after an extended period of thermal stratification.

and bass, and these in turn are succeeded by coarse fishes like bullheads and carp.

The sketch in Figure 5-8 shows how overfertilization afflicts a large lake. Even a relatively mild algal bloom can result in accumulation of substantial decaying scum along the windward shoreline because of the lake's vast surface area. Gentle winds passing over the lake can pick up fishy odors from algae blooms. Algal growth developed in the sunlit epilimnion can settle into the hypolimnion, which is dark and stagnant during lake stratification. Bacterial decomposition of these cells and organic bottom muds deplete dissolved oxygen in the bottom zone. Reduced dissolved oxygen has been associated with the reduction of commercial fishing in some eutrophic lakes, and treatment of water supplies is complicated by removal of tastes and odors created by algal blooms in the epilimnion and anaerobic conditions in the hypolimnion. Perhaps the most devastating aspect of eutrophication is that the process appears to be difficult to retard. Once a lake has become eutrophic it remains

so, at any rate for a very long time, even if nutrients from point sources are reduced. In part, this results from the long water renewal time (detention time) that reduces the rate of flushing.

Small lakes and reservoirs used for recreation often exhibit symptoms of eutrophication. The source of extra nutrients may be wastewaters and, frequently, fertile land drainage. In clear-water impoundments (Figure 5-9a) fertilization leads to abundant plant growth and large populations of tolerant fish species; unfortunately, many of the desired game fish species find the environment unsuitable. Swimming and water skiing may be impaired by weed beds in shallow water. Small bodies of water that are light-limited by soil turbidity support neither heavy aquatic plant growth nor dense algal blooms (Figure 5-9b). Photosynthesis is retarded by the lack of sunlight, even though adequate plant nutrients are available. Although limiting productivity may have some aesthetic advantages, the high turbidity is detrimental to reproduction of game fishes.

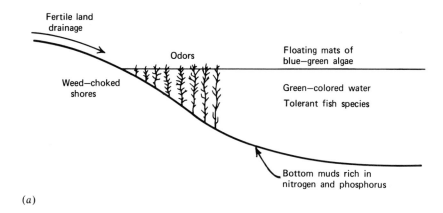

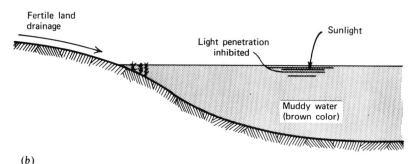

Figure 5-9 The character of small, eutrophic lakes and reservoirs is often defined by clarity of the water. (a) Fertile, clear-water ponds with low soil turbidity reflect eutrophic conditions. (b) Turbid, light-inhibiting waters prevent excessive plant growth by limiting photosynthesis.

Preventing Eutrophication

Macronutrients for plant growth are carbon dioxide, inorganic nitrogen, and phosphate; a variety of trace elements, such as iron, are also needed for growth. The key to controlling rate of lake eutrophication lies in limiting plant nutrients. Natural waters contain sufficient carbon in the bicarbonate alkalinity system to provide carbon dioxide in excess of growth needs. In the majority of lakes investigated, either phosphorus or nitrogen was the limiting factor. Although this idea is not universally accepted, most authorities believe that as a lake becomes more eutrophic the growth-limiting element becomes phosphorus. This concept is reinforced by the fact that blue-green algae can fix atmospheric nitrogen. Because of

this, control of fertilization by limiting the nitrogen supply is highly questionable. At present, emphasis is being placed on phosphorus reduction to control the extent of plant growth in lakes.

Only small amounts of nitrogen and phosphorus are transported to surface waters in a simple agrarian economy, principally in runoff from cultivated farmland. However, in a complex urban economy, such as the United States, phosphate rock is mined and processed into fertilizers, detergents, animal feeds, and other chemicals. In addition to land runoff, animal excreta, and decaying vegetation, a significant amount of human-generated phosphate ultimately reaches surface waters. Approximately 60 percent of the phosphorus in domestic wastewater is from builders used in synthetic detergents. Contributions from agricultural land drainage result from fertilizer additives used to increase crop yields. Our high standard of living has resulted in the use and disposal of large quantities of phosphorus that had previously been underground.

The amounts of inorganic nitrogen and phosphorus needed to produce abundant algae and rooted aquatic weeds are relatively small. The generally accepted upper concentration limits for lakes free of algal nuisances are 0.3 mg/l of ammonia plus nitrate nitrogen and 0.02 mg/l of orthophosphate phosphorus at the time of spring overturn. Lakes with annual mean total nitrogen and phosphorus concentrations greater than 0.8 mg/l and 0.1 mg/l, respectively, exhibit algal blooms and nuisance weed growths during most of the growing season. A detailed discussion of allowable nutrient loadings on lakes is presented by Hammer and Mac Kichan.[2]

The rate of eutrophication of a lake can be retarded by reducing nutrient input. One method used to limit nutrient concentrations has been diversion of nutrient-rich wastewaters. Construction of a pipeline to convey the treated wastewater from the city of Madison, Wisconsin, around a chain of lakes was completed in 1959. Although these lakes still receive nutrients from other uncontrolled sources, such as farmland drainage, the rate of eutrophication has been slowed with no further deterioration since the diversion. Algal blooms are less frequent and less intense because of the appreciable decrease in nitrogen and phosphorus content. Another case history in diversion of wastewaters to retard eutrophication involves Lake Washington in Seattle. Between 1950 and 1958, a rapid increase in algae and a corresponding decline in water transparency observed in the lake were correlated to an increase in nitrogen and phosphorus content originating from wastewater discharges. A comprehensive metropolitan sewerage plan was adopted to intercept all wastewaters entering Lake Washington and to pipe them to two major wastewater treatment plants on Puget Sound. The multimillion-dollar construction work was completed between 1963 and 1968 when the last flow of effluent into the lake was stopped. Significant improvement in water quality occurred within ten years as a result of the flushing action of the nutrient-poor river water entering the lake; algal growth and water transparency have returned to their original, natural levels.

Advanced treatment can be used to remove nutrients from wastewater where diversion is not possible. For instance, all wastewaters flowing into the Great Lakes are processed to reduce the phosphorus concentration to 1.0 mg/l or less, which represents a removal during treatment of about 90 percent. Emphasis is being placed on phosphorus removal, since phosphorus is considered the limiting nutrient and methods are available to precipitate wastewater phosphate chemically in wastewater treatment. Nitrogen compounds are more difficult to eliminate and techniques for nitrogen removal are very costly. Because of the high cost of nutrient removal, diversion should be considered, and, of course, the necessity of nutrient reduction to preserve water quality should be verified. For example, if the majority of nitrogen and phosphorus entering a lake or reservoir originates from land drainage, treatment of point sources may be of limited value. Unfortunately, no ready solution exists for the removal of nutrients from land runoff, whether this be from untilled land or cultivated fields. Land management to control soil erosion and loss of fertilizers are related to local weather, for example, rainfall-runoff patterns during thunderstorms that reduce the success of soil and water conservation practices.

Temporary Remedial Measures

Although the only effective means of preventing or reversing eutrophication is nutrient control, several temporary controls have been used to reduce the nuisance effects in enriched lakes and reservoirs including artificial mixing, harvesting of plants and algae, chemical control, and flushing. Artificial destratification by pumping cold water from the bottom and discharging it at the surface has been effective in improving quality in water-supply reservoirs. Mixing adds dissolved oxygen to the hypolimnion and lowers the temperature of the epilimnion. The latter appears to shift algal populations from the less desirable blue-greens, generally associated with nuisance tastes and odors, to green algae, which are less obnoxious. This change appears to be related to cooling of the epilimnion waters, producing an environment less desirable for the blue-green algae. In lakes where loss of hypolimnion dissolved oxygen is a serious problem during thermal stratification, deep-water diffused aeration can be used in lieu of pumping. One common system is to lay perforated tubing on the bottom to convey and diffuse compressed air.

Power-driven underwater weed cutters with optional barge loading conveyers are available for harvesting aquatic plants. Although costly, weed cutting has been favored for clearing some boating and swimming areas rather than applying herbicides. Harvesting of plants should not be considered a feasible means of removing nutrients from eutrophic lakes. Two tons of aquatic plants by wet weight contain only about one pound of phosphorus and ten pounds of nitrogen. Nutrient removal in fish production is also limited, since fish flesh is only 0.2 percent phosphorus and about 2.5 percent nitrogen. Removal of algae has been suggested as a means for reducing nuisance species and withdrawing nutrients. However, this procedure exists in theory only, since construction of a separator or microstrainer to filter out the algae is prohibitively expensive.

Copper sulfate is commonly used for control of algae in water supply reservoirs. This algicide has several shortcomings as a eutrophication control device. Copper sulfate poisons fish when used in excessive concentrations and has been demonstrated to accumulate in bottom muds of lakes following application over a period of several years. In fertile lakes, the copper sulfate must be applied at intervals throughout the growing season to ensure effective algal control; blooms must be anticipated and treated before they occur. This process is very expensive in both labor and chemical costs. A large number of herbicides are available to provide relief from aquatic weeds, both preemergent and emergent.

Dissolved salts and nutrients bound in algae are washed out of a lake or estuary in the water passing through the system. The rapid recovery of Lake Washington is attributed to the flushing action of relatively pure mountain runoff and rapid renewal of the lake water because of the short detention time of about five years. Artificial flushing can be performed by introducing a large flow of nutrient-poor water to a small eutrophic lake; however, rarely is such a supply of water available for dilution purposes. In the case of a reservoir with multiple ports at various depths for discharge of water, consideration should be given to releasing that water which contributes most directly to a eutrophication problem.

5-9 GROUNDWATER CONTAMINATION

Approximately one half of the population in the United States depends on groundwater as a source of drinking water, with about 30 percent delivered by community systems and another 20 percent from domestic wells. Since groundwater in many regions is a high-quality economical source, future demand is expected to increase for domestic use.

Groundwater contamination commonly results from human activities where pollutants, susceptible to percolation, are stored and spread on or beneath the land surface. Almost every known instance of groundwater contamination has been discovered only after a drinking water supply was affected. Typical pollutant sources are industrial wastewater impoundments, sanitary landfills, storage piles, absorption fields following household septic tanks, improperly constructed wastewater disposal wells, and application of chemicals on agricultural lands.

The amount of water available for infiltration, either from precipitation or the wastewater itself, is a primary factor in carrying pollutants down through a soil profile. Water from the surface passes downward through the unsaturated zone and disperses in an aquifer in a manner depending on site conditions. Figure 5-10 illustrates idealized flow from a wastewater pond overlying an unconfined aquifer forming a recharge mound on the water table that flows laterally outward. Dispersion of a contaminant is influenced both physically by soil porosity and hydraulically by the rate of water movement.

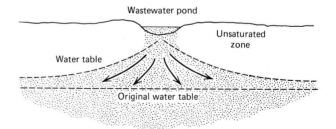

Figure 5-10 Percolation of contaminants from a wastewater pond to an underlying unconfined aquifer.

Natural Contamination

The principal natural chemicals found in groundwater are dissolved salts, iron and manganese, fluoride, arsenic, radionuclides, and trace metals. Both geologic and climatic conditions influence mineral composition. In arid regions with limited water recharge, slow percolation results in mineralized, poor-quality water high in sodium chloride. In humid climates, weathering of sedimentary rock releases calcium and magnesium, creating excessive hardness and often dissolved iron and manganese. Fluoride is a constituent of mineral fluorite found in sedimentary, igneous, and metamorphic rocks. In some regions, high concentrations of fluoride in groundwater result in fluorosis (mottling of teeth) and, in extreme cases, bone damage.

Arsenic can be a significant problem in aquifers of volcanic deposits where concentrations are higher than the maximum contaminant level. These aquifers should be avoided for public water supplies. Natural radioactive substances, while present in most rocks, are generally at very low levels of activity. Geologic formations containing radioactive minerals often contain large amounts of organic matter in the form of lignite or phosphates occurring as phosphatized bone or shell material. Small amounts of metals, such as selenium, cadmium, lead, copper, and zinc, are found in rocks and unconsolidated deposits. Despite this, groundwater generally contains only traces of these metallic elements, and their presence is rarely a water-quality problem.

Changes in water quality resulting from withdrawal of groundwater or channelization of rivers, although induced by human activities, are often viewed as natural contamination. Under undisturbed conditions, groundwater moves toward a river, lake, or the sea (Figure 5-11a). Pumping of nearby wells can reverse this flow, causing migration of polluted surface waters into the aquifer (Figure 5-11b). An example of this is lateral intrusion of salty water in coastal areas where wells overpump the inland, freshwater aquifers. Strict controls over diversion of groundwater in coastal plains can be effective in eliminating saltwater intrusion. In cases where seawater can enter only through limited narrow geologic gaps, freshwater injection wells have been installed parallel to the seashore across the gap. The resulting water mound acts as a barrier, reversing the hydraulic gradient in the aquifer so flow moves toward the sea again.

Contamination from Domestic Wastewaters

A serious source of contamination can be subsurface disposal of domestic wastewater through septic tank–absorption fields. Approximately 30 percent of the population in the United States, representing about 20 million single housing units, use on-site disposal systems. The purpose of an absorption field, which consists of perforated pipe laid in gravel-lined trenches, is to allow the water to seep down into the soil profile. With a high density of these units, serious contamination of the

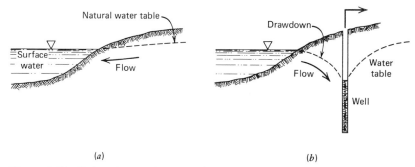

(a)   (b)

Figure 5-11 Induced groundwater recharge resulting from lowering of the water table adjacent to a surface watercourse. (a) Under natural conditions, groundwater normally moves toward flowing or impounded surface waters. (b) Withdrawal from a nearby well can induce recharge, reversing the flow of groundwater.

groundwater can occur. If the soil is permeable, shallow water wells are potential sources of mobile pollutants like detergents, chloride and nitrate ions, and viruses.

Treated wastewaters, if discharged to a dry streambed, can increase infiltration of dissolved salts. Yet this natural percolation of wastewater effluents is generally not viewed with concern since adequate dilution by rainfall, except in an arid climate, reduces significant buildup of salts contributed by the wastewater. In land application of wastewaters for irrigation in arid regions, the groundwater is often not accessible or is naturally saline, making it unsuitable for use; thus, contamination is not a primary concern. Where the quality of infiltration is a major concern, the best cover crop is a perennial grass and the application of wastewater is limited to the amount of nitrogen assimilated by the grass.

Burial of solid wastes can result in degradation of subsurface water through the generation of leachate caused by water percolating through the refuse fill. Leachate is highly mineralized water containing such constituents as sodium chloride, nitrate, trace metals, and a variety of organic compounds. Pollution problems are more likely to occur in humid climates where rainfall exceeds the absorption capacity of the disposal area. Where industrial wastes were included in the landfill, hazardous substances like cyanide, cadmium, and chlorinated hydrocarbons may be present. In the past, dumps were often located on land considered to be of little value at sites susceptible to groundwater pollution, such as marshlands, abandoned sand and gravel pits, old strip mines, and limestone sinkholes. Modern sanitary landfills are designed to reduce environmental damage by managing leachate production. The key factors are proper siting, reduction of infiltration by site grading, selection of a relatively impervious soil cover, and planting of surface vegetation. The base of the refuse can be separated from the water table using a compacted clayey fill.

Industrial and Commercial Sources

The majority of all hazardous wastes from manufacturing are disposed of on the land, mainly because this method is usually the cheapest waste management option. In order of frequency of application, nonreclaimable industrial wastewaters and sludges are discharged to impoundments, buried in landfills, or injected in deep saline aquifers. Impoundments including lagoons, basins, pits, and ponds are a potential source of contamination because of the possibility for leakage of hazardous substances. Since the stored wastewater tends to percolate into the ground, mobile substances are likely to be carried to groundwater either by accident or by design to reduce the liquid in storage. Solid wastes and sludges may also be buried with other refuse in a sanitary landfill. Stringent regulations now control the disposal of hazardous wastes.

Injection into deep wells is environmentally acceptable and often an economical method for disposal of hazardous wastewaters. For toxic chemicals, the boreholes usually penetrate saline aquifers over a thousand feet below the surface. Ideally, in addition to adequate porosity to receive the flow, the stratum should be overlain by impervious layers to prevent upward migration into freshwater zones.

All forms of mining create wastes and changes in hydrogeological conditions that can contribute to degradation of groundwater. The principal impurities from mining wastes are acidity, dissolved solids, metals, and radioactive materials. With both surface and underground mining, refuse piles and slurry lagoons are potential sources of contamination. The open pits from strip mining of ore near the ground surface provide a path for water to reach aquifers. Underground mining forms tunnels and shafts that open new routes for water movement between different water-bearing strata. Procedures for abatement of groundwater pollution are to control infiltration of surface and groundwater into a mine and to reduce the levels of contaminants in mining wastes.

Accidental spills of toxic fluids, gasoline, and oil can migrate through the unsaturated soil zone to groundwater. Hydrocarbons are, by far, the most prevalent contaminants reported in spills and leaky or ruptured buried pipelines and storage tanks. They can persist underground for decades, creating foul tastes at trace levels in groundwater pumped from an aquifer. Most recorded cases of accidental contamination could have been prevented by good management through providing barriers, preventing indiscriminate dumping, and properly cleaning up spills.

Agricultural Sources

Dissolved salts are transported to groundwater by the portion of rainfall or irrigation water that filters through the surface soils of agricultural land. Because of evapotranspiration, the concentration of minerals in the percolate from irrigation is often two or three times greater than in the applied water. To rid the plant root zone of salt deposits, irrigators usually apply more water than is consumed by evapotranspiration and direct percolation to groundwater. The practice results in either overland runoff or subsurface seepage to surface watercourses from which the irrigation return flows are reused in subsequent irrigation cycles. This process unavoidably increases the concentration of dissolved salts with each water reuse, and, as a consequence in arid regions, groundwater quality deteriorates from infiltration of irrigation return flows.

Fertilizers and pesticides can also migrate into the groundwater under cultivated land, except in the case of clayey soils that inhibit infiltration. While some pesticides are persistent, many are readily degraded in the soil environment and do not pose a threat. Furthermore, these chemicals are applied only in limited quantities, reducing the probability of serious groundwater contamination. The most troublesome health-related contaminant from agriculture is the nitrate ion, which is readily carried by water percolating down through unsaturated soil to the saturated zone. Irrigation and application of inorganic-nitrogen fertilizer appear to have contributed to the rapid rise in nitrate levels in many agricultural areas. A significant portion of the nitrogen fertilizer applied to shallow-rooted crops like corn is not used by plant growth. In poorly drained soils with high water tables, much of the leached nitrate may be lost as gaseous nitrogen through denitrification, while drain pipes installed under moderately well-drained soils collect a major portion of the percolated nitrogen. Sandy soils, however, allow nearly all of the nitrate transported deeper than the root zone to contaminate the underlying groundwater. Obviously, the amount of fertilizer applied, soil permeability, and rate of water infiltration are all key factors influencing the movement of nitrate to groundwater.

The problem of increasing nitrate levels in groundwaters is occurring in many regions of arable land. Changes in water management and farming practice can reduce the rate of increase, but nitrate contamination under cultivated land in these regions cannot be prevented without dramatically reducing fertilization or converting to deep-rooted crops like alfalfa or forage grasses. One of the few ways the agriculturists can influence the natural processes governing plant growth is through the use of synthetic fertilizers. Since the cost of ammonia nitrogen is only a fraction of the total cost of farming, applications are often in excess of the amount removed by the crop in order to ensure maximum yield. Nitrate in groundwater in excess of the maximum contaminant level is a serious problem where small towns use groundwater as a drinking water source and where former cultivated areas are urbanized and wells installed for public supplies. To make the predicament worse, water treatment for the removal of nitrate, although technically possible, is generally considered infeasible because of high cost.

Cattle feedlots are also potential sources of nitrate and dissolved salts. Transmission of these or other contaminants is related to local conditions of soils, drainage, rainfall, and surface conditions in the lots. In the case of nitrogen from animal wastes, nitrification-denitrification appears to take place in the mixture of manure and surface soils, reducing the rate of nitrate infiltration. Burial of agricultural solid wastes, including crop residues, dead animals, and manure, can produce leachates, but these sources are small in comparison to surrounding agricultural sources.

Prevention and Control

Prevention is the key to management of groundwater quality. After contamination, remedial actions are usually ineffective since cleanup of most aquifers is not technically or economically feasible. Natural purification requires decades, and if the pollutants do not decay or are not flushed out of the aquifer by groundwater flow, natural cleansing may never occur. Knowledge of potential

MANAGEMENT OF GROUNDWATER QUALITY

POINT SOURCES   DIFFUSE SOURCES

PREVENTION
(Groundwater Protection)

Siting, design,   Site preparation,
construction, and   and control of
operation.   chemical application.

MONITORING
(Early Warning)

Testing of groundwater   Testing of groundwater
from wells installed   from observation, irrigation,
around the site.   and water-supply wells.

ABATEMENT
(Elimination of Source)

Reconstruction of site,   Changing method of
or modifying or   application, or restricting
abandoning operation.   or banning chemical usage.

(Examples: wastewater ponds,   (Examples: Application of
landfills, refuse piles, buried   pesticides or fertilizers
storage tanks, and deep   on agricultural land and
injection wells.)   land application of
                                                                                                    wastewater.)

Figure 5-12   Generalized schemes for controlling groundwater contamination from point
and diffuse sources of pollution.

sources and a comprehensive understanding of the hydrogeology of an area are both essential to preventing contamination. The generalized scheme of prevention, monitoring, and abatement for point and diffuse pollutant sources is outlined in Figure 5-12. Effective prevention from point sources, such as wastewater ponds, landfills, refuse piles, buried storage tanks, and deep injection wells, is based on site selection, controlled design, proper construction, and careful operation. Strategically located monitoring wells around the site for testing of groundwater are recommended for early warning. With the onset of contamination, abatement action must be instituted to stop further damage by eliminating the cause through reconstruction or modifying operations. Diffuse sources from the application of chemicals on the land surface can generally be controlled only by regulatory actions. Agricultural use of pesticides and fertilizers and spreading of highway deicing salts are examples of potential pollutant sources. Prevention is through controlling the rate and method of application. Monitoring can be conducted by testing groundwater samples taken from wells in the area, for example, observation wells for recording groundwater levels, irrigation wells, and private and public groundwater supplies. Abatement is often difficult because private and regional economies may rely in part on the use of chemicals. Reducing applications or changing operational methods may be acceptable or, in extreme cases, the chemicals may be banned from use.

Groundwater is a common resource subject to degradation in quality or reduction in quantity for economic gain by individuals or special-interest groups. In uncontrolled exploitation, each user maximizes his or her benefit at the cost of all others. The remorseless, inherent logic is that abusing a common resource is clearly to an individual's advantage; yet, if all individuals follow this strategy, the resource is diminished to the detriment of all users. Therefore, regulation by a central authority is generally viewed as the only method of preventing exploitation, which in this case results in contamination of groundwater.

REFERENCES

1. *National Interim Primary Drinking Water Regulations,* U.S. Environmental Protection Agency, Office of Water Supply, EPA 570/9-76-003, 1976.
2. Hammer, M. J., and K. A. Mac Kichan. *Hydrology and Quality of Water Resources.* New York: John Wiley, 1981.

PROBLEMS

5-1 What are diffuse sources of water pollution?

5-2 Coliform testing for a town with a population of 800 is performed by the membrane filter technique. How many water samples must be analyzed each month? If the test on a 100-ml sample results in the growth of three coliform colonies, does the drinking water meet the standard for microbiological contaminants?

5-3 What is the health hazard of nitrate in drinking water?

5-4 Write the chemical formulas for the five trihalomethanes that have been found in drinking water. What is the source of trihalomethanes?

5-5 Why was the MCL for turbidity established at a monthly average of one turbidity unit?

5-6 What beneficial use of surface water has the most strict coliform standard and why?

5-7 What are the principal purposes for establishing surface-water quality standards for public water supplies?

5-8 A reactive chemical disposed of by dilution in a river dissipates at a rate proportional to the remaining concentration of unreacted chemical. For this first-order kinetics reaction, given by Eq. 2-23, the $k = 1.15$ per day. The wastewater discharge is 5.0 mgd containing 20 mg/l, while the stream has zero concentration with a flow above the sewer outfall of 105 cu ft/sec. If the mean velocity in the river is 0.50 ft/sec, calculate the flow distance required for the chemical concentration to decrease to a level of 0.20 mg/l. (*Answer* 14 miles)

5-9 A wastewater effluent of 38 cu ft/sec contains a chemical concentration of 2.5 mg/l. The dilutional flow in the receiving stream is 100 cu ft/sec with no chemical residual. Based on a first-order kinetics reaction, calculate the rate constant if the concentration in the streamflow decreases to 0.20 mg/l after 5.0 hr of travel time downstream.

5-10 The limit on increase in streamflow temperature is $4° F$. The critical low flow in the river is 40,000 cu ft/sec when the temperature is $74° F$. Calculate the maximum allowable temperature of 2000 cu ft/sec of cooling water withdrawn and discharged back to the river, without any loss in quantity. (*Answer* $154° F$)

5-11 A 20-mgd domestic wastewater discharge has a concentration of 22 mg/l of ammonia nitrogen. If the receiving stream contains 1.5 mg/l and the maximum allowable concentration is 3.5 mg/l, compute the dilutional streamflow required in cubic feet per second.

5-12 Solve Example 5-1 using a $k_1 = 0.05$ at $20° C$. Assume all other parameters are the same. (*Answer* 6.2 mg/l)

5-13 The following treated effluent is discharged to a stream: $Q = 1.0$ cu ft/sec, DO = 2.0 mg/l, 5-day BOD = 40 mg/l, $k_1 = 0.10$/day, and $T = 20° C$. Upstream from the outfall the watercourse has the following characteristics: $Q = 9.0$ cu ft/sec, DO = 8.0 mg/l, 5-day BOD = 2.0 mg/l, and $T = 20° C$. The stream channel has a $k_2 =$

0.30/day. Calculate the critical dissolved oxygen concentration downstream, and the distance from the outfall to this point assuming a mean velocity of 2.0 ft/sec in the river. (*Answers* 7.0 mg/l, 1.2 days, and 40 miles)

5-14   The characteristics of a river after discharge and mixing of a wastewater are $L_0 = 12.0$ mg/l, $D_0 = 1.0$, $T = 24°$ C, $k_1 = 0.15$/day at $24°$ C, and $Q = 1150$ cu ft/sec. The mean velocity in the river is 0.6 ft/sec and the average depth equals 4.0 ft. Compute the dissolved oxygen level after a time of passage equal to 2.0 days. (*Answer* 5.5 mg/l)

5-15   A wastewater effluent of 50 cu ft/sec with a BOD = 30 mg/l, $k_1 = 0.05$ per day, and DO = 7.2 mg/l enters a slow-moving river where the flow is 50 cu ft/sec with a BOD = 8.0 mg/l and DO = 8.2 mg/l. The downstream flow (100 cu ft/sec) has an average velocity of 0.70 ft/sec, depth of 2.4 ft, and temperature of $23°$ C. The reaeration formula is as given by Eq. 5-6. Calculate the minimum dissolved oxygen level and its distance downstream.

5-16   A stream with a flow of 3.0 cu ft/sec, recovering from a wastewater discharge several miles upstream, merges with a tributary drainage of 1.0 cu ft/sec. The combined flow of 4.0 cu ft/sec travels for another 1.6 miles where it enters a major river. At the point were the stream and tributary join, the ultimate BOD and dissolved oxygen in the stream are 8.0 mg/l and 6.0 mg/l, respectively; and the values in the drainage flow are an ultimate BOD of 2.0 mg/l and DO equal to 8.0 mg/l. The combined flow in the channel to the river is uniform with a water depth of 2.0 ft, mean velocity of 0.20 ft/sec, water temperature of $20°$ C, and k_1 equal to 0.10 per day. What is the lowest dissolved oxygen concentration that exists in the stream reach between the entrance of the tributary and the confluence with the river?

5-17   A wastewater discharge enters a stream 20 miles above a backwater pond. The wastewater characteristics are $Q = 4.0$ mgd, 5-day BOD = 50 mg/l, $k_1 = 0.10$ per day, DO = 2.0 mg/l, and $T = 20°$ C. The stream above the outfall has $Q = 50.0$ cu ft/sec, 5-day BOD = 4.0 mg/l, $k_1 = 0.10$ per day, DO = 8.0 mg/l, and $T = 20°$ C. The 20-mile river reach has an average velocity of 0.50 ft/sec and $k_2 = 0.38$ per day. The backwater is one-quarter mile long with an estimated $k_2 = 0.050$ per day and time of passage (retention time) = 12 hr. Compute the dissolved oxygen concentration at the discharge (downstream) end of the backwater.

5-18   Calculate the five-day BOD load to critical dissolved oxygen drop (ratio of BOD in the stream water to drop in DO level at critical time) for $k_1 = 0.05$ and $k_2 = 0.30$. Assume an initial condition where $D_0 = 0$. (*Answer* BOD/D = 3.8/1.0)

5-19   What is the best way to prevent, or retard the rate of, eutrophication of a lake?

5-20   How does thermal stratification (Section 4-13) influence the quality of water impounded in a eutrophic lake?

5-21   In the disposal of hazardous industrial wastes, what are the most common improper methods of disposal which result in groundwater contamination?

5-22   Describe the circumstances that can result in groundwater contamination with nitrate in agricultural areas.

5-23   Why does thermal stratification of a eutrophic lake worsen the adverse effects of eutrophication?

chapter 6

Water Distribution Systems

The objectives of a municipal water system are to provide safe, potable water for domestic use, adequate quantity of water at sufficient pressure for fire protection, and industrial water for manufacturing. A typical waterworks consists of a source-treatment-pumping and distribution system (Figure 6-1). Sources for municipal supplies are deep wells, shallow wells, rivers, lakes, and reservoirs. About two thirds of the water for public supplies comes from surface-water sources. Large cities generally use major rivers or lakes to meet their high demand, whereas the majority of towns use well water if available. Often groundwater is of adequate quality to preclude treatment other than chlorination and fluoridation. Wells can then be located at several points within the municipality, and water can be pumped directly into the distribution system. However, where extensive processing is needed, the well pumps, or low-lift pumps from the surface water intake, convey the raw water to the treatment plant site. A large reservoir of treated water (clear-well storage) provides reserves for the high demand periods and the equalizing of pumping rates. The high-lift pumps deliver treated water under high pressure through transmission mains to distribution piping and storage.

The distribution consists of a gridiron pattern of water mains to deliver water for domestic, commercial, industrial, and fire fighting purposes. Elevated storage tanks, or underground reservoirs with booster pumps, reserve water for peak periods of consumption and fire demand. A short lateral line connects each fire hydrant to a distribution main. Shutoff valves are located at strategic points throughout the piping system to provide control of any section or service outlet, including hydrants. These valves are used to isolate units requiring maintenance and to insure that main breaks affect only a small section. A service connection to a residence includes a corporation stop tapped into the water main, a service line to a shutoff valve at the curb, and the owner's line into the dwelling, which incorporates a water meter and a pressure regulator or relief valve if necessary.

6-1 WATER QUANTITY AND PRESSURE REQUIREMENTS

The amount of water required by a municipality depends on industrial use, climate, and economic considerations. Although industries in the rural countryside frequently maintain private water systems, major plants in urban areas rely on the municipal waterworks. Approximately two thirds of the water withdrawn in the United States is for nonconsumptive industrial uses; more than 90 percent is cooling water returned to the source.

Municipal water use in the United States averages 600 gpd (2270 l/d) per metered service including residential, commercial, and industrial customers. For residential customers, water consumption in eastern and southern areas is 210 gpd (790 l/d) and in central states 280 gpd (1060 l/d), while western regions use 460 gpd (1740 l/d) per household service. Only a small amount of water is sprinkled on lawns where the rainfall exceeds 40 in./yr (1000 mm/y), while in semiarid climates lawns and gardens are maintained by irrigation. A

Figure 6-1 Sketch of a typical waterworks system that includes source, treatment, pumping, storage, and distribution. (Reprinted by permission of Mueller Co.)

typical city dweller uses about 90 gpd (340 l/d) for personal use.

Lawn sprinkling may have a striking influence on water demand in areas with large residential lots—50 to 75 percent of the total daily volume may be attributed to landscape irrigation. Other climatological factors, for instance, water-chilled air conditioning and swimming pools influence water depletion.

Although water rates have increased significantly during the past decade, residential water use has continued to rise approximately 1 percent/yr. Most new houses have more water fixtures, modern appliances, spacious lawns, and other conveniences that consume larger volumes of water. A flat-rate residential water charge using a fixed fee per dwelling, rather than a rate based on water consumed, results in wasting of water by residents. For example, water may be allowed to run from a faucet continuously to have an instant cold supply. Metering of individual dwellings and establishing water rates on quantity of flow results in decreased waste and, consequently, in reduced water consumption. Municipalities that are entirely metered use approximately 60 percent of the amount that would be consumed based on flat-rate revenues. Increasing water rates have, in some instances, resulted in a decreasing water use by industry, particularly for cooling. Sewer use fees established in many cities bill industries for the quantity of wastewater discharged in municipal sewers. This has resulted in reduced wastewater production and, hence, in reduced water consumption by the industry.

Residential water use varies seasonally, daily, and hourly. Typical daily winter consumption is about 80 percent of the annual daily average, while summer use is 30 percent greater. Variations from these commonly quoted values for a particular community may be significantly greater depending on seasonal weather changes. Maximum daily demand can be considered to be 180 percent of the average daily, with values ranging from about 120 to more than 400 percent. Maximum hourly figures have been observed to range from about 1.5 to more than 10 times the average flow in extreme cases; a mean for the maximum hourly rate is 300 percent. Table 6-1 summarizes variations in residential water consumption for domestic and public uses only.

Table 6-1. Variations in Residential Water Consumption in Gallons per Capita per Day[a]

	Range	Average
Yearly average consumption	100 to 130	110
Mean winter consumption	50 to 130	100
Mean summer consumption	130 to 260	170
Maximum daily use	160 to 520+	230
Maximum hourly use	200 to 1300+	390

[a] 1.0 gal/capita/day = 3.78 liters/person · day

Water flows used in waterworks design depend on the magnitude and variations in municipal water consumption and the reserve needed for fire fighting. Quantities of water required for fire demand, as detailed in Section 6-2, are of significant magnitude and frequently govern design of distribution piping, pumping, and storage facilities. Water intakes, wells, treatment plant, pumping, and transmission lines are sized for peak demand, normally maximum daily use where hourly variations are handled by storage. Standby units in the source-treatment-pumping system may be installed for emergency use, for convenience of maintenance, or to serve as capacity for future expansion. The required design flow of maximum daily consumption plus fire flow frequently determines the size of distribution mains and results in additional pumping capacity and a need for storage reserves, in addition to that required to equalize pumping rates. If the maximum hourly consumption exceeds the maximum daily plus fire fighting demand, it may be the controlling criterion in sizing some units.

The recommended water pressure in a distribution system is 65 to 75 psi (450 to 520 kPa), which is considered adequate to compensate for local fluctuations in consumption. This level of pressure can provide for ordinary consumption in buildings up to ten stories in height, as well as sufficient supply for automatic sprinkler systems for fire protection in buildings of four or five stories. For a residential service connection, the minimum pressure in the water distribution main should be 40 psi (280 kPa). Pressures in excess of 100 psi (690 kPa) are undesirable, and the maximum allowable pressure is 150 psi (1030 kPa). At excessive levels, leaks occur in domestic plumbing requiring pressure reducers in service connections, and undue stress is placed on mains in the ground. Pipe and fittings used in ordinary water distribution systems are designed for a maximum working pressure of 150 psi.

6-2 MUNICIPAL FIRE PROTECTION REQUIREMENTS

The Insurance Services Office (ISO)[1] has developed a standard schedule for the grading of municipalities with regard to their fire defenses and physical conditions. Fire defenses are weighted for evaluation on the basis of 39 percent for water supply, 39 percent for fire department, 13 percent for fire safety control, and 9 percent for fire service communications. In the evaluation of a municipality, deficiency points are assigned for deviations from the criteria published by the Insurance Services Office.[2] Reliability and adequacy of the following major water supply items are considered in the schedule: water supply source, pumping capacity, power supply, water supply mains, distribution mains, spacing of valves, and location of fire hydrants. These are all essential components for fire fighting facilities of a municipality.

Required Fire Flow

This is the rate of flow needed for fire fighting purposes to confine a major fire to the buildings within a block or other group complex. Determination of this flow depends on size, construction, occupancy, and exposure of buildings within and surrounding the block or group complex. The required fire flow is computed at appropriate locations in each section of the city. The minimum amount is 500 gpm, and the maximum for a single fire is 12,000 gpm. Where local conditions indicate that consideration must be given to simultaneous fires, an additional 2000 to 8000 gpm is required. A municipality will have domestic and commercial water demands at the time fires occur; therefore, an adequate system must be able to deliver the required fire flow for the specified duration with municipal consumption at the maximum daily rate. The maximum daily consumption is defined by the Insurance

Services Office as the greatest total amount of water used during any 24-hr period in the past three years. This maximum daily rate, expressed in gallons per minute (liters per second) is the mean usage during the day of peak delivery. In cases where actual use figures are not available, the maximum consumption is estimated on the basis of use in other cities of similar character and climate. Such estimates are to be at least 50 percent greater than the average daily consumption, which is defined as the mean daily usage during a one-year period.

An estimate of fire flow required for a given fire area is calculated by the formula

$$F = 18C(A)^{0.5} \qquad (6\text{-}1)$$

where F = required fire flow, gallons per minute (answer is rounded off to the nearest 250 gpm).

C = coefficient related to type of construction: 1.5 for wood-frame construction, 1.0 for ordinary construction, 0.8 for noncombustible construction, and 0.6 for fire-resistive construction.

A = total floor area including all stories in the building, but excluding basements, square feet. For fire-resistive buildings, the six largest successive floor areas are used if the vertical openings are unprotected; but where the vertical openings are properly protected, only the three largest successive floor areas are included.

The fire flow formula, Eq. 6-1, expressed in SI metric units is

$$F = 3.7C(A)^{0.5} \qquad \text{(SI units)} \quad (6\text{-}2)$$

where F = required fire flow, liters per second
A = total floor area, square meters

Regardless of the calculated value, the fire flow shall not exceed 8000 gpm (500 l/s) for wood-frame or ordinary construction, or 6000 gpm (380 l/s) for noncombustible or fire-resistive buildings. For a normal one-story building of any type, however, it may not exceed 6000 gpm. The fire flow shall not be less than 500 gpm (32 l/s). For groupings of single-family and small two-family dwellings not

Table 6-2. Required Fire Flows for Single-Family and Two-Family Residential Areas Not Exceeding Two Stories in Height

Distance Between Dwelling Units (ft)[b]	Required Fire Flows[a] (gpm)[c]
Over 100	500
31 to 100	750 to 1000
11 to 30	1000 to 1500
10 or less	1500 to 2000
Continuous buildings	2500

Source: Guide for Determination of Required Fire Flow, Insurance Services Office, December 1974.

[a] Where wood shingle roofs could contribute to spreading fires, add 500 gpm.
[b] 1.0 ft = 0.305 m
[c] 1.0 gpm = 0.0631 l/s

Table 6-3. Approximate Fire Flow Requirements for the High-Value Districts in Small Communities

Population	Fire Flow (gpm)[a]	Duration (hr)
1000	1500	2
2500	2500	2
5000	3500	3
10,000	4000	4

[a] 1.0 gpm = 0.0631 l/s

exceeding two stories in height, the fire flows in Table 6-2 may be used.

The value obtained by Eq. 6-1 (or Eq. 6-2) may be reduced up to 25 percent for occupancies having a light fire loading, or it may be increased up to 25 percent for high fire loading. Light fire loadings are occupancies of low hazard, such as all forms of housing, churches, hospitals, schools, offices, museums, and other public buildings. However, after credit is applied, the fire flow cannot be less than 500 gpm (32 l/s). High fire hazard loadings encompass all commercial and industrial activities that involve processing, mixing, storage, or dispensing of flammable and combustible materials. Chemical works, explosives, oil refineries, paint shops, and solvent extracting are examples.

Additional adjustments may be applicable to the fire flow as modified for occupancy. Completely automatic sprinkler protection may reduce the required flow up to 75 percent, but structures within 150 ft (45 m) of the fire area increase the required fire flow. The magnitude of increase for separation depends on the open distance, number of sides exposed, type of construction, occupancy, and other factors. The charge for one exposed side generally does not exceed a maximum of 25 percent, and the total penalty for all sides shall not exceed 75 percent.

After these final corrections, the fire flow shall not exceed 12,000 gpm (760 l/s) or be less than 500 gpm (32 l/s).

The *Guide*[3] was prepared for use by municipal survey and grading personnel of the Insurance Services Office and other fire insurance rating organizations. Although it is available to others as an aid in estimating fire flow requirements, considerable knowledge and experience in fire protection engineering are necessary for detailed application of the guidelines. For example, judgment is used for businesses and industries not specifically mentioned in the *Guide*. A thorough understanding of fire fighting operations is essential when considering the influences of accessibility and configuration of buildings.

Prior to the current grading schedule, municipal fire protection requirements were specified for high-value districts of a community based on population as given in Table 6-3. These requirements are still useful as rule-of-thumb values; however, they should be used only as guidelines in the absence of specific data for calculating the required fire flows. Many changes in community developments over recent years have led to the current ISO schedule. For instance, many towns and cities now have large shopping or office areas located in suburban districts away from the previous "downtown" high-value district.

Duration

The required duration for fire flow is given in Table 6-4. Major components of a water system, on which the reliability of fire flow depends—such as, pumping capacity, power source, supply mains, and treatment works—

Table 6-4.   Required Duration for Fire Flow

Required Fire Flow (gpm)[a]	Required Duration (hr)
10,000 and greater	10
9500	9
9000	9
8500	8
8000	8
7500	7
7000	7
6500	6
6000	6
5500	5
5000	5
4500	4
4000	4
3500	3
3000	3
2500 or less	2

Source: Grading Schedule for Municipal Fire Protection,
Insurance Services Office, 1974.

[a] 1.0 gpm = 0.0631 1/s

must have the ability to deliver the maximum daily consumption rate for several days plus the required fire flow for the number of hours specified at anytime during this interval. The period may be five, three, or two days depending on the system component under consideration and the anticipated out-of-service time needed for maintenance and repair work.

Pressure

The pressure in a distribution system must be high enough to permit pumpers of the fire department to obtain adequate flows from hydrants. In general, a minimum residual water pressure of 20 psi (140 kPa) is required during flow to overcome friction loss in the hydrant and suction hose. Higher pressure is needed where pumpers are not used; a residual pressure of not less than 75 psi permits effective use of streams direct from hydrants that are spaced close enough to allow short hose lines.

Sustained high pressures are of value in permitting direct supply to automatic sprinkler systems, and building standpipe and hose systems.

Water-Supply Capacity

In evaluating a system, the ability to maintain the maximum daily consumption rate plus fire flow in the municipality, at minimum pressure, is considered with one or two pumps out of service. To have no insurance grading deficiency, the capacity remaining with the two most important pumps out of service, in conjunction with storage, must provide this flow for the specified duration any time during a five-day maximum consumption period. Some deficiency is charged against a system that can meet the requirement with only one inoperative pump. Where the capacity remaining, alone or with storage, does not equal the maximum daily use rate, only the amount that is available at required pressure may be considered.

Storage is frequently used to equalize pumping rates into the distribution system as well as to provide water for fire fighting. Since the volume of stored water fluctuates, only the normal minimum daily amount maintained is considered available for fire fighting. In determining the fire flow from storage, it is necessary to calculate the rate of delivery during a specified period. Even though the amount available in storage may be great, the flow to a hydrant cannot exceed the carrying capacity of the mains, and the residual pressure at the point of use cannot be less than 20 psi (140 kPa).

Although a gravity system, that is, delivering water without the use of pumps, is desirable from a fire protection standpoint because of reliability, well-designed and properly safeguarded pumping systems can be developed to such a high degree that no distinction is made between the reliability of gravity-fed and pump-fed systems by the Insurance Services Office. Where electrical power is used, the supply should be so arranged that a failure in any power line or repair of a transformer, or other power device, does not prevent delivery of required fire flow. Underground power lines laid directly from a substation of the power utility to the water plant and pumping stations reduce the probability of power failure caused by adverse weather conditions.

Distribution System

Proper layout of supply mains, arteries, and secondary distribution feeders is essential for delivering required fire flows in all built-up parts of the municipality with consumption at the maximum daily rate. Consideration must be given to the greatest effect that a break, joint separation, or other main failure could have on the supply of water to a system. With the most serious failure, no deficiency is considered if the remaining mains from the source of supply and storage can provide the fire flow for the specified duration, during a period of three days with consumption at the maximum daily rate.

Supply mains, arteries, and secondary feeders should extend throughout the system properly spaced—about every 3000 ft (910 m)—and looped for mutual support and reliability of service. The gridiron pattern of small distribution mains supplying residential districts should consist of mains at least 6 in. (150 mm) in diameter. Where long lengths are necessary, exceeding about 600 ft (180 m), 8-in. (200-mm) or larger intersecting mains should be used. In new construction, 8-in. or larger pipes are used where dead ends and poor gridiron are likely to exist for a considerable time during development, or because of layout of streets and topography. Hydrants for fire protection should never be located on the dead end of 6-in. or smaller mains. In commercial districts, the minimum size main should be 8-in. with intersecting lines in each street, with 12-in. (300-mm) or larger mains used on principal streets and for all long lines that are not connected to other mains at intervals close enough for mutual support.

A distribution system is equipped with a sufficient number of valves located so that a pipeline break does not affect more than $\frac{1}{4}$ mile of arterial mains, 500 ft (150 m) of mains in commercial districts, or 800 ft (240 m) of mains in other districts.

Spacing of fire hydrants is based on required fire flow with the average area served not exceeding that given in Table 6-5. Hydrants should have at least two outlets; one

Table 6-5. Standard Hydrant Distribution

Fire Flow Required (gpm)[a]	Minimum Average Area per Hydrant (sq ft)[b]
1000 or less	160,000
1500	150,000
2000	140,000
2500	130,000
3000	120,000
3500	110,000
4000	100,000
4500	95,000
5000	90,000
5500	85,000
6000	80,000
6500	75,000
7000	70,000
7500	65,000
8000	60,000
8500	57,500
9000	55,000
10,000	50,000
11,000	45,000
12,000	40,000

Source: Grading Schedule for Municipal Fire Protection, Insurance Services Office, 1974.

[a] 1.0 gpm = 0.0631 l/s
[b] 1.0 sq ft = 0.0929 m^2

must be a pumper outlet and others must be at least $2\frac{1}{2}$-in. nominal size. The street connection, not less than 6 in. in diameter, is provided with a gate valve to allow ease of maintenance. The shutoff valve in a hydrant should be designed to remain closed if the barrel is broken.

EXAMPLE 6-1

A three-story wood-frame building with a ground floor area of 7300 sq ft is adjacent to a five-story building of ordinary construction with 9700 sq ft per floor. Determine the fire flow and duration required for each building and the complex assuming the units are connected.

Solution

Using Eq. 6-1, for the three-story building,

$$F = 18 \times 1.5(3 \times 7300)^{0.5} = 4000 \text{ gpm}$$

From Table 6-3, the required duration is 4 hr. For the five-story building,

$$F = 18 \times 1.0(5 \times 9700)^{0.5} = 4000 \text{ gpm}$$

The duration is 4 hr.

In considering the buildings as communicating, F is calculated by proportioning the $C(A)^{0.5}$ values where A is the total floor area equal to 70,400 sq ft (3 × 7300 + 5 × 9700). The three-story building is 31 percent of the total area, and the five-story building is 69 percent

$$F = 18[0.31 \times 1.5(70,400)^{0.5} + 0.69 \times 1.0(70,400)^{0.5}]$$
$$= 5500 \text{ gpm}$$

From Table 6-3, the required duration is 5 hr.

EXAMPLE 6-2

Estimate the fire flow for a 60,000-m², single-story building of ordinary construction.

Solution

Using Eq. 6-2,

$$F = 3.7 \times 1.0(60,000)^{0.5} = 900 \text{ l/s}$$

However, since this is a normal single-story building, the maximum required fire flow is 380 l/s.

6-3   WELL CONSTRUCTION

Water in the voids of underground sand or gravel beds can be tapped for municipal supply by using a drilled well, as shown in Figure 6-2. The main components of a well are solid casing, sealed in the upper soil profile; screen set into the aquifer, allowing water to enter the casing while precluding sand and gravel; and turbine pump suspended in the casing on a column pipe. Groundwater flows toward the well, enters the casing through screen openings, and is lifted for discharge by the pump.

Naturally developed and artificially gravel-packed wells are the two basic types of construction in water-bearing sand and gravel formations. A natural well uses a screen, the same diameter as the well casing, designed with slot openings to permit a predetermined portion of the sand to be removed during development. The finer portion of the formation is drawn through the screen and is removed, leaving the coarser fraction as a natural gravel pack. This highly permeable zone increases the effective diameter of the well and assures freedom from further sand pumping. Construction of an artificially gravel-packed well involves placing an envelope of graded gravel around the well screen to provide a zone of high porosity. Grading of pack material is selected relative to the size of the water-bearing sand, so that the completed well delivers sand-free water after development. Artificial packing may be favored by economics or aquifer conditions. Fine uniform sand formations, thick aquifers that are highly stratified, and loosely cemented sandstone formations are best handled by placed gravel packs.

Cable tool percussion and rotary drilling are the two methods used to advance the hole for a well. With cable tool drilling, the pullback method is frequently the simplest and best way to set the well screen. Casing is sunk to the full depth, the screen is lowered inside, and the pipe is then pulled back a sufficient amount to expose the screen to the water-bearing layer. The pullback method of screen installation is also used in rotary-drilled wells. A double-casing method can be used in constructing gravel-packed wells. After the outer casing is sunk to full depth, an inner pipe with well screen is gravel packed inside and the outer casing is lifted, leaving a gravel-packed screen in the aquifer. When the pullback method is impractical, a screen is set in an open hole drilled below the casing after it is cemented into position. Gravel can be placed around the telescoped screen, which is then sealed to the bottom of the casing. Figure 6-3 illustrates the steps in this type of construction. Although these methods are common for installing wells, the particular techniques used in any region depend on local conditions.

Several different screens are manufactured for use in municipal wells. Three common types are wound wire (Figure 6-3), shutter (Figure 6-2), and perforated or slotted pipe. The screen openings and grain size of the aquifer

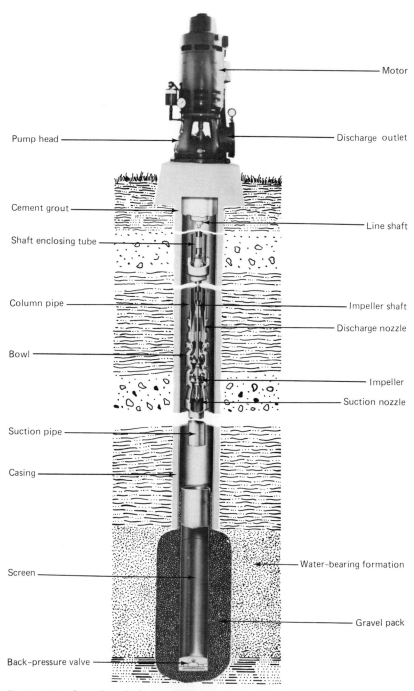

Motor

Pump head

Discharge outlet

Cement grout

Line shaft

Shaft enclosing tube

Column pipe

Impeller shaft

Discharge nozzle

Bowl

Impeller

Suction nozzle

Suction pipe

Casing

Water-bearing formation

Screen

Gravel pack

Back-pressure valve

Figure 6-2 Gravel-packed water well in a sand aquifer equipped with a two-stage vertical turbine pump. (Courtesy of Layne-Western Company, Inc.)

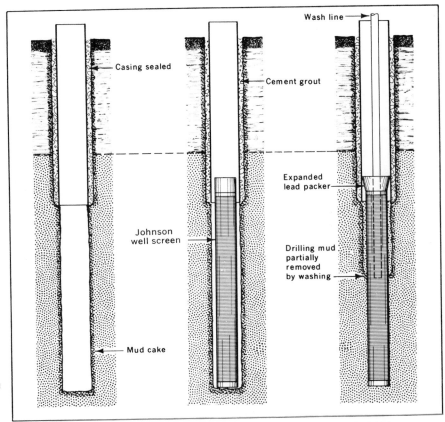

Figure 6-3 Method of placing telescoping screen: (1) After setting casing, hole is extended below by rotary drilling; (2) telescope-size screen is lowered into hole and gravel packed, if necessary; (3) space between top of screen and casing is sealed. The well is then developed to remove mud (clay suspension) used as the drilling fluid to hold the unsupported hole open during construction. (Courtesy of Johnson Division, UOP Inc.)

media dictate the grading of the gravel-pack material. The gravel must be carefully placed, by mechanical or hydraulic means. Merely dumping the material around the casing and screen results in separation of the grain sizes and in bridging, creating poor hydraulic characteristics and possible pumping of sand.

Well development is performed by vigorous hydraulic action through pumping, surging, or jetting. Mechanical surging is done by operating a plunger up and down in the casing like a piston in a cylinder. A heavy baler can be used to produce the surging action, but it is less effective. Air surging is accomplished by lowering an air pipe into the screen section and purging with compressed air. In high-velocity jetting, a tool is slowly rotated inside the screen, shooting water out through the openings. Development is intended to correct any damage or clogging of the water-bearing formation that may have occurred as a side effect of drilling: to increase the porosity and permeability of the natural formations surrounding the screen, and to stabilize the aquifer so that the water is free of sand. The net effect is reduction of water drawdown in the well and higher quality water.

The common well pump is a multistage vertical-turbine pump illustrated in Figure 6-2. The four major components are a motor drive that is mounted at ground level or is submerged below the pump, a discharge pipe column, pump bowls with enclosed impellers, and suction pipe. The pumping units are bottom-suction centrifugal impellers mounted on a vertical shaft that carry one or more stages submerged below the water level is the well casing.

6-4 SURFACE-WATER INTAKES

An intake structure is required to withdraw water from a river, lake, or reservoir. In the latter, it is often built as an integral part of the dam. Typical intakes are towers, submerged ports, and shoreline structures. Their primary functions are to supply the highest quality water from the source and to protect piping and pumps from damage or clogging as a result of wave action, ice formation, flooding, or floating and submerged debris. Towers (Figure 6-4) are common for lakes and reservoirs with fluctuating water levels, or variations of water quality with depth. Ports at several depths permit selection of the most desirable water quality any season of the year. For example, in a eutrophic lake during the summer the surface layer is warm and supports abundant growths of algae, while bottom waters can be devoid of dissolved oxygen, causing foul tastes and odors. On the other hand, during the winter, water immediately under an ice cover may be of highest quality. Submerged ports also have the advantage of being free from ice and floating debris. Underflows of sediment-laden water have been observed during spring runoff in reservoirs, making bottom waters undesirable; hence, a tower with upper entry ports is advantageous. Location, height, and selection of port levels must be related to characteristics of the water body. Therefore, a limnological survey should precede design to define degree and depth of stratification, water currents, sediment deposits, undesirable eutrophic conditions, and other factors.

A submerged intake consists of a rock-filled crib or concrete block supporting and protecting the end of the withdrawal pipe (Figure 6-5). Because of low cost, underwater units are widely used for small river and lake intakes. Although they have the advantage of being protected from damage by submergence, if repair is needed they are not readily accessible. In reservoirs and lakes the distinct disadvantage of bottom intakes is the lack of alternate withdrawal levels to choose the highest quality available throughout the year.

Shore intakes located adjacent to a river must be sited with consideration for water currents that might threaten safety of the structure, location of navigation channels, ice floes, formation of sandbars, and potential flooding. In addition, water-quality considerations and distances from pumping station and treatment plant are important. Figure 6-6 illustrates the type of screening equipment generally employed in shoreline intakes. A coarse screen of vertical steel bars, having openings of 1 to

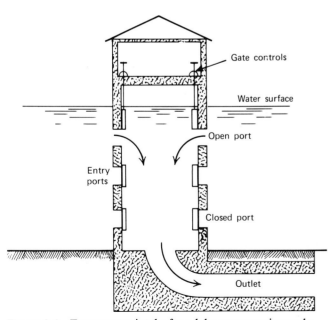

Figure 6-4 Tower water intake for a lake or reservoir supply.

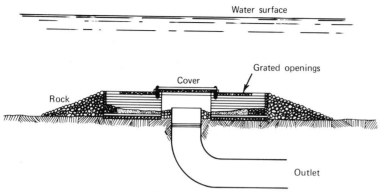

Figure 6-5 Submerged crib intake used for both river and lake sources.

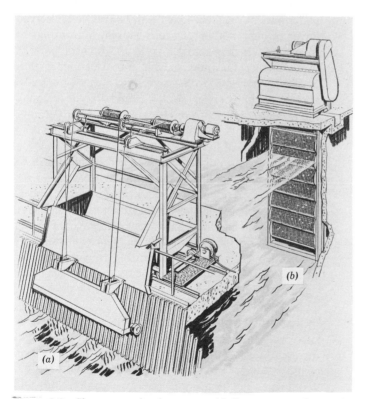

Figure 6-6 Shore water intake screens. (*a*) Coarse screen (bar rack) with power rake and trash hopper for cleaning. (*b*) Traveling water screen with automatic cleaning mechanism to remove debris that passes through bar rack. (Courtesy of Envirex Inc., a Rexnord Company.)

3 in. placed in a near vertical position, is used to exclude large objects. It may be equipped with a trash rack rake to remove accumulated debris. Leaves, twigs, small fish, and other material passing through the bar rack are removed by a finer screen having $\frac{3}{8}$-in. openings. A traveling screen consists of wire mesh trays that retain solids as the water passes through. Drive chain and sprockets raise the trays into a head enclosure, where the debris is removed by means of water sprays. Heat may be required in winter to prevent ice formation. Travel is intermittent and is controlled by the amount of accumulated material.

Low-lift pumping stations transporting raw water from the source to treatment are located as close as is practical to the intake. In the case of a tower or submerged intake, the station is situated on the near shore of the lake or reservoir. Pumps for a shoreline intake are frequently housed in the same structure as the screens. A typical pumping station consists of a suction well located behind the intake screens, or at the discharge end of the withdrawal pipe, pumping equipment, and associated motors, valves, piping, and control systems. With manual controls, an operator must be available to start and stop pumping equipment, to open and close valves, and to monitor other controls. Highly automated pumping stations can be supervised from a central control panel located in the station, or remotely in the treatment plant. Standby pumps and duplicate equipment are provided as is necessary to ensure uninterrupted water supply during maintenance and repair, and to meet emergency water demands.

Selection of low-lift pumps depends on station capacity, the height to which the water must be lifted, the number of pumps required, the anticipated method of operation, and costs. The centrifugal pumps common in low-lift pumping at water intakes are vertical turbine pumps (Figure 6-7) and axial flow pumps (Figure 6-8). These units have liquid flow parallel to the axis of the pump and are designed to operate efficiently under low-head discharge. In the axial flow pump, head is produced by the vortex and lifting action of the rotating propeller, as well as by centrifugal force. The vertical turbine is a centrifugal pump with one or more impellers located in stages. As the water leaves the discharge of the first stage, it enters the suction of the second stage, and continues in

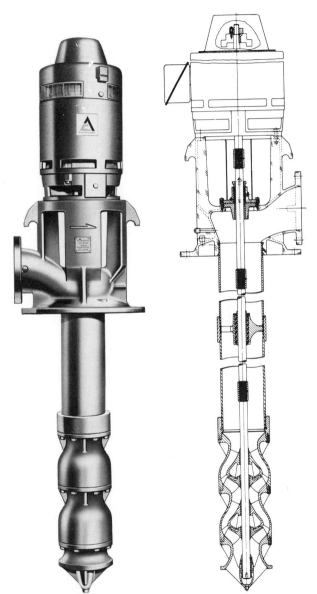

Figure 6-7 Vertical turbine pump. (Courtesy of Allis-Chalmers Corp.)

this manner through subsequent stages. Each stage increases the rate of flow and pressure. No priming is needed for turbine or axial flow pumps since they are installed with the impellers submerged.

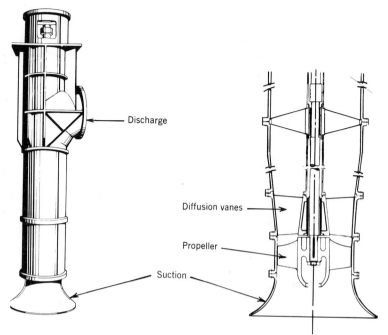

Figure 6-8   Axial flow (vertical propeller) pump. (Courtesy of Allis-Chalmers Corp.)

6-5   PIPING NETWORKS

A municipal water distribution system includes a network of mains with storage reservoirs, booster pumping stations (if needed), fire hydrants, and service lines. Arterial mains, or feeders, are pipelines of larger size that are connected to the transmission lines that supply the water for distribution. All major water demand areas in a city should be served by a feeder loop; where possible the arterial mains should be laid in duplicate. Two moderately sized lines a few blocks apart are preferred to a single large main. Parallel feeder mains are cross-connected at intervals of about one mile, with valving to permit isolation of sections in case of a main break. Distribution lines tie to each arterial loop, forming a complete gridiron system that services fire hydrants and domestic and commercial consumers.

The gridiron system, illustrated in Figure 6-9a, is the best arrangement for distributing water. All of the arterials and secondary mains are looped and interconnected, eliminating dead ends and permitting water circulation such that a heavy discharge from one main allows drawing water from other pipes. When piping repairs are necessary, the area removed from service can be reduced to one block if valves are properly located. Shutoff valves for sectionalizing purposes should be spaced at about 1200-ft (370-m) intervals, and at all branches from arterial mains. At grid intersections no more than one branch, preferably none, should be without a valve. Feeder mains are usually placed in every second or third street in one direction and every fourth to eighth street in the other. Sizes are selected to furnish the flow for domestic, commercial, and industrial demands, plus fire flow. Distribution piping in intermediate streets should not be less than 6 in. (150 mm) in diameter.

The dead-end system, shown in Figure 6-9b, is avoided in new construction and can often be corrected in existing systems by proper looping. Trunk lines placed in the main

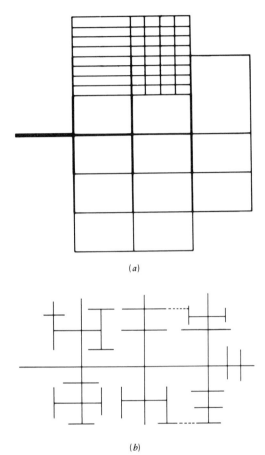

Figure 6-9 Pipe networks. (*a*) Gridiron distribution system consisting of an arterial pipe network with a superimposed system of distribution mains. (*b*) Undesirable dead-end distribution system.

piping, may not be feasible in all cases. For example, in communities located on steep hillsides, elevations may vary so greatly that two or more separate systems are needed. Pressures in a system with large elevation differences should preferably be in the range of 50 to 100 psi (340 to 690 kPa) under average flow conditions. Since a pressure differential of 50 psi (340 kPa) is equivalent to an elevation difference of 115 ft (35 m), the desirable elevation change between the high and low points should be limited to this value. Frequently, separate mains serve different elevation zones with either no direct connections or pressure control valves inserted in the joining lines. Ideally, piping in each zone should be a gridiron system with multiple feeder mains; however, for economic reasons this may not always be feasible. The alternate arrangement is a single arterial main with branches at right angles and subbranches between them such that a grid of interconnected pipes surrounds the central feeder.

Service Connections

A typical service installation consists of a pipe from the distribution main to a turnoff valve located near the property line (Figure 6-10*a*). The pipe is generally attached to the main by means of a corporation stop that can be inserted by using a special tapping machine while the main is in service under pressure. Occasionally, outlets are provided in the main at the time of installation. Access to the curb stop is through a service box extending from the valve to the ground surface. A number of materials are used for service lines, the most popular being copper, plastic, and cast iron. Copper pipe is generally viewed as the standard. However, plastic materials have proved to be just as durable and less expensive. Cast-iron, and sometimes asbestos cement pipe, is used for larger services, generally in a 2-in. size or larger. The water meter, associated shutoff valve, and pressure regulator, if needed, are normally in the basement of the dwelling (Figure 6-10*b*). In some municipalities the meter is placed outside in a box. However, this has the disadvantage of exposing the meter and piping to possible freezing and burial under snow cover. Modern water meters are equipped with remote readers extended to the outside of

streets supply submains, which are extended at right angles to serve individual streets without interconnections. Consequently, if a pipe break occurs, a substantial portion of the community may be without water. Under some conditions, the water in dead-end lines develops tastes and odors from stagnation. To prevent this, dead ends may require frequent flushing where houses are widely separated.

The ideal gridiron system, with duplicating transmission lines, arterial mains, and fully looped distribution

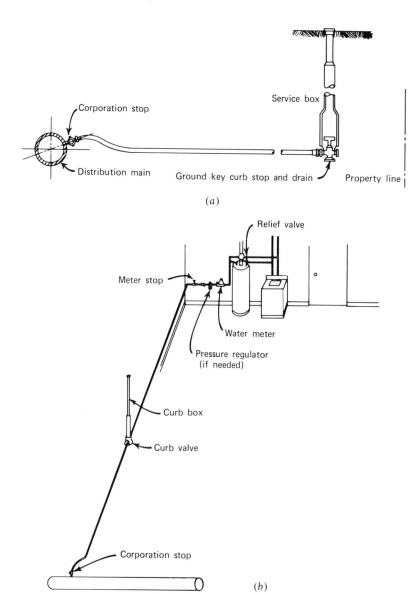

Figure 6-10 Typical residential service connection. (*a*) Service line and meter box installation. (*b*) Complete residential water service.

dwellings so that water utility personnel do not have to enter buildings to record water usage.

The maximum instantaneous residential consumption rate has increased in recent years because of the larger number of fixture units. The estimated rate for a typical house having two bathrooms, full laundry, kitchen, and one or two hose bibbs, is 15 gpm. With the possible exception of lawn sprinkling, a pressure of 15 psi is normally adequate for the operation of any fixture. If one allows 10 psi (23 ft) to overcome static lift from the basement to upper floors, the generally accepted minimum standard of a service is 25 psi at 15 gpm (170 kPa at 0.95 l/s) on the customer side of the meter. For a typical residential service consisting of 40 ft of $\frac{3}{4}$-in. copper pipe and $\frac{5}{8}$-in. disk meter, the friction loss for simultaneous fixture use (6 to 12 gpm) is in the range of 5 to 19 psi. Therefore, if 25 psi is to be available to the customer, the pressure in the main must be approximately 40 psi (280 kPa).

6-6 KINDS OF PIPE

Pipe used for distributing water under pressure includes ductile iron, plastic, asbestos cement, concrete, and steel. Small-diameter pipes for house connections are usually either copper or plastic. For use in transmission and distribution systems, pipe materials must have the following characteristics: adequate tensile and bending strength to withstand external loads that result from trench backfill and earth movement caused by freezing, thawing, or unstable soil conditions; high bursting strength to withstand internal water pressures; ability to resist impact loads encountered in transportation, handling, and installation; smooth, noncorrosive interior surface for minimum resistance to water flow; an exterior unaffected by aggressive soils and groundwater; and a pipe material that can be provided with tight joints and is easy to tap for making connections.

Ductile iron is noted for long life, toughness, imperviousness, and ease of tapping as well as the ability to withstand internal pressure and external loads. Ductile iron is produced by introducing a carefully controlled amount of magnesium into a molten iron of low sulfur and phosphorus content and treating with a silicon base

Table 6-6. Sizes and Thickness Classes of Ductile-Iron Pipe, American National Standards Institute

Size (in.)	Thickness Class					
	1	2	3	4	5	6
	(Nominal Thickness in in.)					
3		0.28	0.31	0.34	0.37	0.40
4		0.29	0.32	0.35	0.38	0.41
6		0.31	0.34	0.37	0.40	0.43
8		0.33	0.36	0.39	0.42	0.45
10		0.35	0.38	0.41	0.44	0.47
12		0.37	0.40	0.43	0.46	0.49
14	0.36	0.39	0.42	0.45	0.48	0.51
16	0.37	0.40	0.43	0.46	0.49	0.52
18	0.38	0.41	0.44	0.47	0.50	0.53
20	0.39	0.42	0.45	0.48	0.51	0.54
24	0.41	0.44	0.47	0.50	0.53	0.56

alloy. This kind of pipe is stronger, tougher, and more elastic than gray cast iron. Iron pipes are available in sizes ranging from 2- to 48-in. diameter (50 to 1200 mm), with several thickness classes in each size (Table 6-6). The selection of a particular thickness class depends on design internal pressure, including allowance for water hammer; external load due to trench backfill and superimposed wheel loads; allowance for corrosion and manufacturing tolerance; and design factor of safety. The procedure for thickness design of water pipes has been simplified by the use of charts and tables that are published by pipe associations and manufacturers. For example, Figure 6-11 is a diagram that can be used to determine the thickness required for ductile-iron pipe under common conditions encountered in construction. The total thickness is dependent on the method of pipe laying, in this case a flat-bottom trench with untamped backfill, depth of cover, pipe diameter, and internal working pressure. The value selected from the chart includes a factor of safety of 2.5, the standard corrosion allowance of 0.08 in., and the specified factory tolerance.

Although ductile iron is resistant to corrosion, aggressive waters may cause pitting of the exterior or tuberculation on the interior. Outside protection is usu-

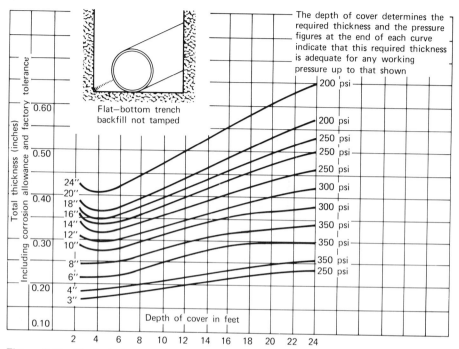

The depth of cover determines the required thickness and the pressure figures at the end of each curve indicate that this required thickness is adequate for any working pressure up to that shown

Figure 6-11 Typical chart to determine the thickness design of ductile iron pipe under common field conditions that include flat-bottomed trench, untamped backfill, factor of safety of 2.5, corrosion allowance of 0.08 in., and a maximum deflection of the pipe of 2 percent. (Courtesy of Clow Corporation.)

ally provided by coating the pipe with a bitumastic tar applied by spray nozzles. In corrosive soils, a tube of polyethylene can be used to encase the pipeline. A section of plastic tubing is placed over each section of pipe before it is lowered into the trench. While the connection is being made, the tubing is pulled over the joint to overlap the adjoining section and then taped in position. The interior of the pipe is covered by a thin coating of cement mortar about $\frac{1}{8}$ in. thick. This lining, placed while the pipe is rotated at a high speed to compact the mortar, adheres closely to the iron surface such that cutting and tapping the pipe will not cause separation.

Pipe lengths, normally 18 ft, may be joined together by several types of joints; the most common are illustrated in Figure 6-12. One of the earliest was the bell and spigot connection in which the plain end of one pipe was inserted into a flared end of another, and then was sealed with the caulking material, such as lead. This development led to the compression-type joint, also referred to as push on or slip joint, which is the most popular type used in water distribution piping today. The beveled spigot end pushes into the bell, which has a specially designed recess to accept a rubber ring gasket. When compressed, it produces a tight joint locking the pipes in place against further displacement. The chief advantages of the compression joint are ease of installation, water tightness, and flexibility. The joint permits about 3 percent deflection, making it possible to install pipe on a gradual curve. The mechanical joint utilizes the principle of the stuffing box to provide a fluid tight, flexible connection for distribution piping; however, it is not as popular as the compression union. Flanged joints are made by threading

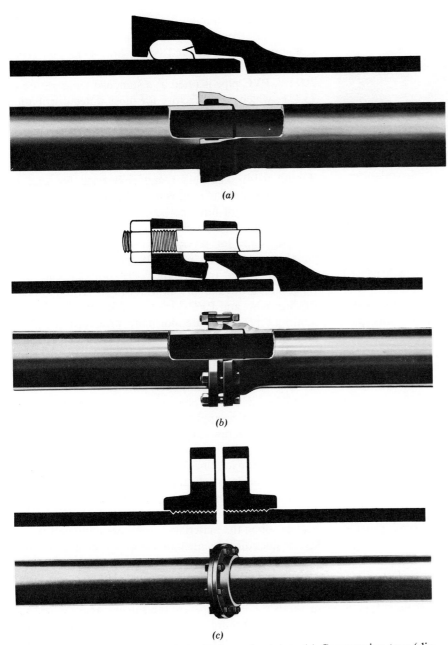

(a)

(b)

(c)

Figure 6-12 Common kinds of ductile iron pipe joints. (*a*) Compression-type (slip joint). (*b*) Mechanical joint. (*c*) Flanged joint. (Courtesy of Clow Corporation.)

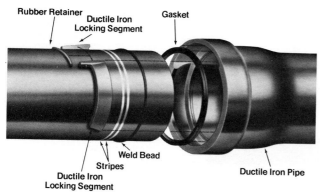

Figure 6-13 Boltless restrained joint for ductile iron pipe. (Courtesy of U.S. Pipe & Foundry Co., U.S. Patent No. 3684320.)

plain-end pipe and screwing on flanges that are then faced and drilled to permit bolting together. Flanged connections are normally used for interior piping in water plants. In addition to these three common joints, particular types are available for special installations. For example, underwater pipelines or other applications requiring a very flexible connection use boltless, or bolted, flexible ball-and-socket joints.

A recent development of a boltless restrained pipe joint for installing distribution system piping is illustrated in Figure 6-13. After a gasket is placed in the socket end of one pipe, the spigot end of the next section is pushed in to make a conventional compression joint. Next, locking segments are inserted on the right and left sides in a slot and slid into position. Finally, a rubber retainer is inserted between the locking segments and wedged into position. The locking segments behind the weld bead provide positive restraint of the joint.

Plastic, or more correctly thermoplastic, pipe is not subject to corrosion or deterioration by electrolysis, chemicals, or biological activity and is exceptionally smooth, minimizing friction losses in water flow. Pipe is manufactured by an extrusion process, and fittings are formed in injection molds. The product is slowly cooled and further shaped through sizing devices to ensure precise dimensions. PVC (polyvinyl chloride) is the plastic preferred for water distribution piping because of its strength and resistance to internal pressure. Yet PVC has

a significantly lower modulus of tensile elasticity than iron and is therefore less resistant to pressure surges. Commonly manufactured sizes are 4–12 in. (100–300 mm), with internal pressure classes of 100, 150, and 200 psi. PVC pipe is rated at a standard temperature of 73.4° F, which is satisfactory for most water distribution systems. At warmer temperatures, the pressure resistance decreases until the critical point is reached near 150° F. For joining PVC water mains, the compression joint uses a rubber seal that fits into a recess formed in the belled end (Figure 6-14). The spigot end is beveled for ease of installation, and the joint is made by simply pushing the lubricated spigot into the bell, compressing the rubber gasket for a pressure-tight fit.

Plastic pipe for service connections and household plumbing systems includes ABS (polymers of acrylonitrile, butadiene, and styrene), PE (polyethylene), and PVC. Sizes are available over wide ranges: ABS, $\frac{1}{2}$ to 12 in.; PE, $\frac{1}{2}$ to 6 in.; and PVC, $\frac{1}{2}$ to 16 in. ABS and PVC, being semirigid products, are normally produced in 20- to 39-ft lengths. Flexible PE pipe is supplied in coils of 100 to 500 ft in length. PE is often used in service connections, and ABS is primarily for drainage, waste and vent fittings, and piping for interior applications.

Joining of plastic pipe is done by several methods varying with size, application, and material. ABS may be connected by screw-threaded couplings, solvent weld, or slip couplings. PE is joined by using insert fittings, flaring, or compression jointing. The insert-fitting method consists of placing an internal support tube inside the pipe ends and holding the connection by outside clamping. PE pipe can be flared by using special tools with the application of heat to soften the material. This allows joining of plastic service lines directly into conventional curb and corporation stops. Yet another method uses internal metal stiffener sleeves and O-rings to form a watertight compression joint. The simplest and most commonly applied joining methods for PVC pipe are the solvent-weld system and the belled-end coupling. Polyvinyl chloride can be chemically bonded through the use of solvents that dissolve the plastic. The solvated plastic surfaces mix when pressed together and leave a monolithic joint on evaporation of the solvent. Belled-end PVC joints may employ a solvent weld or rubber gasket. In the former, the plain pipe end is

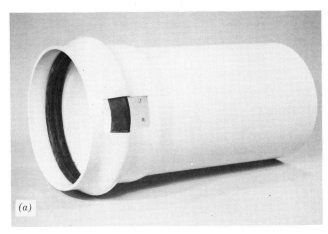

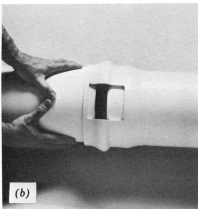

Figure 6-14 PVC bell-and-spigot compression-type joint employing a rubber gasket seal. (Courtesy of Certain-teed Products Corporation.)

cemented into a straight bell formed on the opposite end of a second pipe length.

Asbestos cement pipe is composed of a mixture of asbestos fiber, portland cement, and silica sand. A slurry of these ingredients is formed under high pressure, is heat dried, and then is cured to provide necessary strength and rigidity. Although not as strong as iron pipe, it has the advantages of being a nonconductor immune to electrolysis, resistant to corrosive soil and smooth, providing minimum surface resistance to hydraulic flow.

Most manufacturers produce asbestos cement water pipe in sizes ranging from 4- to 36-in. (100 mm to 900 mm) diameter in laying lengths of 13 ft (4 m). These are provided for three working pressures of 100, 150, and 200 psi, referred to as classes 100, 150, and 200. Pipe lengths are joined by special couplings with rubber sealing rings (Figure 6-15). The O-rings lock into machined grooves when the coupling and pipe ends are pressed together. This joint provides a watertight seal with some degree of flexibility; pipe ends are separated in the coupling to permit expansion and contraction. The class of pipe chosen for application in a water system depends on internal pressure, external load, and safety. Figure 6-16 is a sample selection curve for various sized asbestos cement pipes based on calculations of the combined loading factors for different pipe-laying conditions, internal pressures, and external loads. Pipes selected from this curve have an overall factor of safety of 2.5 for internal pressure in combination with an external backfill plus a 1000-lb wheel load.

Three types of reinforced concrete pipes are used for pressure conduits: steel cylinder, prestressed with steel cylinder, and noncylinder reinforced that is not prestressed. Concrete pipe has the advantages of durability, watertightness, and low maintenance cost. It is particularly applicable in larger sizes that can be manufactured at or near the construction site by using local labor and materials, insofar as they are available. Nonprestressed steel-cylinder pipe is a welded steel pipe surrounded by a cage of steel reinforcement covered inside and out with concrete. The concrete protects the steel from corrosion, provides a smooth interior surface, and contributes high compressive strength to resist stresses from external loading. Applicability is for internal pressures ranging from 40 to 260 psi. For higher pressures, 50 to 350 psi, steel-cylinder pipe is prestressed by winding wire directly on a core consisting of either a steel cylinder lined with concrete, or a pipe of concrete imbedded with a steel cylinder. The exterior of the prestressed core is then finished with a coating of mortar. Reinforced concrete pipe using steel bars or wire fabric, without a steel cylinder, is applicable for low-pressure water transmission requirements, not exceeding 45 psi. Reinforced concrete pipe comes in a variety of diameters ranging from 16 to 144 in. Laying lengths vary with size, the

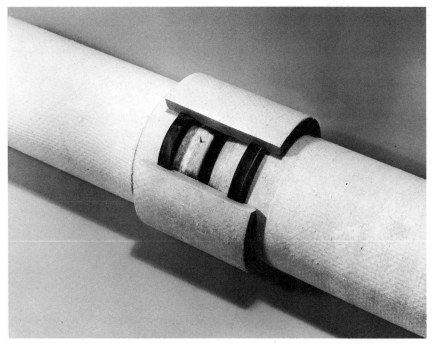

Figure 6-15 Coupling assembly for asbestos cement pipe. Rubber O-rings seal the joint and provide a degree of flexibility. (Courtesy of Certain-teed Products Corporation.)

common manufactured lengths being 8, 12, or 16 ft. The usual concrete pipe joint is a modified bell and spigot with steel rings and a sealing element including a rubber seal. A gasket is placed in the outside groove on the steel-faced spigot ring and then is forced into the steel bell ring of another pipe. The rubber gasket provides a watertight seal by filling the groove in a tightly compressed condition. The narrow space between the bell and pipe surfaces is then sealed with a flexible, adhesive plastic gasket, and the exposed steel joint faces are painted.

Steel pipe, used in transmission lines, exhibits the characteristics of high strength, ability to yield without breaking, and resistance to shock, but careful protection against corrosion is absolutely necessary. A common outside protective coating includes paint primer, coal tar enamel, and wrapping. The particular materials applied

depend on the corrosive environment of the pipe and placement, above ground or underground. Interior linings may be either coal tar enamel or cement mortar. The two types of steel pipe fabricated are electrically welded and mill type. Pipe ends are manufactured in a variety of ways for field connection: plain or beveled ends for field welding, flanges for bolting, bumped or lap ends for riveting, and bell and spigot ends with rubber gasket.

6-7 DISTRIBUTION PUMPING AND STORAGE

High-lift pumps move processed water from a basin at the treatment plant into the distribution system. Different sets may be needed to pump against unequal pressures to separate and distinct service areas. If so, some pumps

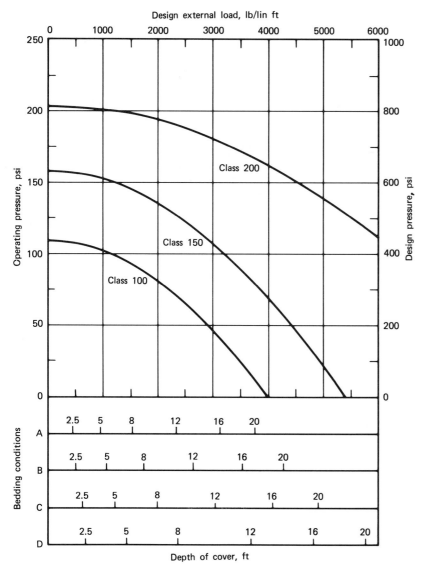

Figure 6-16 Selection curve for 6-in. asbestos cement pipe, applying a safety factor of 4.0 to internal pressures and 2.5 to external loads. Pipe trench bedding conditions: *A*, gravel or sand base, tamped backfill; *B*, same as *A*, backfill not tamped; *C*, pipe barrel resting on flat trench bottom with excavated coupling holes; and *D*, same as *C*, backfill not tamped. (Reprinted from *Standard Practice for the Selection of Asbestos-cement Water Pipe*, AWWA C401-64, by permission of the Association. Copyrighted 1964 by the American Water Works Association, Inc.)

connect directly for main service in lower areas, while booster units are used to reach high elevations in the system. The most common types of pumps in high service are vertical turbine and horizontal split-case centrifugal because of good efficiency and capability to deliver water against high discharge heads. The model illustrated in Figure 6-17 is a double suction design such that water is drawn into both sides of the double volute case, ensuring both radial and axial balance for minimum pressure on bearings. The impeller discharges liquid to the ever-increasing spiral casing, gradually reducing the velocity head while producing an increase in pressure head. This type of pump operates at a range of capacities from design flow to shutoff without excessive loss of discharge pressure or efficiency. Centrifugal pump hydraulics and system characteristics are discussed in Sections 4-4 and 4-5.

Distribution storage may be provided by elevated tanks, standpipes, underground basins, or open reservoirs. Elevated steel tanks are manufactured in a variety of ellipsoidal and spherical shapes and in capacities ranging from 50,000 to 3.0 mil gal (Figure 6-18). The advantage of elevated storage is the pressure derived from holding water higher than surrounding terrain. Steel tanks are normally protected from corrosion by painting the exterior and installing cathodic protection equipment to safeguard interior surfaces. Where gravity water pressure either is not necessary or is provided by booster pumping, ground-level standpipes or reservoirs are pro-

Figure 6-17   Horizontal split-case double-suction centrifugal pump applicable for high-service pumping in waterworks. (Courtesy of Fairbanks Morse Pump Div., Colt Industries.)

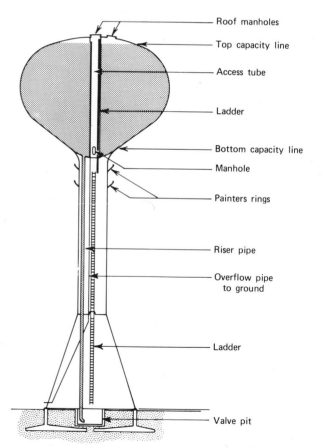

Figure 6-18   Elevated spheroid-shaped water storage tank, available in capacities from 0.20 to 1.25 mil gal and specified pedestal height. (Courtesy of Chicago Bridge and Iron Co.)

vided. Steel standpipes (Figure 6-19) are usually available in sizes up to 5 mil gal. In general, the term standpipe applies where the height of a tank exceeds its diameter, while one that has a diameter greater than its height is referred to as a reservoir.

Concrete reservoirs may be constructed above or below ground. Recent designs in circular, prestressed concrete reservoirs have increased watertightness and reduced maintenance costs not previously available with plain reinforced concrete basins. Although early reservoirs were constructed without covers, current practice is to enclose storage basins, reducing the possibility of pollution and deterioration of the interior surfaces. Exposure to the atmosphere permits airborne contamination, penetration

Figure 6-19 Ornamented steel standpipe for ground-level water storage, designed and built in any capacity. (Courtesy of Chicago Bridge and Iron Co.)

of sunlight which promotes the growth of algae, and, in colder climates, the freezing of the water surface. Most open, paved earth-embankment reservoirs have been abandoned and replaced with other storage facilities, since the cost of covering them is exceedingly high. Clear-well storage at a treatment plant is commonly located under the filter beds that provide a roof structure. Storage capacity at a treatment plant provides allowance for differences in water production rates and high-lift pump discharge to the distribution system. Additional clear-well volume may be supplied to act as distribution storage.

The choice between elevated and ground storage in water distribution depends on topography, size of community, reliability of water supply, and economics. Ground storage facilities on hills high enough to provide adequate pressures are preferred; however, seldom are hilltop locations in a suitable position. Therefore, elevated tanks, or ground-level reservoirs with booster pumping to provide required water pressure, are needed. The economy and desirability of ground storage with booster pumping, as compared with elevated storage, must be determined in each individual area. In general, elevated tanks are most economical and are recommended for small water systems. Reservoirs and booster pumping facilities are often less expensive in large systems where adequate supervision can be provided. Reliable instrumentation and automatic controls are available for remote operation of automatic booster stations.

The principal function of distribution storage is to permit continuous treatment and uniform pumping rates of water into the distribution system while storing it in advance of actual need at one or more locations. The major advantages of storage are the demands on source, treatment, transmission, and distribution are more nearly equal, reducing needed sizes and capacities; the system flow pressures are stabilized throughout the service area; and reserve supplies are available for contingencies, such as fire fighting and power outages. In determining the amount of storage needed, both the volume used to meet variations in demand and the amount related to emergency reserves must be considered. The storage to equalize supply and demand is determined from hourly variations in consumption on the day of maximum water usage. Reserve capacity for fire fighting is computed from required fire flow and duration. Example 6-3 illustrates common techniques in calculating the amount of storage required for a community. Capacity to balance supply and demand, at a constant input pumping rate, is usually in the range of 15 to 20 percent of the daily consumption.

The location of distribution storage as well as capacity and elevation are closely associated with water demands and their variation throughout the day in different parts of the system. Normally, it is more advantageous to provide several smaller storage units in different locations

than an equivalent capacity at a central site. Smaller distribution pipes are required to serve decentralized storage, and more uniform water pressures can be established throughout the system. In normal operating service, some stored water should be used each day to ensure recirculation, and on peak days the drawdown should not consume fire reserves.

EXAMPLE 6-3

Calculate the distribution storage needed for both equalizing demand and fire reserve based on the following information. Hourly demands on the day of maximum water consumption are given in Table 6-7. Listed are hourly consumption expressed in gallons per minute, gallons consumed each hour of the day, and cumulative consumption starting at 12 midnight. Fire flow requirements are 6000 gpm for a duration of 6 hr for the high-value district with 2000 gpm from storage.

Solution

Figure 6-20 is a diagram of the consumption rate-time data given in Table 6-7. When the consumption rate is less than the 1860-gpm pumping rate, the reservoir is filling. When it is greater than 1860 gpm, it is emptying. The area under the emptying, or filling, curve is the storage volume needed to equalize demand for the average 24-hr pumping rate of 1860 gpm.

Since the areas on Figure 6-20 are difficult to measure, a mass diagram is commonly used to determine equalizing storage. Figure 6-21 is a plot of the cumulative flow, column 4 in Table 6-7, versus time. A straight line connecting the origin and final point of this mass curve is the cumulative pumpage necessary to meet the consumptive demand; the slope of this line is the constant 24-hr pumping rate. To find the required storage capacity, construct lines that are parallel to the cumulative pumping rate tangent to the mass curve at the high and low points. The vertical distance between these two parallels is the required tank capacity, in this case, 500,000 gal.

However, sometimes it is not expedient to pump water into the distribution system at a constant rate throughout

Table 6-7. Peak Water Consumption Data on Day of Maximum Water Usage for Example 6-3

Time	Hourly Consumption (gpm)	Hourly Consumption (gal)	Cumulative Consumption (gal)
12 P.M.	0	0	0
1 A.M.	866	52,000	52,000
2	866	52,000	104,000
3	600	36,000	140,000
4	634	38,000	178,000
5	1000	60,000	238,000
6	1330	80,000	318,000
7	1830	110,000	428,000
8	2570	154,000	582,000
9	2500	150,000	732,000
10	2140	128,000	860,000
11	2080	125,000	985,000
12	2170	130,000	1,115,000
1 P.M.	2130	128,000	1,243,000
2	2170	130,000	1,373,000
3	2330	140,000	1,513,000
4	2300	138,000	1,651,000
5	2740	164,000	1,815,000
6	3070	184,000	1,999,000
7	3330	200,000	2,199,000
8	2670	160,000	2,359,000
9	2000	120,000	2,479,000
10	1330	80,000	2,559,000
11	1170	70,000	2,629,000
12	933	56,000	2,685,000
Average = 1860			

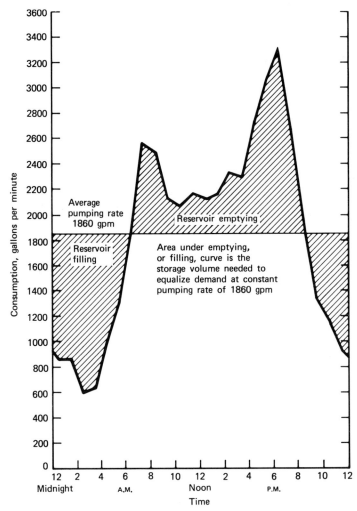

Figure 6-20 Plot of hourly water consumption rates, from Table 6-7 for Example 6-3, to determine storage needed to equalize demand at constant pumping rate.

the day and night. For example, a small community may limit treatment plant operations to daylight hours, or to operation of pumps during off-peak periods when power rates are low. Assume in this instance that the lowest power rates may be obtained during the 8-hr period between 12 midnight and 8 A.M. The graphical procedure for determining storage requirements using this 8-hr

pumping period is illustrated in Figure 6-21 using dashed lines. An accumulated pumping line is drawn from the origin, 0 flow and time of 12 P.M., to the end of the pumping period at 8 A.M. on the maximum cumulative consumption for the day. Storage required is then equal to the vertical distance at 8 A.M. between the accumulated demand line and the maximum daily pumpage. (If a

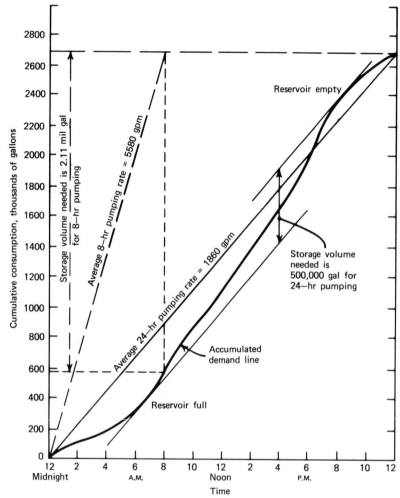

Figure 6-21   Mass diagram of water consumption, from Table 6-7 for Example 6-3, to determine storage needed to equalize demand at constant pumping rates.

starting time for pump operation other than 12 P.M. is to be considered, the data in Table 6-7 must be shifted to the selected time, cumulative consumption values must be recomputed, and Figure 6-21 will have to be redrawn with an origin at the new time.)

Storage required to provide entire fire reserve is equal to flow rate times duration:

$$2000\frac{\text{gal}}{\text{min}} \times 60\frac{\text{min}}{\text{hr}} \times 6 \text{ hr} = 720{,}000 \text{ gal}$$

The total storage capacity required for equalizing demand for a continuous 24-hr pumping rate plus fire protection is equal to

$$0.50 + 0.72 = 1.22 \text{ mil gal}$$

The total considering an 8-hr pumping period plus fire reserve equals

$$2.11 + 0.72 = 2.83 \text{ mil gal}$$

EXAMPLE 6-4

Consider a water-supply system serving a city with the following demand characteristics: average daily demand, 4.0 mgd (2780 gpm); maximum day, 6.0 mgd (4170 gpm); peak hour, 9.0 mgd (6250 gpm); and required fire flow, 7.2 mgd (5000 gpm), resulting in a maximum 5-hr rate of 13.2 mgd (9170 gpm) maximum daily demand plus fire flow. Assume that the minimum pressure to be maintained in the main district is 50 psi (115 ft) except during fire flow, and that the piping system is equivalent to a 24-in.-diameter main with a $C = 100$. Consider the system without storage and with storage beyond the load center.

Solution

Effect of no storage. At each given demand rate the pumping station discharge head must be sufficient to overcome system losses and to maintain a hydraulic gradient with a minimum of 115 ft of head at the load center. Thus, at the average daily demand, the pumping head required is 115 ft plus the head loss in 29,000 ft of 24-in. pipe at 4.0 mgd (using Figure 4-7 for head loss):

$$115 + (0.9 \times 29) = 140 \text{ ft}$$

At maximum daily rate (6.0 mgd):

$$115 + (1.9 \times 29) = 170 \text{ ft}$$

At peak hourly demand (9.0 mgd):

$$115 + (4.0 \times 29) = 230 \text{ ft}$$

And, at maximum daily rate plus fire flow (13.2 mgd):

$$115 + (8.2 \times 29) = 350 \text{ ft}$$

These results are plotted in Figure 6-22.

Storage beyond load center. In the arrangement shown in Figure 6-23, 1.0 mil gal of storage is provided 10,000 ft beyond the load center, 39,000 ft from the pumping station at an elevation of 120 ft. When no water is being taken from storage, the pumping head must be sufficient to pump against the head at the tank and to overcome losses between pumping station and load center. When part of the demand is being supplied from storage,

however, the pumping head need only be sufficient to discharge against the head at load center and to overcome losses in the pipeline between the pumping station and load center.

At the average daily demand, the required pumping rate is 4.0 mgd with no water taken from storage. The hydraulic gradient at the load center is thus identical to that at the tank, namely, 120 ft. The pumping head required is equal to the hydraulic gradient at the load center plus the head loss in 29,000 ft of pipe,

$$120 + (0.9 \times 29) = 146 \text{ ft}$$

During the maximum day, the required pumping rate is 6.0 mgd with no water taken from storage. The pumping head required equal to the hydraulic gradient at the load center plus the head loss in 29,000 ft at 6.0 mgd is

$$120 + (1.9 \times 29) = 176 \text{ ft}$$

At the peak hourly demand, consider supplying 3.0 mgd from storage and the remaining 6.0 mgd from pumping. The hydraulic gradient at the load center is that at the tank minus the head loss in 10,000 ft of pipe between the tank and load center at the storage discharge rate of 3.0 mgd:

$$120 - (0.5 \times 10) = 115 \text{ ft}$$

The pumping head is then

$$115 + (1.9 \times 29) = 171 \text{ ft}$$

Maximum daily plus fire flow. If the 1.0 mil gal in storage is supplied at a uniform rate for the required fire duration of 5 hr, the flow is 3330 gpm (4.8 mgd) with a resulting hydraulic head at the load center of

$$120 - (1.2 \times 10) = 108 \text{ ft}$$

The pumping rate needed is $13.2 - 4.8 = 8.4$ mgd, and the pumping head required is

$$108 + (3.5 \times 29) = 210 \text{ ft}$$

In this analysis the 1.0 mil gal storage supplied about 35 percent of the fire flow with the pumping station providing the balance. This is realistic, since about one half of the storage is to equalize supply and demand. Considering

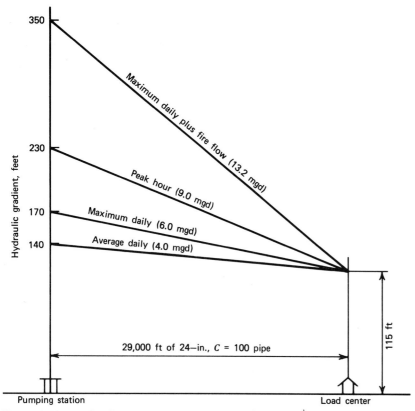

Figure 6-22 Hydraulic gradients with no storage for Example 6-4.

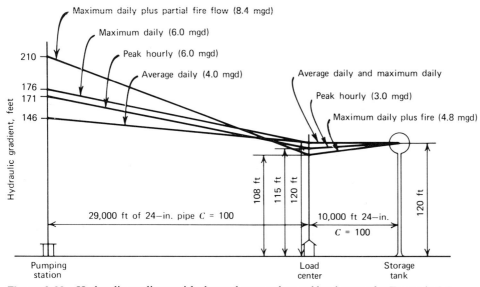

Figure 6-23 Hydraulic gradients with elevated storage beyond load center for Example 6-4.

only 500,000 gal is available for fire demand, the head at the load center is 116 ft and the pumping head increases to 275 ft.

A comparison of Figures 6-22 and 6-23 illustrates the benefits of distribution storage. If no storage is provided, significantly greater pumping heads are required to furnish peak demands. The pumping rate at maximum daily plus fire flow more than doubles the power needed. The same is true at peak hourly pumpage: without storage 9.0 mgd at 230 ft compared to 6.0 mgd at 171 ft. During average and maximum daily demands, the pumping heads are approximately the same.

6-8 VALVES

Valves are installed throughout water systems in treatment plants, pumping stations, and pipe networks, as well as at storage reservoirs. Their purpose is to control the magnitude or direction of water flow. In order to regulate flow, all valves have a movable part that extends into the pipeline for opening or closing the interior passage. The four basic kinds of valves are slide, rotary, globe, and swing; others less common are sphere, diaphragm, sleeve, and vertical-lift disk. Valves are also commonly classified by operating purpose (for example, shutoff and altitude) and function (by-pass and flow control) without regard to the kind of device used. Since the primary purpose of this discussion is to describe the use of valves in water systems, the method of presentation is by function with illustrations showing a commonly used kind of valve. Other kinds may often be used for the same purpose.

The means of operating the movable element of a valve are by screw, gears, or water pressure. Screw stems are common in gate, globe, and needle valves that can be opened or closed by a manually operated handwheel or by a powered operator. In some designs the screw stem rises as the shutoff element closes, and in others the element rides up the screw inside the body of the valve as the stem turns; the latter nonrising stem is more common. In large valves, where water pressure prevents use of a screw, a gear train can be employed to allow the shutoff element to be moved slowly by applying a minimum of torque to the stem. The gear system may be operated manually or by electric, hydraulic, or pneumatic power operators. The kinds of valves that may be equipped with a geared operator include butterfly, gate, globe, and ball. Water pressure can open or close some kinds of valves by direct pressure on the movable element. The simplest is a hinged swing gate that opens under water pressure and closes under the influence of gravity or back pressure. Another example is the automatic globe valve with a body design that controls movement of the shutoff element by system water pressure.

Shutoff Valves

Valves to stop the flow of water through a pipeline are the most abundant valves in a water system. Pipe networks are sectionalized by installation of shutoff valves so that any area affected by a main break or pipe repair can be isolated with a minimum reduction in service and fire protection. Depending on the district within the city and size of the water mains, valve spacings range from 500 to 1200 ft (150 to 370 m). Ideally, a minimum of three of the four pipes connected at a junction are valved and arterial mains have a shutoff every 1200 ft. The pipe connecting a fire hydrant to a distribution main contains a valve to facilitate hydrant repair. In treatment plants and pumping stations, shutoff valves are installed in inlet, outlet, and by-pass lines so that valves and pumps can be removed for maintenance and repair. Gate valves are the popular shutoffs; however, rotary butterfly valves may be installed in large-diameter pipes.

A gate valve consists of a sliding, flat, metal disk that is moved at right angles to the flow direction by a screw-operated stem. When installed in a pipeline, the disk is drawn up into the housing of the case, permitting free flow of the water through the valve opening. The gate is lowered into snugly fitting side channels to block the water flow. The most common valve installed in distribution mains is the double-disk, parallel-seat, cast-iron valve with a nonrising stem (Figure 6-24a). Normally, underground valves in mains are provided with a valve box extending to the street surface. The valve is opened, or closed, by using an extension rod to reach down into the enclosure and turning the nut located on top of the valve stem. Larger gate valves may be placed in vaults or manholes to facilitate operation and maintenance. They are generally equipped with small by-pass valves to reduce pressure differentials on opening and closing (Figure 6-24b) and frequently have spur or beveled gearing to reduce the force required for operation.

A butterfly valve (Figure 6-25) has a movable disk that rotates on a spindle or axle set in the shell. The circular disk rotates in only one direction from full closed to full open, and seats against a ring in the casing. The main disadvantages of a butterfly valve result from the disk always being in the flow stream, restricting the use of pipe-cleaning tools. On the other hand, the advantages of this valve are tight shutoff, low head loss, small space requirement, and throttling capabilities. The latter is one of the most popular applications of butterfly valves, for example, use in a rate of flow controller to regulate the rate of discharge from gravity filters in a water treatment plant. Recently, and more often in the larger sizes, rubber-seated butterfly valves are being used in distribution systems. Because pressure differences across a butterfly valve disk tend to close the valve, a mechanical operator must be employed to overcome the torque in opening the valve, and to resist this force during closing to prevent slamming. For treatment plant applications, the operator is often a hydraulic cylinder and piston rod assembly used to hold the valve disk in any intermediate position, as well

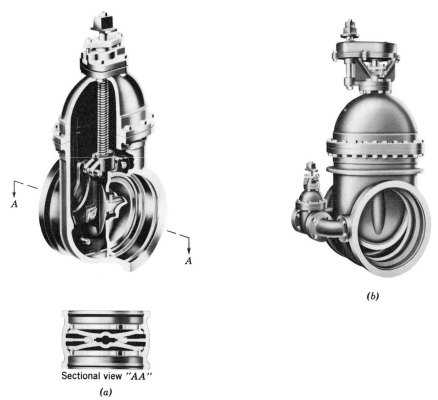

(b)

Sectional view "*AA*"

(a)

Figure 6-24 Typical gate valves used in distribution mains. (a) Cutaway view of a double-disk, parallel-seat gate valve. (b) Larger gate valve equipped with bypass valve and spur gearing for easy operation. (Courtesy of U.S. Pipe and Foundry Co.)

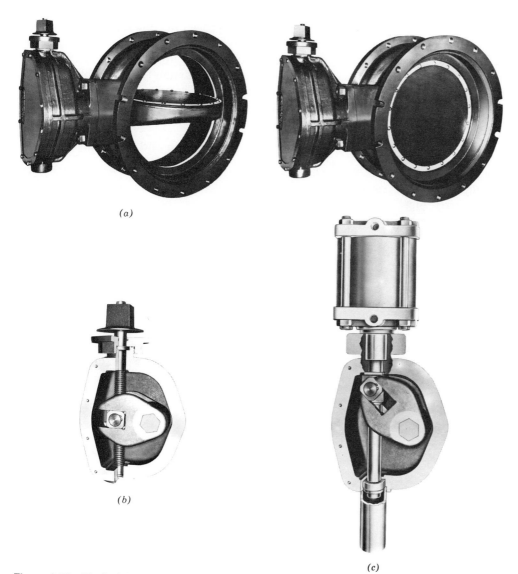

(a)

(b)

(c)

Figure 6-25 Typical butterfly valve used in both water plant and distribution piping. (a) Views of a butterfly valve in open and closed positions. (b) Cutaway view of an operator with a threaded stem for manual operation. (c) Cylinder operator for butterfly valve used in throttling applications to control rate of flow. (Courtesy of Dresser Industries, Inc.)

as in opening and closing action. Operators for large valves may also be motor driven. Manual operation for valves in distribution mains is done by using reducing gears or, as is illustrated in Figure 6-25, a threaded stem that operates a lever arm attached to the valve disk.

Check Valves

A check valve is a semiautomatic device that permits water flow in only one direction. It opens under the influence of pressure and closes automatically when flow ceases. Usual installations are in the discharge pipes of centrifugal pumps to prevent backflow when the pump is not operating, and in conjunction with altitude valves in connections between storage reservoirs and the distribution network.

The two basic configurations of check valves in water system applications are lift check and swing check. A lift check has a flat disk that moves vertically within the valve body, opening with vertical flow and closing by gravity or spring action when flow stops. Of the swing checks, the kind illustrated in Figure 6-26 with a closed disk that rests at right angles to the direction of water flow and where the

disk can be lifted to provide full flow is the most common. Closure may be by gravity or assisted by a counterweight arm attached to the disk. For tight closing, the swing check is manufactured to fit tightly over a rubber or metal seat. The function of the cushion chamber is to prevent hammering action by controlling the closing speed.

Small Pressure-Reducing and Pilot Valves

The function of these valves is to reduce high inlet pressure to a predetermined lower outlet pressure. The applications of small pressure-reducing valves, manufactured in sizes from 0.5 to 2 in., are to protect house plumbing from excessive main pressures and as pilot valves to control the action of large automatic valves that are operated by water pressure. By control of a variable spring, flow through the valve is regulated causing a headloss resulting in a pressure differential between the inlet and outlet. For the unit depicted in Figure 6-27, vertical movement of the stem, predetermined by the position of the diaphragm attached on top, controls the size of the opening through the valve. Outlet pressure, transmitted through the opening connecting the valve

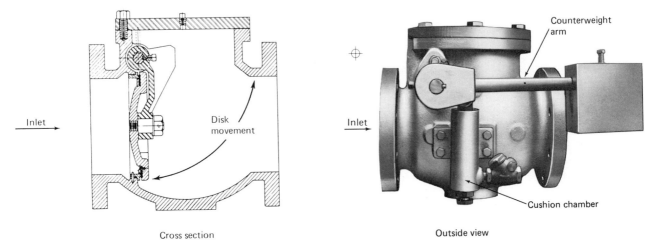

Cross section   Outside view

Figure 6-26   Counterbalanced, swing check valve with a cushion chamber to control the rate of closing. (Courtesy of GA Industries Inc., Golden-Anderson Valve Div.)

chamber to the space under the diaphragm, tends to close the valve by moving the valve disk upward. The force of the spring above the diaphragm pushes downward in an attempt to open the valve. If the pressure on the outlet side decreases below the desired level, the change is transmitted to the diaphragm and the spring forces the valve disk downward to a more open position, thus, allowing greater flow and higher water pressure through the valve. Conversely, if the outlet pressure increases, the diaphragm pulls the stem upward, partially closing the valve and reducing outlet pressure. By turning the handwheel, the force of the spring against the diaphragm changes to allow adjustment of the outlet pressure.

Automatic Control Valves

An automatic valve uses water pressure in the system to operate the movable element to close or open the valve. In large valves, an external pilot valve or a spring-loaded diaphragm assembly directs water pressure to the proper side to set the movable element in the valve body. Sensing devices for signalling valve operation may be hydraulic,

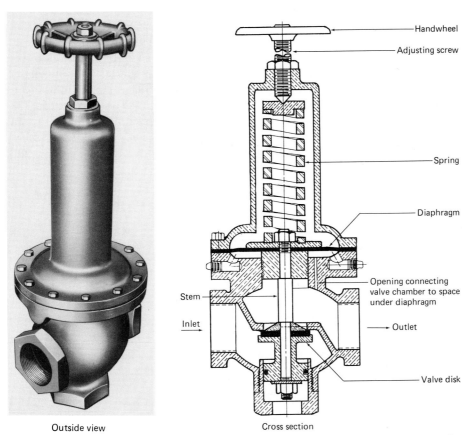

Outside view   Cross section

Figure 6-27   Pressure-reducing valve available in sizes from 0.5 to 2.0 in. for use in service connections (Figure 6-10) and as a pilot valve to control the action of automatic valves (Figures 6-29 and 6-30). (Courtesy of GA Industries Inc., Golden-Anderson Valve Div.)

electric, pneumatic, or solenoid. Of the many applications for automatic valves, the common units in a water system are pressure-reducing, altitude, controlled check, surge relief, and shutoff valves.

The operating principle of one type of automatic globe valve is shown schematically in Figure 6-28. The movable element in the upper chamber of the valve is shaped like a piston, with the top surface area greater than the bottom area. When the chamber above the piston is exhausted to the atmosphere, the piston is held open by inlet water pressure in the body of the valve (Figure 6-28a). Changing the position of the three-way valve on top closes the exhaust and allows inlet water to enter the chamber above the piston. The same magnitude of pressure exerted against both the top and bottom of the piston creates a greater downward force, since the top area of the piston is greater than the bottom area. Thus, when inlet water is allowed to enter the upper chamber, the valve automatically closes (Figure 6-28b). Air moving in or out of the piston chamber passes through a small opening to slow the movement of the piston and prevent water hammer. The three-way valve on top is usually controlled by a hydraulic or solenoid pilot valve.

Pressure-Reducing Valves

Pressure-reducing valves are automatic valves operated by pilots to maintain a preset outlet pressure against a higher inlet pressure. In water distribution systems, the common application is installation in a main connecting separate pipe networks located on two different elevations. As the water flows through the valve from the higher zone to the lower zone, the water pressure is reduced to prevent excessive pressure in the pipe network at the lower elevation.

A pressure-reducing valve is illustrated in Figure 6-29. The inlet pressure of the automatic valve is transmitted to the top of the valve piston, which tends to close the valve, while outlet pressure is transmitted to the top of the piston through the pilot valve. The automatic valve's inlet pressure under the diaphragm of the pilot valve presses upward trying to close the pilot. This force is opposed by the spring pushing downward on the diaphragm trying to open the valve. By turning the handwheel, the force of the spring is adjusted, presetting the reduced outlet pressure of the automatic valve to the desired value. If the pressure at the inlet of the automatic valve is relatively low, the

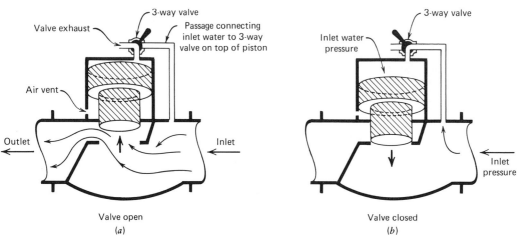

Figure 6-28 Diagrams illustrating the operation of an automatic globe valve. (*a*) The valve is held open by inlet water pressure when the chamber above the piston is vented to the atmosphere. (*b*) The valve is held closed by inlet pressure exerting a greater force on the top of the piston. (Courtesy of GA Industries Inc., Golden-Anderson Valve Div.)

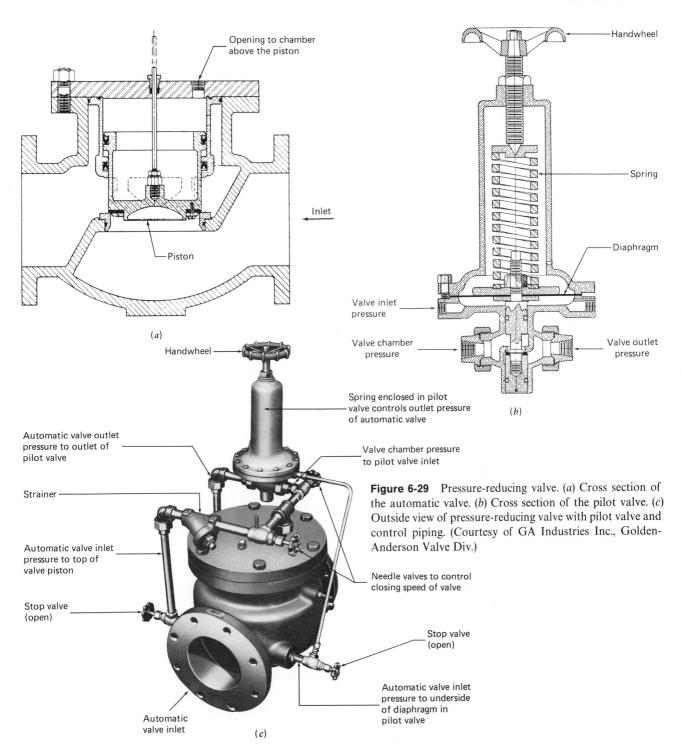

Figure 6-29 Pressure-reducing valve. (a) Cross section of the automatic valve. (b) Cross section of the pilot valve. (c) Outside view of pressure-reducing valve with pilot valve and control piping. (Courtesy of GA Industries Inc., Golden-Anderson Valve Div.)

pressure under the diaphragm of the pilot is lowered, thus opening the pilot valve, which, in turn, relieves the pressure above the valve piston and allows increased flow through the automatic valve. Conversely, a high inlet pressure entering the automatic valve reduces flow through the pilot valve, which increases the pressure on top of the valve piston, reducing flow through the automatic valve. Therefore, the outlet pressure is held constant even though the inlet pressure varies.

Pressure-reducing valves can be equipped with two pilot valves. By this means, the automatic valve can function as both a pressure-reducing and a sustaining valve. It maintains a preset outlet pressure, provided the inlet pressure is above a predetermined value, and closes when the inlet pressure drops below a preset sustaining pressure to protect water services on the inlet side of the valve. Other control arrangements can be used to operate an automatic valve as a pressure-reducing and relief valve or as a pressure-reducing and check valve. A relief pilot opens the automatic valve to equalize inlet and outlet pressures if the inlet pressure drops below a present value. In the pressure-reducing and check valve, backflow is prevented by closing the valve if the inlet pressure drops below the outlet pressure. The type of inlet pressure control selected for a pressure-reducing valve depends on the desired operation for a given installation.

Altitude Valves

Altitude valves are used to automatically control the flow into and out of an elevated storage tank or standpipe to maintain desired water-level elevations. These valves are usually placed in a valve pit adjacent to the tank riser. An altitude valve designed as a double-acting sequence valve (Figure 6-30) is installed in the pipe connecting the tank to the system. The valve automatically closes when the tank is full to prevent overflow. When the pressure on

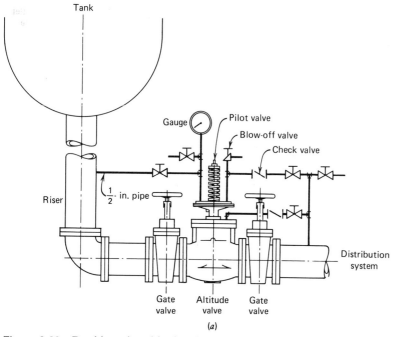

Figure 6-30   Double-acting altitude valve to automatically control water flow into and out of an elevated storage tank. (*a*) Piping arrangement. (*b*) Outside view of the valve. (*c*) Pilot valve operation when water is flowing out of the tank. (*d*) Pilot valve when the tank is full and automatic valve is closed. (Courtesy of GA Industries Inc., Golden-Anderson Valve Div.)

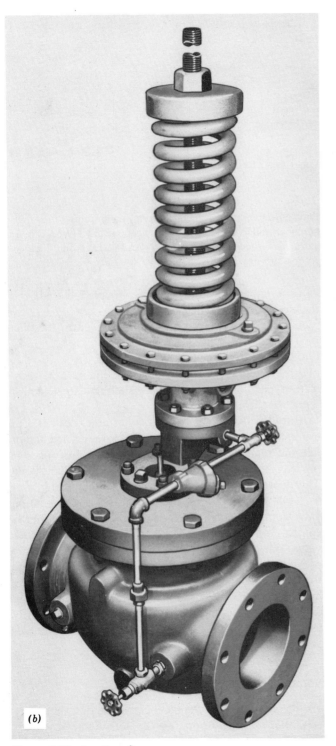

(b)

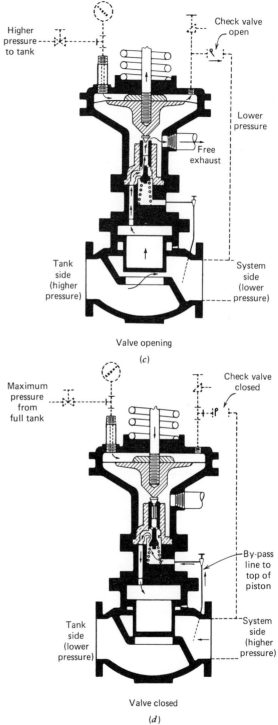

Higher pressure to tank

Check valve open

Lower pressure

Free exhaust

Tank side (higher pressure)

System side (lower pressure)

Valve opening

(c)

Maximum pressure from full tank

Check valve closed

By-pass line to top of piston

Tank side (lower pressure)

System side (higher pressure)

Valve closed

(d)

Figure 6-30 (continued)

the distribution side of the valve is less than on the tank side, it opens allowing water to enter (or leave) the tank.

The piping arrangement for a double-acting valve is shown schematically in Figure 6-30a. The altitude valve is installed in the inlet-outlet pipe between two isolating gate valves. The pilot valve is connected through two small pipes to the tank riser on one side and to the main from the distribution system on the other. An outside view of a double-acting altitude valve is pictured in Figure 6-30b. Speed of valve closing is controlled by adjusting the needle valve in the pipe between the valve body and the base of the pilot. The pilot valve on top is a diaphragm-operated, spring-loaded, three-way valve that transmits water pressure to the top of the piston in the main valve. By turning the nut threaded onto the stem, the force of the spring on the diaphragm can be adjusted to preset the automatically maintained water level in the storage tank. Figure 6-30c illustrates operation of the pilot valve, allowing stored water to flow out of the tank. Water drains out of the chamber above the pilot diaphragm through an open check valve in the small-diameter pipe to the low-pressure system side of the valve. This permits the stem of the pilot to be forced upward when the piston is pushed up by high-pressure water entering the valve from the tank. Therefore, water flows out of the tank through the open valve into the system. During filling of the tank, the direction of flow is reversed, passing from the system into the tank. The pilot does not close the automatic valve during this reverse flow, since the check valve in the pipe to the pilot is closed as a result of the higher pressure from the system side preventing the force of the spring from lowering the diaphragm in the pilot. Thus, in these pilot positions, water is free to flow in and out of the tank and the tank is said to be "floating" on the system. If water continues to enter the tank, eventually the water level rises to the top and the water pressure plus force from the spring lowers the diaphragm, closing the valve between the chamber under the diaphragm and the top of the valve piston. Water can now flow through a by-pass line from the valve body to the top of the piston from the system side, as shown in Figure 6-30d, and the automatic valve closes. (The by-pass pipe is the one attached to the valve body, as pictured in Figure 6-30b.) A globe altitude

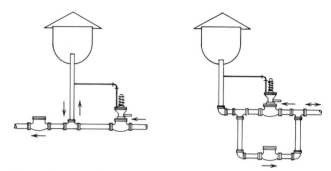

Figure 6-31 Typical arrangements for installing a single-acting altitude valve to automatically control water level in an elevated storage tank. (Courtesy of GA Industries Inc., Golden-Anderson Valve Div.)

valve is either fully opened or completely closed, and therefore does not act as a throttling valve.

A single-acting altitude valve is installed in the inlet pipe to control only water flow into a storage tank. Outflow is through another pipe with a check valve to prevent system water from entering. Figure 6-31 illustrates typical arrangements for installing a single-acting sequence valve.

Altitude valves come in a variety of other designs. For example, a differential altitude valve closes when a tank is full and remains closed until the tank level drops to a predetermined water level before it opens to refill the tank. This operation ensures renewal of the stored water after each filling. An altitude valve may also be a combination valve composed of a swing check built into a butterfly control vane assembly. Outflow is through the check valve. When the water level in the tank drops below a preset elevation, the butterfly vane in the connecting pipe to the system opens. Refilling of the tank occurs when the butterfly control vane is open and the system pressure exceeds tank pressure. When the tank is full, the butterfly vane closes; thus, this kind of valve provides a double-acting operating sequence.

Solenoid Pilot Valves

The force required to change the position of a three-way valve that controls the operation of an automatic valve is very small and can be produced by a solenoid. A solenoid is a coil of wire wound in a helix so that when electric current flows through the wire a magnetic field is established within the helix. The force created by this magnetic field is sufficient to open or close a pilot valve. Figure 6-32 illustrates the operation of a solenoid pilot mounted on top of an angle valve. When the solenoid

pilot is open, the chamber above the valve piston is vented to the atmosphere; simultaneously, a valve in the pipe from the inlet side of the angle valve is closed. This allows the piston in the angle valve to be held open by inlet water pressure. When the solenoid pilot is closed, the vent is sealed and pressure from the inlet water pushing on the

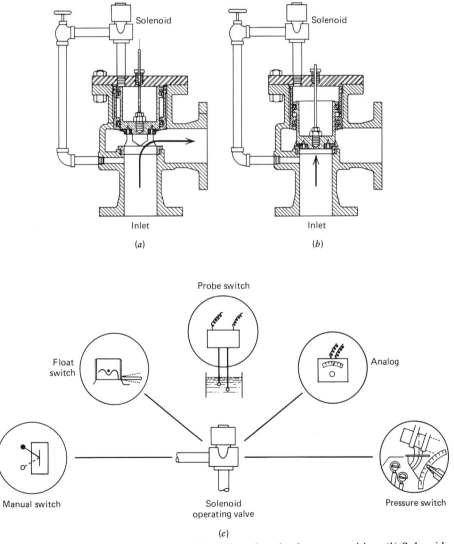

Figure 6-32 Solenoid pilot valve. (*a*) Solenoid angle valve in open position. (*b*) Solenoid angle valve in closed position. (*c*) Several kinds of electrical controls that may be used to operate a solenoid valve. (Courtesy of GA Industries Inc., Golden-Anderson Valve Div.)

larger top area of the piston forces it downward, closing the angle valve.

Several kinds of electrical controls may be used to energize a solenoid pilot valve. All of these allow operation of a valve from a remote location, which is an essential feature for telemetry and automation of a water distribution system. The simplest electrical control is a manual on-off switch. Automatic control can be achieved by using sensing devices to transmit a signal to the solenoid valve. Discreet-level sensors provide only on or off, open or close control. Included in this category are storage-tank-level controls such as float, mercury, and on-off pressure switches.

Analog-level sensors continuously measure the water level over a predetermined range. Examples of analog sensors are pressure-measuring diaphragms, long probes, and sonic signals that reflect back from the water surface. Of these, pressure-measuring devices using diaphragm deformation to measure pressure on a load cell, or some other principle of measurement, are particularly advan-

tageous since they can be used with both altitude valves and pressure-reducing valves. Analog-level sensors can also be used to control the operation of pumps.

Air-Release Valves

Air can enter a pipe network from a pump drawing air into the suction pipe, through leaking joints, and by entrained or dissolved gases being released from the water. Air pockets increase the resistance to the flow of water by accumulating in the high points of distribution piping, in valve domes and fittings, and in discharge lines from pumps. Air-release valves are installed at these locations to discharge the trapped air. The common air-release valve shown in Figure 6-33 contains a ball that floats at the top of a cylinder, sealing a small opening. When air accumulates in the valve chamber, the ball drops away from the outlet orifice, allowing the air to escape. With return of the water level, the ball reseals the outlet.

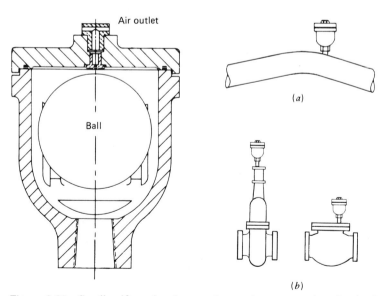

Figure 6-33 Small-orifice air-release valve and common locations of installation in a water distribution system (*a*) at the high points in pipelines and (*b*) on the domes of valves. (Courtesy of GA Industries Inc., Golden-Anderson Valve Div.)

6-9 BACKFLOW PREVENTERS

The water in a distribution system must be protected against contamination from backflow through customer service lines and other system outlets. A cross connection refers to an actual or potential connection between a potable water supply and an industrial or residential source of contamination. By practicing cross-connection control through enforcement of a plumbing code and inspection of backflow preventers in service connections, a municipal utility or private water purveyor can ensure against distribution contamination under foreseeable circumstances. Of greatest concern is backflow of toxic chemicals and wastewaters that may contain pathogens. The risk of back siphonage is reduced by maintaining adequate pressure in the supply mains to prevent reversal of flow. In undersized water distribution networks, low pressures can result from undersized piping, inadequate pumping capacity, or excessive peak water consumption.

Back siphonage is backflow resulting from negative or reduced pressure in the supply piping. Back siphonage can also result from a break in a pipe, repair of a water main at an elevation lower than the service point, and reduced pressure from the suction side of booster pumps.

In contrast, back pressure causes reversal of flow when the pressure in a customer's service connection exceeds the pressure in the distribution main supplying the water. An example of backflow resulting from back pressure is chemically treated water from a malfunctioning boiler system being forced back through the feed line into the potable water supply.

The simplest method of preventing backflow is to provide an air space between the free-flowing discharge end of a supply pipe and an unpressurized receiving vessel. To have an acceptable air gap, the end of the discharge pipe has to be at least twice the diameter of the pipe above the highest rim of the receiving vessel, but in no case should this distance be less than 1 in. (Figure 6-34a). Although air gaps are common in small installations, the loss of water pressure as a result of the physical separation in the supply line to a large facility requires repressurizing the system. Installation of a storage tank and pumps is often more costly than installing a mechanical backflow preventer that transmits supply pressure. In many countries, household water supplies are isolated from the public water-supply system by air gap separation using a storage tank in the attic or on the roof of the dwelling. Flow into the tank

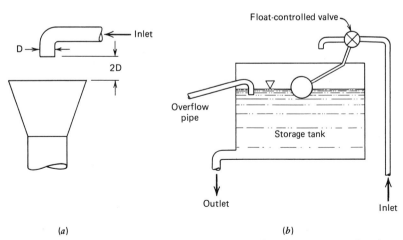

Figure 6-34 Air-gap separation to prevent backflow into the supply pipe. (a) Minimum recommended air gap is twice the diameter of the supply pipe. (b) Storage tank with air-gap separation and overflow pipe to prevent overfilling the tank if supply valve fails to close tightly.

is controlled by a float valve in the inlet pipe, and an emergency overflow pipe ensures that failure of the inlet valve does not result in flow over the tank rim (Figure 6-34*b*). The disadvantage of this arrangement is that a direct pressure connection is necessary for supplying adequate pressure to operate some appliances, like an automatic washing machine, and for lawn sprinkling.

The four kinds of mechanical backflow preventers are the atmospheric-vacuum breaker, pressure-vacuum breaker, double check valve, and reduced-pressure-principle device. The one selected depends on the type of installation and the hazard involved if backflow occurs. For direct water connections subject to back pressure, only the reduced-pressure-principle backflow preventer is considered adequate as an alternate to an air gap separation. Still, air gaps are recommended in industrial applications where toxic chemicals are being mixed with the water, for example, vats containing acids and solutions for metal plating. For applications not subject to back pressure, vacuum breakers can be installed. The atmospheric-vacuum breaker is used in flush valve toilets, and either an atmospheric- or a pressure-vacuum breaker may be connected to a lawn sprinkler system.

Vacuum and Pressure-Vacuum Breakers

An atmospheric-vacuum breaker has an inside moving element that prevents water from discharging through the top of the breaker during flow and drops down to provide a vent opening when flow stops. As diagramed in Figure 6-35*a*, the check-float element is open, raised by the pressure of water flowing up through the valve. When flow stops, the element drops onto the valve seat by the force of gravity. This prevents back siphonage by allowing air to enter and break the siphon in the elevated pipe

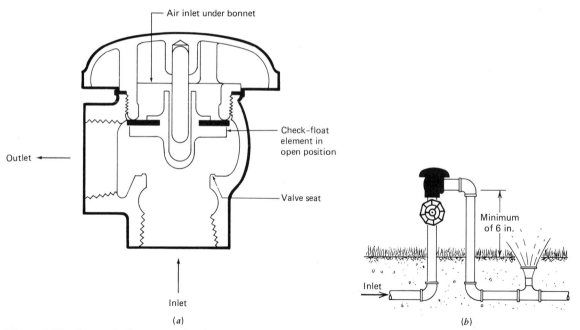

Figure 6-35 Atmospheric vacuum breaker. (*a*) The check-float element seals the air inlet with water flow through the unit. When flow stops, the element drops to cover the inlet and allows air to enter to break the siphon. (*b*) Typical installation on a sprinkler system. (Courtesy of Febco Sales, Inc.)

loop. A typical installation for lawn sprinklers is shown in Figure 6-35b. Another application of an atmospheric-vacuum breaker is installation on the overhead pipe of a water filling station where water can be added to mobile chemical tanks to mix fertilizer, herbicide, or other chemical solutions. Since the hose attached to the end of the fill pipe can be submerged, back siphonage of the solution is possible unless the system is designed to prevent backflow. A vacuum breaker on the top of the riser pipe will break the siphon in the overhead pipe when the water supply is shut off, thus, preventing backflow. An atmospheric breaker should not remain under pressure for long periods of time and cannot have a shutoff valve installed on the discharge side of the breaker.

A pressure-vacuum breaker contains an assembly consisting of a spring-loaded check valve and a spring-loaded air valve, as shown in Figure 6-36. The check valve prevents backflow, and the air valve opens to admit air when the pressure within the body of the breaker approaches atmospheric pressure. If the check valve does not close tight because of interference from foreign matter, air is drawn in through the automatically operating vent valve to preclude backflow. The breaker body is fitted with two test cocks for checking valve tightness against reverse flow and a shutoff valve on each side of the breaker. Pressure-vacuum breakers are installed where low inlet water connections are not subject to back pressure; however, they may be attached to direct water connections subject to backflow provided the cross connection is with water containing only nontoxic substances.

Double Check Valve Assembly

A check valve backflow preventer is composed of two single, independently acting check valves. An installation also has two tightly closing shutoff valves located on each

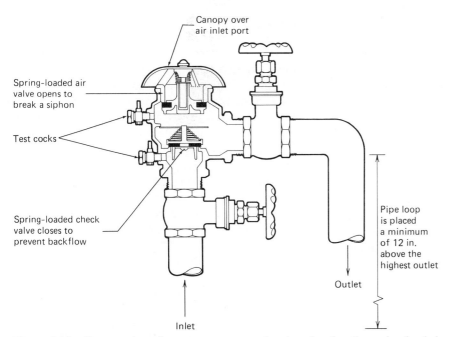

Figure 6-36 Cross section of a pressure-vacuum breaker showing the spring-loaded check valve over the inlet and the spring-loaded air valve over the air inlet port under the canopy. (Courtesy of Febco Sales, Inc.)

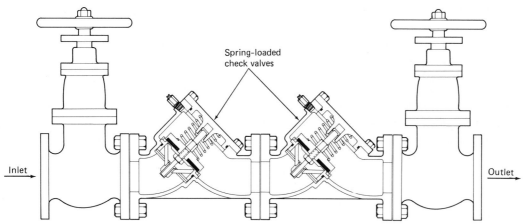

Spring-loaded
check valves

Inlet

Outlet

Figure 6-37 A double-check valve backflow preventer consists of two spring-loaded check valves that operate independently. (Courtesy of Febco Sales, Inc.)

side of the four test cocks for checking tightness of the valves against backflow. In Figure 6-37, both check valves are closed. As with pressure-vacuum breakers, double check valves are used to protect against backflow of nontoxic substances in direct water connections subject to backflow and to protect against backflow of nontoxic and toxic substances in water connections not subject to back pressure.

Reduced-Pressure-Principle Backflow Preventer

This backflow device consists of two independent, spring-loaded check valves with an automatically operating, pressure differential, relief valve located between the two checks. The two isolating gate valves are for testing the operation of the three interior valves and to facilitate removal of the unit for maintenance. The four modes of operation described below are illustrated by diagrams in Figure 6-38.

1. **Normal Flow.** Both check valves are open allowing flow through the preventer. Head loss through the first check valve reduces the pressure in the zone between the two valves 5 to 11 psi below the supply pressure. Controlled by differential pressure, the relief valve is held closed by the greater supply pressure.

2. **Static Pressure.** With no flow through the unit, the check valves are held closed by force of the valve springs. The relief valve is held closed by the pressure difference of 5 to 11 psi between the supply and the intermediate (reduced-pressure) zone.

3. **Back Siphonage.** With a negative supply pressure, both the check valves are closed. As a result of pressure equalization (loss of pressure differential) between the supply and the intermediate zone, the relief valve opens allowing discharge of the water trapped in the intermediate zone. Thus, if any water seeps through the second check valve from the discharge side, it drains out through the relief valve.

4. **Back Siphonage and Back Pressure with Both Checks Fouled.** As in diagram 3, the negative supply pressure causes the relief valve to open. Now, assume the worst condition of both check valves held partly open with pieces of pipe scale or other material. The condition permits a negative pressure in the intermediate zone and allows potentially contaminated water to enter this zone. Since the relief valve is double seated, backflow entering through the second check valve is drained out, while the other half of the relief valve

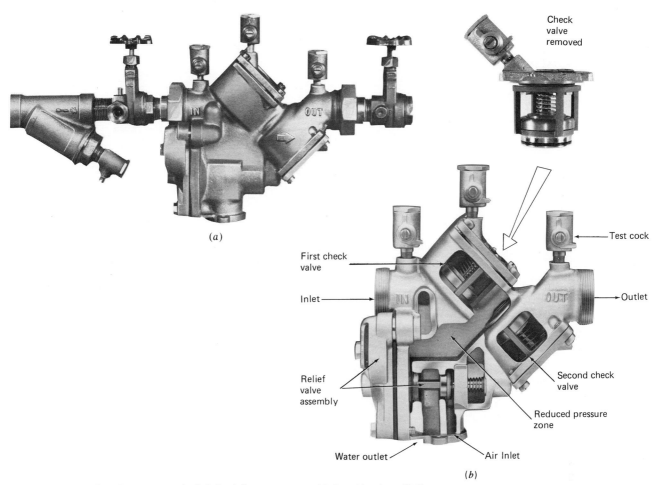

Figure 6-38 Reduced-pressure-principle backflow preventer. (*a*) Outside view. (*b*) Cutaway section showing the three interior valves. (*c*) Illustrations of the four modes of operation. (Courtesy of Watts Regulator Co.)

allows air to be drawn in by the negative supply pressure to enter the intermediate zone above the first check valve.

Reduced-pressure-principle backflow preventers are comparable in safety to air gap separation. In addition, they have the advantage of transmitting water pressure to the user's piping system. This allows operation of water fixtures with main pressure, including an automatic sprinkler system for fire protection. Common preventer installations are in service connections to industries, chemical plants, hospitals, mortuaries, and irrigation systems.

All mechanical backflow preventers must be inspected periodically to ensure proper operation of the interior valves. The procedure involves attaching a differential pressure gauge to selected test cocks and closing one of the isolating shutoff valves to apply backflow water pressure to the discharge side of an internal valve. Then, by release of the water pressure from the inlet side of the valve, tightness of the closure is checked by the ability of the valve to hold backflow pressure without leaking.

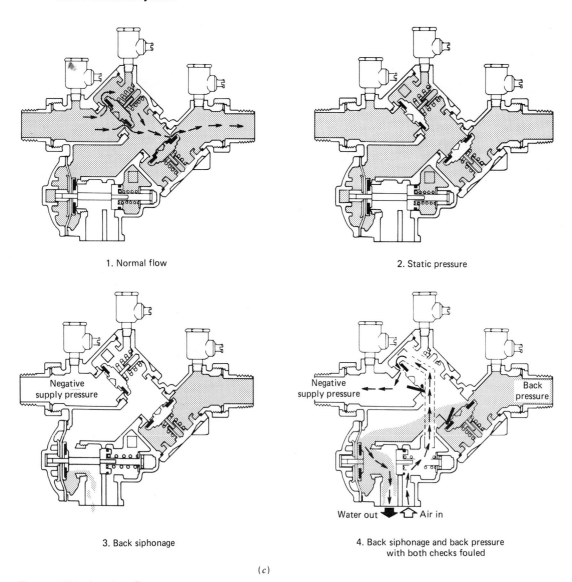

1. Normal flow

2. Static pressure

3. Back siphonage

4. Back siphonage and back pressure
with both checks fouled

(c)

Figure 6-38 (*continued*)

6-10 FIRE HYDRANTS

Hydrants provide access to underground water mains for the purposes of extinguishing fires, washing down streets, and flushing out water mains. Figure 6-39 shows the principal parts of a hydrant; the cast-iron barrel is fitted with outlets on top and a shutoff valve at the base is operated by a long valve stem that terminates above the barrel. A typical unit has two $2\frac{1}{2}$-in.-diameter hose nozzles and one $4\frac{1}{2}$-in. pumper outlet for a suction line. The barrel and valve stem are designed such that accidental breaking of the barrel at ground level will not unseat the valve, thus

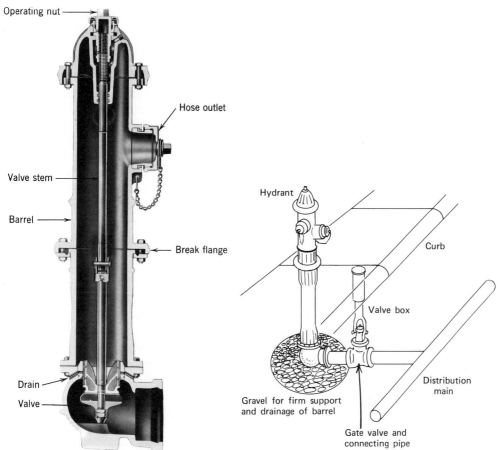

Operating nut

Hose outlet

Valve stem

Barrel

Break flange

Drain

Valve

Hydrant

Curb

Valve box

Distribution main

Gravel for firm support and drainage of barrel

Gate valve and connecting pipe

Figure 6-39 Cutaway section of a fire hydrant and typical installation adjacent to a street. (Reprinted by permission of Mueller Co.)

preventing water loss. Hydrants are installed along streets behind the curb line a sufficient distance, usually 2 ft, to avoid damage from overhanging vehicles. The pipe connecting a hydrant to a distribution main is normally not less than 6 in. (150 mm) in diameter and includes a gate valve allowing isolation of the hydrant for maintenance purposes. A firm gravel or broken-rock footing is necessary to prevent settling and to permit drainage of water from the barrel after hydrant use. In cold climates the water remaining in a hydrant can freeze and break the barrel. Where groundwater stands at levels above the hydrant drain, generally the drain is plugged at the time of

installation and, for service in cold climates, is pumped out after use. The Insurance Services Office recommends a minimum fire hydrant spacing based on required fire flow (Table 6-5).

6-11 DESIGN LAYOUTS OF DISTRIBUTION SYSTEMS

The arrangement of a water system is dictated by the source of water supply, topography of the distribution area, and variations in water consumption. If water is supplied at only one point in a pipe network, elevated storage or ground-level storage with booster pumping is required in remote areas to maintain residual pressures. Also, the arterial mains between supply and storage must be large enough to transmit water without excessive head loss. On the other hand, if water enters a distribution network at several points from individual wells, the storage capacity can be reduced and the pipe sizes can be smaller because the water supply does not have to travel as far before consumption. Topography can be the major factor in system design. Consider the two extremes, one where the water is supplied at a high elevation and flows by gravity through the pipe network and the other where the supply enters at a low elevation and must be pumped up into the pipe network and storage tanks. Many distribution systems have independent pipe networks at different elevations for pressure control; these may be completely separated or connected by pipes with pressure-reducing valves or through a common reservoir that acts as storage on the lower network and pumping suction for the upper system. The pattern for water consumption is directly a function of industrial, commercial, and residential demands. For municipalities, land-use planning and zoning are commonly employed to control variations of water consumption in the distribution network. Climate has a definite effect, since lawn watering creates a major demand in residential areas in semiarid regions.

Each water distribution system is unique and is influenced by local conditions. Since pumping capacity, sizes of network pipes, and volume of storage are all interrelated, increase in one of these can compensate for deficiencies in the others. Although founded on economic considerations, the options for expansion or modification of a system are governed to a considerable extent by the arrangement of existing facilities. Furthermore, the costs of construction and operation vary with time. For example, as a result of increasing energy costs, larger pipes to reduce pumping pressures and bigger storage reservoirs to equalize pumping rates are desirable options. In all cases, however, the primary engineering objectives are to provide a stable hydraulic gradient pattern for maintaining adequate pressure throughout the service area and sufficient

pumping and storage capacities to meet fire demands and emergencies, such as main breaks and power failures.

The following simplified distribution systems illustrate the basic principles of design. In Figure 6-40a, an arterial pipe network for a small system is shown with a high-service pumping station and an elevated storage tank. The ground-level storage at the pumping station receives water after treatment or directly from wells. By providing larger mains between the pumping station and the elevated tank, an adequate quantity of stored water can be maintained to supply peak demands in the area of the system around the elevated tank. The hydraulic gradient over the system, produced by the high-service pumps, is supported by the water level in the elevated storage tank. In Figure 6-40b, wells distributed throughout the pipe network pump water directly into the system at several locations, allowing the installation of smaller-diameter pipes. The primary functions of the storage tank are to stabilize pressures and equalize pumping rates. Wells are operated as needed to provide water, and standby power is installed at a sufficient number of locations to meet demand in emergencies. Some wells are used only during part of the year, for example, during a dry summer season when consumption reaches a maximum. The hydraulic gradient is supported by the operating wells and the level in the elevated storage tank.

6-12 EVALUATION OF DISTRIBUTION SYSTEMS

Quantity

The supply source plus storage facilities should be capable of yielding enough water to meet both the current daily demands and the anticipated consumption ten years hence. To quantitatively evaluate water usage, a record of average daily, peak daily, and peak hourly rates of consumption for the past ten years should be recorded. Projected future needs can be estimated from these values and other factors relating to community growth.

The minimum amount of water available from a supply source should always be sufficient to insure uninterrupted service. Consideration must be given to the probability of

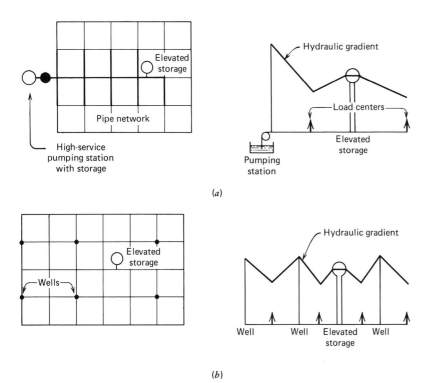

Figure 6-40 Simplified layouts of distribution systems. (*a*) The high-service pumping and elevated storage tank support the hydraulic gradient over the distribution system. (*b*) Wells located throughout the pipe network support the hydraulic gradient.

a succession of drought years equal to the previous worst drought experience, and the possibility of lowering groundwater levels. For surface supplies, the tributary watershed should be able to yield the estimated maximum daily demand ten years in the future. As a general rule, the storage capacity of an impounding reservoir should be equal to at least 30 days maximum daily demand five years into the future. Ideally, for well supplies there should be no mining of water, that is, neither the static groundwater level nor the specific capacity of the wells (gpm/ft of drawdown), should decrease appreciably as demand increases. Preferably these values should be constant over a period of five years except for minor variations that correct themselves within one week.

Intake Capacity

A surface-water intake must be large enough to deliver sufficient water to meet municipal consumption and treatment plant needs (e.g., filter backwashing) during any day of peak demand. With respect to fire flow requirements, if no storage is available the intake capacity must be large enough to meet fire demand, maximum hourly flow, and inplant process needs simultaneously. On the other hand, where the quantity of distribution storage is sufficient to meet all fire flow requirements, the intake capacity involves comparing total storage of the system to the maximum amount of water needed, for both present and future demands. Intake facilities should be sized to meet maximum needs projected for at least five years into the future.

A water intake system must be reliable; it must be located, protected, or duplicated such that no interruption of service to customers or to fire protection occurs by reason of floods, ice, or other weather conditions, or for reasons of breakdown, equipment repair, or power failure. In other words, intake facilities should be so reliable that no conceivable interruption in any part of the facility could cause curtailment of water supply service to a community.

Pumping Capacity

In a typical surface-water supply system, low-lift pumps draw water from the source and transport it to the treatment plant. After processing, high-lift pumps deliver the water from clear-well storage to the distribution system. In the case of a groundwater supply, the well pumps deliver the raw water to treatment. However, if processing is not needed, the wells may discharge water directly to the transmission mains. In large communities, or in areas with widely varying elevations in topography, booster pumping stations may be needed to increase pressure in the distribution network and to extend the system for greater distances from the main pumping station.

With due consideration for the amount of water storage available, a pumping system must have sufficient capacity to provide the amount of water at pressures and flow rates needed to meet both daily and hourly peak demand with required fire flow. Pumping facilities must also be able to meet demands, taking into account common system failures and maintenance requirements. For example, the specified maximum pumping capacity should be obtainable with the two largest pumps out of service. In many instances system storage, either elevated or ground level with booster pumping, is a component part of the pumping system to meet peak hourly demands.

Pumping facilities should be sufficiently reliable through duplication of units, standby equipment, and alternate sources of power such that no interruptions of service occur for any reason. In the event of power outage, standby power sources must be capable of sufficiently quick response to prevent exhaustion of water available in distribution storage during peak hourly demands including required fire flow. Where unattended automatic booster stations exist, the control system should report back to a central station both the condition of operation and any departure from normal.

Piping Network

Arterial and secondary feeder mains should be designed to supply water service for 40 or more years after installation. Actual useful service life of mains under normal conditions is 50 to 100 years. Submains should be at least 6 in. in diameter in residential districts, and the minimum size in important districts should be 8 in. in diameter with intersecting mains 12 in. Distribution lines are laid out in a gridiron pattern, avoiding dead ends by proper looping. An adequate number of valves should be inserted to permit shutoff in case of break so that no more than one block is out of service. The distribution of hydrants is based on Insurance Services standards.

REFERENCES

1. *Grading Schedule for Municipal Fire Protection*, Insurance Services Office, 160 Water Street, New York, N.Y. 10038, 1974; *Guide for Determination of Required Fire Flow*, Insurance Services Office, December 1974.
2. *Grading Schedule for Municipal Fire Protection.*
3. *Guide for Determination of Required Fire Flow.*

PROBLEMS

6-1   In what region of the United States does the average residential customer consume the greatest amount of water? Why?

6-2   A city with a population of 130,000 has an average consumption of 12 mgd during December, January, and February. The maximum daily use is 42 mgd, which usually occurs in July. Compute the water consumptions per capita and compare these to the data given in Table 6-1. (*Answers* 92 gpcd, 320 gpcd)

6-3 Calculate the fire flow and duration required for a 200-ft by 400-ft five-story building. The structure is of fire-resistive construction with all interior vertical openings protected to preclude spread of flames, but without automatic sprinkler protection. (*Answer* 5250 gpm, 5 hr)

6-4 Calculate the required fire flow and duration for a fire-resistive building that has four floors each with 45,000 sq ft. The vertical openings are protected, and the entire building is equipped with automatic sprinklers.

6-5 Compute the required fire flow and duration for a five-story apartment building with 280 m^2 per floor constructed of ordinary materials with vertical openings protected. (*Answers* 110 l/s, 2 h)

6-6 Calculate the required fire flow and duration for a two-story, wood-frame building with an area of 400 m^2 on each floor and protected vertical openings.

6-7 What are the fire flow requirements for a residential area of single-story houses of wood-frame construction, asphalt roofing, and 2000 sq ft floor area? The minimum distance between dwellings is 20 ft. What is the required standard hydrant distribution (spacing) in this residential area?

6-8 If hose streams for projecting water onto a fire require a minimum of 75 psi, why is the minimum residual hydrant pressure required for fire flow only 20 psi?

6-9 Describe the procedure for drilling and placing the casing and screen in a water well.

6-10 What kind of water intake structure is recommended (a) for a reservoir and (b) along a river shore?

6-11 Why is a gridiron pipe network the best arrangement for distributing water?

6-12 List the major components of a house water service connection.

6-13 An 8-in. ductile-iron water main is buried 14 ft below the ground surface. Based on Figure 6-11, what classification of pipe is recommended and what is the maximum allowable working pressure? (*Answers* Thickness Class 2, 350 psi)

6-14 What are the common joints used to connect ductile iron pipe in water distribution systems?

6-15 Describe the common jointing methods for asbestos cement and plastic pipe.

6-16 Discharge pressure and flow measurements were recorded under a variety of pumping conditions at a city's high-lift station in order to gather data to draw pump head-discharge curves. The following is a summary of the results:

Number of Pumps Operating	Discharge Head (ft)	Pumping Rate (mgd)
3	210	3.0
3	175	4.6
3	125	6.0
4	200	5.0
4	150	6.8
5	220	6.0
5	200	7.2
5	150	9.0

(a) Plot these data as three pump head-discharge curves on graph paper, as discussed in Section 4-4. (b) Assume these curves are for the pumping station in Figure 6-23. Plot a system head-discharge curve for the conditions shown in Figure 6-23 on the same graph, as discussed in Section 4-5. (c) At maximum daily flow (6.0 mgd), how many pumps are operating and what is the discharge pressure? (d) Assume the station is operating with four pumps at a discharge head of 175 ft and 6.0 mgd. What happens to the discharge flow and pressure conditions if a fifth pump is turned on? If one of the four pumps is turned off?

6-17   Discharge pressure and flow measurements were recorded at a city's high-lift station under a variety of pumping conditions at different system demands and varying water levels in the elevated storage tank. Based on these measurements, the system head-discharge curves were defined as follows:

Consumption	System Discharge (1/s)	System Heads at Different Water Levels in Elevated Storage		
		Low (m)	Mid-Height (m)	Full (m)
(Static pressure)	0	30	35	40
Winter average	140	37	42	47
Annual average	180	40	45	50
Summer average	225	45	50	55
Maximum day	270	50	55	60
Maximum day and fire flow	350	60	65	70

The pumping station has a total of six constant-speed pumps. Numbers 1, 2, and 3 are identical with a best efficiency point (bep) at 100 l/s and 45 m; 60 percent bep is 60 l/s and 60 m; and, 120 percent bep is 120 l/s and 30 m. Numbers 4, 5, and 6 are identical with a bep at 150 l/s and 60 m; 60 percent bep is 90 l/s and 80 m; and 120 percent bep is 180 l/s and 35 m. Draw the pump head-discharge curves for the following combinations of pumps operating in parallel: pump 1 alone; pumps 1 and 2; pumps 1, 2, and 3; pump 4 alone; pumps 1 and 4; pumps 4 and 5; and pumps 4, 5, and 6. Draw the system head-discharge curves on the same graph.

What pumps would normally be operated during days when the consumption is at (a) winter average, (b) annual average, (c) summer average, and (d) the maximum day? Assume that the volume of the storage tank is adequate to equalize the pumping rate on the maximum day with reserve storage for fire demand. Is the pumping capacity of the station adequate?

6-18   From the results of Example 6-3, the storage volume needed to equalize the 24-hr pumping rate on the day of maximum water usage is 500,000 gal. What percentage of the total consumption for that day is this required storage capacity? How does this value compare to the rule-of-thumb percentages given in Section 6-7 for balancing pumping?

6-19   The water usage on the day of maximum consumption is as follows:

Time	Hourly Rate (gpm)	Time	Hourly Rate (gpm)
12 P.M.	—	1 P.M.	6600
1 A.M.	2200	2	6400
2	2100	3	6300
3	1800	4	6400
4	1400	5	6400
5	1300	6	6700
6	1200	7	7400
7	2000	8	9200
8	3500	9	8400
9	5000	10	5000
10	6000	11	3200
11	6400	12	2800
12	7000		

Plot a consumption-time curve like that shown in Figure 6-20. Calculate hourly cumulative consumption values and plot a mass diagram as illustrated in Figure 6-21. What is the constant 24-hr pumping rate and required storage capacity to equalize demand over the 24-hr period? Calculate the required storage capacity as a percentage of the total cumulative consumption.

6-20   The two principal functions of water storage in a distribution system are to equalize pumping rates and provide a reserve supply for fire fighting. In analysis of a municipality for fire protection, what amount of the storage capacity is considered available for fire fighting?

6-21   The water system for a town with a population of 900 has four wells that have yields of 400 gpm, 400 gpm, 600 gpm, and 800 gpm and an elevated storage tank with a capacity of 100,000 gal. The annual average water consumption is 120,000 gpd, and the maximum daily usage is 280,000 gal. The required fire flow for the commercial district is 1500 gpm with a duration of 2 hr. (a) Assuming this town has no large industrial water users, do the average and maximum daily consumptions appear to be reasonable? (b) Assume the piping in the distribution network is adequate to convey water from the wells and elevated storage to the commercial district under conditions of peak demand without excess pressure losses. Are the well pumps in combination with distribution storage able to meet the specified fire demand with no insurance deficiency?

6-22   The internal valve of a fire hydrant was pulled from its seat when the hydrant was struck by a truck. After water gushed from the broken barrel for several hours, the flow from the hydrant was stopped by closing shutoff valves in the piping network, which also resulted in several hundred residents being without water while the hydrant was replaced. What are the deficiencies of this distribution system?

6-23   Refer to the pressure-reducing valve in Figure 6-27. Explain the hydraulic action of the valve that results in reduction of water pressure.

6-24   Refer to the pressure-reducing valve in Figure 6-29. Assume that this automatic valve is installed in a pipeline and is adjusted to reduce pressure from 90 psi to 70 psi. In what direction should the handwheel be turned to de-crease the outlet pressure to 60 psi? Explain the resulting change in operation of the valve.

6-25   The pilot valve on top of the double-acting altitude valve illustrated in Figure 6-30 is a three-way valve. With reference to both Figure 6-28 and Figure 6-30, explain how the pilot valve automatically opens and closes the globe valve to maintain the water level in the elevated storage tank below preset maximum elevation.

6-26   The operation of high-service pumps can be controlled by the water level in an elevated storage tank in the distribution system. As the water level in the tank lowers below predetermined elevations, additional high-service pumps can be turned on; conversely, as the water level rises above preset elevations, selected pumps can be turned off. How can this kind of control be achieved?

6-27   What kinds of backflow preventers are usually installed on lawn sprinkler and landscape irrigation systems?

6-28   A large house is being converted into a mortuary. What special provisions should be considered in modifying the current residential water service connection?

6-29   Explain why the backflow preventer illustrated in Figure 6-38 is referred to as a reduced-pressure-principle backflow preventer.

chapter 7
Water Processing

The objective of municipal water treatment is to provide a potable supply—one that is chemically and microbiologically safe for human consumption. For domestic uses, treated water must be aesthetically acceptable—free from apparent turbidity, color, odor, and objectionable taste. Quality requirements for industrial uses are frequently more stringent than for domestic supplies. Thus, additional treatment may be required by the industry. For an example, boiler feed water must be demineralized to prevent scale deposits.

Common water sources for municipal supplies are deep wells, shallow wells, rivers, natural lakes, and reservoirs. Well supplies normally yield cool, uncontaminated water of uniform quality that is easily processed for municipal use. Processing may be required to remove dissolved gases and undesirable minerals. The simplest treatment illustrated in Figure 7-1a is disinfection and fluoridation. Deep-well supplies are chlorinated to provide residual protection against potential contamination in the water distribution system. In the case of shallow wells recharged by surface waters, chlorine both disinfects the groundwater and provides residual protection. Fluoride is added to reduce the incidence of dental caries. Dissolved iron and manganese in well water oxidize when contacted with air, forming tiny rust particles that discolor the water. Removal is performed by oxidizing the iron and manganese with chlorine or potassium permanganate, and removing the precipitates by filtration (Figure 7-1b). Excessive hardness is commonly removed by precipita- tion softening, shown schematically in Figure 7-1c. Lime and, if necessary, soda ash are mixed with well water, and settleable precipitate is removed. Carbon dioxide is applied to stabilize the water prior to final filtration. Aeration is a common first step in treatment of most groundwaters to strip out dissolved gases and add oxygen.

Pollution and eutrophication are major concerns in surface-water supplies. Water quality depends on agricultural practices in the watershed, location of municipal and industrial outfall sewers, river development such as dams, season of the year, and climatic conditions. Periods of high rainfall flush silt and organic matter from cultivated fields and forest land, while drought flows may result in higher concentrations of wastewater pollutants from sewer discharges. River temperature may vary significantly between summer and winter. The quality of water in a lake or reservoir depends considerably on season of the year. Municipal water quality control actually starts with management of the river basin to protect the source of water supply. Highly polluted waters are both difficult and costly to treat. Although some communities are able to locate groundwater supplies, or alternate less polluted surface sources within feasible pumping distance, the majority of the nation's population draw from nearby surface supplies. The challenge in waterworks operation is to process these waters to a safe, potable product acceptable for domestic use.

The primary process in surface-water treatment is

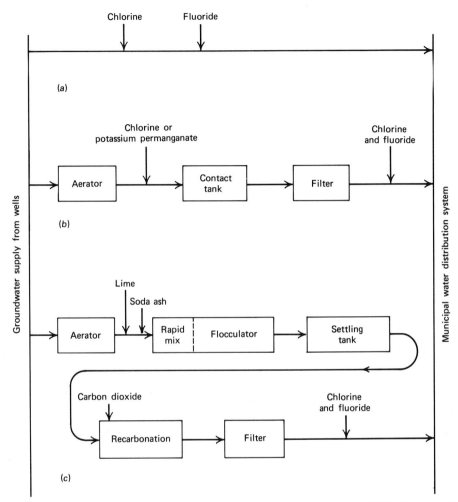

Figure 7-1 Flow diagrams of typical groundwater treatment systems. (*a*) Disinfection and fluoridation. (*b*) Iron and manganese removal. (*c*) Precipitation softening.

chemical clarification by coagulation, sedimentation, and filtration, as illustrated in Figure 7-2. Lake and reservoir water has a more uniform year-round quality and requires a lesser degree of treatment than river water. Natural purification results in reduction of turbidity, coliform bacteria, color, and elimination of day-to-day variations. On the other hand, growths of algae cause increased turbidity and may produce difficult-to-remove tastes and odors during the summer and fall. The specific chemicals applied in coagulation for turbidity removal depend on

the character of the water and economic considerations. The most popular coagulant is alum (aluminum sulfate). As a flocculation aid, the common auxiliary chemical is a synthetic polymer. Activated carbon is applied to remove taste- and odor-producing compounds. Chlorine and fluoride are post-treatment chemicals. Prechlorination may be used for disinfection of the raw water only if it does not result in formation of trihalomethanes. River supplies normally require the most extensive treatment facilities with greatest operational flexibility to handle the

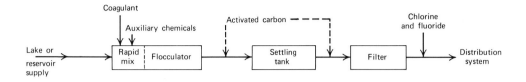

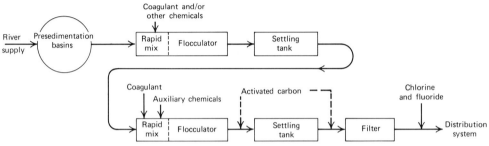

Figure 7-2   Schematic patterns of typical surface-water treatment systems.

day-to-day variations in raw water quality. The preliminary step is often presedimentation to reduce silt and settleable organic matter prior to chemical treatment. As illustrated in Figure 7-2, many river-water treatment plants have two stages of chemical coagulation and sedimentation to provide greater depth and flexibility of treatment. The units may be operated in series, or by split treatment with softening in one stage and coagulation in the other. As many as a dozen different chemicals may be used under varying operating conditions to provide a satisfactory finished water.

The two primary sources of waste from water treatment processes are sludge from the settling tank, resulting from chemical coagulation or softening reactions, and wash water from backwashing filters. These discharges are highly variable in composition, containing concentrated materials removed from the raw water and chemicals added in the treatment process. The wastes are produced continuously, but are discharged intermittently. Historically, the method of waste disposal was to discharge to a watercourse or lake without treatment. This practice was justified from the viewpoint that filter backwash waters and settled solids returned to the watercourse added no

new impurities, but merely returned material that had originally been present in the water. This argument is no longer considered valid, since water quality is degraded to the extent that a portion of the water is withdrawn, and chemicals used in processing introduce new pollutants. Therefore, more stringent federal and state pollution control regulations have been enacted requiring treatment of waste discharges from water purification and softening facilities.

The situation of each waterworks is unique and controls to some extent the method of ultimate disposal of plant wastes. For example, they may be piped to a municipal sewer for processing, or may be discharged to lagoons, provided that sufficient land area is available. Ultimate disposal by landfill, or barging to sea, requires thickening for economical handling and hauling. A variety of alternative processing methods are available; however, because of unique characteristics of each plant's waste, no specific process can be universally applied. Figure 7-3 is a typical system for dewatering alum sludge. Filter backwash water is discharged to a clarifier-surge tank. Overflow is recycled to the raw-water inlet of the treatment plant while settled solids are discharged, along

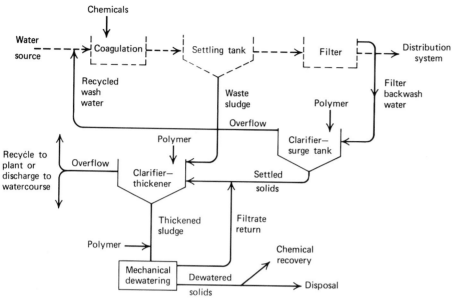

Figure 7-3 Sketch of a dewatering system for alum sludge from a water treatment plant.

with waste sludge from the settling tank, to a clarifier-thickener. Supernatant from this unit may be recycled to the head of the plant or may be discharged to a watercourse. Thickened sludge is mechanically dewatered, usually by centrifugation or filtration. Dewatered solids may be processed for recovery of chemicals or may be discharged to drying beds, to land burial, or by barging to sea.

7-1 MIXING AND FLOCCULATION

Chemical reactors in water treatment and biological aeration basins in wastewater processing are designed as either completely mixed or plug-flow basins. In an ideal completely mixed unit, the influent is immediately dispersed throughout the volume, and the concentration of reactant in the effluent is equal to that in the mixing liquid (Figure 7-4a). For steady-state conditions the first-order reaction kinetics is given by Eq. 7-1. Applications of complete mixing in water treatment are rapid (flash or quick) mix tanks used to blend chemicals into raw water

for coagulation, and the mixing and reaction zone in flocculator-clarifiers. In wastewater treatment, high-rate activated sludge and extended aeration basins are completely mixed.

$$t = \frac{V}{Q} = \frac{1}{k}\left(\frac{C_0}{C_t} - 1\right) \tag{7-1}$$

where t = detention time
V = volume of basin
Q = quantity of flow
k = rate constant for first-order kinetics
C_0 = influent reactant concentration
C_t = effluent reactant concentration

In an ideal plug-flow system, the water flows through a long chamber at a uniform rate without intermixing. The concentration of reactant decreases along the direction of flow, remaining within the imaginary plug of water moving through the basin (Figure 7-4b). For steady-state conditions the relationship between detention time and concentration, applying first-order kinetics, is given in Eq. 7-2. In practice, ideal plug flow is very difficult to

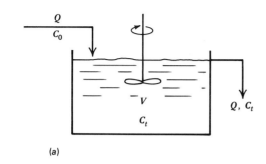

(a)

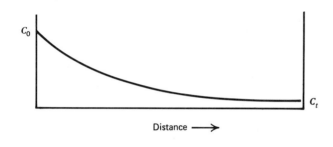

Distance ⟶

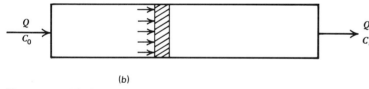

(b)

Figure 7-4 Ideal completely mixed and plug-flow tanks. Symbols are defined in Eq. 7-1. (a) Completely mixed. (b) Plug flow.

achieve because of short-circuiting and intermixing caused by frictional resistance along the walls, density currents, and turbulent flow. In wastewater treatment, flow through long narrow aeration tanks approaches plug flow. Flocculation basins in chemical water processing are designed as plug-flow units using baffles to reduce short-circuiting.

$$t = \frac{V}{Q} = \frac{L}{v} = \frac{1}{k}\left(\log_e \frac{C_0}{C_t}\right) \qquad (7\text{-}2)$$

where t, V, Q, C_0, and C_t are same as Eq. 7-1
   L = length of rectangular basin
   v = horizontal velocity of flow
   k = rate constant for first-order kinetics

The in-line mixing and flocculation unit shown in Figure 7-5 has a propeller to provide rapid mixing of chemicals with the raw water, and horizontal paddle units for gentle, slow mixing to build settleable floc. The redwood paddle flocculators (Figure 7-5c) are separated by baffles and have variable-speed drives so that the peripheral speed of the paddles can be set between 0.6 and 1.5 ft/sec for optimum flocculation.

The *Recommended Standards for Water Works, Great Lakes–Upper Mississippi River Board of State Sanitary Engineers* (*GLUMRB*)[1] recommend that quick mixing, for rapid dispersion of chemicals throughout the water, be performed with mechanical mixing devices with a detention time not less than 30 sec.

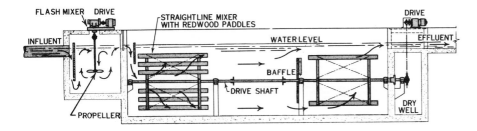

Figure 7-5 Straightline rapid mix and flocculation basin used in chemical treatment of water. (*a*) Longitudinal section of flash and horizontal slow mixers. (*b*) Propeller mixer provides rapid blending of chemicals into water. (*c*) Slow, horizontal-paddle mixer allows flocculation. (Courtesy of FMC Corp., Material Handling Systems Div.)

These recommendations are made for flocculation basins:

1. Inlet and outlet design shall prevent short-circuiting and destruction of floc.
2. Minimum flow-through velocity shall not be less than 0.5 nor greater than 1.5 ft/min (2.5 to 7.5 mm/s) with a detention time for floc formation of at least 30 min.
3. Agitators shall be driven by variable speed drives with the peripheral speed of paddles ranging from 0.5 to 2.5 ft/sec (0.15 to 0.75 m/s). Flocculation and sedimentation basins shall be as close together as possible. The velocity of flocculated water through conduits to settling basins shall not be less than 0.5 nor greater than 1.5 ft/sec (0.15 to 0.45 m/s). Allowances must be made to reduce turbulence at bends and changes in direction.

4. Baffling (rather than mechanical mixers) may be used to provide flocculation in small plants, but only after consultation with the reviewing authority.

EXAMPLE 7-1

Based on laboratory studies, the rate constant for a chemical coagulation reaction was found to be first-order kinetics with a k equal to 75 per day. Calculate the detention times required in completely mixed and plug-flow reactors for an 80 percent reduction, $C_0 = 200$ mg/l and $C_t = 40$ mg/l.

Solution

Using Eq. 7-1 for complete mixing,

$$t = \frac{\text{day}}{75} \times 1440 \frac{\text{min}}{\text{day}} \left(\frac{200 \text{ mg/l}}{40 \text{ mg/l}} - 1 \right) = 77 \text{ min}$$

For plug flow, substituting into Eq. 7-2,

$$t = \frac{1440}{75}\left(\log_e \frac{200}{40}\right) = 19.2(2.3 \log 5)$$

$$= 19.2 \times 2.3 \times 0.7 = 31 \text{ min}$$

7-2   SEDIMENTATION

Sedimentation, or clarification, is the removal of particulate matter, chemical floc, and precipitates from suspension through gravity settling. The common criteria for sizing settling basins are detention time, overflow rate, weir loading and, with rectangular tanks, horizontal velocity. Detention time, expressed in hours, is calculated by dividing the basin volume by average daily flow, Eq. 7-3.

$$t = \frac{V \times 24}{Q} \tag{7-3}$$

where t = detention time, hours
      V = basin volume, million gallons (cubic meters)
      Q = average daily flow, million gallons per day (cubic meters per day)
     24 = number of hours per day

The overflow rate (surface loading) is equal to the average daily flow divided by total surface area of the settling basin, expressed in units of gallons per day per square foot, Eq. 7-4.

$$V_0 = \frac{Q}{A} \tag{7-4}$$

where V_0 = overflow rate (surface loading), gallons per day per square foot (cubic meters per square meter per day)
      Q = average daily flow, gallons per day (cubic meters per day)
      A = total surface of basin, square feet (square meters)

Most settling basins in water treatment are essentially up-flow clarifiers where the water rises vertically for discharge through effluent channels; hence, the ideal basin shown in Figure 7-6 can be used for explanatory purposes. Water entering a settling basin is forced to the bottom

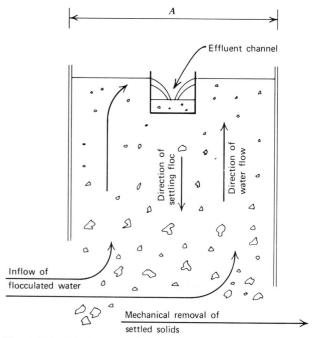

Figure 7-6   Ideal sedimentation tank. Symbols are defined in Eq. 7-4.

behind a baffle wall, and then rises vertically, overflowing the weir of a discharge channel at the tank surface. Flocculated particles settle downward, in a direction opposite to the flow of water, and are removed from the bottom by a continuous mechanical sludge removal apparatus. The particles with a settling velocity v greater than the overflow rate Q/A are removed while lighter flocs, with settling velocities less than the overflow rate, are carried out in the basin effluent.

Weir loading is computed by dividing the average daily quantity of flow by the total effluent weir length, and expressing the results in gallons per day per foot (cubic meters per meter per day).

Sedimentation basins may be rectangular, circular, or square. They are designed for slow uniform water movement with a minimum of short-circuiting. The rectangular tank in Figure 7-7a contains partitioning baffles to guide the flow vertically to collecting troughs that extend across, and around the periphery, of the clarifier. Scrapers

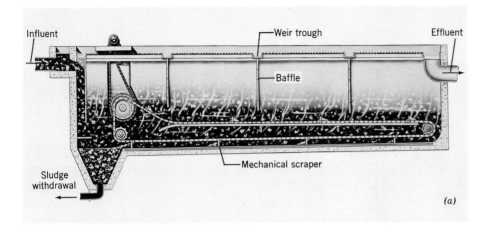

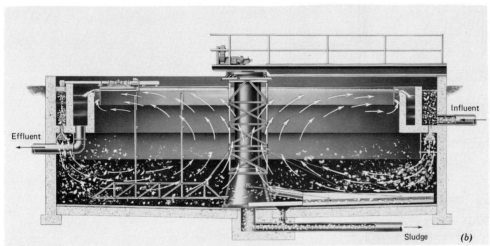

Figure 7-7 Cross-sectional diagrams of rectangular and circular sedimentation tanks used in water treatment. (*a*) Rectangular, baffled clarifier with a conventional plow-type scraper mechanism. (*b*) Circular rim-flow clarifier with a rotating, hydraulic sludge pickup arm. (Courtesy of Envirex Inc., a Rexnord Company.)

drawn by endless chains slowly move the settled solids to a sludge hopper at the inlet end. Figure 7-7*b* illustrates a circular clarifier where the influent enters and the effluent leaves from channels around the periphery of the tank. The sludge removal arm, which revolves slowly around a center shaft, is a tapered tube with openings spaced along the length. Instead of a scraping action, the settled solids are picked up by a slight suction action and are conveyed through the arm to be discharged from the bottom of the tank at the center. Figure 7-8 illustrates an in-line flash

mixer, a flocculator with paddle units set perpendicular to the direction of flow, and a square center-feed clarifier. Water enters the clarifier through a center column and is directed downward by an influent well (circular baffle). The flow radiates out from the center and overflows a peripheral weir. Settled solids are scraped to a hopper for discharge by a center-driven rake arm equipped with corner blades.

Presedimentation basins for settling muddy river water are generally circular with hopper bottoms, equipped

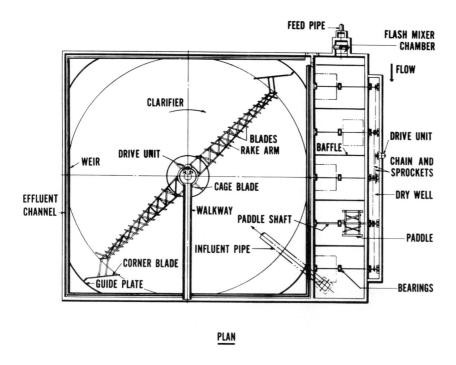

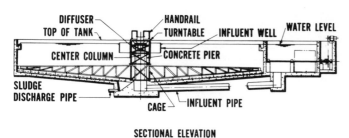

PLAN

SECTIONAL ELEVATION

Figure 7-8 Flocculator and square sedimentation tank for water clarification. (Courtesy of Dorr-Oliver, Inc.)

with scraper-type sludge removal arms. *GLUMRB Standards* recommend a detention time of not less than 3 hr. For basins following chemical flocculation, the *Standards* recommend the following: minimum detention time of 4 hr, maximum horizontal velocity through the settling basin of 0.5 ft/min (2.5 mm/s), and maximum weir loading of 20,000 gpd/ft (250 m^3/m·d) of weir length. Overflow rates generally fall in the range of 500 to 800 gpd/sq ft (20 to 33 m^3/m^2·d).

The actual detention times of individual particles of water in a sedimentation tank vary with the physical features of the tank. As shown by the flow arrows in Figure 7-7, flow routes through the sedimentation basins differ, resulting in nonideal conditions. In an ideal plug-flow tank, all of the particles remain in the tank for the same length of time as defined by Eq. 7-3. The hydraulic character of a clarifier is defined by the residence time distribution of individual particles of the water flow-

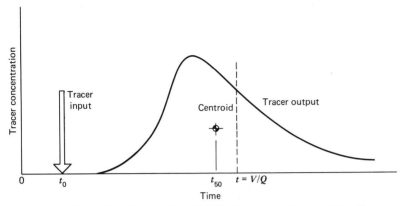

Figure 7-9 General residence time-distribution curve for particles of water passing through a sedimentation tank determined by measuring the concentration of a tracer in the effluent resulting from a pulse tracer input.

ing through the tank. Evaluation of an actual tank is performed by introducing a pulse of dye tracer in the influent during steady-state flow and measuring the concentration in the effluent over an extended period of time. As sketched in Figure 7-9, the tracer output curve of dye concentration in the effluent rises sharply and decreases slowly. The mean residence time t_{50} (time to the centroid) is to the left of the ideal detention time t as a result of flow dispersion with backmixing and short-circuiting of flow past the water held in stagnant pockets. The mean residence time, calculated from experimental data, equals the summation of time-concentration values divided by the summation of the concentrations.

EXAMPLE 7-2

The in-line system shown in Figure 7-8 has the following sized units: flash mixing chamber with a volume of 1500 gal; a flocculator 15 ft wide, 70 ft long, and 8.0 ft liquid depth; and a settling tank 75 ft square, 12 ft liquid depth, and 300 ft of effluent weir. Calculate the major parameters used in design of these units based on a water flow of 3.0 mgd.

Solution

$$Flow = 3.0 \text{ mgd} = 2080 \text{ gpm} = 401{,}000 \text{ cu ft/day}$$
$$= 278 \text{ cu ft/min}$$

Detention time in flash mixer

$$t = \frac{1500 \text{ gal}}{2080 \text{ gal/min}} \times 60 \frac{\text{sec}}{\text{min}}$$
$$= 43 \text{ sec} \quad (>30 \text{ sec } GLUMRB)$$

Detention time and horizontal velocity in flocculator

$$t = \frac{15 \text{ ft} \times 70 \text{ ft} \times 8.0 \text{ ft}}{278 \text{ cu ft/min}}$$
$$= 30 \text{ min} \quad (=30 \text{ min } GLUMRB)$$
$$v = \frac{Q}{A} = \frac{278 \text{ cu ft/min}}{15 \text{ ft} \times 8.0 \text{ ft}}$$
$$= 2.3 \text{ ft/min} \quad (>1.5 \text{ ft/min})$$

Settling time, weir loading, and overflow rate in clarifier

$$t = \frac{75 \text{ ft} \times 75 \text{ ft} \times 12 \text{ ft}}{278 \text{ cu ft/min}} \times \frac{\text{hr}}{60 \text{ min}}$$
$$= 4.0 \text{ hr} \quad (=4.0 \text{ hr } GLUMRB)$$
$$\text{Weir loading} = \frac{3{,}000{,}000 \text{ gal/day}}{300 \text{ ft}}$$
$$= 10{,}000 \text{ gpd/ft} \quad (<20{,}000 \text{ gpd/ft})$$
$$V_0 = \frac{3{,}000{,}000 \text{ gal/day}}{75 \text{ ft} \times 75 \text{ ft}} = 530 \text{ gpd/sq ft}$$

7-3 FLOCCULATOR-CLARIFIERS

Flocculator-clarifiers, also referred to as solids contact units or up-flow tanks, combine the processes of mixing, flocculation, and sedimentation in a single compartmented tank. One such unit, shown in Figure 7-10, introduces coagulants, or softening chemicals, in the influent pipe and mixes the water under a central cone-shaped skirt where a high-floc concentration is maintained. Flow passing under the hood is directed through the sludge blanket at the bottom of the tank, to promote growth of larger conglomerated clusters, where the heavier particles have settled. Overflow rises upward in the peripheral settling zone to radial weir troughs, or circular inboard troughs suspended at the surface. These units are particularly advantageous in lime softening of groundwater, since the precipitated solids help seed the floc, growing larger crystals of precipitate to provide a thicker waste sludge. Recently, flocculator-clarifiers are receiving wider application in the chemical treatment of industrial wastewaters amd surface-water supplies. The major advantages promoting their use are reduced space

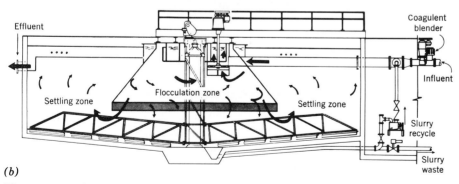

Figure 7-10 Flocculator-clarifier provides rapid mixing, flocculation, and sedimentation in a single compartmented tank. (*a*) Photograph of a flocculator-clarifier in a water softening plant. (Metropolitan Utilities District, Omaha, Nebraska.) (*b*) Cross-sectional view of unit pictured above. (Courtesy of Walker Process Corp.)

requirements and less costly installation. However, the unitized nature of construction generally results in a sacrifice of operating flexibility.

The *GLUMRB Standards* stress that solids contact units are most acceptable for clarification in combination with softening for the treatment of waters with relatively uniform characteristics and flow rates. They recommend the following in sizing units: flocculation and mixing time of not less than 30 min; minimum sedimentation time of 2 hr for turbidity removal in treatment of surface waters and 1 hr for precipitation in lime-soda ash softening of groundwaters; weir loadings not exceeding 10 gpm/ft (180 m³/m·d) for turbidity removal units and 20 gpm/ft (360 m³/m·d) for softening units; and upflow rates not to exceed 1.0 gpm/sq ft (60 m³/m²·d) and 1.75 gpm/sq ft (100 m³/m²·d), respectively. The volume of sludge removed from these units should not exceed 5 percent of the water treated for turbidity removal or 3 percent for softening.

7-4 FILTRATION

The granular-media gravity filter is the most common type used in water treatment to remove nonsettleable floc remaining after chemical coagulation and sedimentation. A typical filter bed (Figure 7-11), is placed in a concrete box with a depth of about 9 ft. The granular media, about 2 ft deep, are supported by a graded gravel layer over underdrains. During filtration, water passes downward through the filter bed by a combination of water pressure from above and suction from the bottom. Filters are cleaned by backwashing (reversing the flow) upward through the bed. Wash troughs suspended above the filter surface collect the backwash water and carry it out of the filter box.

The common flow scheme for municipal water processing incorporates flocculation with a chemical coagulant and sedimentation prior to filtration (Figure 7-2). Instead of separate in-line tanks, flocculator-clarifiers can be used for mixing and sedimentation, particularly in treatment of groundwaters. Filtration rates following flocculation and sedimentation are in the range of 2 to 10 gpm/sq ft (1.4 to 6.8 l/m²·s), with 5 gpm/sq ft (3.4 l/m²·s) normally the maximum design value.

Direct filtration, which does not include sedimentation, can be used to treat surface waters with low turbidity and color. The typical flow scheme is rapid mixing for blending a coagulant into the raw water, flocculation for greater than 30 min, addition of a polymer coagulant aid, and filtration at a rate of 1 to 6 gpm/sq ft (0.7 to 4.1 l/m²·s). Floc removed from the water is collected and stored in the filter, which is subsequently cleaned by backwashing. In general, the turbidity of the raw water should be less than 5 units and the color below 40 units. A turbidity consistently over 15 units is likely to cause operational problems in direct filtration that could have been avoided by providing prior sedimentation. The success of direct filtration is based on use of a coarse-to-fine dual-media filter for a greater capacity to store the impurities removed, backwashing systems using mechanical or air agitation to clean the media, and selection of a suitable polymer coagulant aid.

Flow Control

Traditional gravity systems regulate the rate of filtration by controlling the rate of discharge from the filter underdrain. Influent entering each filter box is controlled automatically, or manually, to equal outflow so that a constant water level is maintained above the filter bed.

The following description of filter operation follows the valve numbering in Figure 7-12. Initially valves 1 and 4

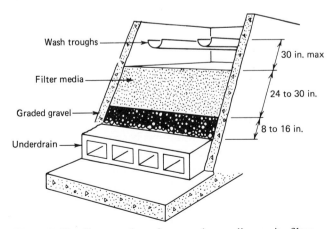

Wash troughs

Filter media

Graded gravel

Underdrain

30 in. max

24 to 30 in.

8 to 16 in.

Figure 7-11   Cross section of a granular-media gravity filter.

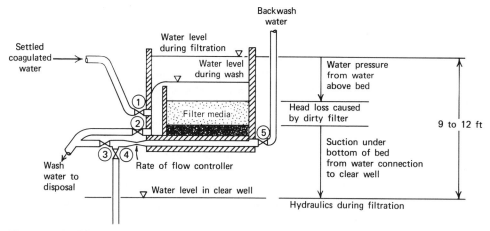

Figure 7-12 Diagrammatic sketch showing operation of a gravity filter with a rate of flow controller regulating the rate of filtration.

are opened, and 2, 3, and 5 are closed for filtration. Overflow from the settling basin applied to the filter passes through the bed and underdrain system to the clear well underneath. The depth of water above the filter surface is between 3 and 4 ft. The underdrain pipe is trapped in the clear well to provide a liquid connection to the water being filtered, thus preventing backflow of air into the underdrain. The maximum head available for filtration is equal to the difference between the elevation of the water surface above the filter and the level in the clear well; this is commonly 9 to 12 ft. When the filter is clean, flow through the bed must be regulated to prevent an excessive filtration rate. A rate-of-flow controller, consisting of a valve controlled by a venturi meter, throttles flow in the discharge pipe restraining the rate of filtration. As the filter collects impurities, the resistance to flow increases and the flow controller valve opens wider to maintain a preset rate.

A filter is cleaned by backwashing when the measured head loss through the media is approximately 8 ft. Valves 1 and 4 are closed (3 remaining closed) and 2 is opened. This valving arrangement drains away the water from above the filter bed down to the elevation of the outlet-channel wall. If air agitation is used, compressed air is introduced through the underdrain to loosen accumulated impurities from the grains of media by rubbing them against each other in the turbulence caused by the

air mixing. If rotary washer units are employed, they are placed in operation prior to starting the water backwash. By opening valve 5, clean water flows into the filter underdrain and passes up through the bed hydraulically expanding the stationary bed depth about 50 percent. Dirty wash water is collected by troughs and conveyed to disposal or for treatment to remove the impurities before recycling to the plant influent for reuse (Figure 7-3). The first few minutes of filtered water at the beginning of the next run may be wasted to flush the wash water remaining in the bed out through the drain. This is accomplished by opening valve 3 when valve 1 is opened to start filtration (valves 2, 4, and 5 are shut). Opening valve 4 while closing 3 permits filtration to proceed again.

Filter Media

The action that takes place in a granular-media filter is extremely complex, consisting of straining, flocculation, and sedimentation. Gravity filters do not function properly unless the applied water has been chemically treated and, if necessary, settled to remove the large floc. Coagulant carryover is essential in removing microscopic particulate matter that would otherwise pass through the bed. If an excessive quantity of large floc overflows the settling basin, a heavy mat forms on the filter surface by straining action and clogs the bed. However, the impurities in improperly coagulated water may penetrate too far into the bed and can be flushed through before being trapped, causing a turbid effluent. Optimum filtration occurs when nonsettleable coagulated floc are held in the pores of the bed and produce "in-depth" filtration.

The ideal filter media possesses the following characteristics: coarse enough for large pore openings to retain large quantities of floc yet sufficiently fine to prevent passage of suspended solids; adequate depth to allow relatively long filter runs; and graded to permit effective cleaning during backwash. The earliest filter medium used was a fine sand of nearly uniform size. After backwashing, a sand bed becomes hydraulically graded with the largest grains on the bottom and the finest on the top. This forms a bed where most of the removal takes place at or near the surface, which is undesirable. The earliest modification to this single-medium filter was to increase the sand size to allow longer filter runs and greater depth of penetration while reducing the undesirable effect of surface straining. Dual-media beds of coarser anthracite overlying the sand filter provide an upper layer of increased porosity to reduce surface plugging. Another problem of early sand filters was poor cleaning during hydraulic backwashing.

Mud balls and other heavy impurities, formed during filtration of certain waters, would drop down to the gravel underlayer when the bed was expanded rather than being washed out in the overflow. Higher backwash rates in an attempt to resolve this problem often resulted in disturbance of the graded gravel underdrain, or in carryover of sand in the wash water. Several devices were developed to improve backwashing by increasing the scrubbing action in the expanded bed. Most current installations provide some kind of washing aid like a mechanical agitator or air scrubbing before water backwash. A mechanical surface agitator consists of a rotating arm, placed just above the sand and beneath the wash troughs, with nozzles directing water jets at right angles to the bed surface. After the sand layer has been expanded by introducing backwash, water under pressure is introduced to the agitator, which rotates, increasing turbulence for better washing action. An alternate system consists of fixed nozzles that spray water into an expanded bed.

Granular media are specified by effective size and uniformity coefficient. Effective size is the 10 percentile diameter, which means that 10 percent of the filter grains by weight are smaller than this diameter. Uniformity

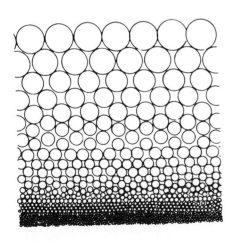

(a)

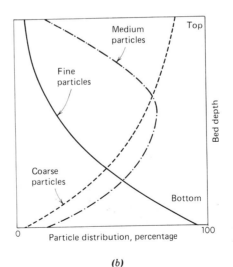

(b)

Figure 7-13 A mixed-media filter using coal, sand, and garnet approaches the ideal distribution for a filter bed. (a) Ideal filter bed with decreasing particle size from top to bottom. (b) Particle size distribution in a mixed-media filter. (Courtesy of Neptune Microfloc Inc.)

coefficient is the ratio of the 60 percentile diameter to the 10 percentile diameter. Common ranges in effective size and uniformity coefficient for single-medium sand filters are 0.40 to 0.55 mm and 1.3 to 1.7, respectively. The depth of the sand layer is 24 to 30 in. (610 to 760 mm).

A coal-sand dual-media filter permits use of a relatively coarse anthracite medium with a specific gravity of 1.4 to 1.6 over a finer sand layer with a specific gravity of 2.6. The effective size and uniformity coefficient of the coal medium are 0.9 to 1.1 mm and less than 1.7, respectively. Both media layers are generally 12 in. (300 mm) in thickness. The upper layer of coarser anthracite has voids about 20 percent larger than the sand, and thus a coarse-to-fine grading of media is provided in the direction of flow. After backwashing, the bed stratifies with the heavier sand on the bottom and the lighter, coarser coal medium on top. Larger floc particles are adsorbed and trapped in the surface coal layer, while finer material is held in the sand filter; therefore, the bed filters in greater depth, preventing premature surface plugging.

Mixed-media beds using coal, sand, and garnet very closely approach the ideal filter (Figure 7-13). Garnet, the finest medium, has a specific gravity of about 4.2, which is greater than that of the sand or coal. The three media are sized so that intermixing of these materials occurs after backwashing with no discrete interface between the three. This eliminates stratification and more closely approximates the idea of a uniform decrease in pore space with increasing filter depth. A typical filter has a particle size gradation decreasing from about 2 mm at the top to 0.2 mm at the bottom.

Filter Underdrains

Several different kinds of manufactured filter bottoms are available, including porous plates, filter blocks, and nozzles. The filter bottom in Figure 7-14 is constructed of vitrified tile blocks having extruded upper and lower laterals and punched orifices. The gravel layer placed between the block and the filter media is 8 to 12 in. in thickness with a gradation between $\frac{3}{32}$ and $\frac{3}{4}$ in. Wash water enters the tile underdrain through the feeder lateral, rises through control orifices (holes) to the compensating lateral, and finally proceeds through the dispersion orifices. By this arrangement, the pressure of the wash water is equalized to produce uniform upward flow through the orifices along an entire row of blocks. The coarser gravel on top of the block ($\frac{3}{4}$ to $\frac{1}{8}$ in.) disperses the wash-water flow, and the finer overlying gravel ($\frac{1}{8}$ to $\frac{3}{32}$ in.) serves as a barrier to prevent intermixing of the gravel and filter sand. The dual-arm agitator consists of two independently rotating arms equipped with wash-water nozzles. The surface agitator is located approximately 2 in. above the compacted bed, and the subsurface arm is 6 in. above the interface of the sand and anthracite layers, which is about 18 in. above the supporting gravel layer. Prior to starting the upward flow of water from the filter bottom, the surface arm is rotated counter-clockwise for 2 min to break up the impurities that have accumulated on the surface of the filter bed. Water backwash is then started. When the bed expands to about 10 percent, the subsurface agitator starts to rotate clockwise. Dual agitation continues for 5 to 6 min with the full bed expansion of up to 50 percent. The agitators are then stopped while backwash continues, to allow the media mixture to reclassify with the anthracite above the sand. Backwash is terminated, permitting the media to settle down. Water consumption by the agitators is about 2 gpm/sq ft, and the backwash rate, including agitators and underflow, is usually 15 gpm/sq ft (10 l/m²·s) for 5 to 10 min. Filters are commonly cleaned every 24 hr. The amount of water used in backwashing varies from 2 to 4 percent of the filtered water.

The underdrain system in Figure 7-15 consists of plastic nozzles placed through holes in a steel underdrain plate. Each nozzle is constructed of wedge-shaped segments assembled with stainless steel bolts to form media-retaining slots that are tapered larger in the direction of water flow during filtration. The recommended medium for this filter nozzle is a 16-in. gravel layer graded from $1\frac{1}{2}$ in. to $\frac{3}{32}$ in. supporting 10 in. of sand and 20 in. of anthracite media. After the water in the filter box is lowered, air is introduced into the filter bottom. The nozzle tail pipes have air holes sized to maintain a 3.5-in. air cushion under the plate to ensure equal nozzle pressure and uniform air distribution to the bottom of the bed. After the media are mixed and scoured by aeration

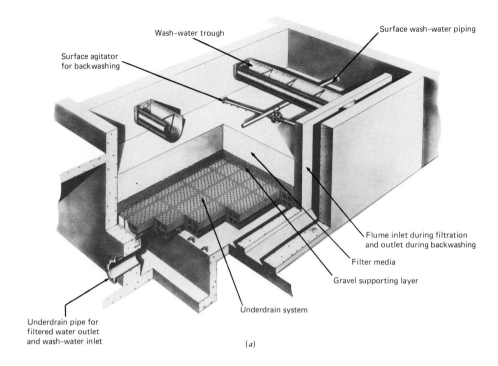

Surface agitator
for backwashing

Wash-water trough

Surface wash-water piping

Flume inlet during filtration
and outlet during backwashing

Filter media

Gravel supporting layer

Underdrain system

Underdrain pipe for
filtered water outlet
and wash-water inlet

(a)

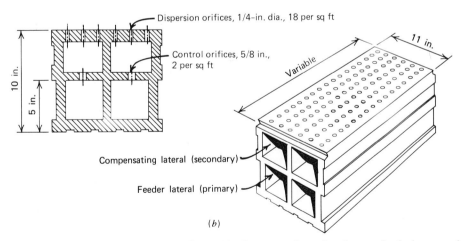

Dispersion orifices, 1/4-in. dia., 18 per sq ft

Control orifices, 5/8 in.,
2 per sq ft

10 in.

5 in.

11 in.

Variable

Compensating lateral (secondary)

Feeder lateral (primary)

(b)

Figure 7-14 Dual-media gravity filter. (a) Cutaway view showing underdrain, gravel supporting layer, filter media, surface agitator, and wash troughs. (b) Detail of vitrified clay underdrain block. (Courtesy of The F. B. Leopold Co. Inc., a subsidiary of Mueller Co.)

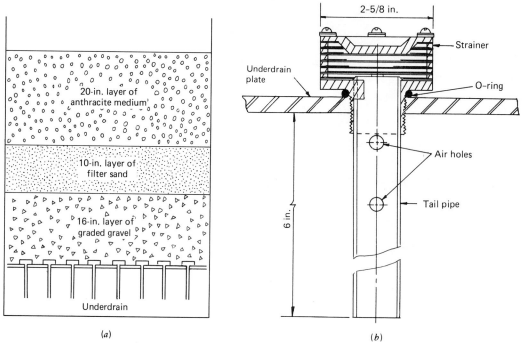

Figure 7-15 Underdrain system for air scouring and water backwashing of a granular-media filter. (*a*) Cross section of the filter. (*b*) Detail of the air-water nozzle. (Courtesy of General Filter Co., Ames, Iowa.)

for about 3 min, wash water flows upward through the bed for 3 to 4 min at 15 gpm/sq ft (10 l/m²·s) to purge out the impurities and restratify the media.

Filter Control

The control console for each filter unit has a head loss gauge, flow meter, and rate controller. A filter run is normally terminated when head loss through the filter reaches a prescribed value between 6 and 9 ft, or when effluent turbidity exceeds the amount considered acceptable. As suspended matter collects in the pores of the filter media and head loss increases through the bed, the lower portion of the filter is under suction from the water column connected through the piping to the clear well (Figure 7-12). This partial vacuum permits release of dissolved gases that tend to collect in the filter pores and, as a result, cause air binding and a reduced rate of

filtration. In addition to shortening filter runs, the gases accumulated in a bed may tend to discharge violently during the start of the backwash cycle, potentially causing disruption of the gravel base.

The rate controller prevents rapid changes in filtration rates and controls the velocity of flow through a clean bed. Continuous-flow photoelectric turbidimeters are used for monitoring filter effluent quality. By detecting turbidity in the filtered water, the operator can terminate a filter run if the quality deteriorates prior to reaching the maximum head loss.

Other Types of Filters

Pressure filters have media and underdrains contained in a steel tank. Media are similar to those used in gravity filters, and filtration rates range from 2 to 4 gpm/sq ft. Water is pumped through the bed under pressure, and the unit is backwashed by reversing the flow, flushing out impurities. Pressure filters are not generally employed in large treatment works because of size limitations; however, they are popular in small municipal water plants that process groundwater for softening and iron removal. Their most extensive application has been treating water for industrial purposes.

Diatomaceous earth filters generally have been limited to swimming pools, portable field units, industrial applications, and installations for small towns. The filter medium is supported on a fine metal screen, porous ceramic material, or synthetic fabric. The filtration cycle consists of precoating the medium, filtering water to a predetermined head loss, and removing the filter cake. During the precoat operation, a slurry of diatomite in water is used to deposit a filter on the screen. Raw water with a small amount of diatomaceous earth (body feed) is drawn through the filter formed by the initial layer. Body feed continually builds the filter so that surface clogging is minimized. When maximum pressure drop, or minimum filtration rate is reached, the filter medium is washed by reversing the flow discharging the dirty cake to waste. The cycle is then repeated. Common rates of filtration range from 1 to 5 gpm/sq ft.

Microstrainers are mechanical filtering devices with a medium normally consisting of finely woven, stainless steel fabric. The pore openings of the fabric are fine enough to remove microscopic organisms and debris. A common design consists of a rotating drum with the fabric mounted on the periphery; raw water passes from the inside to the outside of the drum. Backwashing is usually continuous using wash-water jets. These units have found greatest application in treatment of industrial waters and final polishing filtration of wastewater effluents.

Slow sand filters are limited to low-turbidity waters not requiring chemical treatment. Filtration rates are in the range of 0.05 to 0.10 gpm/sq ft (0.03 to 0.07 l/m²·s). Filtering action is a combination of straining, adsorption, and biological extraction. The biological growths and impurities that accumulate on the surface of the bed are removed periodically by scraping off the upper inch or two of sand. Eventually, the filter has to be rebuilt with clean sand. Use of slow sand filters has declined because of high cost of construction, large filter area needed, and

unsuitability for processing waters with high-turbidity or polluted waters requiring chemical coagulation.

EXAMPLE 7-3

The filter unit illustrated in Figure 7-14 is 15 ft by 30 ft. After filtering 2.5 mil gal in a 24-hr period, the filter is backwashed at a rate of 15 gpm/sq ft for 12 min. Compute the average filtration rate and the quantity and percentage of treated water used in backwashing.

Solution

$$\text{Filtration rate} = \frac{2,500,000 \text{ gal/day}}{15 \text{ ft} \times 30 \text{ ft} \times 1440 \text{ min/day}}$$

$$= 3.9 \frac{\text{gpm}}{\text{sq ft}}$$

$$\text{Quantity of wash water} = \frac{15 \text{ gal}}{\text{min} \times \text{sq ft}} \times 12 \text{ min}$$

$$\times 15 \text{ ft} \times 30 \text{ ft}$$

$$= 81,000 \text{ gal}$$

$$\frac{\text{Wash water}}{\text{Treated water}} = \frac{81,000 \text{ gal}}{2,500,000 \text{ gal}} \times 100 = 3.2 \text{ percent}$$

7-5 CHEMICAL FEEDERS

A chemical feeder is a mechanical device for measuring a quantity of chemical and applying it to a water at a preset rate. Liquid feeders apply chemicals in solutions or suspensions; dry feeders apply them in granular or powdered forms. Some chemicals, such as ferric chloride, polyphosphates and sodium silicate, must be fed in solution form while others—ferrous sulfate and alum—are fed dry. If a chemical does not dissolve readily, it may be applied dry or in suspension provided that the solution is continuously stirred.

Solution feeders are small positive-displacement pumps that deliver a specific volume of liquid for each stroke of a piston or rotation of an impeller. Diaphragm and plunger metering pumps are the two general types used for delivery against pressure. The former, illustrated in Figure 7-16, contains a flexible diaphragm driven by a mechanical linkage that converts the rotary motor input to a reciprocating push-rod motion. Inlet and outlet valves are

Figure 7-16 Solution feeder (diaphragm metering pump) for applying chemical solutions and slurries in water treatment. (Courtesy of Wallace & Tiernan Division, Pennwalt Corp.)

operated by suction and pressure created in the diaphragm chamber. Stroke length, and consequently rate of feed, may be controlled manually at a fixed rate, or with a pneumatic control unit that provides stroking in proportion to water flow or an automatic water-quality feedback signal.

The plunger pump has a reciprocating piston that alternately forces solution out of a chamber and then, on its return stroke, refills the chamber by pulling solution from a reservoir. In addition, several types of rotary pumps qualify as positive-displacement feeders, including gear, swinging vane, sliding vane, oscillating screw, eccentric, and cam pumps.

The criteria used in selecting a solution feeder are accuracy, durability, capacity, corrosion resistance, and pressure capability. Most manufacturers have separate designs for very corrosive chemicals, milder chemicals, and slurries.

The two types of dry feeders are volumetric and gravimetric, depending on whether the chemical is measured by volume or weight. Volumetric dry feeders are simpler, less expensive, and slightly less accurate than gravimetric units. Several types are available with a variety of feed mechanisms that include rotating roller, disk, or screw, star wheel, moving belt, vibratory pan, and oscillating hopper. The unit shown schematically in Figure 7-17a feeds material from a hopper by a rotating screw that moves back and forth. The feed screw is designed so that chemical discharges first from one end of the screw and then from the other. Rate of feed is controlled by regulating the speed of the helical screw.

Gravimetric dry feeders are extremely accurate, available in large sizes to deliver high feed rates, readily adaptable to recording and automatic control, and more expensive than volumetric feeders. There are two general types: One weighs the material on a section of moving belt, and the other operates on loss-in-weight of the feeder hopper. The extreme accuracy is accounted for

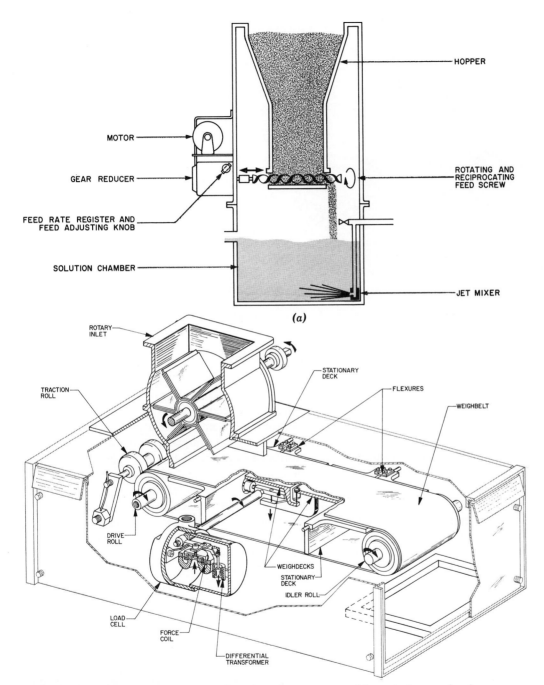

Figure 7-17 Dry feeders, volumetric and gravimetric types, are used for metering powdered chemicals. (*a*) Screw-feed volumetric dry feeder. (*b*) Belt-type gravimetric feeder. (Courtesy of Wallace & Tiernan Division, Pennwalt Corp.)

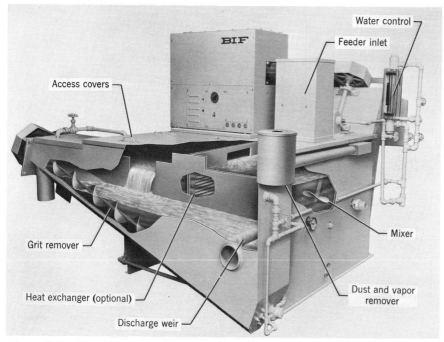

Figure 7-18   Lime slaker mixes powdered lime with water, forming a slurry that is fed by a solution pump in treatment of water. (Courtesy of BIF.)

by the fact that material is dispensed by weight, removing the variables of density and consistency. The latter do affect the feed rate of volumetric units. Figure 7-17b is an operational diagram of a belt-type feeder. Chemical enters on a weigh belt over a stationary deck. Movement of the belt conveys the material over a weigh deck section, which continuously transmits the quantity to a loading beam scale. As the amount of material on the platform varies, the scale beam moves, actuating a pneumatic beam balancing cylinder that, in turn, controls belt speed and loading to maintain the preset feed rate.

Applying lime requires a special type of unit that is referred to as a slaker. (*Slake* means to cause to heat and crumble by treatment with water.) The chemical equation for slaking is

$$CaO \text{ (lime)} + H_2O = Ca(OH)_2 \text{ (lime slurry)} \quad (7\text{-}5)$$

A gravimetric feeder applies powdered lime to a mixing chamber where dilution water is supplied to bring the slurry to desired strength (Figure 7-18). The slaker proportions lime feed to water maintaining the correct concentration; pH controls are available to signal small changes in dosage to compensate for variations in water composition and purity of the lime. The ratio of water to lime is about 5 to 1 and the process time is 30 min. Regulators and alarms protect against excessive temperatures caused by the chemical reaction. An automatic grit separator removes coarse, inert material from the slurry before feeding. The lime is pumped from the holding tank to a slurry feeder that is controlled by an automatic pH and flow control system.

Figure 7-19 shows chemical feeders in a water treatment plant. Each is supplied from a hopper-bottom, chemical storage tank that extents through the ceiling to the floor above. Mechanical vibrators attached to the hoppers keep the chemicals flowing freely out into the feeders. After the dry chemicals are measured into continuously stirred dissolving chambers, the solutions flow either by gravity into an open flume or are transferred by pumps into a pressure main.

Figure 7-19 Photograph of gravimetric feeders in a water treatment plant. The unit on the right is a lime slaker; the one in the center is feeding alum.

7-6 CHLORINATION

Chlorine is used in water and wastewater treatment for disinfection to destroy pathogens and control nuisance microorganisms, and for oxidation. As an oxidant, it is used in iron and manganese removal, destruction of taste and odor compounds, and elimination of ammonia nitrogen.

Chlorine Chemistry

Chlorine is a heavier than air, greenish-yellow-colored, toxic gas. One volume of liquid chlorine confined in a container under pressure yields about 450 volumes of gas. It is a strong oxidizing agent reacting with most elements and compounds. Moist chlorine is extremely corrosive; consequently, conduits and feeder parts in contact with chlorine are either special alloys or nonmetal. The vapor is a respiratory irritant that can cause serious injury if exposure to a high concentration occurs.

Hypochlorites are salts of hypochlorous acid (HOCl). Calcium hypochlorite ($Ca(OCl)_2$) is the predominant dry form used in the United States. High-test calcium hypochlorites, available commercially in granular, powdered or tablet forms, readily dissolve in water and contain about 70 percent available chlorine. Sodium hypochlorite (NaOCl) is commercially available in liquid form at concentrations between 5 and 15 percent available chlorine. Most water treatment plants in the United States use liquid chlorine, since it is less expensive than hypochlorites. The latter are used in swimming pools, small waterworks, and emergencies.

Chlorine combines with water forming hypochlorous acid, which, in turn, can ionize to the hypochlorite ion. Below pH 7 the bulk of the HOCl remains unionized, while above pH 8 the majority is in the form of OCl^-, Eq. 7-6.

$$Cl_2 + H_2O \rightarrow HCl + HOCl \underset{pH < 7}{\overset{pH > 8}{\rightleftharpoons}} H^+ + OCl^- \quad (7\text{-}6)$$

Hypochlorites added to water yield the hypochlorite ion directly, Eq. 7-7.

$$Ca(OCl)_2 + H_2O = Ca^{++} + 2OCl^- + H_2O \quad (7\text{-}7)$$

Chlorine existing in water as hypochlorous acid and hypochlorite ion is defined as free available chlorine.

Chlorine readily reacts with ammonia in water to form chloramines as follows:

$$HOCl + NH_3 = H_2O + NH_2Cl \qquad \text{(monochloramine)}$$
$$(7\text{-}8)$$

$$HOCl + NH_2Cl = H_2O + NHCl_2 \quad \text{(dichloramine)}$$
$$(7\text{-}9)$$

$$HOCl + NHCl_2 = H_2O + NCl_3 \qquad \text{(trichloramine)}$$
$$(7\text{-}10)$$

The reaction products formed depend on pH, temperature, time, and initial chlorine to ammonia ratio. Monochloramine and dichloramine are formed in the pH range of 4.5 to 8.5. Above pH 8.5 monochloramine generally exists alone, but below pH 4.4 trichloramine is produced. Chlorine existing in chemical combination with ammonia nitrogen or organic nitrogen compounds is defined as combined available chlorine. (The laboratory test procedure for available chlorine is given in Section 2-11.)

When chlorine is added to water containing ammonia, the residuals that develop yield a curve similar to that shown in Figure 7-20. The straight line from the origin is the concentration of chlorine applied, or the residual chlorine if all of that applied appeared as residual. The curved line represents chlorine residuals, corresponding to various dosages, remaining after a specified contact time, such as 20 min. Chlorine demand at a given dosage is measured by the vertical distance between the applied and residual lines. This represents the amount of chlorine reduced in chemical reactions and, therefore, the amount that is no longer available. With molar chlorine to ammonia-nitrogen ratios less than 1 to 1, monochloramine and dichloramine are formed with the relative amounts depending on pH and other factors. Higher dosages of chlorine increase the chlorine to nitrogen ratio and result in oxidation of the ammonia and reduction of

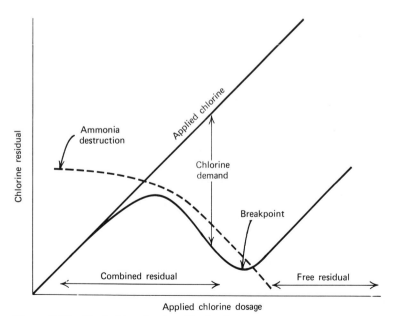

Figure 7-20 Typical breakpoint chlorination curve.

the chlorine. Theoretically, three moles of chlorine react with two moles of ammonia to drive nitrogen off as gas and to reduce chlorine to the chloride ion,

$$2NH_3 + 3Cl_2 = N_2 + 6HCl \qquad (7\text{-}11)$$

Chloramine residuals decline to a minimum value referred to as the breakpoint. Dosages in excess of this breakpoint produce free chlorine residuals. The breakpoint curve is unique for each water tested, since chlorine demand depends on the concentration of ammonia, presence of other reducing agents, contact time between chlorine application and residual testing, and other factors.

Disinfection

The most common application of chlorination is disinfection of drinking water to destroy microorganisms that cause diseases in humans. The bactericidal action of chlorine results from a chemical reaction between HOCl and the bacterial or viral cell structure, inactivating required life processes. Rate of disinfection depends on the concentration and form of available chlorine residual, time of contact, pH, temperature, and other factors. Hypochlorous acid is more effective than hypochlorite ion; therefore, the power of free chlorine residual decreases with increasing pH. Bactericidal action of combined available chlorine is significantly less than that of free chlorine residual.

Information is unavailable to provide specific application rates in chlorination to achieve 100 percent kill of all microorganisms of sanitary significance for the variety of water supplies being treated for domestic use. The water source selected should, of course, be the least polluted available, and regulations should be set up to protect its quality. Current disinfection practice is based on establishing a given kind and amount of chlorine residual during treatment and, then, maintaining an adequate residual to the customer's faucet. Thus, a major part of quality control is testing water in the distribution system for chlorine residual. Effectiveness of the disinfection process is determined by testing for the coliform group as indicators of water quality. The sensitivity of bacteria to chlorination is well understood, while the effect on protozoans and viruses has not been clearly delineated. Protozoal cysts and enteric viruses are more resistant to chlorine than are coliforms and other enteric bacteria. On the other hand, very little evidence exists to indicate that current water treatment practices are not adequate. No outbreaks of waterborne viral or protozoal infections have been documented in public water supplies after proper treatment by chemical coagulation, filtration, and chlorination. Furthermore, waterborne diseases attributed to these pathogens are not common in the United States. (Refer to Section 3-7 for a discussion of waterborne diseases.)

Recommended minimum bactericidal chlorine residuals are listed in Table 7-1. These values are based on destruction of coliform bacteria at water temperatures between 20 and 25°C after 10-min contact for free chlorine and a 60-min contact for combined available chlorine. Minimum residuals for virus inactivation and protozoal cyst destruction are considerably greater.

Table 7-1. Recommended Minimum Bactericidal Chlorine Residuals for Disinfection

pH Value	Minimum Free Available Chlorine Residual after 10-Min Contact (mg/l)	Minimum Combined Available Chlorine Residual after 60-Min Contact (mg/l)
6.0	0.2	1.0
7.0	0.2	1.5
8.0	0.4	1.8
9.0	0.8	>3.0
10.0	0.8	>3.0

Therefore, recommended treatment of surface waters includes coagulation, sedimentation, and filtration to reduce the density of viruses and physically remove cysts, followed by chlorination to establish a free residual. With this water processing, establishing a free residual as recommended in Table 7-1 has proven to be satisfactory for protection of public water supplies. This requires breakpoint chlorination if the surface water contains dissolved ammonia.

Occasionally, a combined chlorine residual is established to control bacterial growths in a treated water. Compared to free-residual chlorination, the advantages are that chloramines are less reactive and a residual can be maintained for a longer period of time without rechlorination. For instance, a combined residual can be applied to a treated water before it is pumped through a long pipeline to a municipal distribution system. If insufficient natural ammonia is present in the water, gaseous anhydrous ammonia is applied by feeding equipment similar to that used for chlorine.

Oxidation

Chorine is a much stronger oxidizing agent for manganese than is dissolved oxygen. Therefore, one of the treatment processes for iron and manganese removel (Figure 7-1b) uses chlorine to remove these metals from solution.

Hydrogen sulfide present in groundwater can be rapidly converted to the sulfate ion using chlorine.

$$H_2S + 4Cl_2 + 4H_2O = H_2SO_4 + 8HCl \quad (7\text{-}12)$$

Breakpoint chlorination in the treatment of surface waters may be used to destroy objectionable tastes and odors and to eliminate bacteria, minimizing biological growths on filters and aftergrowths in the distribution system. Breakpoint chlorination of polluted surface waters, however, can result in formation of trihalomethanes (Section 7-7).

Distribution System Chlorination

A new main after installation should be pressure tested, flushed to remove all dirt and foreign matter, and disinfected by one of the following methods prior to being placed in service. The continuous-feed method involves supplying water to the new main with a chlorine concentration of at least 50 mg/l. Either a solution-feed chlorinator or a hypochlorite feeder injects chlorine into the water that fills the main. The chlorinated water should remain in the pipe for a minimum of 24 hr while all valves and hydrants along the main are operated to ensure their disinfection. At the end of 24 hr, no less than 25 mg/l chlorine residual should be remaining. In the slug method, a continuous flow of water is fed to the main with a chlorine concentration of at least 300 mg/l. The rate of flow is set so that the column of chlorinated water contacts the interior surfaces of the main for a period of at least 3 hr. As the slug passes other connections, the valves are operated to ensure disinfection of appurtenances. This method is used principally for large-diameter mains where the continuous-feed technique is impractical. The tablet method, although commonly used for small-diameter mains, is least satisfactory, since it precludes preliminary flushing. Calcium hypochlorite tablets are placed in each section of the pipe, hydrants, and other appurtenances during construction. The new main is then slowly filled with water to dissolve the tablets without flushing them to one end of the pipe. The final solution, with a residual of at least 50 mg/l, should remain in contact for a minimum of 24 hr. Following disinfection by any of these methods, the chlorinated water should be flushed to waste by using potable water. Microbiological tests should then be conducted before placing the main in service.

A broken main is isolated by closing the nearest valves. The first step in repair involves flushing the broken section to remove contamination while pumping the discharge out of the trench. Minimum disinfection includes swabbing the new pipe sections and fittings with a 5 percent hypochlorite solution before installation, and flushing the main from both directions before returning the system to service. Where conditions permit, the repaired section should be isolated and disinfected by the procedures prescribed for new mains.

Tanks and reservoirs before being placed into service or following inspection and cleaning should be disinfected. One method is to add chlorine directly to the filling water, either using hypochlorite or a portable solution-feed chlorinator. Standard procedure is a 50 mg/l residual maintained for a minimum of 6 hr. An alternate method, where convenient, is to spray the walls and other surfaces with a solution containing about 500 mg/l of available chlorine.

After the construction or repair of wells, disinfection is performed by using a 50- to 100-mg/l free chlorine residual for 12 to 24 hr. A new well should be operated until the water is practically free of turbidity. A quantity of chlorine solution equal to at least twice the volume of water in the well is added through a pipe or hose. The pump cylinder and drop pipe are then washed with a strong chlorine solution as the assembly is lowered into the casing. After a minimum contact of 12 hr, the chlorinated water is pumped to waste.

Equipment and Feeders

Liquid chlorine is shipped in pressurized steel cylinders. The most popular sizes are 100 or 150 lb, but 1-ton containers may be used in large installations. Storage areas should be cool, well ventilated, and protected from corrosive vapors and continuous dampness. Gas is withdrawn from a container through a valve connection at the top. Liquid chlorine in the cylinder vaporizes to allow continuous gas withdrawal. Design and operation of all facilities should minimize hazards associated with connecting, emptying, and disconnecting containers. The characteristic odor of chlorine provides warning of leaks. Since it reacts with ammonia to form dense white fumes, a leak can be readily detected by holding a cloth swab saturated with strong ammonia water near the suspected area. Calcium hypochlorite is relatively stable under normal conditions; however, reactions may occur with organic substances. Preferably it should be stored in a location segregated from other chemicals and materials.

The most essential unit in a chlorine gas feeder is the variable orifice inserted in the feed line to control rate of flow out of the cylinder. Its operation is somewhat analogous to a water faucet with a constant supply pressure behind it. The amount of water discharged can be governed by opening the faucet, and if supply pressure does not change, a constant rate of flow is maintained for any given setting. The orifice shown in detail in Figure 7-21 consists of a grooved plug sliding in a fitted ring.

Feed rate is adjusted by varying the V-shaped opening. However, since chlorine cylinder pressure varies with temperature, the discharge through such a throttling valve does not remain constant without frequent adjustments of the valve setting. Furthermore, the condition on the outlet side may vary as a result of pressure changes at the point of application. To ensure that these variable conditions do not affect control, a pressure-regulating valve is inserted between the cylinder and the orifice, with a vacuum compensating valve on the discharge side. A safety pressure-relief valve is held closed by vacuum. If vacuum is lost and chlorine under pressure passes the inlet valve, the relief valve opens and the chlorine gas is vented outside the building. The visible flow meter (rotameter), pressure gauges, and feeder rate adjusting knob are located on the front panel of the chlorinator cabinet.

Direct feed of chlorine gas into a pipe or channel has certain limitations; for example, a considerable safety problem would exist if gas piping was extended throughout the treatment plant. To overcome the difficulties of conveyance and direct introduction of chlorine gas, an injector is employed to permit solution feed. Water flowing through the injector creates a vacuum that draws in chlorine gas from the feeder and mixes it with the water supply. This concentrated solution is relatively stable and can be safely piped to various points in the treatment plant for discharge into an open channel, closed pipeline, or suction end of a pump.

Chlorine feeders can be controlled manually or automatically based on flow or chlorine residual or both. Manual adjustment establishes a continuous feed rate. This type of regulator is satisfactory only where chlorine demand and flow are reasonably constant, and where an operator is available to make adjustments as necessary. Automatic proportional control equipment adjusts the feed rate to provide a constant preestablished dosage for all rates of flow. This is accomplished by metering the main flow and using a transmitter to signal the chlorine feeder. Automatic residual control uses an analyzer downstream from the point of application to pace the chlorinator. Combined automatic flow and residual control (Figure 7-22) maintains a preset chlorine residual in the water independent of demand and flow variations. The feeder is responsive to signals from both the flow

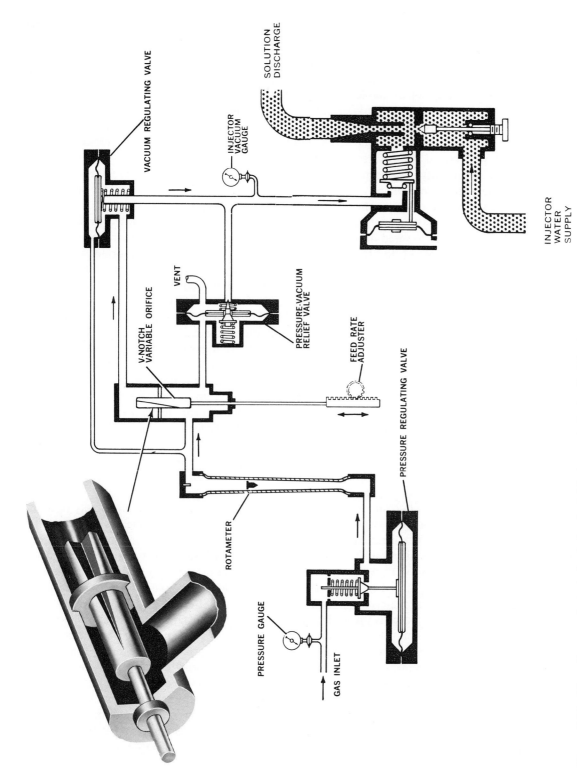

Figure 7-21 Flow diagram of a typical solution feed chlorinator and a detail of the V-notch variable orifice. (Courtesy of Wallace & Tiernan Division, Pennwalt Corp.)

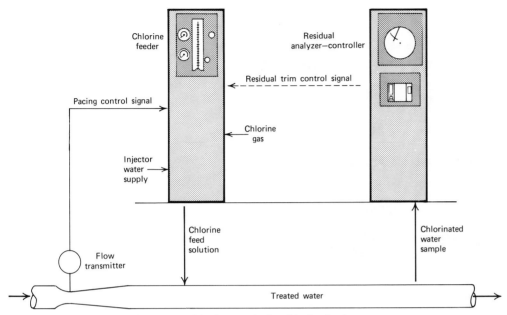

Figure 7-22 Automatic flow and residual control of a chlorine feeder.

meter transmitter and the chlorine residual analyzer. This system is most effective when flow pacing is the primary chlorine feed regulator, and residual monitoring is used to trim the dosage.

Positive-displacement diaphragm pumps, mechanically or hydraulically actuated, are used for metering hypochlorite solutions. The hypochlorinator in Figure 7-23 consists of a water-powered pump paced by a positive-displacement water meter. The meter register shaft rotates in proportion to the main line flow and controls a cam-operated pilot valve. This, in turn, regulates the flow of water behind the power diaphragm to produce a discharge of hypochlorite proportional to the main flow. Admitting main pressure behind the pumping diaphragm balances the water pressure in the pumping head. The advantage of this is that only a small force is needed for the discharge stroke, placing minimum stress on the diaphragm. Another advantage is that the pump does not require electrical power. Hypochlorite dosage can be manually adjusted by changing the stroke length setting of the pump.

EXAMPLE 7-4

Chlorine usage in the treatment of 5.0 mgd of water is 17.0 lb/day. The residual after 10 min contact is 0.20 mg/l. Compute the dosage in milligrams per liter and chlorine demand of the water.

Solution

$$\text{Dosage} = \frac{17.0 \text{ lb/day}}{5.0 \text{ mil gal/day}} \times \frac{\text{mg/l}}{8.34 \text{ lb/mil gal}}$$

$$= 0.41 \text{ mg/l}$$

$$\text{Chlorine demand} = 0.41 - 0.20 = 0.21 \text{ mg/l}$$

EXAMPLE 7-5

A new water main is disinfected using a 50 mg/l chlorine dosage by applying a 2.0 percent hypochlorite solution. (a) How many kilograms of dry hypochlorite powder, containing 70 percent available chlorine, must be dissolved in 100 liters of water to make a 2.0 percent

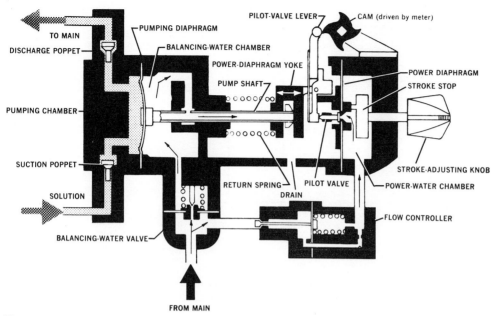

Figure 7-23   Hydraulically actuated diaphragm (water operated) hypochlorinator. (Courtesy of Wallace & Tiernan Division, Pennwalt Corp.)

(20,000 mg/l) solution? (b) At what rate should this solution be applied to the water entering the main to provide a concentration of 50 mg/l? (c) If 34,000 liters of water are used to fill the main at a dosage of 50 mg/l, how many liters of hypochlorite solution are used?

Solution

(a)   Kilograms of hypochlorite powder
 for 2.0 percent solution

$$= \frac{100\ l \times 1.0\ kg/l \times 0.02}{0.70}$$

$$= 2.86\ kg/100\ l$$

(b)   Feed rate for 50 mg/l $= \dfrac{50\ mg/l}{20,000\ mg/l}$

$$= \frac{1 \text{ volume of } 2.0\% \text{ solution}}{400 \text{ volumes of water}}$$

(c)   Solution usage $= 34,000\ l \times \dfrac{50\ mg/l}{20,000\ mg/l} = 85\ l$

7-7   CONTROL OF TRIHALOMETHANES

Chlorination of surface waters containing humic substances from natural sources, such as decaying vegetation, produce trihalomethanes (THMs) with chloroform and bromodichloromethane being the most common chemical forms. THMs are suspected carcinogens and have a maximum contaminant level of 0.10 mg/l in drinking water. (Section 5-4, Drinking Water Standards, discusses trihalomethanes under the heading Organic Chemicals.)

The practice of breakpoint chlorination as an initial step in processing of surface waters for disinfection and control of taste and odor contributes to the formation of trihalomethanes. Also, sustained high free chlorine residual increases the concentration of THMs with time. The following modifications to existing methods of water treatment are recommended as ways of reducing the level of THMs in treated water: application of chlorine after chemical coagulation and sedimentation; improvement of clarification processes; use of alternate disinfectants, such as chlorine dioxide; and application of powdered activated carbon to adsorb trihalomethanes and humic substances.

Applying chlorine to later stages in treatment of surface waters is the easiest method for reducing trihalomethane formation. Chlorine can be added after coagulation and sedimentation or after filtration, if prior control of microbial populations is not necessary. The benefits of delayed chlorination are a reduction in the required dosage and removal of humic substances before applying chlorine. If necessary, powdered activated carbon can be applied during early treatment stages to adsorb humic substances, and improved coagulation can be implemented to provide better removal of organic matter prior to filtration.

Alternative disinfectants that do not form trihalomethanes are chloramines, chlorine dioxide, and ozone. Use of chloramines, which are weak disinfectants for bacteria and much less effective in inactivating viruses, increases the risk of pathogens in treated drinking water. Despite this, they can be used as a secondary disinfectant to establish a combined residual in the water entering the distribution system. Chlorine dioxide has had limited application as a water disinfectant, although it is used for taste and odor control. The advantage is production of a persistent residual without forming chloramines. As a disadvantage, the possible health effects of products formed when chlorine dioxide reacts with organic substances are uncertain. Ozone is an effective disinfectant that forms no known products that could be detrimental to human health. The disadvantages of ozone oxidation are high cost and lack of residual. If used as a primary disinfectant, chlorine can be added as a secondary disinfectant to provide a protective residual.

Trihalomethanes are difficult to remove from water. Aeration can reduce the total THM level, but practical problems with aeration are purification of the air to prevent water contamination and enclosure to reduce cooling. The effectiveness of adsorbing trihalomethanes by powdered activated carbon and granular activated carbon is variable. Bromine trihalomethanes are more readily adsorbed than chloroform, which is the most common trihalomethane. The preferred approach to control of trihalomethanes is prevention rather than removal after formation.

7-8 FLUORIDATION

During the past three decades, hundreds of studies have been undertaken to correlate the concentration of waterborne fluoride and the incidence of dental caries. Examination of children's teeth consuming water with natural fluoride led to the following three distinct relationships: An excessive fluoride level increases the occurrence and severity of dental fluorosis (mottling of teeth) while decreasing the incidence of decaying, missing, or filled teeth; based on climate, the optimum occurs at a concentration of 0.6 to 1.2 mg/l, resulting in maximum reduction in caries with no esthetically significant mottling; and, below optimum some benefit occurs but decay reduction is not as great and decreasing fluoride levels relate to increasing incidence of caries. Controlled fluoridation in water treatment to bring the natural content up to optimum level achieves the same beneficial results. Recommended limits given in Table 5-1 are based on air temperature, since this influences the amount of water ingested by people. Recent studies have indicated that fluoride also benefits older people in reducing prevalence of osteoporosis and hardening of the arteries. Without doubt, fluoride in drinking water prevents dental caries and controlled fluoridation is an acceptable public health measure. Over one half of the nation's population consumes water containing near optimum fluoride content with the bulk of these residing in communities that deliberately add a chemical compound to provide fluoride ion.

The three most commonly used fluoride compounds in water treatment are sodium fluoride, sodium silicofluoride, and fluosilicic acid, also known as hydrofluosilicic, hexafluosilicic, or silicofluoric acid. Table 7-2 lists some of the characteristics of these compounds. Sodium fluoride, although one of the most expensive for the amount of available F, is one of the most widely used chemicals. The crystalline type is preferred when manual handling is involved, since the absence of fine powder results in a minimum of dust. Fluosilicic acid is a colorless, transparent, fuming, corrosive liquid having a pungent odor and an irritating action on the skin. The major source of acid is a by-product from phosphate fertilizer manufacture. Sodium silicofluoride, the salt of fluosilicic acid, is the most widely used compound

Table 7-2. Characteristics of the Three Commonly Used Fluoride Compounds

	Sodium Fluoride	Sodium Silicofluoride	Fluosilicic Acid
Formula	NaF	Na_2SiF_6	H_2SiF_6
Fluoride ion, percent	45	61	79
Molecular weight	42	188	144
Commercial purity, percent	90 to 98	98 to 99	22 to 30
Commercial form	Powder or crystal	Powder or fine crystal	Liquid
Dosage in pounds per million gallons Required for 1.0 mg/l at indicated purity	18.8 (98%)	14.0 (98.5 %)	35.2 (30%)

primarily because of its low cost. The white, odorless, crystalline powder is commercially available in various gradations for optimum application by various feeders.

No one specific type of fluoridation system is applicable to all water treatment plants. Selection is based on size and type of water facility, chemical availability, cost, and type of operating personnel available. For small utilities, some type of solution feed is almost always selected and batches are manually prepared. A simple system consists of a solution tank placed on a platform scale, for the convenience of weighing during preparation and feed, and a solution metering pump with appropriate piping from the tank to the water main for application. If fluosilicic acid is used, it may be diluted with water in the feed tank or may be applied at full strength directly from the shipping drum. When sodium fluoride is used, the feed solution may be prepared to a desired strength or as a saturated solution in a dissolving tank. Sodium fluoride has a maximum solubility of 4.0 percent (18,000 mg F/l), which is essentially independent of water temperature. Saturators are commercially available to prepare a saturated solution by allowing water to trickle through a bed containing a large excess of sodium fluoride. Water used for dissolution should be softened whenever the hardness exceeds 75 mg/l. While NaF is quite soluble, calcium and magnesium fluorides form precipitates that can scale and clog feeders and lines.

Fluoride solution feed must be paced to water flow. If a pump has a fixed delivery rate, the feeder can be tied

electrically to turn on and off with pump operation, and the fluoride application can be adjusted to the rate of water discharge. For small water plants, a water meter contactor can be used to pace the feeder. Essentially, a contactor is a switch that is geared to the water meter movement so it makes contact at specific volumetric discharges. A pulse from the meter contactor can be used to energize the feeder in proportion to the meter's response to flow.

Large waterworks use gravimetric dry feeders to apply sodium silicofluoride, or solution feeders to apply fluosilicic acid directly. Automatic control systems use flow meters and recorders to adjust feed rate.

Fluoride must be injected at a point where all the water being treated passes. If no single common point exists, separate fluoride feeding installations are required for each water facility. In a well-pump system, application can be in the discharge line of each pump, or in a common line leading to a storage reservoir. Fluoride applied in a water plant can be introduced in a channel or main coming from the filters, or directly to the clear well. Whenever possible it should be added after filtration to avoid losses that may occur as a result of reactions with other chemicals. Of particular concern are coagulation with heavy alum doses and lime-soda softening. Fluoride injection points should be as far away as possible from any chemical that contains calcium so as to minimize loss by precipitation.

Surveilance of water fluoridation involves testing both

the raw and treated water for fluoride. The concentration in treated water should be the amount recommended by drinking water standards. Records of the weight of chemical applied and the volume of water treated should be kept to confirm that the correct amount of fluoride is being added. The amount added to the water should equal the quantity calculated from the increase in concentration and quantity of water treated.

Defluoridation

When water supplies contain excess fluorides, the teeth of most consumers over a period of several years become mottled with a permanent brown to gray discoloration of the enamel. Children who have been drinking water containing 5 mg/l develop fluorosis to the extent that the enamel is severely pitted, resulting in loss of teeth.

Treatment methods for defluoridation use either activated alumina or bone char. Water is percolated through insoluble, granular media to remove the fluorides. The media are periodically regenerated by chemical treatment after becoming saturated with fluoride ion. Regeneration of bone char consists of backwashing with a 1 percent solution of caustic soda and then rinsing the bed. Reactivation of alumina also involves backwashing with a caustic solution.

Removal of excess fluoride from public water supplies is a sound economic investment when related to the increased cost of dental care and loss of teeth. Despite the obvious need, some communities have not installed defluoridation units, allegedly because of excessive costs of construction and operation. A few communities have been able to seek alternate sources of water to solve their fluoride problem without special treatment.

EXAMPLE 7-6

A liquid feeder applies a 4.0 percent saturated sodium fluoride solution to a water supply, increasing the fluoride concentration from the natural level of 0.4 mg F/l to 1.0 mg F/l. The commercial NaF powder contains 45 percent F by weight. (a) How many pounds of NaF are required per million gallons treated? (b) What is the dosage of 4.0 percent NaF solution per million gallons?

Solution

(a) $\dfrac{\text{NaF required}}{\text{Million gallons}} = \dfrac{(1.0 \text{ mg F/l} - 0.4 \text{ mg F/l})}{0.45 \text{ mg F/mg NaF}}$

$$\times\ 8.34 \dfrac{\text{lb/mil gal}}{\text{mg/l}} = 11.1 \text{ lb}$$

A 4.0% NaF solution contains 40,000 mg NaF/l, and $0.45 \times 40,000 = 18,000$ mg F/l.

(b) $\dfrac{\text{Solution dosage}}{\text{Million gallons}}$

$$= \dfrac{1,000,000 \text{ gal} (1.0 \text{ mg/l} - 0.4 \text{ mg/l})}{18,000 \text{ mg/l}} = 33.3 \text{ gal}$$

EXAMPLE 7-7

A dry feeder recorded a weight loss of 10 kg of sodium silicofluoride in the treatment of 7000 m³ of water. From Table 7-2, commercial powder is 98 percent pure Na_2SiF_6 and 61 percent of the pure compound is F. Calculate the concentration of fluoride ion added to the treated water.

Solution

Fluoride concentration

$$= \dfrac{10 \text{ kg} \times 1000 \text{ g/kg} \times 0.98 \times 0.61}{7000 \text{ m}^3} = 0.85 \text{ mg/l}$$

7-9   IRON AND MANGANESE REMOVAL

Ferrous iron (Fe^{++}) and manganous manganese (Mn^{++}) are soluble, invisible forms that may exist in well waters or anaerobic reservoir water. When exposed to air, these reduced forms slowly transform to insoluble, visible, oxidized ferric iron (Fe^{+++}) and manganic manganese (Mn^{++++}). The rate of oxidation depends on pH, alkalinity, organic content, and presence of oxidizing agents. If not removed in treatment, the brown-colored oxides of iron and manganese create unaesthetic conditions and may interfere with some water uses.

Preventive measures may sometimes be used with reasonable success. Addition of sodium hexametaphosphate, while not preventing oxidation of the metal ions, may keep them in suspension and thus moving through the system without creating accumulations that periodically cause badly discolored water. Success of this treatment is very difficult to predict, since it depends on the concentrations of iron and manganese, the level of chlorine residual established for disinfection, and the time of passage through the distribution system. The latter is established by the extent of the distribution system, pipe sizes in the network, and location and volume of storage reservoirs.

Reduced iron in water promotes the growth of autotrophic bacteria in distribution mains (Section 3-1). Periodic flushing of small distribution pipes may be effective in removing accumulations of rust particles; however, elimination of iron bacteria is generally difficult and expensive. Biological growths are particularly obnoxious when they decompose in the piping system, releasing foul tastes and odors. Heavy chlorination of isolated sections of water mains followed by flushing has been effective in some cases. The only permanent solution to iron and manganese problems is removal by treatment of the water.

Aeration, Sedimentation, and Filtration

The simplest form of iron oxidation in treatment of well water is plain aeration. A typical tray-type aerator has a vertical riser pipe that distributes water on top of a series of trays from which it then drips and splatters down through the stack. The trays frequently contain coke or stone contact beds that develop and support oxide coatings that speed up the oxidation reactions.

$$Fe^{++}(\text{ferrous}) + \text{oxygen} \rightarrow FeO_x(\text{ferric oxides}) \quad (7\text{-}13)$$
$$\underset{\text{Soluble iron}}{\phantom{Fe^{++}}} \qquad\qquad\qquad \underset{\text{Insoluble iron}}{}$$

Manganese cannot be oxidized as easily as iron, and plain aeration is generally not effective. Sometimes, increasing the pH to about 8.5 with lime or soda ash enhances the oxidation of manganese on coke beds coated with manganic oxides; however, high removals are not assured. Inefficient removal of manganese can cause serious problems with post-chlorination for establishing a residual in the distribution system. Manganese not taken out by the aeration-filtration process may be oxidized in the injector mechanism of a solution-feed chlorinator and may clog the unit; or in the distribution piping, the oxidizing effect of the chlorine residual may create a staining water.

Aeration, Chemical Oxidation, Sedimentation, and Filtration

This is the sequence of processes, illustrated in Figure 7-1b, that is the common method for removing iron and manganese from well water without softening treatment. Preliminary aeration strips out dissolved gases and adds oxygen. Iron and manganese are chemically oxidized by free chlorine residual, Eq. 7-14, or by potassium permanganate, Eqs. 7-15 and 7-16, at rates of oxidation much greater than dissolved oxygen. When chlorine is used, a free available residual is maintained throughout the treatment processes. The specific dosage required depends on the concentration of metal ions, pH, mixing conditions, and other factors. Potassium permanganate is a dark purple crystal or powder available commercially at 97 to 99 percent purity. Theoretically, 1 mg/l of potassium permanganate oxidizes 1.06 mg/l of iron or 0.52 mg/l of manganese. In actual practice, the amount needed is often less than this theoretical requirement. Permanganate oxidation may be advantageous for certain waters, since its rate of reaction is relatively independent of pH.

$$Fe^{++} + Mn^{++} + \text{oxygen} \xrightarrow[\text{residual}]{\text{free chlorine}}$$
$$\underset{\text{Soluble ions}}{\phantom{Fe^{++}}}$$

$$FeO_x\downarrow + MnO_2\downarrow \quad (7\text{-}14)$$
$$\underset{\text{Insoluble metal oxides}}{}$$

$$\underset{\substack{\text{Ferrous}\\\text{bicarbonate}}}{Fe(HCO_3)_2} + \underset{\substack{\text{Potassium}\\\text{permanganate}}}{KMnO_4} \longrightarrow$$

$$\underset{\substack{\text{Ferric}\\\text{hydroxide}}}{Fe(OH)_3\downarrow} + \underset{\substack{\text{Manganese}\\\text{dioxide}}}{MnO_2\downarrow} \quad (7\text{-}15)$$

$$\underset{\substack{\text{Manganous}\\\text{bicarbonate}}}{Mn(HCO_3)_2} + \underset{\substack{\text{Potassium}\\\text{permanganate}}}{KMnO_4} \longrightarrow \underset{\substack{\text{Manganese}\\\text{dioxide}}}{MnO_2\downarrow} \quad (7\text{-}16)$$

Effective filtration following chemical oxidation is essential, since a significant amount of the flocculent metal oxides are not heavy enough to settle by gravity.

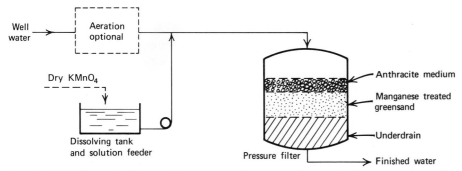

Figure 7-24 Removal of iron and manganese from groundwater by the manganese zeolite process employing a dual-media pressure filter with manganese-treated greensand.

Iron and manganese, carried over to the filter, coat the media with oxides that enhance filtration removal. Practice has shown that new filters pass manganese until the grains are covered with oxide that develops naturally during filtration of manganese-bearing water.

Manganese Zeolite Process

Manganese zeolite is a natural greensand coated with manganese dioxide that removes soluble iron and manganese from solution. After the zeolite becomes saturated with metal ions, it is regenerated using potassium permanganate. A continuous flow system is illustrated in Figure 7-24. Permanganate solution is applied to the water ahead of a pressure filter that contains a dual-media anthracite and manganese zeolite bed. The iron and manganese oxidized by the permanganate feed are removed by the upper filter layer. Any ions not oxidized are captured by the underlying manganese zeolite layer. If surplus permanganate is inadvertently applied to the water, it passes through the coal medium and regenerates the greensand. When the bed becomes saturated with metal oxides, it is backwashed to remove particulate matter from the surface layer and to regenerate the zeolite with potassium permanganate.

Water Softening

Precipitation softening using lime takes iron and manganese out of solution. If split treatment is employed in softening a groundwater containing iron and manganese, potassium permanganate can be used to oxidize metal ions in the water by-passing the first-stage lime treatment. Lime treatment has been used for the principal purpose of removing organically bound iron and manganese in processing surface waters.

7-10 PRECIPITATION SOFTENING

Hardness in water is caused by calcium and magnesium ions resulting from water coming in contact with geological formations. Public acceptance of hardness varies, although many customers object to water harder than 150 mg/l. The maximum level considered for public supply is 300 to 500 mg/l. A moderately hard water is generally defined as 60 to 120 mg/l. Hardness interferes with laundering by causing excessive soap consumption, and may produce scale in hot water heaters and pipes. To a considerable extent these disadvantages have been overcome by the use of synthetic detergents and the lining of pipes in small hot-water heaters. Industries generally pretreat boiler water to prevent scaling.

Precipitation softening uses lime (CaO) and soda ash (Na_2CO_3) to remove calcium and magnesium from solution. In addition, lime treatment has the incidental benefits of bactericidal action, removal of iron, and aid in clarification of turbid surface waters. Carbon dioxide can be applied for recarbonation after lime treatment to lower the pH by converting the excess hydroxide ion and carbonate ion to bicarbonate ion.

Lime is sold commercially in the forms of quicklime and hydrated lime. Quicklime, available in granular form, is a minimum of 90 percent CaO with magnesium oxide being the primary impurity. A slaker is used to prepare quicklime for feeding in a slurry containing approximately 5 percent calcium hydroxide. Powdered, hydrated lime contains approximately 68 percent CaO, and may be prepared by fluidizing in a tank containing a turbine mixer. Lime slurry is written as $Ca(OH)_2$ in chemical equations. Soda ash is a grayish-white powder containing at least 98 percent sodium carbonate.

Carbon dioxide is a clear, colorless gas used for recarbonation in stabilizing lime-softened water. The gas is produced by burning a fuel, such as coal, coke, oil, or gas. The ratio of fuel to air is carefully regulated in a CO_2 generator to provide complete combustion. Gas under pressure from the combustion chamber is forced through diffusers immersed in a treatment basin. Several manufacturers produce recarbonation systems for generating and feeding carbon dioxide.

The chemical reactions in precipitation softening are

$$CO_2 + Ca(OH)_2 = CaCO_3\downarrow + H_2O$$
$$(7\text{-}17)$$

$$Ca(HCO_3)_2 + Ca(OH)_2 = 2CaCO_3\downarrow + 2H_2O$$
$$(7\text{-}18)$$

$$Mg(HCO_3)_2 + Ca(OH)_2 = CaCO_3\downarrow$$
$$+ MgCO_3 + 2H_2O$$
$$MgCO_3 + Ca(OH)_2 = CaCO_3\downarrow + Mg(OH)_2\downarrow$$

$$Mg(HCO_3)_2 + 2Ca(OH)_2 = 2CaCO_3\downarrow$$
$$+ Mg(OH)_2\downarrow + 2H_2O$$
$$(7\text{-}19)$$

$$MgSO_4 + Ca(OH)_2 = Mg(OH)_2\downarrow + CaSO_4$$
$$(7\text{-}20)$$

$$CaSO_4 + Na_2CO_3 = CaCO_3\downarrow + Na_2SO_4$$
$$(7\text{-}21)$$

Lime added to water reacts first with any free carbon dioxide forming a calcium carbonate precipitate, Eq. 7-17. Next, the lime reacts with any calcium bicarbonate present, Eq. 7-18. In both of these equations, one equivalent of lime combines with one equivalent of either CO_2 or $Ca(HCO_3)_2$. Since magnesium precipitates as $Mg(OH)_2$ ($MgCO_3$ being soluble), two equivalents of lime are needed to remove one equivalent of magnesium bicarbonate, Eq. 7-19. Noncarbonate hardness (calcium and magnesium sulfates or chlorides) requires addition of soda ash for precipitation. Equation 7-21 shows that one equivalent of soda ash removes one equivalent of calcium sulfate. However, magnesium sulfate needs both lime, Eq. 7-20, and soda ash, Eq. 7-21.

The calcium ion can be effectively reduced by the lime additions defined in the equations that raise the pH of the water to approximately 10.3. But precipitation of the magnesium ion demands a higher pH and the presence of excess lime in the amount of about 35 mg/l CaO (1.25 meq/l) above the stoichiometric requirements. The practical limits of precipitation softening are 30 to 40 mg/l of $CaCO_3$ and 10 mg/l of $Mg(OH)_2$ as $CaCO_3$, while the theoretical solubility of these compounds is considerably less, being approximately one fifth of these limits.

A major advantage of precipitation softening is that the lime added is removed along with the hardness taken out of solution. Thus, the total dissolved solids in the water are reduced. When soda ash is added, Eq. 7-21, the sodium ions remain in the finished water along with the accompanying anion, either sulfate or chloride; however, noncarbonate hardness requiring the addition of soda ash is generally a small portion of the total hardness in a raw water. The chemical reactions of lime-soda ash softening can also be used to estimate the quantity of solids produced in waste sludge.

Recarbonation is used to stabilize lime-treated water, reducing its scale-forming potential. Carbon dioxide neutralizes excess lime, precipitating it as calcium carbonate. Further recarbonation converts carbonate to bicarbonate.

$$Ca(OH)_2 + CO_2 = CaCO_3\downarrow + H_2O \qquad (7\text{-}22)$$
$$CaCO_3 + CO_2 + H_2O = Ca(HCO_3)_2 \qquad (7\text{-}23)$$

Excess Lime Softening

Excess lime treatment is used to remove calcium and magnesium hardness to the practical limit of about 40 mg/l. Lime and soda ash dosages are estimated by using the chemical reactions, plus the excess lime addi-

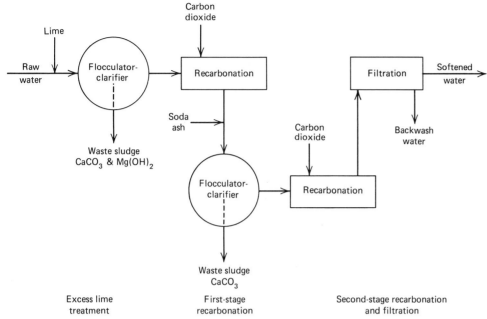

Figure 7-25   Schematic flow diagram for a two-stage excess lime softening plant.

tion needed to precipitate the magnesium. A general flow scheme for two-stage excess lime softening is given in Figure 7-25. After excess lime addition, the water is flocculated and settled to remove $CaCO_3$ and $Mg(OH)_2$ precipitates. Subsequent recarbonation is done in two stages. In the first step, carbon dioxide is added to lower the pH to approximately 10.3 and convert the dissolved excess lime into solid $CaCO_3$ (Eq. 7-22) for removal by flocculation and sedimentation. Soda ash, if needed, is also applied in this stage to precipitate noncarbonate calcium hardness. In the second step, the pH is further reduced to the range of 8.5 to 9.5 to convert most of the remaining carbonate ion to bicarbonate ion (Eq. 7-23) to stabilize the water against scale formation during filtration and in the distribution piping.

Selective Calcium Carbonate Removal

Selective calcium carbonate removal may be used to soften a water low in magnesium hardness, less than 40 mg/l as $CaCO_3$. A magnesium hardness greater than

40 mg/l is not recommended because of the possible formation of hard magnesium silicate scale in high temperature (180° F) services. The usual processing scheme is lime clarification with single-stage recarbonation followed by filtration, as shown in Figure 7-26. Enough lime is added to precipitate the desired reduction in calcium hardness without adding excess lime for magnesium removal. Soda ash may or may not be required depending on the extent of noncarbonate hardness. If the $CaCO_3$ precipitate does not settle satisfactorily, a small amount of polymer or alum can be added to aid in flocculation. Recarbonation is usually performed to reduce scale formation on the filter media and to produce a stable softened water.

Split-Treatment Softening

Split treatment consists of dividing the raw water into two portions for softening in a two-stage system illustrated by the flow diagram in Figure 7-27. The larger portion is given excess lime treatment in the first stage by using a

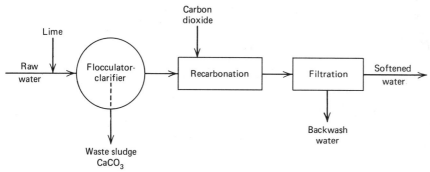

Figure 7-26 Schematic flow diagram for a single-stage calcium carbonate softening plant.

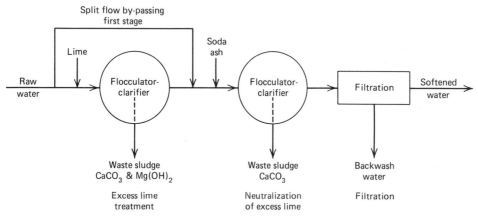

Figure 7-27 Schematic flow diagram for a split-treatment softening plant.

flocculator-clarifier, or in-line mixing and sedimentation basins. Soda ash is added to the second stage where the split flow is blended with the treated water. Excess lime used to force precipitation of magnesium in the first stage now reacts with the calcium hardness that was by-passed around the lime treatment. Thus, the excess lime is used in the softening process instead of being wasted at the expense of carbon dioxide neutralization. Recarbonation is not customarily required, but it may be desirable in the treatment of some waters for stabilization. Since hardness levels of 80 to 100 mg/l are generally considered acceptable, split treatment can result in considerable chemical savings. Lime and recarbonation costs are lower than excess lime treatment,

while the advantage of reducing magnesium hardness to less than 40 mg/l is still possible. The amount of magnesium in the finished water can be calculated by multiplying the by-pass flow times the magnesium concentration in the raw water plus the lime-treated flow times 10 mg/l divided by the total raw water flow. The quantity of flow split around the first stage is often determined by the level of magnesium desired in the softened water.

Softening surface waters may not permit by-passing flow without treatment to remove turbidity. In this case, the split leg is frequently treated by using a coagulant, such as alum. Additionally, coagulant aids or activated carbon or both may be applied to second-stage blending.

EXAMPLE 7-8

Water defined by the following analysis is to be softened by excess lime treatment. Assume that the practical limit of hardness removal for $CaCO_3$ is 30 mg/l, and that of $Mg(OH)_2$ is 10 mg/l as $CaCO_3$.

CO_2 = 8.8 mg/l
Ca^{++} = 40.0 mg/l
Mg^{++} = 14.7 mg/l
Na^+ = 13.7 mg/l

$Alk(HCO_3^-) = \dfrac{135 \text{ mg/l}}{\text{as } CaCO_3}$
$SO_4^=$ = 29.0 mg/l
Cl^- = 17.8 mg/l

(a) Sketch a meq/l bar graph and list the hypothetical combinations of chemical compounds in solution.
(b) Calculate the softening chemicals required, expressing lime dosage as CaO and soda ash as Na_2CO_3.
(c) Draw a bar graph for the softened water before and after recarbonation. Assume that one half the alkalinity in the softened water is in the bicarbonate form.

Solution

Component	mg/l	Equivalent Weight	meq/l
CO_2	8.8	22.0	0.40
Ca^{++}	40.0	20.0	2.00
Mg^{++}	14.7	12.2	1.21
Na^+	13.7	23.0	0.60
Alk	135	50.0	2.70
$SO_4^=$	29.0	48.0	0.60
Cl^-	17.8	35.0	0.51

(a) From the meq/l bar graph drawn in Figure 7-28a, the hypothetical combinations are $Ca(HCO_3)_2$, $Mg(HCO_3)_2$, $MgSO_4$, Na_2SO_4, and NaCl. Calcium hardness = $2.00 \times 50 = 100$ mg/l as $CaCO_3$ and magnesium hardness = $1.21 \times 50 = 60.5$ mg/l.

Component	meq/l	Applicable Equation	Lime meq/l	Soda Ash meq/l
CO_2	0.40	7-15	0.40	0
$Ca(HCO_3)_2$	2.00	7-16	2.00	0
$Mg(HCO_3)_2$	0.70	7-17	1.40	0
$MgSO_4$	0.51	7-18 & 7-19	0.51	0.51
			4.31	0.51

(b) Required lime dosage equals the amount needed for the softening reactions plus 35 mg/l CaO of excess lime to precipitate the magnesium.

Dosage = $4.31 \times 28 + 35 = 156$ mg/l CaO

Dosage of soda ash = $0.51 \times 53 = 27$ mg/l Na_2CO_3

(c) A hypothetical bar graph for the water after addition of softening chemicals is shown in Figure 7-28b. The dashed box to the left of zero is the excess lime (35 mg/l CaO = 1.25 meq/l) added to increase the pH high enough to precipitate the $Mg(OH)_2$. The 0.6 meq/l of Ca^{++} (30 mg/l as $CaCO_3$) and 0.2 meq/l of Mg^{++} (10 mg/l as $CaCO_3$) are the practical limits of hardness reduction. The 1.11 meq/l of Na^+ is the sum of the sodium originally in the water (0.60 meq/l) plus the amount increased by soda ash addition (0.51 meq/l). Alkalinity consists of 0.20 meq/l of OH^- associated with $Mg(OH)_2$, and the remainder is 0.60 meq/l carbonate ion. The $SO_4^=$ and Cl^- meqs/l are unchanged by the softening process.

Recarbonation converts the excess lime to calcium carbonate precipitate, Eq. 7-22, which is removed by second-stage sedimentation and filtration. Further carbon dioxide addition converts $CO_3^=$ to HCO_3^-, Eq. 7-23. Figure 7-28c illustrates the composition of the stabilized, finished water with a total hardness of 40 mg/l.

EXAMPLE 7-9

Consider selective calcium carbonate removal softening of a raw water with a bar graph as drawn in Figure 7-29a. Calculate the required lime dosage as CaO, and sketch the softened water bar graph after recarbonation and filtration.

Solution

The only hypothetical combination involving calcium is 2.0 meq/l of $Ca(HCO_3)_2$; therefore, no soda ash is needed and the lime required is 2.0 meq/l (Eq. 7-18), which equals 56 mg/l of CaO.

The softened water bar graph has 0.6 meq/l of calcium hardness (the practical limit of 30 mg/l), and the total alkalinity is 0.8 meq/l, which is 0.6 meq/l from the practical limit and 0.2 meq/l associated with magnesium in the raw water bar graph. The degree of recarbonation

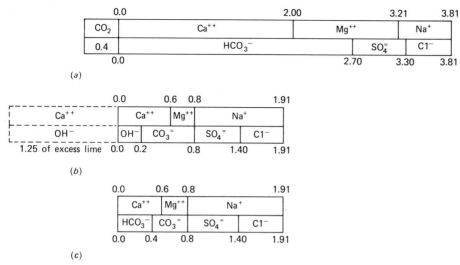

(a)

(b)

(c)

Figure 7-28 Bar graphs of raw and treated waters in milliequivalents per liter for Example 7-8. (a) Raw water prior to any treatment. (b) Water after excess lime softening, but before recarbonation and filtration. (c) Lime–soda ash softened water following recarbonation and filtration.

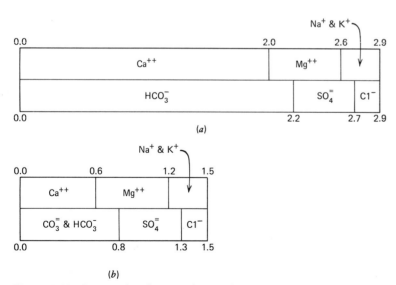

(a)

(b)

Figure 7-29 Bar graphs of raw and treated waters in milliequivalents per liter for Example 7-9. (a) Raw water prior to any treatment. (b) Water after selective calcium carbonate removal, recarbonation, and filtration.

determines the relative amounts of carbonate and bicarbonate anions. The other ions in the softened water are the same as in the raw water.

EXAMPLE 7-10

Consider precipitation softening by split treatment of the raw water described in Example 7-8. Assume that 40 percent of the flow is by-passed, and that 60 percent is given excess lime treatment in the first stage. Compute the chemicals required and the hardness of the finished water.

Solution

Flow passing through the first stage is processed in the same fashion as the excess lime treatment described in the solution to Example 7-8. However, the chemical additions are reduced to 60 percent, since only that fraction of the raw water is being treated.

$$\text{Lime dosage} = 0.60 \times 156 = 94 \text{ mg/l CaO}$$
$$\text{Soda ash} = 0.60 \times 27 = 16 \text{ mg/l Na}_2\text{CO}_3$$

Figure 7-28a is the bar graph of the by-passed water, and Figure 7-28b is the analysis of the first-stage effluent. These two are blended in the second stage where the excess lime reacts with the untreated water. The amount of excess hydroxide in the mixed flow streams is equal to

$$0.60(1.25 + 0.20) = 0.87 \text{ meq/l}$$

The other components of interest in the blended water are

$$\text{CO}_2 = 0.40 \times 0.40 = 0.16 \text{ meq/l}$$
$$\text{Ca(HCO}_3)_2 = 0.40 \times 2.00 = 0.80 \text{ meq/l}$$

First, the carbon dioxide is eliminated by the excess lime

$$0.87 \text{ meq/l} - 0.16 \text{ meq/l} = 0.71 \text{ meq/l}$$

The balance of the hydroxide ion reacts with calcium bicarbonate reducing it to

$$0.80 \text{ meq/l} - 0.71 \text{ meq/l} = 0.09 \text{ meq/l}$$

Final calcium hardness equals this remainder plus the limit of calcium carbonate removal,

$$(0.09 \text{ meq/l} + 0.60 \text{ meq/l})50 = 35 \text{ mg/l as CaCO}_3$$

Magnesium hardness in the finished water is

$$(0.40 \times 1.21 \text{ meq/l} + 0.60 \times 0.20 \text{ meq/l})50$$
$$= 30 \text{ mg/l as CaCO}_3$$

Total hardness in the softened water is $35 + 30 = 65$ mg/l.

7-11 ION EXCHANGE SOFTENING AND NITRATE REMOVAL

The common applications of ion exchange in water treatment are softening and removal of selected contaminants, such as the nitrate ion. Also, demineralization can be used to produce an ion-free water for laboratory and industrial purposes. By the exchange process, ions of a particular species in solution are replaced by ions of a different species attached to an insoluble resin. In practice, the water being treated is passed through a filter bed of ion exchange material that releases the substitute ions for the ions adsorbed from the water. When the capacity for exchanging ions has been depleted, a regenerating solution with a high concentration of the released ions is passed through the bed. This regeneration process displaces the contaminant ions and rejuvenates the exchange resin.

Cation Exchange Softening

The common cation exchange material in water softening is a polystrene resin manufactured chemically in the form of small spheres approximately 0.5 mm in diameter. Equation 7-24 is the reaction that takes place with R representing the anionic component of the resin. Calcium and magnesium cations are adsorbed, and an equivalent amount of sodium ions is released into solution. When the exhausted bed is regenerated, the reverse exchange takes place. As shown in Eq. 7-25, a strong sodium chloride solution displaces the bivalent hardness ions and regenerates the resin with sodium.

$$\begin{matrix} \text{Ca}^{++} \\ \text{Mg}^{++} \end{matrix} + \text{Na}_2R \rightarrow \begin{matrix} \text{Ca}R \\ \text{Mg}R \end{matrix} + \text{Na}^+ \qquad (7\text{-}24)$$

$$\begin{matrix} \text{Ca}R \\ \text{Mg}R \end{matrix} + \text{NaCl} \xrightarrow[\text{NaCl}]{\text{excess}} \text{Na}_2R + \begin{matrix} \text{Ca}^{++} \\ \text{Mg}^{++} \end{matrix} \qquad (7\text{-}25)$$

The milliequivalents bar graph of a water after cation exchange softening has the same number of milliequivalents as the raw water with all of the Ca^{++} and Mg^{++} ions replaced by Na^+ ion.

Anion Exchange for Nitrate Removal

The most effective exchanger for nitrate removal is a strongly basic, anion resin that uses sodium chloride for regeneration. All anion exchange resins preferentially remove divalent anions; therefore, both sulfate and nitrate ions are removed and replaced by chloride ions. This constitutes a major difficulty in processing water for nitrate removal, since many of the groundwaters contaminated with nitrate also have a significant sulfate concentration. When the capacity for extracting nitrate ions is depleted, a regenerating solution with a high salt content is passed through the bed to displace the sulfate and nitrate ions (Eq. 7-26).

$$RCl + \frac{SO_4^=}{NO_3^-} \underset{\substack{\text{regeneration} \\ \text{with NaCl}}}{\overset{\text{nitrate removal}}{\rightleftharpoons}} \frac{RSO_4}{RNO_3} + Cl^- \quad (7\text{-}26)$$

The quantity of waste brine from regeneration is about 5 percent of the water processed. This amount varies with the sulfate content of the water being treated and the exchange capacity of the resin, which decreases with usage. The major disadvantages of anion exchange treatment for nitrate removal are high operating costs and the problem of brine disposal.

Demineralization

This process uses both cation and anion exchangers in single- or dual- process tanks to completely remove all metal and nonmetal ions from solution. A strong-acid resin replaces cations with hydrogen ions, and a weak-base resin exchanges hydroxide ions for the removed anions. Deionization is not applied in municipal water treatment because of high cost; however, it is extensively used in industry to produce water of extremely high quality.

7-12   TURBIDITY REMOVAL

Surface waters generally contain suspended and colloidal solids from land erosion, decaying vegetation, microorganisms, and color-producing compounds. Coarser materials, such as sand and silt, can be eliminated to a considerable extent by plain sedimentation, but finer particles must be chemically coagulated to produce larger floc that is removable in subsequent settling and filtration. Destabilization of colloidal suspensions is discussed in Section 2-8, and typical flow schemes for surface-water treatment plants are diagrammed in Figure 7-2.

Coagulation and flocculation are sensitive to many variables, for instance, the nature of the turbidity-producing substances, type and dosage of coagulant, pH of the water, and the like. Of the many variables that can be controlled in plant operation, pH adjustment appears to be most important. Generally, the type of coagulants and aids available are defined by the plant process scheme; of course, dosages of these substances can be regulated to meet changes in raw-water quality. Also, mechanical mixing can be adjusted by varying the speed of the flocculator paddles.

Jar tests are widely used to determine optimum chemical dosages for treatment. This laboratory test (Section 2-11) attempts to simulate the full-scale coagulation-flocculation process and can be conducted for a wide range of conditions. The interpretation of test results involves visual and chemical testing of the clarified water. Ordinarily, a treatment plant gives better results than the jar test for the same chemical dosages. Of course, to confirm optimum chemical treatment, the water should be tested at various processing stages including the finished effluent. One common monitoring technique is analysis of the filtered water for turbidity; another is to record the length of time between backwashing of filters.

Coagulants

Commonly used metal coagulants in water treatment are (1) those based on aluminum, such as aluminum sulfate, sodium aluminate, potash alum, and ammonia alum; and (2) those based on iron, such as ferric sulfate, ferrous sulfate, chlorinated ferrous sulfate, and ferric chloride. The description that follows gives some of the relevant properties and chemical reactions of these substances in the coagulation of water. The latter are presented in the form of hypothetical equations with the understanding that they do not represent exactly what happens in water. Studies have shown that the hydrolysis of iron and aluminum salts is far more complicated than these formulas would indicate; however, they are useful in approximating the reaction products and quantitative relationships.

Aluminum sulfate, $Al_2(SO_4)_3 \cdot 14.3H_2O$, is by far the most widely used coagulant; the commercial product is commonly known as alum, filter alum, or alumina sulfate. Filter alum is a grayish-white crystallized solid containing approximately 17 percent water-soluble Al_2O_3 and is available in lump, ground, or powdered forms as well as concentrated solution. Ground alum is commonly measured by a gravimetric-type feeder into a solution tank from which it is transmitted to the point of application by pumping. Amber-colored liquid aluminum sulfate contains about 8 percent available Al_2O_3.

The hydrolysis of aluminum ion in solution is complex and is not fully defined. In pure water at low pH, the bulk of aluminum appears as Al^{+++} while in alkaline solution complex species such as $Al(OH)_4^-$ and $Al(OH)_5^=$ have been shown to exist. In the hypothetical coagulation equations, aluminum floc is written as $Al(OH)_3$. This is the predominant form found in a dilute solution near neutral pH in the absence of complexing anions other than hydroxide. The reaction between aluminum and natural alkalinity is given in Eq. 7-27. If lime or soda ash is added to the water with the coagulant, the theoretical reactions are as shown in Eqs. 7-28 and 7-29.

$$Al_2(SO_4)_3 \cdot 14.3H_2O + 3Ca(HCO_3)_2$$
$$= 2Al(OH)_3\downarrow + 3CaSO_4 + 14.3H_2O + 6CO_2 \quad (7\text{-}27)$$

$$Al_2(SO_4)_3 \cdot 14.3H_2O + 3Ca(OH)_2$$
$$= 2Al(OH)_3\downarrow + 3CaSO_4 + 14.3H_2O \quad (7\text{-}28)$$

$$Al_2(SO_4)_3 \cdot 14.3H_2O + 3Na_2CO_3 + 3H_2O$$
$$= 2Al(OH)_3\downarrow + 3Na_2SO_4 + 3CO_2 + 14.3H_2O \quad (7\text{-}29)$$

Based on these reactions, 1.0 mg/l of alum with a molecular weight of 600 reacts with 0.50 mg/l natural alkalinity, expressed as $CaCO_3$; 0.39 mg/l of 95 percent hydrated lime as $Ca(OH)_2$, or 0.33 mg/l 85 percent quicklime as CaO; or 0.53 mg/l soda ash as Na_2CO_3. When lime or soda ash is reacted with the aluminum sulfate, the natural alkalinity of the water is unchanged. Sulfate ions added with the alum remain in the finished water. In the case of natural alkalinity and soda ash, carbon dioxide is produced. The dosages of alum used in water treatment are in the range of 5 to 50 mg/l, with the higher concentrations needed to clarify turbid surface waters. Alum coagulation is generally effective within the pH limits of 5.5 to 8.0.

Sodium aluminate, $NaAlO_2$, may be used as a coagulant in special cases. The commercial grade has a purity of approximately 88 percent and may be purchased either as a solid or as a solution. Because of its high cost, sodium aluminate is generally used to aid a coagulation reaction instead of being the primary coagulant. It has been found effective for secondary coagulation of highly colored surface waters, and as a coagulant in the lime–soda ash softening process to improve settleability of the precipitate.

Ferrous sulfate, $FeSO_4 \cdot 7H_2O$ (also commonly known as copperas), is a greenish-white crystalline solid that is obtained as a by-product of other chemical processes, principally from the pickling of steel. Although available in liquid form from processed spent pickle liquor, the common commercial preparations are granular. Ferrous iron added to water precipitates in the oxidized form of ferric hydroxide; therefore, the addition of lime or chlorine is generally needed to provide effective coagulation. Ferrous sulfate and lime coagulation, shown in Eq. 7-30, is effective for clarification of turbid water and other reactions conducted at high pH values, for example, in lime softening.

$$2FeSO_4 \cdot 7H_2O + 2Ca(OH)_2 + \tfrac{1}{2}O_2$$
$$= 2Fe(OH)_3\downarrow + 2CaSO_4 + 13H_2O \quad (7\text{-}30)$$

Chlorinated copperas is prepared by adding chlorine to oxidize the ferrous sulfate. The advantage of this method, relative to lime addition, is that coagulation can be obtained over a wide range of pH values, 4.8 to 11.0. Theoretically, according to Eq. 7-31, each milligram per liter of ferrous sulfate requires 0.13 mg/l of chlorine, although additional chlorine is generally added to ensure complete reaction.

$$3FeSO_4 \cdot 7H_2O + 1.5Cl_2$$
$$= Fe_2(SO_4)_3 + FeCl_3 + 21H_2O \quad (7\text{-}31)$$

Ferric sulfate, $Fe_2(SO_4)_3$, is available as a commercial coagulant in the form of a reddish-brown granular material that is readily soluble in water. It reacts with the natural alkalinity of water according to Eq. 7-32, or with added alkaline materials, such as lime or soda ash, Eq. 7-33.

$$Fe_2(SO_4)_3 + 3Ca(HCO_3)_2$$
$$= 2Fe(OH)_3\downarrow + 3CaSO_4 + 6CO_2 \quad (7\text{-}32)$$
$$Fe_2(SO_4)_3 + 3Ca(OH)_2$$
$$= 2Fe(OH)_3\downarrow + 3CaSO_4 \quad (7\text{-}33)$$

In general, ferric coagulants are effective over a wide pH range. Ferric sulfate is particularly successful when used for color removal at low pH values; at high pH, it may be used for iron and manganese removal and as a coagulant in precipitation softening.

Ferric chloride, $FeCl_3 \cdot 6H_2O$, is used primarily in the coagulation of wastewater and industrial wastes, and it finds only limited use in water treatment. Normally, it is produced by chlorinating scrap iron, and is available commercially in solid and liquid forms. Being highly corrosive, the liquid must be stored and handled in corrosion-resistant tanks and feeders. The reactions of ferric chloride with natural and added alkalinity are similar to those of ferric sulfate.

Coagulant Aids

Difficulties with coagulation often occur because of slow-settling precipitates, or fragile flocs that are easily fragmented under hydraulic forces in basins and sand filters. Coagulant aids benefit flocculation by improving settling and toughness of flocs. The most widely used materials are polyelectrolytes; others are activated silica, adsorbent-weighting agents, and oxidants.

Synthetic polymers are long-chain, high-molecular-weight, organic chemicals commercially available under a wide variety of trade names. Polyelectrolytes are classified according to the type of charge on the polymer chain. Those possessing negative charges are called anionic, those positively charged are called cationic; and those carrying no electric charge are nonionic. Anionic or nonionic are often used with metal coagulants to provide bridging between colloids to develop larger and tougher floc growth (Figure 2-7c). The dosage required as a flocculent aid is generally in the order of 0.1 to 1.0 mg/l. In the coagulation of some waters, polymers can promote satisfactory flocculation at significantly reduced alum dosages. The potential advantages of polymer substitution are in reducing the quantity of waste sludge produced in alum coagulation, and in changing the character of the sludge such that it can be more easily dewatered.

Cationic polymers have been used successfully in some waters as primary coagulants for clarification. Although the unit cost of cationic polymers is about 10 to 15 times higher than the cost of alum, the reduced dosages required may nearly offset the increased cost of chemicals. Furthermore, unlike the gelatinous and voluminous aluminum hydroxide sludges, polymer sludges are relatively dense and easier to dewater for subsequent handling and disposal. Sometimes cationic and nonionic polymers may be used together to provide an adequate floc, the former being the primary coagulant and the latter a coagulant aid. Although significant strides have been made in the application of polyelectrolytes in water treatment, their main application is still as an aid rather than as a primary coagulant. Many waters cannot be treated by using polymers alone, but require aluminum or iron salts. Jar tests and actual plant operation must be used to determine the effectiveness of a particular proprietary polyelectrolyte in flocculation of a given water.

Activated silica is sodium silicate that has been treated with sulfuric acid, aluminum sulfate, carbon dioxide, or chlorine. A dilute solution is partially neutralized by an acid and then is allowed to age for a period of time up to

2 hr; additional aging runs the risk of solidification of the entire solution by gelation. The aged silicate solution is generally further diluted before being applied to the water. As a coagulant aid, it offers the advantages of increased rate of chemical reaction, reduced coagulant dose, extended optimum pH range, and production of a faster settling, tougher floc. One of the disadvantages of activated silica, relative to polyelectrolytes, is the precise control required in preparation and feeding. Activated silica is normally used with aluminum coagulants at a dosage between 7 and 11 percent of the alum dose, expressed as milligrams per liter of SiO_2.

Bentonitic clays may be used in treating waters containing high color, low turbidity, and low mineral content. Iron or aluminum floc produced under these conditions is frequently too light to settle readily. Addition of clay results in a weighting action that improves settleability. Clay particles may also adsorb organic compounds, improving treatment. Although exact dosage must be determined by testing, 10 to 15 mg/l frequently results in formation of a good floc. Other weighting agents that have been used besides clay are powdered silica, limestone, and activated carbon; the latter possess the additional advantage of high adsorptive capacity.

Acids and alkalies can be added to adjust the pH for optimum coagulation. The common acid used to lower the pH is sulfuric acid. Increase in pH is done by the addition of lime, soda ash, or sodium hydroxide. The latter is purchased and fed as a concentrated solution. The characteristics of lime and soda ash are discussed in Section 7-10.

EXAMPLE 7-11

A dose of 50 mg/l of alum is used in coagulating a turbid surface water. (a) How much natural alkalinity is consumed? (b) What changes take place in the ionic character of the water? (c) How many milligrams per liter of aluminum hydroxide are produced?

Solution

(a) From the text, 1.0 mg/l of alum reacts with 0.5 mg/l of natural alkalinity; therefore,

$$50 \text{ mg/l (alum)} \frac{0.50 \text{ mg/l (alk)}}{1.0 \text{ mg/l (alum)}}$$

$$= 25 \text{ mg/l of alkalinity as } CaCO_3$$

(b) 50 mg/l of alum is equivalent to 0.50 meq/l; therefore, based on Eq. 7-27, 0.50 meq/l of sulfate ion is added to the water. The aluminum ions precipitate out of solution, the calcium content is not affected, and 0.50 meq/l of bicarbonate is converted to carbon dioxide.

(c) From Eq. 7-27, one mole of alum (600 g) reacts to produce two moles of aluminum hydroxide ($2 \times 78 = 156$ g). Therefore,

$$\frac{156}{600} = \frac{\text{mg/l of } Al(OH)_3 \text{ produced}}{50 \text{ mg/l alum dose}},$$

$$\text{or } Al(OH)_3 = 13 \text{ mg/l}$$

EXAMPLE 7-12

A surface water is coagulated by adding 30 mg/l of ferrous sulfate and an equivalent dosage of lime. How many pounds of coagulant are used per million gallons of water processed? How many pounds of lime are required at a purity of 80 percent CaO?

Solution

Consumption of ferrous sulfate equals

$$30 \text{ mg/l} \times 8.34 \frac{\text{lb/mil gal}}{\text{mg/l}} = 250 \text{ lb/mil gal}$$

One equivalent weight of ferrous sulfate (139) reacts with one equivalent of 80 percent CaO (28/0.80 = 35). Therefore,

$$\text{Lime dosage} = 250 \frac{\text{lb}}{\text{mil gal}} \times \frac{35}{139} = 63 \text{ lb/mil gal}$$

7-13 TASTE AND ODOR CONTROL

One of the objectives in water treatment is to produce a palatable water that is aesthetically pleasing. Repeated presence of objectionable taste, odor, or color in a water supply may cause the public to question its safety for consumption. Flavor may be affected by inorganic salts or metal ions, a variety of organic chemicals found in nature or resulting from industrial wastes, or products of biological growths. Algae are the most frequent cause of taste and odor problems in surface supplies. Their metabolic activities impart odorous compounds that produce fishy, grassy, musty, or foul odors. Actinomycetes are moldlike bacteria that create an earthy odor.

Problems related to the palatability of water are generally unique in each system, and they must be individually studied to determine the best approach for prevention and cure. In groundwater treatment, aeration is frequently effective, since the odor compounds are often dissolved gases that can be stripped from solution. However, aeration is rarely effective in processing surface waters where the odor-producing substances are generally nonvolatile.

First consideration should be given to preventative measures in surface water supplies. If the interference can be traced to an industrial waste discharge, the source may be removed. In a eutrophic reservoir or lake, regular copper sulfate applications to the impounded water are effective in suppressing algal blooms that cause taste and odor, and filter clogging.

Carbon adsorption is usually the most effective chemical for reducing the level of taste and odor in the treatment of water. Activated carbon can be prepared from hardwood charcoal, lignite, nut shells, or other carbonaceous materials by controlled combustion to develop adsorptive characteristics. Each particle is honeycombed with thousands of molecular-sized pores that adsorb odorous substances. Activated carbon is available in powdered and granular forms. Granular carbon adsorption beds have not been used extensively in municipal water processing because of economic considerations; however, they are used by food and beverage industries in purification of product water. Carbon in municipal treatment is a finely ground, insoluble black powder that can be applied either through dry feed machines or as a slurry. It can be introduced in any stage of processing before filtration where adequate mixing is available to disperse the carbon, and where the contact time is 15 min or more before sedimentation or filtration. The optimum point of application is generally determined

by trial and error, and previous experience. Carbon adsorbs chlorine, and therefore these two chemicals should not be applied simultaneously, or in sequence without an appropriate time interval.

Oxidation of taste and odor compounds can be done by chlorination, chlorine dioxide, potassium permanganate, or ozone. Heavy chlorination may not be desirable because of trihalomethane formation (Section 7-7). Chlorine dioxide, produced by combining solutions of sodium chlorite and chlorine, has approximately the same oxidative power as chlorine without forming trihalomethanes. While oxidizing some aromatic organic compounds, it is ineffective for many other odorous substances. Potassium permanganate is a strong chemical effective in oxidizing a variety of taste- and odor-producing compounds. Since one of the reactions in water produces manganese dioxide (Eq. 7-15), it must be applied in processing before filtration. The use of ozone is limited in the United States.

7-14   REMOVAL OF DISSOLVED SALTS

Processes for separating salts from water in potable water treatment include distillation, reverse osmosis, and electrodialysis. Distillation and reverse osmosis are the common methods for desalinization of sea water. Reverse osmosis and electrodialysis are used to desalt brackish groundwater or to reduce the concentration of contaminants that are hazardous to human health, such as nitrate, fluoride, and radionuclides. In addition to high energy consumption, a significant problem at inland desalting plants is the disposal of reject brine. Common methods are evaporation ponds, deep-well injection, and piping to the ocean depending on the volume of reject brine, site geography, and climate.

Sea water has a salinity of 35,000 mg/l or more that is mostly sodium chloride. Desalinization by distillation is feasible, since the process operates nearly independent of dissolved solids concentration. Moreover, a product purity of less than 100 mg/l is easily attained to satisfy the quality standards of less than 100 mg/l of sodium and 200 mg/l of chloride in drinking water. The process involves heating sea water to the boiling point and converting it into steam to form water vapor that is then condensed, yielding a salt-free water. In order to increase energy efficiency, a series of 15 to 25 stages of evaporation-condensation units are employed to obtain maximum utilization of the heat energy in the applied steam.

Brackish groundwater, containing a few thousand milligrams per liter of dissolved solids, can be desalted by electrodialysis. In this process, ions are separated from the water by attracting them through selective ion-permeable membranes using an electrical potential. The unit consists of membranes interposed in the path of a direct current generated by end electrodes. The selectivity of the membranes alternates so that those permeable to anions, such as Cl^-, are placed between membranes permeable to cations like Na^+. With the saline water flowing between the membranes, the direct current passing through the water draws anions toward the anode from one compartment while cations are drawn in a similar manner in the opposite direction toward the cathode. From this movement of ions, the water in alternate compartments is reduced in ionic concentration by increasing the salinity of the water in the other compartments. The two flows become the desalted product water and the reject brine.

Reverse Osmosis

Reverse osmosis is the forced passage of water through a membrane against the natural osmotic pressure to accomplish separation of water and ions. The process of osmosis is illustrated in Figure 7-30a, where a thin membrane of cellulose acetate (0.10 to 0.15 mm thick) separates two salt solutions. Water from the side of lower salt concentration flows through the membrane to the solution of high concentration, attempting to equalize the salt content, while the membrane allowing water flow blocks passage of salt ions. If pressure is applied to the side of higher salt content, flow of water can be prevented; this pressure, at no net flow, is termed the osmotic pressure. The osmotic pressure of sea water is about 350 psi; brackish groundwater pressure, having a lower salt concentration, is significantly less. If pressure is increased (Figure 7-30c), the water flow is reversed and passes from salt water to the fresh water; in this manner the salts are separated from solution. Reverse-osmosis operating pressures vary between 350 and 1500 psi with a typical range of 600 to 800 psi. The rate of water transfer depends primarily on the difference in salt concentration between the solu-

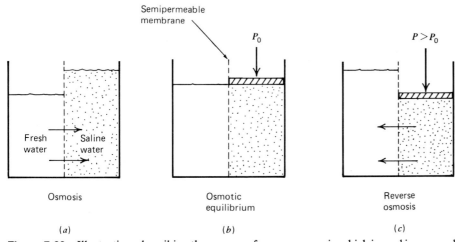

Figure 7-30 Illustrations describing the process of reverse osmosis, which is used in removal of dissolved salts from water. (a) Normal osmotic flow of water through a semipermeable membrane is from the low to high concentration salt solution. (b) The pressure at which no net flow occurs is referred to as the osmotic pressure, P_0. (c) In reverse osmosis, water is forced by high pressure from the salty solution through the membrane that blocks the passage of ions.

tions, characteristics of the membrane, and magnitude of the applied pressure.

The spiral-wound module, shown in Figure 7-31, is constructed of large membrane sheets covering both sides of a porous backing material that collects the permeate (product water). The membranes are sealed in pairs on the two long edges and one end to form an envelope enclosing the permeate collector. The other end of the membrane envelope is connected to a central perforated tube, which receives and carries away the permeate from the collectors. Several of these membrane envelopes with mesh spacers between them for brine flow are rolled up to form a spiral-wound module. Saline water enters the end of the module through voids between the membrane envelopes

provided by the mesh spacers. Under high pressure, water is forced from the brine in the spacer voids through the membranes and conveyed by the enclosed porous permeate collectors to the perforated tube in the center of the module. Reject brine discharges from the spacer voids at the outlet end of the module.

A second common membrane is manufactured of aromatic polyamide in the form of hollow fibers that are finer than a human hair with an outside diameter of 85 to 100 μm and an inside diameter of 42 μm. The pressure vessel of a hollow-fiber module contains a very large number of microfiber membranes densely packed in a U-bundle with their openings secured in an end block of the module. Saline water enters the module through a central

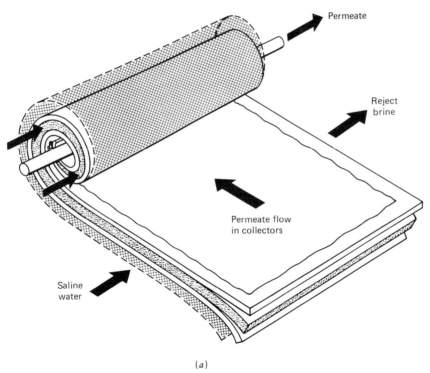

(a)

Figure 7-31   Spiral-wound module for reverse osmosis. (a) Schematic of unwound module. (b) Cutaway view of wound module. (c) Detail of layered construction. (Courtesy of Fluid Systems Division, UOP.)

perforated feed tube and flows radially through the fiber bundle toward the outer shell of the cylinder. Under high pressure, water from the surrounding brine enters the hollow fibers and exits from their open ends at the outlet end of the module. Reject brine arriving at the outer shell of the pressure vessel is collected by a flow screen and conveyed out of the module.

A basic reverse-osmosis system consists of pretreatment units, pumps to provide high operating pressure, post treatment, tanks and appurtenances for cleaning and flushing, and a disposal system for reject brine. The saline water fed to RO modules must be clear of suspended solids and free of organic matter and excessive hardness, iron, and manganese to prevent fouling and scaling of the membranes. As diagramed in Figure 7-32 and pictured in Figure 7-33, pretreatment commonly includes granular-media and cartridge (25 μm mesh size) filtration, acidification, and addition of a scale inhibitor. In large plants, the granular-media filtration may be preceded by chemical coagulation for turbid waters or precipitation softening for hard waters. If a water contains dissolved organic substances, the common pretreatment is filtration through granular activated carbon followed by chlorination to inhibit biological growths. Reducing the pH to 6 or less by feeding sulfuric acid converts a portion of the bicarbonate ion to carbon dioxide gas, making the water corrosive to reduce the probability of calcium carbonate scale and oxidation of iron and

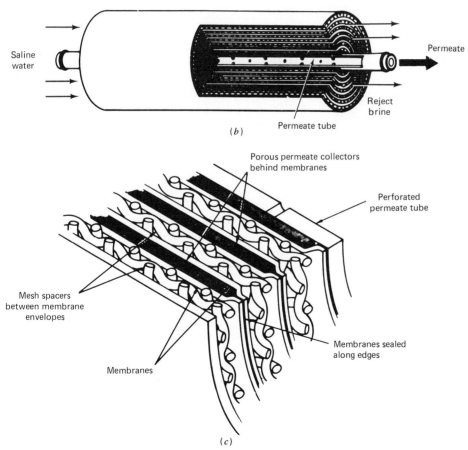

(b)

(c)

Figure 7-31 (*continued*)

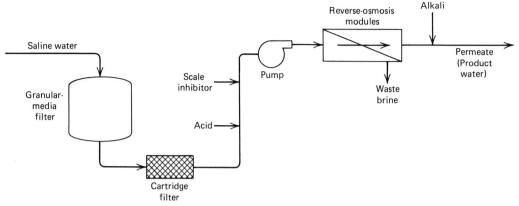

Figure 7-32 Flow diagram of a basic reverse-osmosis system.

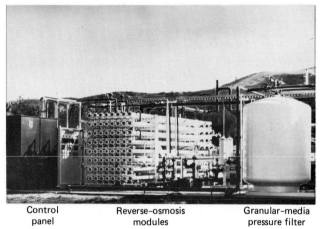

Figure 7-33 Typical small reverse-osmosis installation. (Courtesy of Fluid Systems Division, UOP.)

manganese. Sodium hexametaphosphate, or some other inhibitor, is applied to avoid chemical scaling. Post treatment is necessary to stabilize the permeate since the carbon dioxide formed in acidification can pass through the membranes with the product water. Aerators are used to remove the carbon dioxide, and lime or soda ash added to provide final pH adjustment. Periodic cleaning of the membrane surfaces is required to maintain high water transfer efficiency. Modules are flushed with acid rinses and cleaning agents to remove

any buildup of metal ions, salt precipitates, organic matter, or biological growths.

The quantity of reject brine varies from 10 to 30 percent of the saline-water feed depending on method of operation. A typical plant that is operating on a feed of 2000 mg/l total dissolved solids can economically achieve about 75 percent conversion; that is, 100 gal of feed produces 75 gal of water with a brine rejection of 25 gal containing 6000 to 7000 mg/l of dissolved solids. A higher percentage of recovery is possible at reduced hydraulic loading or higher pressures; however, the higher dissolved solids content in the rejected water has a greater potential for forming scale on the membranes. The costs involved in disposing of a large volume of reject water are a critical economic and environmental problem associated with reverse-osmosis treatment at inland sites.

7-15 CORROSION CONTROL

Ferrous metal when placed in contact with water or moist soil results in an electric current caused by the reaction between the metal surfaces and existing chemicals in the earth or water. The basic electrode reactions are represented by Eqs. 7-34 and 7-35. At the anode, iron goes into solution as the ferrous ion, and at the cathode, when dissolved oxygen is present, hydroxide ion is formed.

Anode: $Fe \rightarrow Fe^{++} + 2$ electrons   (7-34)

Cathode: 2 electrons $+ H_2O + \frac{1}{2}O_2 \rightarrow OH^-$ (7-35)

Ferrous (Fe^{++}) iron is further oxidized to the ferric (Fe^{+++}) state in the presence of oxygen and precipitates as insoluble ferric hydroxide, or rust, Eq. 7-36.

$$2Fe^{++} + 5H_2O + \tfrac{1}{2}O_2 \rightarrow 2Fe(OH)_3\downarrow + 4H^+ \quad (7\text{-}36)$$

The rate of corrosion is controlled by the concentration and diffusion rate of dissolved oxygen to the surface of the iron. Diffusion of oxygen to the anode is limited somewhat by the physical barrier of rust forming on the surface. The result of pipe corrosion is pitting and scaling of the surface referred to as tuberculation, discoloration of the water, and eventual failure of the pipe.

External corrosion of ductile-iron pipe buried in normal, nonaggressive soils is prevented by a coating of coal tar or enamel on the exterior of the pipe. In corrosive soils, ductile-iron pipelines including the fittings can be encased in plastic tubing to isolate the exterior from the surrounding soil at the time of pipe installation. Plastic pipe and asbestos cement pipe are resistant to corrosion.

The best way to protect a ductile-iron pipe against internal corrosion is by lining with a thin layer of cement mortar placed during manufacture of the pipe. In addition to cement-mortar lining, a calcium carbonate film can be maintained by chemical stabilization of the water to protect the interior of pipes.

Water Stabilization

The formation of a protective film of calcium carbonate on the interior of water pipes depends on control of the following chemical reaction:

$$CO_2 + CaCO_3 + H_2O \rightleftharpoons Ca(HCO_3)_2 \quad (7\text{-}37)$$

Excess free carbon dioxide dissolves $CaCO_3$, while less than equilibrium carbon dioxide forms $CaCO_3$ scale. For stability of a water, the pH must be compatible with carbonate-carbon dioxide equilibrium at the measured water temperature. In actual practice, a calculated oversaturation of 4 to 10 mg/l of $CaCO_3$ is recommended in treatment of a water to eliminate aggressive carbon dioxide. The minimum calcium and alkalinity values should be in the range of 40 to 70 mg/l as $CaCO_3$ depending on the concentrations of other ions in solution. Finally, the water must contain sufficient dissolved oxygen, normally 4 to 5 mg/l, to oxidize the nonprotective ferrous hydroxide to ferric hydroxide. Ferric hydroxide is precipitated along with the calcium carbonate scale.

The most common corrosion index for predicting the stability of a water to deposit or dissolve calcium carbonate is the Langelier saturation index, which is applicable in the pH range of 6.5 to 9.5. This index equates pH, calcium content, alkalinity, and ionic strength as follows:

$$\begin{aligned}
SI &= pH - pH_s \\
&= pH - [(pK'_2 - pK'_s) + pCa^{++} + pAlk] \quad (7\text{-}38)
\end{aligned}$$

where

$$\begin{aligned}
SI &= \text{saturation index, dimensionless} \\
pH &= \text{measured pH of the water} \\
pH_s &= \text{pH at equilibrium } (CaCO_3 \\
&\quad \text{saturation)} \\
(pK'_2 - pK'_s) &= \text{constants based on ionic} \\
&\quad \text{strength and temperature} \\
pCa^{++} &= \text{negative logarithm of the} \\
&\quad \text{calcium ion concentration in} \\
&\quad \text{moles per liter} \\
pAlk &= \text{negative logarithm of the total} \\
&\quad \text{alkalinity in equivalents per liter}
\end{aligned}$$

A positive index signifies that the water is oversaturated and can precipitate calcium carbonate. A negative value signifies a corrosive water that can dissolve calcium carbonate. A saturation index of less than -2.0 indicates a highly corrosive water, and an index between -2.0 and -0.1 indicates a moderately corrosive water. The numerical value, however, does not relate to either the rate at which stability will be attained or the dosage of chemicals necessary to achieve equilibrium.

The value of $pK'_2 - pK'_s$ is based on temperature and ionic strength as determined from Table 7-3. The ionic strength of a water is calculated as

$$\text{Ionic strength} = 0.5(C_1Z_1{}^2 + C_2Z_2{}^2 + \cdots + C_nZ_n{}^2)$$
$$(7\text{-}39)$$

where $C =$ concentration of an ionic species, moles per liter

$Z =$ valence of the individual ion of that species

Table 7-3. Values of pK'_2 and pK'_s at 25° C for Various Ionic Strengths and the Difference ($pK'_2 - pK'_s$) for Various Temperatures

Ionic Strength	Total Dissolved Solids (mg/l)	25° C			$pK'_2 - pK'_s$									
		pK'_2	pK'_s	$pK'_2 - pK'_s$	0° C	10° C	20° C	30° C	40° C	50° C	60° C	70° C	80° C	90° C
0.000		10.26	8.32	1.94	2.45	2.23	2.02	1.86	1.68	1.52	1.36	1.23	1.08	0.95
0.001	40	10.26	8.19	2.07	2.58	2.36	2.15	1.99	1.81	1.65	1.49	1.36	1.21	1.08
0.002	80	10.25	8.14	2.11	2.62	2.40	2.19	2.03	1.85	1.69	1.53	1.40	1.25	1.12
0.003	120	10.25	8.10	2.15	2.66	2.44	2.23	2.07	1.89	1.73	1.57	1.44	1.29	1.16
0.004	160	10.24	8.07	2.17	2.68	2.46	2.25	2.09	1.91	1.75	1.59	1.46	1.31	1.18
0.005	200	10.24	8.04	2.20	2.71	2.49	2.28	2.12	1.94	1.78	1.62	1.49	1.34	1.21
0.006	240	10.24	8.01	2.23	2.74	2.52	2.31	2.15	1.97	1.81	1.65	1.52	1.37	1.24
0.007	280	10.23	7.98	2.25	2.76	2.54	2.33	2.17	1.99	1.83	1.67	1.54	1.39	1.26
0.008	320	10.23	7.96	2.27	2.78	2.56	2.35	2.19	2.01	1.85	1.69	1.56	1.41	1.28
0.009	360	10.22	7.94	2.28	2.79	2.57	2.36	2.20	2.02	1.86	1.70	1.57	1.42	1.29
0.010	400	10.22	7.92	2.30	2.81	2.59	2.38	2.22	2.04	1.88	1.72	1.59	1.44	1.31
0.011	440	10.22	7.90	2.32	2.83	2.61	2.40	2.24	2.06	1.90	1.74	1.61	1.46	1.33
0.012	480	10.21	7.88	2.33	2.84	2.62	2.41	2.25	2.07	1.91	1.75	1.62	1.47	1.34
0.013	520	10.21	7.86	2.35	2.86	2.64	2.43	2.27	2.09	1.93	1.77	1.64	1.49	1.36
0.014	560	10.20	7.85	2.36	2.87	2.65	2.44	2.28	2.10	1.94	1.78	1.65	1.50	1.37
0.015	600	10.20	7.83	2.37	2.88	2.66	2.45	2.29	2.11	1.95	1.79	1.66	1.51	1.38
0.016	640	10.20	7.81	2.39	2.90	2.68	2.47	2.31	2.13	1.97	1.81	1.68	1.53	1.40
0.017	680	10.19	7.80	2.40	2.91	2.69	2.48	2.32	2.14	1.98	1.82	1.69	1.54	1.41
0.018	720	10.19	7.78	2.41	2.92	2.70	2.49	2.33	2.15	1.99	1.83	1.70	1.55	1.42
0.019	760	10.18	7.77	2.41	2.92	2.70	2.49	2.33	2.15	1.99	1.83	1.70	1.55	1.42
0.020	800	10.18	7.76	2.42	2.93	2.71	2.50	2.34	2.16	2.00	1.84	1.71	1.56	1.43

Source: Larson, T. E., and A. M. Buswell. "Calcium Carbonate Saturation Index and Alkalinity Interpretations." *Journal American Water Works Association*, Vol. 34, 1942, p. 1667. Copyright 1942 by the American Water Works Association, Inc.

An unstable water can be treated with chemicals to achieve the proper amounts and balance between calcium and alkalinity at equilibrium pH. Lime is the strongest chemical for treating a corrosive water to increase pH and neutralize free carbon dioxide. If the water is soft, lime has the advantage of adding calcium. Soda ash also increases pH and neutralizes carbon dioxide, but at a higher dosage than lime. Soda ash is often used to increase alkalinity. Addition of carbon dioxide controls a scale-forming water by reducing the pH and increasing alkalinity. Normally only one of these chemicals is needed to adjust pH and stabilize an untreated or treated water. The exception is remineralization of desalinated sea water that requires both lime and carbon dioxide to increase both the concentration of calcium and alkalinity and to establish the proper pH. After chemical treatment of a water to make it slightly scale forming, metaphosphates can be added at a low dosage to sequester the excess calcium and carbonate ions to prevent precipitation. An equilibrium is reached whereby the $CaCO_3$ coating does not dissolve because the water is saturated with calcium and carbonate ions; yet these ions cannot precipitate to form scale because of the sequesting action of the metaphosphate ions. The common chemical used in water treatment is sodium hexametaphosphate, which is available in both powdered and granular forms.

Cathodic Protection

Cathodic protection is a means of counteracting corrosion by preventing the dissolution of iron shown in Eq. 7-34. Either a sacrificial galvanic anode, composed

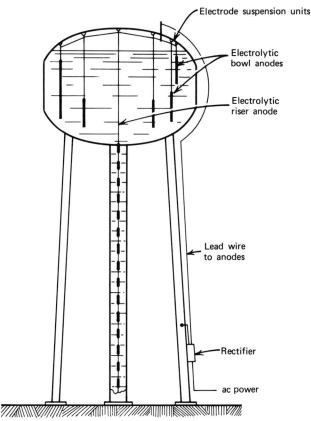

Figure 7-34 Cathodic protection can be used to prevent corrosion of the interior surfaces of a water storage tank.

of a metal higher in the electromotive series, or an electrolytic anode energized by an external source of direct current may be used to protect ferrous structures. Magnesium or zinc galvanic anodes provide a battery action with iron such that they corrode and are sacrificed while the iron structure they are connected to is protected from dissolution. This type of system is common to small hot-water heaters.

Large steel waterworks structures are given cathodic protection using electrolytic anodes that are made anionic by application of a direct current. Anodes may be of many different metals, for instance, graphite, carbon, platinum, aluminum, iron, or steel alloys. They are energized by connecting to the positive terminal of a direct current power supply, usually a rectifier, while the structure being

protected is connected to the negative terminal. The impressed current attracts electrons to the iron, preventing ionization and, hence, corrosion. Figure 7-34 sketches the application of cathodic protection to the interior of an elevated storage tank. Under some conditions, galvanic anodes are employed in lieu of, or in combination with, the rectifier depending on the tank condition and chemical characteristics of the water. Buried tanks are protected from exterior corrosion by placing the anodes in the ground surrounding the tank. Except in unusual cases, cathodic systems are not used to protect water distribution piping because of the high cost.

EXAMPLE 7-13

The chemical analyses of a water after split-treatment softening are listed below. The measured pH was 8.5. Calculate the Langelier saturation index at a water temperature of 25° C.

Solution

Component	mg/l	Molecular Weight	moles/l	Equivalent Weight	meq/l
Ca^{++}	19	40.1	0.00047	20.0	0.95
Mg^{++}	10	24.3	0.00041	12.2	0.82
Na^+	41	23.0	0.00178	23.0	1.78
$CO_3^=$	18	60.0	0.00030	30.0	0.60
HCO_3^-	59	61.0	0.00097	61.0	0.97
$SO_4^=$	76	96.0	0.00079	48.0	1.58
Cl^-	14	35.5	0.00040	35.5	0.40

The ionic strength using Eq. 7-39 is calculated as follows:

$$Ca^{++} = 0.5 \times 0.00047 \times 4 = 0.00094$$
$$Mg^{++} = 0.5 \times 0.00041 \times 4 = 0.00082$$
$$Na^+ = 0.5 \times 0.00178 \times 1 = 0.00089$$
$$CO_3^= = 0.5 \times 0.00030 \times 4 = 0.00060$$
$$HCO_3^- = 0.5 \times 0.00097 \times 1 = 0.00048$$
$$SO_4^= = 0.5 \times 0.00079 \times 4 = 0.00158$$
$$Cl^- = 0.5 \times 0.00040 \times 1 = \underline{0.00020}$$
$$\text{Ionic strength} = 0.00551$$

From Table 7-3, $pK'_2 - pK'_s = 2.2$ at $25°$ C

$$pCa^{++} = -\log 0.00047 = 3.3$$

$$pAlk = -\log\left(\frac{0.60 + 0.97}{1000}\right) = 2.8$$

Substituting into Eq. 7-38, the Langelier saturation index is

$$SI = 8.5 - (2.2 + 3.3 + 2.8) = 0.2$$

The value $+0.2$ signifies a slightly scale-forming water.

The maximum recommended oversaturation of a treated water is 10 mg/l of $CaCO_3$, which amounts to 4 mg/l of Ca^{++} and 6 mg/l of $CO_3^=$. (Refer to the discussion in the text under Eq. 7-37.) The following calculations recompute the saturation index after reducing the calcium concentration to 15 mg/l and the carbonate to 12 mg/l.

$$Ca^{++} = 15/(40.1 \times 1000) = 0.00037 \text{ moles/l}$$

$$pCa^{++} = -\log 0.00037 = 3.4$$

$$CO_3^= = 12/30 = 0.40 \text{ meq/l}$$

$$pAlk = -\log\left(\frac{0.40 + 0.97}{1000}\right) = 2.9$$

The change in ionic strength is not significant.

$$SI = 8.5 - (2.2 + 3.4 + 2.9) = 0$$

Therefore, the stability of the water appears to be satisfactory for distributing in a pipe network.

7-16   SOURCES OF WASTES IN WATER TREATMENT

Wastes originating from water treatment in approximate order of abundance are residues from chemical coagulation, precipitates from softening, filter backwash water, settled solids from presedimentation, oxides from iron and manganese removal, and spent brines from regeneration of ion exchange units. These wastes vary widely in composition, containing the concentrated materials re-moved from raw water and chemicals added in treatment. They are produced continuously and discharged intermittently. Settled floc is allowed to accumulate in clarifiers over relatively long periods of time, while backwashing of filters produces a high flow of wastewater for a few minutes, usually once a day for each filter.

Coagulation Wastes

The chief constituent in coagulation sludge is either hydrated aluminum from alum or iron oxides from iron coagulants. Small quantities of activated carbon and coagulant aids, such as polyelectrolytes and activated silica, may be included. Particulate matter entrained in the floc is mostly inorganic in nature, being principally silt and clay. Since the organic fraction is small, the sludge does not undergo active biological decomposition. Aluminum hydroxide sludges are gelatinous in consistency, which makes them difficult to dewater. Settled sludges have low solids concentrations, usually between 0.2 and 2.0 percent. Iron precipitates are slightly denser than alum sludges. The total solids produced in alum coagulation of a surface water can be estimated by using the following relationship, as illustrated in Example 7-14.

Sludge solids,
  lb/mil gal

$$= 8.34\left(\frac{\text{alum dosage, mg/l}}{4} + \frac{\text{raw-water}}{\text{turbidity}}\right) \quad (7\text{-}40)$$

Softening Sludge

Lime–soda ash softening of groundwaters yields a sludge free of extraneous inorganic and organic matter consisting of calcium carbonate, magnesium hydroxide, and unreacted lime. This residue is generally stable, dense, inert, and relatively pure, allowing recovery of the lime by recalcination where economically feasible. The settled solids concentration in softening sludges is in the range of 2 to 15 percent depending on the ratio of calcium to magnesium precipitates. Calcium carbonate provides a compact sludge easy to handle while magnesium hydroxide, like aluminum hydroxide, is gelatinous and does not consolidate well by gravity settling. Highly turbid, hard surface waters are often treated by both precipitation softening and coagulation. The character and settleability of these combined sludges vary considerably depending on such factors as nature of turbidity in the raw water, ratio of calcium to magnesium in the softening precipitate, type and dosage of metal coagulant, and filter aids used. In general, lime and iron sludges have solids concentrations greater than 10 percent and often greater than 15 percent, while alum-lime sludges have settled solids densities of less than 10 percent. The quantity of sludge solids produced in softening can be estimated, as is shown in Example 7-15, by using the precipitation-softening Eqs. 7-17 to 7-21.

Other Residues

The amount of sludge attributable to iron and manganese removal is relatively small and negligible compared to coagulation or softening process wastes. Where the sole treatment is iron and manganese removal from groundwater, 50 to 90 percent of the hydrated ferric and manganic oxides are trapped in the filters and appear in dilute backwash water. Since the concentration of these metals in raw water is generally low, the amount of solids produced and volume of wastewater are correspondingly small.

Where turbid surface waters are given presedimentation, the accumulated solids consist of fine sands, silts, clays, and organic debris. Sometimes polymers or other chemicals are applied to the raw water to improve settleability; yet, they rarely contribute a measurable amount of solids to the sludge. Presedimentation residue may be hauled to landfill or, in some instances, discharged back to the watercourse.

Spent brine solutions from regeneration of ion exchange softeners contain regenerate sodium chloride plus calcium and magnesium ions displaced from the exchange resin. The character and volume of the waste liquid depend on the amount of hardness removed and operation of the exchange unit; the volume of spent brine is usually 3 to 10 percent of the treated water.

Filter Wash Water

Backwashing of filters produces a relatively large volume of wastewater with low solids concentration in the range of 0.01 to 0.1 percent (100 to 1000 mg/l). The total solids content depends on efficiency of prior coagulation and sedimentation and may be a substantial fraction, say 30 percent, of the residue resulting from treatment. Two to three percent of all water processed is used for filter washing; the exact amount is contingent on the type of treatment system and the filter backwashing technique. Wash water may be discharged to a recovery basin and recycled for processing with the raw water. In the case of a lime softening plant that is treating groundwater, the backwash may be collected, mixed, and returned to the inlet of the plant without solids removal. However, in surface-water plants this often creates a buildup of undesirable solids, for example, algae, that keep cycling through the system. Here, the suspended solids are allowed to settle, often after the addition of a polyelectrolyte to improve flocculation, and only the overflow is returned for reprocessing, as is illustrated in Figure 7-3. Settled sludge is discharged from the bottom of the clarifier-surge tank to a sludge thickener, or dewatering unit, or directly to disposal. Sometimes backwash water is released to a sanitary sewer for final disposition at the wastewater treatment plant; a surge tank may still be required to even out the flow so that the sewer is not hydraulically overloaded. Where lagoons are used to dispose of sludge solids, the wash water may be directed to these ponds, and occasionally the pond water is decanted from near the surface and recycled to the plant.

EXAMPLE 7-14

A surface-water treatment plant coagulates a raw water having a turbidity of 9 units by applying an alum dosage of 30 mg/l. Estimate the total sludge solids production in pounds per million gallons of water processed. Compute the volume of sludge from the settling basin and filter backwash water using 1.0 percent solids concentration in the sludge and 500 mg/l of solids in the wash water. Assume that 30 percent of the total solids are removed in the filter.

Solution

Applying Eq. 7-40,

$$\text{Total sludge solids} = 8.34\left(\frac{30}{4} + 9\right) = 137 \text{ lb/mil gal}$$

$$\text{Solids in sludge} = 0.70 \times 137 = 96 \text{ lb}$$

$$\text{Solids in backwash water} = 0.30 \times 137 = 41 \text{ lb}$$

$$\text{Volume} = \frac{\text{sludge solids (lb)}}{\text{solids fraction} \times 8.34 \text{ (lb/gal)}}$$

$$\text{Sludge volume} = \frac{96}{\dfrac{1.0}{100} \times 8.34} = 1150 \text{ gal/mil gal}$$

$$\text{Wash-water volume} = \frac{41}{\dfrac{500}{1,000,000} \times 8.34}$$

$$= 9830 \text{ gal/mil gal}$$

EXAMPLE 7-15

Calculate the residue produced in the excess lime softening of the water defined in Example 7-8.

Solution

Component in Water			Precipitate Produced	
Formula	meq/l	Applicable Equation	$CaCO_3$ meq/l	$Mg(OH)_2$ meq/l
CO_2	0.40	7-15	0.40	0
$Ca(HCO_3)_2$	2.00	7-16	4.00	0
$Mg(HCO_3)_2$	0.70	7-17	1.40	0.70
$MgSO_4$	0.51	7-18 and 7-19	0.51	0.51
Excess lime	1.25	7-20	1.25	0
			7.56	1.21
Minus the practical limits (solubility)			−0.60	−0.20
			6.96	1.01

$$\text{Residue} = CaCO_3 + Mg(OH)_2$$

$$= 6.96 \times 50 + 1.01 \times 29.2 = 378 \text{ mg/l}$$

7-17 DEWATERING AND DISPOSAL OF WASTES FROM WATER TREATMENT PLANTS

Prior to regulations governing the disposal of processing wastes, settled sludges and backwash waters were discharged to rivers. Lagoons, or sludge drying beds, were used where a watercourse was not conveniently available for waste discharge. In some municipalities, sanitary sewers received the waste flows, and in rare cases softening plants used recalcination to process calcium carbonate precipitate back to lime for reuse. Since residues from each water plant are unique, no specific treatment process for dewatering and disposal will yield the same results. In fact, although a variety of alternative methods are available, only one or two techniques may be applicable to a particular location.

Lagoons

Ponding is a popular and acceptable method for dewatering, thickening, and the temporary storage of waste sludge. Lagoons are constructed by excavation or enclosing of an area by dikes. Drainage may be improved by blanketing the bottom with sand and installing underdrains. An overflow device is normally constructed to decant clear supernatant, particularly if filter wash water is discharged to the lagoon. Although ponding is an inefficient process, where suitable land area is available that is inexpensive, any other disposal method may be difficult to justify. The amount of land needed depends on factors such as type, character, and solids content of the waste sludge; design features including underdrains and decanters; climate; and method of operation.

Sludges from water-softening plants are readily dewatered in lagoons. Slow consolidation of the settled precipitates yields concentrations of about 50 percent solids. This material can be removed by a scraper or dragline; however, lime sludge is considered a poor landfill, and final disposition of lagoon residue can be a serious problem where burial sites are scarce. Alum sludge is more difficult to dewater and, even after a long period of time, consolidates to only 10 to 15 percent solids density.

If the surface of settled alum sludge is exposed, evaporation may provide a hard crust but the remaining depth has the character of a gel that turns into a viscous liquid on agitation. This slurry may be removed by a dragline and dumped on the banks to air dry prior to hauling. As with lime sludge, it does not make a good landfill. In cold climates, freezing can enhance the dewatering character of alum sludge. On thawing, the moisture content decreases and the solids are in the form of small granular particles that dry to a brown powder.

Drying Beds

Sand drying beds are generally applicable to small water plants in areas where land is readily available. A bed consists of 6 to 12 in. of coarse sand underlain by layers of graded gravel. The earth bottom is graded slightly to tile underdrains placed in trenches. A surrounding wall is used to contain the liquid sludge when applied to the bed; filling depths usually range from 2 to 5 ft. Dewatering action is essentially gravity drainage aided by air drying, although the operation may include decanting supernatant. After repeated sludge applications over a period of several months, the sludge cake formed on the surface is removed either by hand shoveling or mechanical means.

Gravity Thickening

Separate thickeners may be used as the first step in sludge processing for volume reduction when consolidation of sludge in settling basins used to clarify coagulated water is inadequate. Polymers or other coagulant aids may be applied to improve the settleability of the sludge solids. Figure 7-35 illustrates a typical gravity thickener. Compared to a conventional settling basin, the tank is deeper to accommodate a greater volume of sludge and has a heavier raking mechanism. Pickets or palings attached to the collector arms stir through the sludge, providing cavities for the release of trapped water. Flow enters from behind an inlet well in the center of the tank and is directed downward. Supernatant overflows a peripheral weir, while underflow of thickened sludge is drawn from a bottom sump in the tank. In waste-

water treatment, thickening is normally considered a continuous-flow process, but in handling water treatment plant or industrial waste sludges, flow to the thickener is often intermittent. While performing clarification, tanks fed intermittently can be designed to allow for storage of accumulated sludge. This holding tank function is often essential to efficient subsequent dewatering by mechanical units, such as filters or centrifuges. Gravity thickeners are often sized on the basis of solids loading per square foot of tank area; values in the range of 5 to 10 lb/sq ft/day are common. However, individual sludges vary widely in their consolidating characteristics, and thickeners should be selected on the basis of local conditions.

Centrifugation

The first centrifuge developed was a basket type that is sketched in Figure 7-36a. It consists of a spinning cylinder creating high centrifugal forces to push solids against the drum wall. Feed slurry enters in the center at the bottom, and clarified liquid discharges over a lip ring at the top. When cake depth in the bowl reaches a preset thickness, the centrifugation process is interrupted to scrape out collected sediment using a skimmer or knife plow. After cake discharge, feed is again started.

The most popular centrifuge in dewatering water and wastewater sludges is the solid bowl scroll type illustrated in Figure 7-36b. The unit consists of a rotating solid bowl in the shape of a cylinder with a cone section on one end, and an interior rotating screw conveyer. Feed slurry, entering from the center, is held against the bowl wall by centrifugal force. Settled solids are moved by the conveyer to one end of the bowl while clarified effluent discharges at the other end. Additional details are shown in the cutaway view in Figure 7-37. The conical bowl shape at the solids discharge end enables the conveyer to move the solids out of the liquid for drainage before being discharged.

A major advantage of the solid bowl centrifuge is operational flexibility. Machine variables include pool volume, bowl speed, and conveyer speed. The depth of liquid held against the bowl wall can be controlled by an adjustable plate dam at the discharge end of the bowl. Pool volume adjustment changes the drainage deck

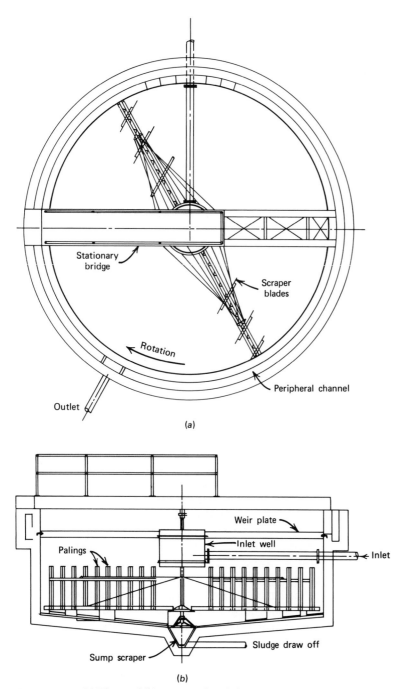

(a)

(b)

Figure 7-35   (a) Plan and (b) cross-sectional view of a gravity sludge thickener. (Courtesy of Infilco Degremont, Inc.)

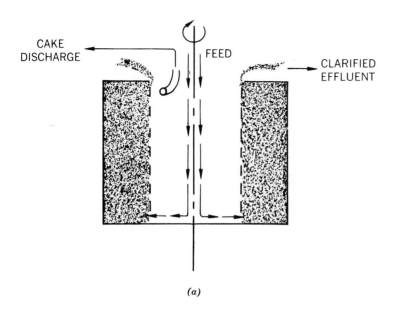

(a)

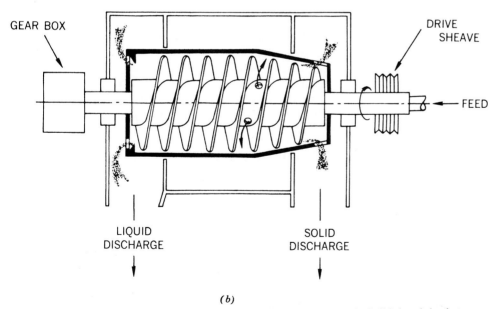

(b)

Figure 7-36 Schematic diagrams for two types of centrifuges. (*a*) Solid bowl basket centrifuge with skimmer. (*b*) Solid bowl scroll centrifuge, also illustrated in Figure 7-37, is the common type used in dewatering waste sludges from water and wastewater treatment. (Courtesy of Sharples-Stokes Div., Pennwalt Corp.)

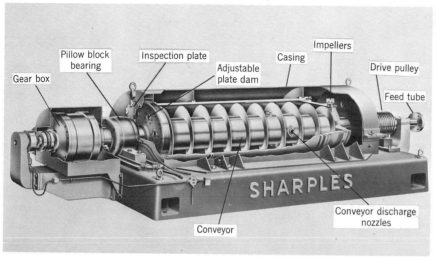

Figure 7-37   Cutaway view of a horizontal, solid bowl scroll centrifuge. (Courtesy of Sharples-Stokes Div., Pennwalt Corp.)

surface area of the solids discharge section. Bowl speed affects gravimetric forces on the settling particles, and conveyer speed controls the solids retention time. The driest cake product results when bowl speed is increased, pool depth is at the minimum allowed, and differential speed between the bowl and conveyer is the maximum possible. Flexibility of operation allows a wide-range moisture content in the solid discharge, varying from a dry cake to a thickened liquid sludge. Feed rates, solids content, and prior chemical conditioning are also variables that influence performance. Removal of solids can be enhanced by adding polyelectrolytes or other coagulants with the slurry feed. The scroll centrifuge also operates with little surveillance.

Centrifuges are being used to dewater treatment plant sludges drawn from settling tanks or gravity thickeners. Lime sludges compact readily, producing a cake with 60 to 70 percent solids (toothpaste consistency). Thus, centrifuges are used to dewater lime-softening sludges in preparation for recalcination. Although aluminum hydroxide sludges do not dewater as readily as lime precipitates, centrifugation with polymer addition to aid flocculation provides a viscous, liquid, solids discharge suitable for either further processing or disposal. Sludge containing about one half aluminum hydroxide can be thickened to 10 to 15 percent solids, while approximately one quarter hydrate slurry can be thickened to 20 to 25 percent. Removal of solids from the feed ranges from 50 to 95 percent depending on operating conditions and polymer dose. Holding all other variables constant, the percentage of solids recovery and density of the thickened discharge are directly related to polyelectrolyte dosage.

Pressure Filtration

The plate-and-frame filter press is the type of pressure filter adopted for dewatering waste chemical sludges. It consists of a series of recessed plates with cloth filters and intervening frames held together to form enclosed filter chambers. These chambers are filled with dewatered solids by pumping in sludge under high pressure, forcing the water out through the cloth filters. At the end of the feed and pressure cycles, the plates are separated to remove the sludge cake. Plate-and-frame filters are capable of producing a dry, rigid cake with no visible water and a clear filtrate very low in suspended solids. The two designs of filter presses are the fixed-volume press and the variable-volume diaphragm press.

Figure 7-38   Fixed-volume plate-and-frame filter press for dewatering waste sludge (Courtesy of Passavant Corporation.)

A fixed-volume plate-and-frame press installation is pictured in Figure 7-38. A view of the plate-and-frame section of this unit is shown in Figure 7-39. Steps in operation are chemical conditioning of the raw sludge if necessary, precoating the filter media, pressure sludge dewatering, and discharge of the sludge cakes. The filter media is precoated by feeding a water suspension of diatomaceous earth. Precoat prevents binding of the filter cloth with sludge particles and facilitates discharge of cake at the end of the filter cycle. Conditioned sludge enters the center feed port immediately after the precoat slurry and is distributed throughout the chambers located between the individual plates. Water is forced through the filter cloth and out through drains between the frames. When the chambers have filled with accumulated solids, a high-pressure cycle consolidates the cakes. Compressed air at about 225 psi (1550 kN/m^2) is applied to the sludge inlet and maintained until the desired density is obtained. After the filter cycle, a core flow valve opens and compressed air blows the feed sludge remaining in the center feed port back to a holding tank. The filter plates are then separated and the dewatered sludge cakes drop into a hopper equipped with a conveyor mechanism. Release of the cakes is assisted by manually separating the

press frames and loosening the layers of cake from the recessed plates with a wooden paddle, if they do not drop out by force of gravity. The entire filter cycle takes from 2 to 3 hr and is completely automated including precoat, filtration, and opening of the press.

The variable-volume diaphragm press is designed for completely automatic operation by forcefully discharging the cakes after the frames open and washing the filter cloths before the press closes for another cycle. Figure 7-40 is an overhead view of a press installation. The automatic operation of a variable-volume diaphragm press is diagramed in Figure 7-41. Sludge is pumped into the recessed chambers at a pressure of 100 psi (690 kN/m^2) for 10 to 20 min allowing filtrate to drain out through the clothes on both sides of the chamber. After sludge feed is shut off, water under high pressure is pumped into the space between the diaphragm and one side of the plate body. This compression is held for 15 to 30 min to dewater the accumulated solids to the desired cake dryness. At the end of the compression period, the sludge feed and filtrate lines are blown out with compressed air. The press then automatically opens and the cakes are released by pulling the filter cloths down around small-diameter rollers. The cloths are washed

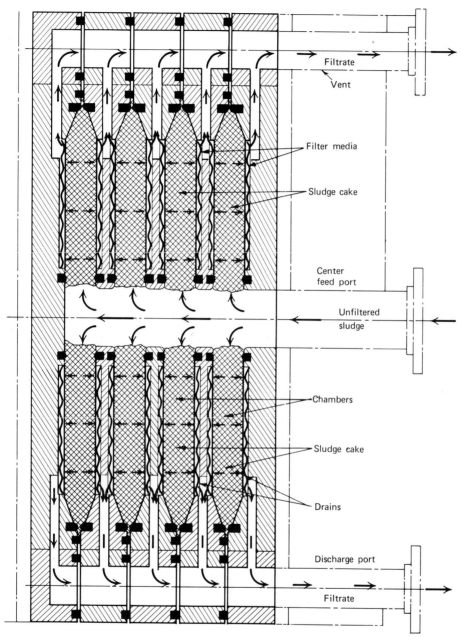

Figure 7-39 Cross-sectional view and flow diagram of the plate section of a fixed-volume filter press. (Courtesy of Passavant Corporation.)

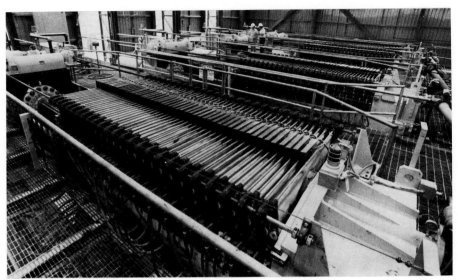

Figure 7-40   Variable-volume diaphragm filter presses for dewatering waste sludge. (Courtesy of Ingersoll-Rand, IMPCO Division as licensed by Ishigaki Mechanical Industry Company Limited.)

on both sides as they retract up to their original position. The cycle time of the diaphragm press is 15 to 25 percent of the cycle time of a fixed-volume press, and the cakes are one half or less in thickness. As a result, the yield for a diaphragm press can be more than twice the yield of a fixed-volume press in terms of mass of solids dewatered per unit area of filter cloth per unit of time.

Pressure filtration appears to be particularly advantageous for thickening alum sludges where a high solids content is most desirable. With lime or polymer conditioning, aluminum hydroxide sludges can be pressed to a solids content of 30 to 40 percent, which is adequate for hauling by truck.

Vacuum Filtration

Vacuum filters are not used extensively in dewatering water plant sludges even though they are extremely popular in processing wastewater treatment sludges; refer to Section 11-18. Where applied to pure lime sludges, belt filters with synthetic cloth have worked rather successfully while coil filters were plagued with encrustation

of the coils. Vacuum filtration of aluminum hydroxide precipitates appears to be both economically and technologically questionable. Even with polymers and precoating it is difficult to consistently achieve a suitable dry cake. Based on preliminary studies, it appears that vacuum filtration will not receive broad application in the dewatering of chemical sludges, since other processes seem to be more economical and to achieve better results.

Recovery of Chemicals

Lime can be recovered from softening sludges by recalcination. The basic steps are gravity thickening; dewatering by centrifugation or pressure filtration after treatment to remove magnesium hydroxide and other impurities; for some processes, flash drying by injecting the dewatered sludge into a stream of hot off-gases from the recalciner; and recalcining, which is accomplished by heating the dried solids in a rotary kiln or incinerator to produce lime and carbon dioxide, Eq. 7-41. An impurity that must be removed to produce a high-quality lime is the magnesium hydroxide precipitate. At present this is

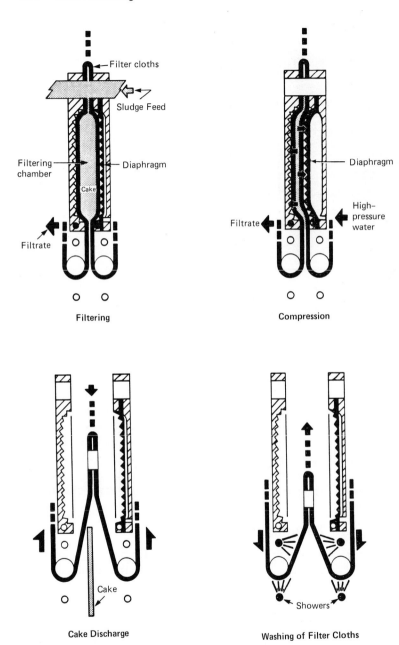

Filter cloths

Sludge Feed

Filtering chamber

Diaphragm

Cake

Filtrate

Filtering

Diaphragm

High–pressure water

Filtrate

Compression

Cake

Cake Discharge

Showers

Washing of Filter Cloths

Figure 7-41   Diagrams of the four steps in automatic operation of a variable-volume diaphragm filter press. (Courtesy of Ingersoll-Rand, IMPCO Division as licensed by Ishigaki Mechanical Industry Company Limited.)

wasted prior to recalcination, although recovery of magnesium carbonate from the magnesium hydroxide sludge is under consideration if it should prove to be a favorable coagulant for reuse in water treatment. Another problem associated with recalcining is interference from clay turbidity where the softening plant treats a surface water. Clay mixed with the calcium carbonate sludge results in buildup of insolubles in the finished lime; currently, there is no feasible means of separating colloidal impurities like clay. Froth flotation is one technique being considered.

$$CaCO_3 \xrightarrow{\text{heat}} CaO + CO_2 \qquad (7\text{-}41)$$

Recovery of alum is rarely practiced because of low-process efficiency and poor quality of the regenerated alum. The key problem is separation of aluminum hydroxide from other impurities in the waste sludge. Where practiced, the sludge is mixed with sulfuric acid and discharged to a gravity thickener for separation of alum solution from settleable insoluble residue. The decanted supernatant contains the bulk of the alum, but separation is incomplete. Carryover of colloidal matter, including organic materials charred by the sulfuric acid, contaminates the recovered alum solution. Reclaimed alum is held in tanks and metered to points of application in the treatment plant. Underflow from the thickener is wasted by treating with lime, dewatered, and hauled away for land burial.

REFERENCE

1. *Recommended Standards for Water Works*, Great Lakes–Upper Mississippi River Board of State Sanitary Engineers, 1976 Edition (Health Education Service, Albany, N.Y.).

FURTHER READINGS

1. Bellack, E., *Fluoridation Engineering Manual*, Environmental Protection Agency, Water Supply Programs Division, EPA-520/9-74-022, 1974.
2. Viessman, W., Jr., and M. J. Hammer. *Water Supply and Pollution Control*, 4th Edition. New York: Harper & Row, 1985.
3. *Water Chlorination Principles and Practices*, American Water Works Association, Manual M20, New York, 1973.

PROBLEMS

7-1 List the impurities typically removed in treating groundwater for a municipal supply. What are the most common pollutants extracted from surface water?

7-2 Why do plants for processing river water generally have the most extensive and flexible treatment systems of any water source?

7-3 Laboratory tests were conducted on the chemical treatment of a water using both a long narrow flocculation tank (plug flow) and a completely mixed unit. The observed rate constants under steady-state conditions were first-order kinetics equal to "12 per day" for plug flow and "36 per day" for complete mixing. For an influent concentration of 80 mg/l, which process requires the shorter retention time to achieve a 95 percent completion of the reaction? (*Answers* plug flow at 6.0 hr, complete mixing 12.7 hr)

7-4 The kinetics of an oxidation reaction were analyzed by the addition of chlorine to the water in laboratory experiments. Samples were withdrawn from the completely mixed reaction vessel at 1-min intervals for determining the concentration of the unoxidized metal ions. The data for one test were as follows: $t = 0$ min, $C_0 = 0.70$ mg/l; $t = 1.0$, $C = 0.47$; $t = 2.0$, $C = 0.32$; $t = 3.0$, $C = 0.26$; and $t = 4.0$ min, $C = 0.19$ mg/l. Plot a graph of $\log_e C_0/C_t$ versus t and determine the value of k.

7-5 A treatment plant has straightline processing with rapid mix chambers, flocculation basins, and settling tanks sized according to the minimum values specified by *GLUMRB*. If the volume of the flocculators is 100,000 gal, compute the design capacity of the plant. What are the volumes of the tanks for applying chemicals to the influent and settling the flocculated water?

7-6 Is the flocculation tank illustrated in Figure 7-5 more representative of complete mixing or plug flow?

7-7 A flocculator designed to treat 10 mgd is 48 ft long, 52 ft wide, and 12 ft deep. Compute the detention time and horizontal flow-through velocity. Do these values satisfy the *GLUMRB Standards*? (*Answers t* = 32 min > 30, yes; *v* = 1.5 ft/min = 1.5, yes)

7-8 A water treatment plant designed for a flow of 10 mgd has units as illustrated in Figure 7-8. The flocculator is 20 ft wide, 120 ft long, and 10 ft in liquid depth; the clarifier is 120 ft by 120 ft, and 15 ft deep. Calculate settling time in hours, flocculation time in minutes, and horizontal velocity through the flocculator in feet per minute.

7-9 Calculate the detention time and overflow rate for a sedimentation basin with a volume of 1.0 mil gal and surface area of 12,500 sq ft treating 8.0 mgd. (*Answers* 3.0 hr, 640 gpd/sq ft)

7-10 Two rectangular clarifiers each 45 ft long, 10 ft wide, and 10 ft deep settle 0.40 mgd following alum coagulation. The effluent weir length is 50 ft. Do these basins meet *GLUMRB Standards*?

7-11 Two rectangular clarifiers each 27 m long, 5.0 m wide, and 3.8 m deep settle 6000 m^3/d. Total effluent weir length is 50 m. Calculate the detention time, overflow rate, and weir loading. (*Answers* 4.1 h, 22 m^3/m^2·d, and 120 m^3/m·d)

7-12 What diameter circular clarifier and side water depth are needed for a 15,000 m^3/d flow based on a maximum overflow rate of 16 m^3/m^2·d and detention time of 4.0 h?

7-13 The flow of water through a long rectangular sedimentation tank was analyzed by injecting a pulse of a dye tracer in the influent. The concentrations of dye in the effluent at time intervals of 20 min starting from injection of the dye were as follows:

Time (min)	Concentration (μg/l)	Time (min)	Concentration (μg/l)
0	0 (start)	120	0.78
20	0.00	140	0.52
40	0.40	160	0.25
60	0.75	180	0.08
80	1.42 (peak)	200	0.02
100	1.20	220	0.00

Draw a residence time distribution curve as shown in Figure 7-9. Locate the mean residence time (centroid), which can be calculated as $t_{50} = \Sigma t_i C_i / \Sigma C_i$. Locate the theoretical detention time equal to 120 min; this value was calculated by dividing the tank volume by the flow during the test.

7-14 The settling velocity of calcium carbonate floc formed during flocculation in a flocculator-clarifier (Figure 7-10) is 2.1 mm/s. If the detention time in the settling zone is 1.0 h and upflow rate is 1.75 gpm/sq ft, what is the minimum depth of water required to ensure removal of the floc by gravity settling?

7-15 What is the function of the rate-of-flow controller in filter operation?

7-16 In a gravity filter, how can the head-loss gauge record a loss of 9 ft if the depth of water above the filter surface is only 4 ft?

7-17 A filter run is terminated as a result of either one of two conditions. State the two conditions.

7-18 Why do granular-filter media have uniformity coefficients of 1.7 or less?

7-19 What is the principal advantage of a dual-media anthracite-sand filter as compared to a single-medium sand filter?

7-20 How does the surface agitator shown in Figure 7-14 function during backwashing?

7-21 Outline the steps used to backwash the filter illustrated in Figure 7-15. What is the function of each step?

7-22 What causes air binding in a gravity filter bed?

7-23 What are the advantages and disadvantages of a gravimetric feeder compared with a volumetric feeder?

7-24 While touring a treatment plant, you visit the room housing the chemical feeders used to apply powdered alum and lime. In looking at the feeders, how would you distinguish the lime feeders from the alum feeders?

7-25 After powdered chemicals are metered from the storage hoppers, as pictured in Figure 7-19, how are they applied to the water being treated?

7-26 Results of a chlorine demand test on a raw water are as follows:

Sample Number	Chlorine Dosage (mg/l)	Residual Chlorine after 10-min Contact (mg/l)
1	0.20	0.19
2	0.40	0.36
3	0.60	0.50
4	0.80	0.48
5	1.00	0.20
6	1.20	0.40
7	1.40	0.60
8	1.60	0.80

Sketch a chlorine demand curve as shown in Figure 7-19. What is the breakpoint dosage, and what is the chlorine demand at a dosage of 1.2 mg/l?

7-27 What is the recommended minimum free chlorine residual to ensure disinfection of a water at pH 8? Minimum combined chlorine residual? (*Answers* 0.4 mg/l, 1.8 mg/l)

7-28 Why are pressure-regulating and vacuum-compensating valves needed on a solution feed chlorinator?

7-29 A dosage of 0.60 mg/l liquid chlorine is needed to maintain a residual of 0.20 mg/l in the distribution system. What dosage rate in pounds per day is needed to treat 6.0 mgd? (*Answer* 30 lb/day)

7-30 In treating 100,000 m^3 of water, 80 kg of liquid chlorine are applied. Compute the dosage in milligrams per liter.

7-31 How many pounds of available chlorine are contained in 1.0 gal of sodium hypochlorite with a strength of 15 percent? How many gallons would be required for a dosage of 0.6 mg/l to 6.0 mil gal of water? (*Answers* 1.25 lb/gal, 24 gal)

7-32 How many grams of dry hypochlorite powder with 70 percent available chlorine must be added to 400 liters of water to make a 1.0 percent solution?

7-33 A small town chlorinates its water supply using powdered calcium hypochlorite with 70 percent available chlorine. At what rate (pounds per hour) should the gravity feeder be set to apply 1.2 mg/l of chlorine for a water flow of 85 gpm?

7-34 A new main is disinfected with water containing 50 mg/l of chlorine by feeding a 1.0 percent solution of chlorine to the water entering the pipe. (a) How many pounds of dry hypochlorite powder, containing 70 percent available chlorine, must be added to 40 gal of water to make a 1.0 percent solution? (b) At what rate should this 1.0 percent solution be applied to the water entering the pipe to provide a concentration of 50 mg/l?

7-35 A new water storage tank with a volume of 950 m^3 must be disinfected by filling with chlorinated water for 6 hr. The chlorine dosage applied to the filling water is to be 70 mg/l. How many kilograms of commercial hypochlorite should be applied to the 950 m^3 of water if the commercial powder contains 70 percent available chlorine?

7-36 A river water has foul tastes and odors when land drainage in the spring of the year washes decaying vegetation from cultivated fields and woodlands into the river. The water processing consists of presedimentation with breakpoint chlorination, split treatment with one portion coagulated with alum and the other softened with lime, blending of the two flows for stabilization with the addition of activated carbon, filtration, and post chlorination. The concentration of total trihalomethanes in the treated water often exceeds the maximum contaminant level of 100 μg/l. What are your recommendations?

7-37 A liquid feeder applies 12 gal of 4.0 percent saturated sodium fluoride solution to 400,000 gal of water. What is the concentration of fluoride added? (*Answer* 0.54 mg/l)

7-38 A fluoridation installation consists of a feed pump that applies fluosilicic acid with 30 percent purity to a water supply directly from the shipping drum placed on a platform scale. The recorded weight loss of the drum is 82 lb in processing 2.3 mil gal of water. Calculate the fluoride ion concentration added to the water in milligrams per liter.

7-39 A saturator (dissolving tank and liquid feeder) is used to apply a 4.0 percent saturated sodium fluoride solution to a water, increasing the fluoride ion content 0.80 mg/l. How many pounds of saturated solution should be fed per million gallons of water?

7-40 A sodium silicofluoride solution is prepared by dissolving 5.0 kg of 98 percent commercial powder in 95 kg of water. Calculate the feed rate of this solution into a water being treated to increase the fluoride ion content 0.80 mg/l. Express your answer in grams of solution applied per cubic meter of water treated.

7-41 A community with a population of 10,000 has a groundwater supply with an average maximum pH of 8.1 and natural fluoride ion concentration of 0.2 mg/l. The annual water consumption is 550 mil gal, and the annual average maximum daily air temperature is 65° F. What minimum free chlorine residual do you recommend maintaining in the distribution system? What fluoride concentration do you recommend? If fluoride ion is applied using commerical fluosilicic acid with a purity of 25 percent, calculate the pounds of commercial acid required to fluoridate the community water supply for one year.

7-42 A chlorinated well water, without additional treatment, supplies a distribution system with plastic piping. Customer complaints include periodic staining water and the emission of foul odors. Explain the origin and formation of both the discolored water and obnoxious odors.

7-43 A groundwater treatment plant, processing a water high in both iron and manganese, consists of tray-type aeration, contact time in a settling basin, sand filtration, and post chlorination. What problem would you anticipate with this system?

7-44 What dosage of potassium permanganate is theoretically needed to oxidize 5.0 mg/l of reduced iron and 1.2 mg/l of manganese? (*Answer* 7.0 mg/l)

7-45 What is the advantage of manganese-treated greensand in a filter bed rather than a plain sand medium following potassium permanganate application for oxidation of iron and manganese?

7-46 The analysis of a water is as follows:

Ca^{++}	= 3.7 meq/l	HCO_3^-	= 4.0 meq/l
Mg^{++}	= 1.0 meq/l	$SO_4^=$	= 1.2 meq/l
Na^+	= 1.0 meq/l	Cl^-	= 1.0 meq/l
K^+	= 0.5 meq/l		

(a) Draw a milliequivalents-per-liter bar graph, list the hypothetical combinations of chemical compounds, and calculate the total hardness. (b) Calculate the chemical doses needed for excess lime softening. (c) Draw a bar graph for the finished water after two-step precipitation softening by excess lime treatment with intermediate and final recarbonation. Assume the practical limits of hardness removal for calcium as 30 mg/l and magnesium as 10 mg/l. (d) Calculate the milligrams per liter of carbon dioxide reacted for neutralization by two-stage recarbonation assuming one half of the final alkalinity is converted to bicarbonate. (e) Calculate the chemical doses required for selective calcium carbonate removal to produce a finished water with a hardness of 120 mg/l. [*Answers* (a) 6.2 meq/l, 235 mg/l; (b) 175 mg/l CaO, 37 mg/l Na_2CO_3; (c) 3.0 meq/l; (d) 41 mg/l CO_2; (e) 81 mg/l CaO]

7-47 Draw a milliequivalents-per-liter diagram and list the hypothetical combinations for the following groundwater data:

Calcium hardness	= 175 mg/l
Magnesium hardness	= 40 mg/l
Sodium ion	= 14 mg/l
Potassium ion	= 4 mg/l
Alkalinity	= 200 mg/l
Sulfate ion	= 29 mg/l
Chloride ion	= 14 mg/l
pH	= 7.7

(a) Calculate the chemical doses needed for excess lime softening. Draw a bar graph for the finished water after two-stage precipitation softening by excess lime treatment with intermediate and final recarbonation. Assume that one half the alkalinity in the finished water is in the bicarbonate form. (b) Calculate the lime dosage required for selective calcium carbonate removal. Draw a bar graph for the finished water. Is this softening process recommended for this water? (c) Calculate the chemical doses for split-treatment softening with 65 percent of the flow given excess lime treatment and 35 percent by-passed to second-stage blending. Estimate the hardness of the finished water.

7-48   A groundwater supply has the following analysis:

Calcium = 94 mg/l	Bicarbonate = 317 mg/l
Magnesium = 24 mg/l	Sulfate = 67 mg/l
Sodium = 14 mg/l	Chloride = 24 mg/l

Calculate the quantities of lime and soda ash for excess lime softening, and the carbon dioxide reacted for neutralization by two-stage recarbonation. Assume the practical limits of hardness removal for calcium as 30 mg/l and magnesium as 10 mg/l, and three quarters of the final alkalinity is converted to bicarbonate. Sketch the bar graphs for the raw and finished waters.

7-49   Compute the quantities of chemicals for split treatment of the water described in Problem 7-48 based on a magnesium hardness in the finished water of 40 mg/l and maximum total hardness of 100 mg/l. (*Answers* excess lime softening of two-thirds of the raw flow; 160 mg/l CaO, 53 mg/l Na_2CO_3; finished hardness = 40(Mg) + 60(Ca) = 100 mg/l)

7-50   The analysis of a river water in a semiarid region is as follows:

Calcium = 78 mg/l	Alkalinity = 126 mg/l
Magnesium = 30 mg/l	Sulfate = 282 mg/l
Sodium = 106 mg/l	Chloride = 96 mg/l
Potassium = 5 mg/l	pH = 8.1
Total dissolved = 680 mg/l
solids

(a) Draw a milliequivalents-per-liter bar graph. What forms of alkalinity are present in the water? (b) The recommended limit of total dissolved solids in drinking water is 500 mg/l. Can precipitation softening reduce the total dissolved solids from 680 mg/l to less than 500 mg/l?

7-51   The ionic character of a groundwater is defined by the following hypothetical combinations:

$Ca(HCO_3)_2$ = 3.0 meq/l	Na_2SO_4 = 0.2 meq/l
$Mg(HCO_3)_2$ = 0.4 meq/l	NaCl = 0.6 meq/l
$MgSO_4$ = 0.8 meq/l	

Draw a milliequivalents-per-liter bar graph. Is the hardness of this water considered excessive? What process do you recommend for precipitation softening of this water?

7-52   Assume cation exchange softening is used to remove the hardness ions from the water described in Problem 7-46. Sketch a milliequivalents-per-liter bar graph for the finished water. What would be the disadvantages of this treatment process for a municipal water supply?

7-53   What is your comment relative to the following statement: "Fluoridation of a municipal water supply is of no benefit to consumers that have water softeners in their homes since the fluoride ion is taken out by the ion-exchange resin."

7-54   What is the health hazard associated with excessive nitrate concentration in a public water supply?

7-55   Assume the anion exchange process is used to remove nitrate from a water with an ionic character as given in Problem 7-50. What would be a major problem of this treatment process?

7-56   Results of a jar-test demonstration on alum coagulation are tabulated below. (Jar testing is discussed in Section 2-11.) Jars 1 through 5 contained a clay suspension in tap water, while jar 6 was a clay suspension in distilled water.

Jar	Alum Dosage	Floc Formation
1	0 mg/l	None
2	10 mg/l	Fair
3	20 mg/l	Good
4	30 mg/l	Heavy
5	40 mg/l	Heavy
6	40 mg/l	None

(a) What is the lowest recommended dosage for treating this water? (b) Why didn't the clay suspension in jar 6 destabilize?

7-57   Alum is applied at a dosage of 20 mg/l in coagulating a surface water. How much natural alkalinity is consumed? What is the change in pH? (*Answers* 10 mg/l, decreases)

7-58   A dosage of 40 mg/l of alum is added in coagulating a water.
(a) How many milligrams per liter of alkalinity are consumed?
(b) What changes take place in the ionic character of the water? State the ions that change forms and express the amounts in milligrams per liter. (c) What is the effect on the pH of the water? [*Answers* (a) 20 mg/l; (b) $SO_4^=$ increases by 19 mg/l, HCO_3^- reduces by 24 mg/l; (c) decrease in pH resulting from CO_2 formed]

7-59   Sodium aluminate and alum can be used in treatment of boiler water. Based on the following balanced reaction, what is the weight ratio of sodium aluminate to alum applied?

$$6NaAlO_2 + Al_2(SO_4)_3 \cdot 14.3H_2O$$
$$= 8Al(OH)_3\downarrow + 3Na_2SO_4 + 2.3H_2O$$

7-60   A dosage of 30 mg/l of alum and a stoichiometric amount of soda ash are added in coagulation of a surface water. What changes take place in the ionic character of the water as a result of this chemical treatment?

7-61   Write the reaction between ferric chloride and lime slurry that results in precipitation of $Fe(OH)_3$. For a dosage of 40 mg/l $FeCl_3$: (a) What is the required stoichiometric addition of lime expressed as CaO? (b) How many pounds of $Fe(OH)_3$ are produced per million gallons of water treated? [*Answers* (a) 20.7 mg/l CaO, (b) 220 lb/mil gal]

7-62   A dose of 35 mg/l of $FeSO_4 \cdot 7H_2O$ and an equivalent amount of lime are added in coagulation. What are the changes that take place in the ionic character of the water as a result of this chemical treatment?

7-63   A surface water is coagulated by applying 30 mg/l of ferric sulfate. How many milligrams per liter of alkalinity are consumed?

7-64   How many milligrams per liter of CaO are needed to react with a ferrous sulfate dosage of 20 mg/l?

7-65   A 20.0 mg amount of sodium carbonate is added to 1.0 l of water resulting in a pH change from 7.2 to 7.4. How much is the alkalinity increased? What is the ionic form of the alkalinity increase?

7-66   In taste and odor control, what is usually the most effective chemical? Why is heavy prechlorination of surface waters likely to be undesirable?

7-67   A river-water supply is treated by the sequence of unit operations and chemical additions listed below. State in a few words the purpose or purposes for each unit process and chemical application.
(a) presedimentation with polymer addition
(b) mixing and flocculation with additions of alum and polymer
(c) addition of activated carbon
(d) sedimentation
(e) granular-media filtration
(f) post chlorination

7-68   The processing scheme for treatment of lake water consists of (1) rapid mixing and flocculation, (2) sedimentation, (3) filtration, and (4) clear-well storage. The chemicals available for treatment are activated carbon, alum, chlorine, fluosilicic acid, and polymers. At what principal and alternate locations in the processing scheme are these chemicals likely to be added?

7-69   Why is the passage of water through a membrane for removal of dissolved salts called *reverse* osmosis?

7-70   Describe how the feed water and permeate (product water) flow through a spiral-wound module for reverse osmosis.

7-71   How does the addition of acid prevent scale formation on reverse-osmosis membranes?

7-72   The quality characteristics of a deep-well water in an arid region with a hot climate are

Calcium hardness	= 270 mg/l
Magnesium hardness	= 180 mg/l
Sodium ion	= 138 mg/l
Iron ion	= 0.40 mg/l
Manganese ion	= 0.15 mg/l
Total dissolved solids	= 830 mg/l
Temperature	= 27° C
pH	= 8.1
Alkalinity	= 120 mg/l
Sulfate ion	= 110 mg/l
Chloride ion	= 344 mg/l
Fluoride ion	= 1.4 mg/l
Nitrate nitrogen	= 15 mg/l

Draw a milliequivalents-per-liter bar graph and compare the concentrations of chemicals with the primary and secondary drinking water standards (Table 5-1). (a) Is the water quality satisfactory for a municipal supply without treatment? Discuss any quality problems with respect to health and esthetic standards. (b) One recommendation is to treat the water by lime–soda ash softening to improve the quality. Will this processing provide a satisfactory quality for a public water supply? Consider each of the quality parameters of concern. What are the advantages and disadvantages of precipitation softening of this water? (c) Propose a treatment scheme that will provide a water quality adequate to meet both health and esthetic standards. Sketch a flow diagram showing the unit processes, chemical additions, and sources of wastes for disposal. (d) Sketch an approximate bar graph of the treated water.

7-73   Calculate the Langelier saturation index for the water analysis given in Problem 7-48 at 25° C assuming a measured pH of 7.7.

7-74   Estimate the sludge solids produced in coagulating a surface water having a turbidity of 12 turbidity units with an alum dosage of 40 mg/l. What is the sludge volume if the settled waste sludge and filter backwash water are concentrated to 1000 mg/l of solids in a clarifier-thickener? (*Answers* 180 lb/mil gal, 22,000 gal/mil gal)

7-75   Lime added to water containing calcium hardness results in precipitation of calcium carbonate. If 126 mg/l of lime with a purity of 78 percent CaO are added to a hard water, how many milligrams per liter of $CaCO_3$ are formed? (*Answer* 350 mg/l)

7-76   Calculate the waste sludge produced in the excess lime treatment described in Problem 7-48 in units of pounds dry solids per million gallons of water processed and gallons per million gallons assuming a settled solids concentration of 10 percent. (*Answers* 5900 lb/mil gal, 70 gal/mil gal)

7-77   Calculate the quantities of waste sludges produced in softening the water described in Problem 7-47. Assume that the solids concentration of the gravity-thickened sludge from excess lime and split-treatment processes is 8 percent, and the solids concentration from selective calcium carbonate removal is 12 percent.

7-78   Extraction of iron and manganese from groundwater is often performed by the process sequence of aeration, chemical oxidation, sedimentation, and filtration. In which unit operation is the majority of the metal oxides removed?

7-79   Which of the following waste sludges is the easiest to dewater: $CaCO_3 + Mg(OH)_2$ precipitate, alum sludge, or $CaCO_3$ slurry?

7-80   Which of the items listed below are advantages of pressure filtration relative to centrifugation?
(a) Filtration is a continuous flow process.
(b) The sludge can be dewatered to a solid cake.
(c) Hydroxide sludges can be dewatered with relative ease.
(d) Filtration is more adaptable for dewatering prior to recalcining.

7-81   Why is the variable-volume diaphragm press more suitable than the fixed-volume filter press for installation at a large treatment plant?

chapter 8

Operation of Waterworks

Management of a water utility requires extensive knowledge in distribution systems and water processing; refer to Chapters 6 and 7, respectively. Water quality (Chapter 5) and testing for both chemical and biological characteristics are also prerequisite to systems control. This chapter discusses inspection and maintenance of water distribution systems, operational control of treatment plants, and management of waterworks.

8-1 DISTRIBUTION SYSTEM INSPECTION AND MAINTENANCE

Fire Hydrants

Hydrant maintenance is the responsibility of the water utility; however, the fire department may assist in hydrant inspections. Use of hydrants as a source of water for street cleaning and construction work must be restricted and controlled. After each use, a hydrant should be inspected to be sure it is in operating condition in case of fire. A stock of repair parts, particularly replacement internal valves, is necessary to make immediate repairs. Although the frequency of periodic inspection varies among utilities, a yearly maintenance schedule is common. The exact inspection procedure depends on the type of hydrant. The most common is a center-stem dry-barrel fire hydrant (Figure 6-39) equipped with a progressive drain. Rather

than opening and closing instantaneously, the drain operates progressively as the main valve opens and closes. During use and inspection, the hydrant should be fully opened so that the drain is fully closed. Water seeping through a partially closed drain can saturate the drain field and lead to washing away the soil around the hydrant. In cold climates, drainage is essential to prevent damage from water freezing and expanding in the barrel.

The first step in inspecting a hydrant is to remove an outlet-nozzle cap and check the barrel for water by inserting and retracting a plumb bob. The presence of water means that either the main valve is leaking or the groundwater level is high. A listening device is used to check for leakage, and if necessary the main valve is repaired. If groundwater is high, a plug can be placed in the drain to preclude groundwater. Water in the barrel must then be pumped out after hydrant use. Drainage from an unplugged barrel can be checked after flushing. Upon closing the main valve, the drainage rate should be rapid enough to create a noticeable suction on a hand placed over the opening of one open nozzle. Leaks from joints and nozzle caps are detected by opening the main valve with the caps in place. To prevent damage, air is vented from the barrel while filling. Compression packings can be lubricated and tightened or gaskets replaced. For the center-stem hydrant, a screw or fitting in the operating nut indicates the need for manual lubrication, since some brands have automatic lubrication. In either case, the lubrication should be checked. Finally, all of the nozzle caps are inspected, threads cleaned and lubricated, and screwed on sufficiently tight to prevent removal by hand.

Valves

Shutoff valves are installed in a distribution system to isolate pipe sections for maintenance or to repair a break. A preventative maintenance program ensures proper functioning of valves when they are needed. Without periodic inspection and operation, they may become inoperable, covered by repaving of streets or paving of driveways, or difficult to locate because of inaccurate mapping. Valves can also be partially or fully closed obstructing water flow. A good maintenance program includes verifying locations, inspecting the condition of valve boxes, and opening and closing the valves. Depending on local conditions, valves in trunk and feeder mains are inspected at least once a year and others in the pipe network once every one to three years. Incrustation of internal moving parts can prevent full closure and, in extreme cases, prevent movement of the disk without risk of damage. Periodic opening and closing loosens the interfering scale.

Preventative maintenance involves location, inspection, and operation. Location of the valve box is confirmed by checking the measurements given on the map. Next, the valve box cover is removed and with a flashlight the interior is inspected. Finally, the valve is turned fully closed and a count of the number of turns required is recorded. By listening with a leak detector, the tightness of closure is checked. To avoid failure of the stem, the applied torque should not exceed the recommended limit specified by the manufacturer. Manual turning, a slow and arduous process, can be done more economically by power equipment. Portable units are powered by compressed air or electricity. Power turning is particularly valuable for repeated operation of a valve to remove internal scale.

The common defects requiring repair are a broken stem, need for packing, and internal incrustation. Repair of a valve is a difficult task requiring isolation of that section of the pipe network, excavation, and removal of the valve. Either the damaged valve can be replaced or, if the overall condition is good, new parts can be installed on site.

Backflow Preventers and Meters

A water utility has the responsibility of protecting its potable supply from contamination resulting from backflow and back siphonage. Therefore, a program for control of cross connections is essential, including inspection of customer facilities that have the potential of contaminating the water supply with hazardous substances and pathogens. Examples are hospitals, mortuaries, chemical industries, and irrigation systems. Backflow preventers (Section 6-9) installed in service lines require periodic testing to ensure proper operation. Since enforcement of a cross-connection-control program involves health agencies, plumbing inspection, water suppliers and owners, inspection of backflow preventers may be done by water utility personnel, a private testing agency, or another organization.

Accurate water metering is basic to customer billing and to provide data for comparing supply with consumption to estimate water loss from the system. Household water meters are ordinarily under the control of the water utility. If underregistration is suspected, or if the homeowner complains of improper registration, the meter is replaced. Many utilities have an ongoing program for repair and replacement of worn meters. Commercial and industrial users are generally obligated to show proof of regular meter inspection and calibration.

Leak Detection

The objectives of leak detection are to locate and repair small defects in a pipe network before failures occur and to reduce water loss from the system. As water under pressure exits a crack or small hole, sound waves in the audible range are emitted by the pipe wall and surrounding soil. Electronic detectors, capable of amplifying sound waves and filtering out unwanted background noise, can locate and isolate leaks. With experience, the person operating the detector can also estimate the rate of leakage. The sound transmitted by the pipe wall can be heard by listening at hydrants, main valves, and curb valves. In conducting a survey, areas of water loss are identified by listening at these direct contact points. An approximate location is established by evaluating the intensity from different locations. Water impacting the soil and circulating in a cavity creates lower frequency waves that have limited transmission through the ground. Through the use of a surface microphone, the leak can be located with greater precision. The nature of the sound emitted by escaping water depends on water pressure in the main, pipe material and size, soil conditions, and configuration of the opening. Consequently, locating and estimating the severity requires operator training and experience.

Pipes and joints are commonly repaired by placing an external cover over the leak. Repair sleeves are cylindrical halves that are fitted around the pipe or joint and bolted together. At a joint, a solid insert may be placed inside with half of the length in each pipe and a split-coupling clamped around the outside.

Pipe Cleaning

Periodic flushing of water mains is a means of removing metal oxides and other precipitates from a pipe network before they are drawn into service connections. To be effective, the flows in the pipes during flushing must be turbulent enough to suspend the impurities and convey them out of the system. Inconvenience to customers can be minimized by advanced planning. In general, the best procedure is to open hydrants to full discharge, continue flushing until the water is clear, and then gradually close the hydrants. The best time is at night between 9 P.M. and 3 A.M. to avoid mealtimes and times when residents wash clothes. Flushing schedules and an explanation of the program should be published, and commercial establishments that depend on a clear water supply—for example, laundries, hospitals, and restaurants—should be given special notification.

Mechanical scraping is required to remove tuberculation and scale from the inside of pipes to reduce resistance to flow and prevent quality deterioration of the conveyed water. The most effective and least costly cleaning method is done by pushing a bullet-nosed cylinder, called a pig, through the pipe with water pressure. The pigs pictured in Figure 8-1a are made of flexible polyurethane rubber foam with a nose of hard polyurethane and a cylinder banded with wire brush scrapers. With the use of a firm but flexible interior, the cylinder can contract and expand

(a)

Figure 8-1 Cleaning of water mains. (a) Polyurethane pigs. (b) Methods of inserting pigs into mains. (c) Diagram of a launcher or receiver. (Courtesy of Polly-Pig by Knapp, Inc.)

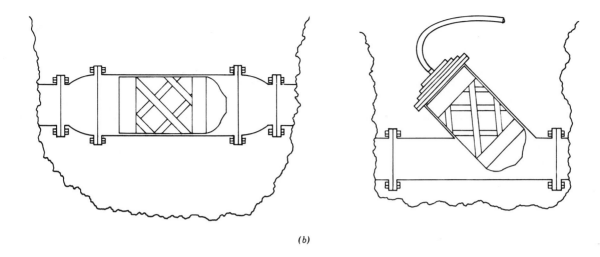

(b)

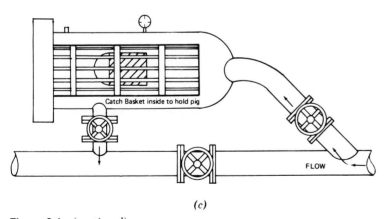

(c)

Figure 8-1   *(continued)*

with variations of the pipe interior. With a single spiral wire brush, the pig is more flexible for pushing through tight bends in the pipeline. For removal of harder deposits, the brush bands are in a crisscross pattern or coat the entire exterior wall. A bare foam rubber swab may be used first to check the direction of flow or to help remove debris after cleaning. Figure 8-1b illustrates two ways of inserting pigs. By cutting out a section of the pipe, a temporary launcher can be inserted and the pig propelled by pressure from the rear. Another method is to install a wye launcher and force the pig into the pipe by external water pressure. For small pipes, a hydrant with the valve removed may be suitable for mounting the launcher. A common procedure at the outlet end is to cut out a section of pipe for an exit and install temporary piping to convey flushing water out of the excavation. Figure 8-1c is an attachment designed to serve as either a launcher or a receiver. After the pig is placed in the chamber, the gate valves are opened to force it into the main. At the receiving end, the pig exits into the basket.

A recommended procedure for cleaning mains is referred to as progressive cleaning. First, the pipeline is isolated by closing boundary valves and fully opening the valves in the line to be cleaned. After a full-sized bare swab is run through to prove the direction of flow, a bare squeegee pig is forced through the main to determine the clear internal diameter of the pipe. Pigs up to 1 in. larger in diameter are progressively forced through with the final pig diameter equal to or slightly greater than the pipe diameter. The usual speed for propelling a pig is 40 to 60 ft/min; however, higher speeds can be used for straight pipelines without restrictions. Flushing after each run and pushing a bare swab through after cleaning remove loose scale.

Tapping of Mains

Employees of a water utility make the service connections to mains for building contractors. A tapping machine drills and taps a hole through the pipe wall under pressure, hence the name "wet" tapping. The machine then either inserts a corporation stop for small-diameter connections or permits closure of an outlet gate valve for large connections. Dry tapping can be performed before a

main is filled with water or if the pipe is removed from service.

The tapping machine for small connections is clamped against a saddle that matches the machine to the outside diameter of the pipe. A gasket is placed under the saddle, and the proper tapping tool is selected for the pipe material. The three diagrams in Figure 8-2 illustrate the three main steps in making a house service connection. First, the hole is drilled into the pipe with the combined drill bit and threading tool, and then the drilled hole is threaded with the same tool. After the tool is retracted within the drill-and-tap unit, it is swung aside placing the corporation stop in position for screwing into the hole.

Large service connections require installation of a sleeve and gate valve. After the pipe exterior has been cleaned, a tapping sleeve in cylindrical halves is clamped around the pipe and sealed by split gaskets held against the sleeve ends by rings placed around the pipe and bolted to the sleeve. One half of the tapping sleeve has a hole with a short piece of flanged pipe. A gate valve is bolted to the sleeve pipe, and a drilling machine connected to the valve. The cutting bit, extending through the open gate valve, drills a hole through the pipe wall. After the hole has been completed, the cutter is retracted and the tapping gate valve closed. The drilling machine is removed and the service line is attached to the flange of the gate valve.

8-2 DISTRIBUTION SYSTEM TESTING

Fire Flow Tests

These tests are important in determining the efficiency and adequacy of a distribution system in transmitting water, particularly during days of high demand, and are used to measure the amount of water available from hydrants for fire fighting. The required rate of flow for fire fighting must be available at a specified residual pressure. The minimum is 20 psi (140 kPa) if fire department pumpers are used to supply hose streams. If hoses are to be connected directly to hydrants, a residual pressure of 50 to 75 psi is needed, depending on local structural conditions. For automatic sprinkler systems, a pressure of 15 psi is usually specified at the top line of sprinklers.

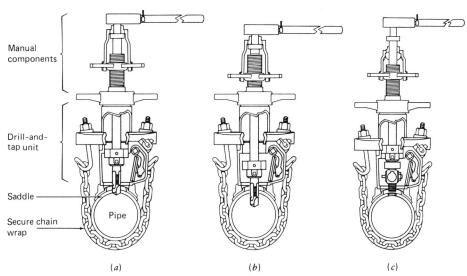

Manual
components

Drill-and-
tap unit

Saddle

Secure chain
wrap

Pipe

(a) (b) (c)

Figure 8-2 Tapping a water main for a service connection. (a) Drilling a hole through the pipe wall. (b) Threading the drilled hole. (c) Inserting the corporation stop. (Courtesy of Mueller Co.)

Flow tests consist of discharging water at a measured rate of flow from one or more fire hydrants and observing the corresponding pressure drop in the mains through another nearby hydrant. Figure 8-3a illustrates a flow test in which hydrants 1, 2, 3, and 4 are discharging while the residual pressure drop is read at hydrant R. The residual hydrant is chosen so that the hydrants flowing water are between it and the larger feeder mains supplying water to the area. The number of hydrants used for discharge, and the rates of flow, should be such that the drop in pressure at the residual hydrant is not less than 10 psi (70 kPa). For single main testing, the layouts and hydrant selection are shown in Figure 8-3b.

The typical test procedure is as follows: One of the outlet caps on the residual hydrant is replaced with a pressure gauge, the hydrant valve is opened to exhaust air from the barrel, and the initial water pressure reading is recorded. The surrounding hydrants are then opened and discharge is measured using pitot gauges. A pitot tube centered in the flow stream from a hydrant nozzle measures the pressure exerted by the velocity of the flowing water. The quantity of discharge from a hydrant nozzle can be calculated from the pitot pressure reading

by using Eq. 8-1. Since it is difficult to establish exactly the rate of discharge needed to produce a specific residual pressure, the flow that would be available at a given residual pressure, for example, 20 psi, must be computed from test data by using Eq. 8-2.

$$Q = 29.8C \, d^2(p)^{1/2} \qquad (8\text{-}1)$$

where Q = discharge, gallons per minute
C = coefficient, normally 0.90
d = diameter of outlet, inches
p = pitot gauge reading, pounds per square inch

$$Q_R = Q_F \frac{H_R^{0.54}}{H_F^{0.54}} \cong Q_F \left(\frac{H_R}{H_F}\right)^{1/2} \qquad (8\text{-}2)$$

where Q_R = computed discharge at the specified residual pressure, gallons per minute
Q_F = total discharge during test, gallons per minute
H_R = drop in pressure from original value to specified residual, pounds per square inch
H_F = pressure drop during test, pounds per square inch

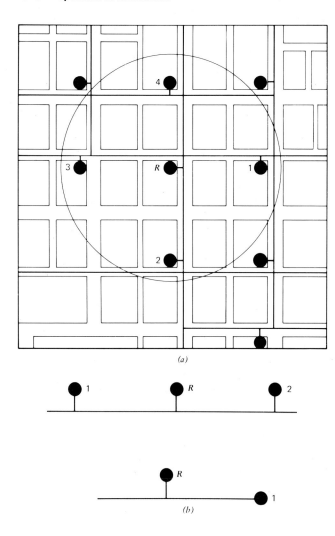

(a)

(b)

Figure 8-3 Diagrams showing selection of fire hydrants for flow tests to determine the amount of water available for fire fighting. Numbered hydrants discharge water with resulting pressure drop measured at hydrant *R*. (*a*) Typical group hydrant flow test. (*b*) Layouts for single main flow tests.

Flow test results show the strength of a distribution system but not necessarily the degree of adequacy of the entire waterworks. Consider a system supplied by pumps at one location and having no elevated storage. If the pressure at the pump station decreases during the test, this is an indication that the distribution system is capable of delivering more than the pumps can provide at their normal operating pressure, and the value for the drop in pressure measured during the test must be corrected. It is equal to the actual drop obtained in the field minus the drop in discharge pressure at the pumping station. If sufficient pumping capacity is available at the station and the discharge pressure can be maintained by operating additional pumps, the water system as a whole is able to deliver the computed quantity. If, however, additional pumping units are not available, the distribution system is capable of delivering the computed amount, but the water system as a whole is limited by the pumping.

The corrections for pressure drops in tests on systems with storage are generally estimated from a study of all the pressure drops observed on recording gauges at the pumping station. The corrections may be very significant for tests near the pumping station while decreasing to zero for remote tests.

Pressure Tests

These tests are generally performed by attaching a pressure gauge to a hose nozzle of a hydrant and opening the valve to fill the barrel with water under residual pressure. Where more than instantaneous readings are desirable, a recording unit can be installed. Operational evaluations normally include monitoring several points continuously for pressure, especially during periods of maximum water consumption.

Hydraulic-Gradient Tests

These tests are conducted to determine the ability of a distribution system to transmit water with adequate residual pressures. Testing is normally performed during periods of peak delivery when the system flows are substantially constant; this often occurs during midmorning or midafternoon. Pressure measurements are taken at various points located along a line from the source of delivery to an outlying section. Pressure observations are plotted vertically along a horizontal distance scale. The curve connecting the plotted values gives a visual picture of the head losses between points as the water is transported through the pipeline. Sections with steep hydraulic gradients represent areas where the greatest friction losses occur.

Coefficient Tests

Occasionally, determining the internal condition of a pipeline with respect to friction loss during water flow is desirable. The procedure is to isolate a pipe section to the greatest extent possible, including closing service connections. The hydraulic gradient (pressure drop) is measured between two points a known distance apart while a measured flow is transmitted through the pipe. Assuming that complete isolation of the pipe section being tested is possible, flow can be measured by hydrant discharges as previously described. Otherwise, the quantity of flow is measured by using a modified pitot meter that can be inserted into the pipeline through a 1-in. corporation stop. With these data, the coefficient of roughness can be computed from the Hazen Williams formula. Then, with the proper coefficient known, head losses during other quantities of flow can be calculated.

EXAMPLE 8-1

Calculate the discharge at a residual pressure of 20 psi based on the following flow test. Hydrants 1, 2, 3, and 4 as illustrated in Figure 8-3a were all discharging the same amount of water through $2\frac{1}{2}$ in.-diameter hose nozzles. The pitot tube pressure reading at each hydrant was 16 psi. At this flow the residual pressure at hydrant R dropped from 90 psi to 58 psi.

Solution

Using Eq. 8-1,

$$Q = (4)(29.8)(0.90)(2.5)^2(16)^{1/2} = 2680 \text{ gpm}$$

From Eq. 8-2,

$$Q_R = 2680 \frac{(90-20)^{0.54}}{(90-58)^{0.54}} = 4090 \text{ gpm}$$

or

$$Q_R = 2680 \left(\frac{90-20}{90-58}\right)^{1/2} = 3970 \text{ gpm}$$

8-3 CONTROL OF WATERWORKS OPERATIONS

Water Treatment Process Control

Quality of the raw-water supply, the chemical processes in treatment, and the physical facilities of flocculation, sedimentation, and filtration determine the complexity of controlling the operation of a treatment plant. The degree of automation, although influenced by these factors, is related primarily to size of the plant. In addition to monitoring the treatment processes, a preventative maintenance schedule is necessary for mechanical and electrical equipment. Chemical feeders, air compressors, sludge pumps, and standby generators are examples of units that require regular inspection and maintenance.

Manual operation is limited to small treatment plants. Most municipal waterworks incorporate some automation to control system hydraulics and chemical dosing. Water level in a clear well can be monitored and the information transmitted to raw-water pumps to vary the rate of flow. Rate of filtration is commonly controlled and the programmed backwash cycle started automatically, or by pressing a button. Treatment processes can be operated by a control loop that continuously monitors a chemical parameter, such as pH or turbidity, and automatically adjusts chemical feed rate. In fully automated plants, a central network links pumping, chemical dosing, and filter operation to a control room that contains a panel and operator's console.

The control panel for a 60-mgd groundwater plant is pictured in Figure 8-4. Displayed from left to right are well pump operating lights; six flocculator-clarifiers and eight filters, with flow and level controllers mounted underneath; warning alarm panel; and chlorine dosage and residual indicators. The edge of the operator's console is in the foreground of the picture. Well pumps can be started and stopped from the console, and running lights on the panel show the pumps that are in service. The capacity of each well is marked on the board, enabling the operator to select the proper combination for any desired pumpage. Flow of raw water into the treatment basins is controlled from the console by rate of flow and level controllers. Each inlet line has a butterfly valve regulated by signals from a flow meter and a bubbler tube for level

Figure 8-4   Control panel and operator's console at an automated groundwater treatment plant. (Courtesy of the Metropolitan Utilities District, Omaha, Nebraska.)

control. Since some imbalance in flow exists because of signal oscillation, two or more basins are set for approximate flows and one for level control. The latter takes minor surges as a result of pressure fluctuations and valve cycling. If all of the basins are placed on automatic flow control, there is a tendency for valves to "hunt" or oscillate as each, in turn, corrects for the slight variation that another makes. Flow passing into each unit is recorded automatically at the control panel, which receives a propeller meter pulse from each influent line.

Water forced into the plant by the well pumps discharges to selected flocculator-clarifiers. Lime slurry is applied by gravity feed to the mixing zones of the basins. Slakers and feeders are capable of automatic pacing at preset feed rates using the same propeller meter pulse that transmits the rate of flow signals to the control panel. Units are also provided for adding a coagulant, potassium permanganate, and fluosilicic acid. Feeder operation is displayed on the control panel by running lights, and is tied to the alarm section. Warning lights and a horn announce interruption of feed caused by blockage in the hopper or other malfunctions.

The chemically treated water is filtered through dual-media beds after clarification. Filter backwash is initiated by the operator either when the head loss increases to 8 ft or after 100 hr of operation. After manual starting of the cycle, the air and water process continues on an automatic program. The bed is scoured by air agitation and then hydraulically backwashed. Timers for both the air and water periods are adjustable at an auxiliary control panel. After cleaning, the filter is held out of service for a period of time to improve the initial filtrate quality. An alternate procedure is to waste the first portion of filtered water, which is likely to show slightly higher turbidity.

Filtered water is routed to clear-well storage or can be pumped directly into the distribution system. Chlorine is applied to filter influent and/or to high-service pump discharge. Chlorine rates and residuals can be read at the main control panel. Treated water is constantly sampled by means of a closed loop, and residuals are adjusted to a preset level. Chlorinators and evaporators are placed in a separate room that is equipped with a detector and automatic blower system for ventilation should an accident occur. The detection unit is also connected to the alarm and warning horn in the control room. Gas masks and compressed air packs are stored just outside the chlorine room for emergency use.

Fluosilicic acid, found to be the most convenient and economical chemical for fluoridation, is metered into the water prior to filtration by using a solution feeder from a tank mounted on a platform scale. A recorder graphs loss of weight of the tank as a function of time. This arrangement allows the operator to constantly confirm feed pump accuracy against the quantity of acid applied. Regular laboratory testing of both raw and finished supplies is used both to establish treatment rate and confirm that the proper dosage is being applied.

Clear-well storage capacity is approximately 10 percent of the plant design flow. High-service pumps are controlled from the main panel, and running lights indicate their status to the operator. Both pumpage and discharge pressure are charted on recorders. While electrically powered pumps require little attention, the natural gas engines need periodic operator observation. An hourly checklist includes running speed, cooling water and lubrication oil temperatures, water jacket temperature and pressure, crankcase vacuum, manifold pressures, and other items. High-service engines require prelubrication before starting and warm-up prior to applying a pumping load.

This water treatment plant with modern equipment and automation permits operation with minimum labor, usually a two-person operating shift. This savings in operating personnel is partially offset by the need for a highly skilled maintenance crew to perform constant inspection and service functions. During the evening and night shifts, operating personnel generally take over some of the maintenance chores, such as machine cleanup, lubrication, and minor adjustments. The alarm panel may summon the shift operator several times in an 8-hr period to handle irregularities in normal treatment and pumping operation. Chemical feeder and chlorinator stoppages most frequently demand the operator's attention. Other problems can be contributed by air compressors, steam boilers, sludge pumps, air handling units, and flow controllers. Pneumatic chemical unloading systems require operator time since truck shipments, especially, may arrive at any time of the day or night.

Operation of a surface-water plant treating a turbid river water requires more direct operator attention because of the variable characteristics of the raw water that can drastically change from season to season and day to day. In contrast to the minor treatment problems encountered in a groundwater plant, the superintendent of a river-water facility may have a few weeks during the year when his problems seem unsurmountable. During critical seasons a laboratory technician is on duty around the clock to run jar tests, and to perform quality surveillance. A delay in modifying chemical treatment for even a few hours, during a period of rapid water-quality deterioration, can lead to taste and odor problems in the finished water. Problems of coagulation, taste and odor removal, and color reduction are intensified when the water quality deteriorates during spring runoff and chemical reactions proceed more slowly at water temperatures close to freezing. Furthermore, disinfection efficiency is likely to suffer because of the higher turbidity in the raw water. Large quantities of activated carbon are necessary to produce a palatable water. Major rivers are subject to pollution from oil spills, industrial accidents that contribute exotic contaminants, and agricultural drainage that increases the organic content as well as producing a variety of pesticide residuals. Therefore, plant personnel can never relax their vigilance in water-quality monitoring.

Water-Quality Control

A quality control program is essential to the operation of any water treatment plant. For surface supplies it is mandatory but, even if processing well water is limited to iron and manganese removal, adequate laboratory control is beneficial to overall operation. Some small plants routinely carry out comprehensive sampling and testing programs, while others rely completely on state or local health agencies for surveillance. But even most small waterworks now find it within their ability to train operators in many of the common chemical and bacteriological tests by using proprietary kits and reagents.

Availability of complete laboratory reports is valuable for public relations and in combating unwarranted criticism of system operations. Industries served by a water system are often interested in the chemical characteristics of the treated water and request monthly reports. A typical analysis includes results of all common physical and chemical tests, and concentrations of major cations and anions. Periodic trace metal analyses may be conducted for arsenic, barium, cadmium, and other heavy metals. With increasing concern for organic contaminants, particularly chlorinated hydrocarbons, comprehensive testing also includes examination for the common pesticides.

All states require analysis of finished water for coliform bacteria; the number of tests required is based on the population served. Fecal coliform counts, while not generally required by regulatory agencies, are not difficult and can give additional information regarding sources of contamination. Chemical standards, such as chlorine residual, turbidity, trihalomethanes, dissolved solids, nitrate, and color, are sometimes specified. Measurement of chlorine residuals in the distribution system is mandatory to determine if the chlorination practice is adequate. Other laboratory testing is related to control of chemical treatment, trouble-shooting problems in the distribution system, and handling customer complaints on quality. Shipments of treatment chemicals, if purchased under specifications, should be routinely tested, and penalties should be assessed for deficiencies. A penalty clause in the contract supported by an approved laboratory analysis can protect the plant against inferior material. Recent water pollution control legislation now requires laboratory monitoring of waste discharge streams, which represents another addition to the duties of the quality control section.

The smallest plant laboratory may contain only a turbidimeter, a chlorine residual comparator, a pH meter, and glassware for hardness and alkalinity determinations. A completely equipped laboratory for a metropolitan area includes infrared, ultraviolet, and atomic absorption spectrophotometers, a gas chromatograph, an amperometric titrator, and a conductance meter, plus the common ovens and glassware. Incubators, autoclave, special glassware, and culture media are needed for bacteriological work. If a nuclear power plant is located upstream, some radiochemical surveillance is necessary and a proportional counter and scintillation counter

may be needed. Laboratories serving large water systems are usually supervised by a graduate chemist or chemical engineer. Additional professionals and two or three laboratory technicians may also be employed, depending on the extent of the testing program. Laboratory personnel may also serve part time or full time in plant supervision. Shift operators can be trained to do routine tests like those for turbidity, alkalinity, and chlorine residual. Standard coliform counts can also be handled by operators who have had the proper training. The advantage of laboratory-trained operators is that operating shifts cover the entire 24-hr day while laboratory personnel typically work only an 8-hr day. Capability for around-the-clock surveillance is important in detecting sudden changes in raw-water quality.

Laboratory jar tests and past experience are both used to determine optimum chemical dosages for coagulation and lime softening processes. Problems may be anticipated by careful testing of the raw water so that variations in the treatment process can be made promptly. Past treatment records are helpful, for example, efficient lime softening can be constantly evaluated by maintaining a record of lime feed to hardness removed. The ratio of CaO applied to hardness precipitated may range from 0.8 to 1.2 depending on water temperatures and retention time, but it should be fairly uniform during any one season of the year.

Taste and odor problems in surface-water supplies can easily be the primary concern in water-quality control for several months of the year. The specific causative compounds are often dissolved organics that are exceedingly difficult, if not impossible, to identify. The standard chlorine demand test gives an excellent indication of overall water quality. Other useful tests include those of potassium permanganate demand and the standard 4-AAP phenols procedure. Determination of threshold odor number is another means of evaluating water quality. The test is conducted by a panel of observers who attempt to detect odor in successively diluted samples. Odor number is calculated from the number of dilutions necessary for the last traces of odor to disappear. Although a useful tool, the test lacks reproducibility, and many instances have been reported in which treated water with a low threshold odor number has been responsible for numerous customer complaints. Odor dilution, it seems, is not always linear, and rather offensive samples may occasionally respond well to only one or two dilutions.

Distribution System Control

Management of small distribution systems is generally performed from the treatment plant where pumpage and pressures can be controlled. As the area served increases, however, system pressures can no longer be accurately controlled, or even determined, from one point. Operations must then be extended into the pipe network to monitor pumps and reservoirs to ensure adequate service to all parts of the system. For a large system, central control provides uniform and accurate data at one location. This allows the supervisor to make the best possible operational decisions. Also, critical conditions can be anticipated and emergencies can be handled rapidly.

The following is a brief description of the surveillance techniques employed by a utility serving more than 100,000 water customers. The system's control center is a data gathering and computing facility for supervising water operations. The control panel pictured in Figure 8-5 allows the operator to enter data into the computer,

interrogate information stored in its memory, control data logging, and set system alarm limits. The overall function of the system is to establish water production rates from the two treatment plants, and to operate nine booster pumping stations and three reservoirs.

Data from pumping stations, reservoirs, and pressure points are telemetered continuously to the center. Flows for each pumping station and consumption in the two pressure zones are calculated by the computer, which also stores the information and hourly prints out the rate and quantity of flow for each pumping station and pressure zone. The controller is warned of low-pressure conditions, depleted water storage, chlorine leaks, and equipment malfunctions by an audible alarm and printout on the teletype. Discharge pressure from three major booster pumping stations is recorded on continuous strip charts, and pump speed for the variable speed units is displayed on an indicator. Several reservoir valves can be remotely controlled to regulate flow into, or out of, storage. Some valves can be positioned to control rate of flow while others are simple open-or-close valves. Pumps at all booster stations are started and stopped by remote control. Pump check valves are hydraulic or pneumatic operated ball valves. On the electric motor-driven pumps, these valves open automatically when the unit is operated. Panel lights indicate valve positions and pump operation.

8-4 RECORDKEEPING

Efficient management of a water utility is founded on detailed records of physical facilities, operation, and maintenance. A distribution system is documented by data describing the pipe network and its hydraulic characteristics, and water treatment records keep data on the operation and maintenance of the plant and water quality. Records of business affairs include meter readings, billing accounts, material and equipment purchases, insurance, and personnel records.

The basic documents for a distribution system are maps showing mains, valves, hydrants, and service connections. In addition to mapping, data for each valve are retained on an individual valve record card. Precise location is described by at least two permanent reference points. Additional data on the card are valve number, type, size,

Figure 8-5 Operator's control and supervisory console for a large distribution system. (Courtesy of the Metropolitan Utilities District, Omaha, Nebraska.)

model, and manufacturer; number of turns and direction of turning to open; and a record of inspection and maintenance of the valve. Each hydrant is recorded on an individual card with the data including number, location, description, inspections, and maintenance. Service connection cards have a drawing for exact location and information on the pipe, meter, and other appurtenances. Also, inventories are kept of installed mains, valves, hydrants, and meters and of materials, supplies, equipment, and spare parts stocked for installation and repair work. Operations of a distribution system are documented by water consumption, pressure gauge records, quantity and pressure of pump station discharges, and levels in storage reservoirs. All of these data are essential for computer modeling of a system.

Treatment plant recordkeeping encompasses water quality and quantity data and plant operations. Some tests on the treated water are required by regulatory agencies to show that the quality meets drinking water standards. Daily tests include coliform analyses and usually one or more chemical parameters. Periodic analyses are performed for inorganic and organic chemicals. For process control, the laboratory testing is defined by the treatment, for example, jar tests for turbidity removal or hardness for softening. Also, the dosages and quantities of chemicals applied in treatment are recorded. Quantity and flow data include the amount of influent water, finished water, filter rates, water loss by backwashing of filters, and sludge wastage. If the source is groundwater, recorded well data are hours of operation, pumping rate, static and pumping water levels, discharge pressure, power consumption, and maintenance.

An annual report summarizes operation, maintenance, business, and cost data. The following are some of the important items discussed in a report. Consumption data are listed giving the minimum, average, and maximum values as total daily quantities and per capita usages. Quality data are compared to drinking water standards. Other topics are major problems and their solutions, consumer complaints and their resolutions, operating costs, personnel, completed improvements, and improvements needed. In addition to providing a permanent record of activities, a good annual report is valuable in consumer relations and to obtain public support for necessary waterworks improvements.

Computers are ideally suited for recordkeeping. The unit may be a terminal attached to the municipal computer or a separate microcomputer consisting of a typewriter-style keyboard and video monitor. A terminal for the latter contains a microprocessor chip, memory chips, power supply, display screen electronics, printed circuitry that routes power and data internally, and ports to connect peripheral devices, such as printers, disk drives, plotters, and telephone communicators. Internal memories allow data to be entered from the keyboard, or an auxiliary device, for storage and retrieval. With removable disks, auxiliary storage capacity can be expanded for recording and transferring data to the computer's working memory when desired. Printers are used to print out both input data and results from program calculations. Prepared programs for recordkeeping can be purchased on prerecorded disks to direct the operation of the computer. Programs are available for scheduling preventative maintenance, inventory control, process control calculations, plotting water consumption data, and other purposes. Since stored data can be printed without difficulty at any time, records are immediately available for analysis and writing reports.

8-5   WATER CONSERVATION

An important economic advantage of reducing water demand is extending the use of a waterworks without expansion of facilities. Lower consumption also reduces the costs of water and wastewater processing and slows depletion of a limited water supply. Conservation programs can be directed toward either consumer consumption or water losses in the distribution system.

Water conservation, oriented toward reducing consumer usage, is based on public education, installation of water-efficient plumbing fixtures, and institution of water rates that encourage conservation. While an essential adjunct, an education program without other incentives is not likely to be effective with passage of time. Plumbing fixtures that use less water are effective; however, the most successful applications have been in new construction where the community has either a limited source of supply or difficulty in wastewater disposal, or both. Many of the temporary measures—for instance, dams placed in the water tanks of conventional toilets and low-flow showerheads—are easily abandoned. In contrast, a new residence with washdown or siphonic toilets that use only about 1 gal per flush, a water-saving washing machine, smaller sinks, and minimum-flow showerheads and faucets can reduce domestic consumption by approximately one half. Charging a higher price for water can reduce consumption but, unless the public understands the need for restricting usage or additional revenue, the utility can lose public support. Domestic consumption is difficult to lower below some minimum amount, as determined by normal household uses, without replacing plumbing fixtures. On the other hand, lawn watering can be reduced by instituting an escalating rate schedule that penalizes the homeowner for overirrigation and wastage. Commercial and industrial users are much more likely to reduce their normal consumption as a result of higher prices. With economic incentive, they can often justify modifying processes to save water and develop reuse systems.

Water conservation by a utility is initiated by conducting a water audit to detect leaks and reduce the quantity of water that cannot be accounted for by metered sales. In addition to conserving water, potential benefits of conducting a water audit are increasing revenue by replacing meters that underregister and preventing serious damage that can result from uncorrected leaks. As a general rule, good performance is when the metered sales are at least 85 to 90 percent of the metered supply. The major phases of a water audit are flow measurement and leak detection.

Survey of a system begins with checking the accuracy of the master meters that measure flow into the distribution system. The common instrument for measuring flow in a pipeline is a pitot tube flow meter and recorder. With an accuracy of ± 2 percent, a wide range of flows can be measured in different size pipes. The meter can be temporarily installed through a 1-in. corporation stop. Although testing of residential meters during an audit is not practical, the municipal program of maintenance and replacement is assessed if underregistration is suspected.

Many communities remove and rebuild meters on a cycle of eight to ten years. Since small meters tend to underregister with age, the metered sales are likely to be less than the actual quantity of water delivered.

The pipe network is then divided into districts, and each district evaluated independently by isolating the area through the closing of all boundary valves except in one pipe for measuring influent flow. Since no storage exists, the recorded flow pattern is the variation of consumption—metered sales plus leakage. If the district is a residential area, the probability of significant leakage is determined by the ratio of minimum flow rate at night to the 24-hr average flow rate. Generally, the quantity of leakage is not significant if the ratio is 35 percent or less. A ratio of greater than 40 percent indicates significant leakage and the district is scheduled for leak detection. The district may be subdivided into smaller areas for flow measurements as part of the leak survey. If a district contains industries, the nighttime usage of water should be considered in evaluating the minimum night ratio. Leak detection surveys are conducted as described in Section 8-1.

8-6 ENERGY CONSERVATION

In waterworks operations, the greatest potential for energy conservation is in distribution system pumping. Not only does this consume the majority of energy, but pumping is also the most likely source of inefficient use of electrical power.

Electricity rate schedules are based on charges for energy consumed, high-demand usage, and power factor, which are all components of the price of power for electric motor-pump operation. The energy charge is determined by the quantity of electricity consumed. The demand charge is a penalty for peak usage for a short period of time. The power factor charge is to compensate for the increased cost of providing power for reactive loads, such as induction motors. These charges are discussed in greater detail in Section 12-6. Before a survey for energy conservation is undertaken, the billing records of the water utility are studied with reference to the rate schedule of the power company supplying the electricity.

Reducing Power Charges

Increasing energy efficiency is a major consideration in expansion and improvement of a distribution system. For example, additional distribution storage can equalize pumping rates and reduce the cost of electricity by reducing the demand charge. Of course, an economic analysis should be performed to justify a major capital expenditure to reduce operating costs. Lowering high service pressures in districts where lower water pressures will not adversely affect fire protection or customer usage can result in reduced power consumption. Repair of leakage can conserve both power and water. Since each distribution system is unique, the best methods for energy conservation differ. The general concepts are reduction in pumping head by lowering pressure, increasing pipe sizes, and increasing storage capacity; reduction in volume of pumpage by water conservation; increasing efficiency of motors and pumps; improvement of the power factor by installing capacitors and changing motor loadings; and reduction of demand charges by using gravity feed from elevated storage, starting engine-driven pumps to meet extreme peak demands, and water pricing to reduce peak consumption. Unfortunately, the institution of many potential energy conservation measures is limited by the existing physical system. A long-term energy management plan (Section 12-6) is essential to an energy conservation program.

Replacing ordinary electric motors with premium or high-efficiency motors results in energy savings. In large sizes, an ordinary industrial motor converts approximately 90 percent of the electrical energy input to mechanical energy output. Premium motors in the same size range have efficiencies of 92 to 93 percent, and they generally have a higher power factor and longer life. Depending on electricity pricing and percentage of time the motor is operated, a cost that is 15 to 25 percent greater than that for an ordinary motor can be paid back by energy savings in the first few years of operation.

A variable-speed drive can improve efficiency of a motor-pump system in an installation where the pump must be throttled to reduce discharge. Changing impeller speed to meet the flow demand increases the efficiency of the pump. In order to realize this benefit, the variable-speed drive must be efficient in converting electrical input to drive-shaft output. Efficiencies of variable-speed drives operating at rated (full) speed range from 80 to 90 percent, which is 5 to 10 percent less than a constant-speed drive. When operating a 50 percent of rated speed, however, different kinds of drives have efficiencies ranging from about 40 to 80 percent. Therefore, in selection of a particular variable-speed drive unit, the efficiency of the overall system must be considered for the normal anticipated operating range.

Pump Station Evaluations

Review of an existing pump station is started by testing the system hydraulics. A performance test is conducted to determine the head-discharge and efficiency curves for each installed pump (Figure 8-6a). Next, station discharge pressures and rates are measured to plot system head-discharge curves. The procedure involves operating selected combinations of pumps, at times of both low and high water consumption, to provide plotting points (Figure 8-6b). Pump suction and water levels in nearby reservoirs should be recorded along with any other data that may influence pump and system curves. (Section 4-4 discusses centrifugal pump characteristics, and Section 4-5 presents system characteristics.)

Improper selection or oversizing of pumps can be a major cause of excessive power consumption. Consider the following simplified example. The system head-discharge curve for a pump station is drawn in Figure 8-7a showing average, normal maximum, and peak discharges. Based on the concept that pumps of equal size would provide easy operation and maintenance, three identical constant-speed pumps were selected of sufficient capacity so that two could meet the peak demand at best operating efficiency, with the third unit as a standby. At average flow, however, when only one pump is needed, the pump is oversized. As sketched in Figure 8-7b, the head on the pump curve at average discharge is H_1 while on the system curve the desired head is H_2. In order to discharge water from the pump station at H_2, the pump discharge must be throttled to decrease the head from H_1 to H_2. This head loss is wasted energy. A better design for energy conservation, as illustrated in Figure 8-7c, would have been to

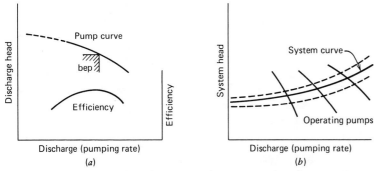

(a)

(b)

Figure 8-6 Pump station data to evaluate operation for an energy conservation survey. (*a*) Pump head-discharge and efficiency curves. (*b*) System head-discharge curve plotted from operating points of selected combinations of operating pumps.

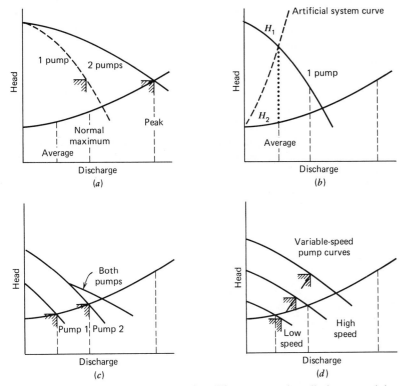

(a)

(b)

(c)

(d)

Figure 8-7 Head-discharge curves for different pump installations supplying water to the same system. (*a*) Two constant-speed pumps of equal size discharging peak demand. (*b*) The discharge of one pump supplying average flow must be throttled to reduce the discharge head from H_1 to H_2. (*c*) Two smaller pumps sized for efficient supply of average and normal maximum discharges. (*d*) A variable-speed pump to operate in the range of average to normal maximum flows.

provide different size pumps with their best efficiency points along the system head-discharge curve. An alternate design would have been to install variable-speed pumps to operate in the range of average to normal maximum discharge. As shown in Figure 8-7*d*, if the operating range is with the low speed near average flow, the variable-speed drive must be selected to provide high efficiency at low speed.

The operating efficiency of an oversized constant-speed pump can be improved by changing the size of the impeller. With a smaller diameter impeller, the characteristic pump curve is lowered so that the best operating point is closer to the system head-discharge curve. The existing impeller can be cut down, or a smaller diameter impeller can be ordered from the pump manufacturer. For some pumps, an alternate narrower impeller is available to reduce discharge pressure for the same flow.

Distribution system expansions and improvements can change the system head-discharge curve, resulting in inefficient pump operation. For example, a system curve can be lowered by installing larger diameter feeder mains that reduce frictional resistance. The effect, drawn in Figure 8-8*a*, is the same as having installed an oversized pump. In contrast, a system curve can be raised by aging of pipes, extension of the network, or installation of elevated storage. As sketched in Figure 8-8*b*, the increased

head at the same discharge results in less efficient pump operation. Forecasting the effects of system modifications on pumping is essential for maintaining an energy efficient system. A calibrated computer model of a pipe network is useful in predicting the influence of proposed system changes.

8-7   WATER RATES

A water utility must receive sufficient revenue to cover the costs of adequate service. Basic revenue requirements are for operation and maintenance expenses, debt service including interest and stipulated reserves, utility extensions and improvements, and plant replacement for perpetuation of the system. The total revenue collected should reflect not only recent cost experience but also should recognize anticipated future costs during the nominal period for which rates are being established, for example, five years.

Contrary to usual opinion, a schedule of water rates can be developed to recover the costs of serving different classes of customers while maintaining reasonable equity. Proper rate design does not take quantity discount into consideration nor does it advocate lower rates simply for water sold in larger amounts, as has been the popular past practice. Modern water rates recognize the costs of

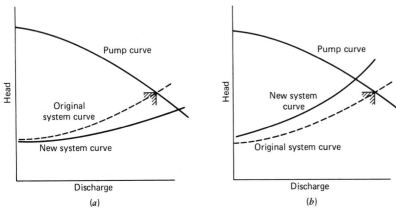

Figure 8-8   Changes in the system head-discharge curve can result in inefficient pump operation. (*a*) Reducing the frictional resistance of the distribution system lowers the system curve. (*b*) Increasing the static head or frictional resistance in the distribution system raises the system curve.

supplying the amount of water consumed, the rate of use or demand for use, and the expense involved in maintaining the customer account. Recognition of different classes of customers is reasonable in measuring costs to serve these diverse groups. Charges should be commensurate with the service rendered. For example, a customer with a high peak rate of use in comparison with the average consumption would require larger capacity pumps, pipes, and other facilities than a customer who has a comparable total consumption but uses water continuously at essentially a constant rate. Accordingly, cost-allocation procedures should recognize the particular service requirements of a user for not only total volume but also peak rates, and other factors.

Development of water rates involves the following major areas of study: determination of the amount of revenue required for the water utility; allocation of this annual revenue requirement to basic cost functions, which in turn are attributed to customer classes in accordance with respective class requirements for service; and design of water rates that recover from each customer class the respective costs of providing service. The latter, of course, requires an evaluation of water use for each class of customer based on local conditions that are influenced by climate, industrial demand, and other factors.

One technique for allocating the cost of service to cost functions is referred to as the commodity-demand method. This separates the three primary cost functions of demand, consumption, and customer costs. Demand costs are associated with providing facilities to meet peak rates of use. They include capital charges and operating costs on the part of the treatment works designed to meet peak requirements. Consumption costs include power, chemicals, and other incremental expenses that vary with the quantity of water produced. Customer expenses are associated with serving consumers irrespective of the amount of water used, such as meter reading, billing, accounting, and maintenance on meters and services.

The next step is to ascertain customer classes on the basis of hourly or daily demand characteristics, although some special categories may be based on unusual annual requirements. Typically, the three principal classes are residential, commercial, and industrial. In any given system, there may be users with unusual water-use characteristics that should be given separate consideration, for example, large institutions such as a university.

System costs for demand, consumption, and customer services are now related to the chosen customer classes to establish cost responsibility. Each customer's demand expense is based on his or her fair share of the total system demand. Consumption costs are allocated according to total annual use, and customer costs are allocated on the basis of number of metered services. Finally, a rate schedule is fixed to collect sufficient revenue to meet expenses of the waterworks.

PROBLEMS

8-1 Why are shutoff valves closed and reopened at least once a year?

8-2 Describe the procedure and the equipment used to clean water pipes.

8-3 Is a water main taken out of service to tap the pipe for a new service connection?

8-4 Calculate the discharge at a residual pressure of 20 psi based on the following flow test data. Hydrants 1, 2, 3, and 4 as situated in Figure 8-3a were discharging water with pitot tube pressure readings of 12 psi, 14 psi, 14 psi, and 16 psi, respectively. The hydrant nozzles were all $2\frac{1}{2}$-in. diameter. During these flows, the residual pressure drop at hydrant R was from 70 psi to 45 psi. (*Answers* 3640 gpm or 3540 gpm)

8-5 A fire flow test was conducted on a section of main in a residential district employing three hydrants with 2.5-in. nozzles, as shown in the upper diagram of Figure 8-3b. The pitot tube pressures measured in the discharges from hydrants 1 and 2 were both 10 psi, while the residual pressure drop at hydrant R was from 50 psi to 35 psi. Compute the discharge at 20 psi.

8-6 What kinds of records are kept for a water utility? What data are essential for computer modeling of a water system?

8-7 Describe the procedure for conducting a water audit.

8-8 How can modifications and expansions of a water system reduce the efficiency of pump operation?

chapter 9

Wastewater Flows and Characteristics

Domestic or sanitary wastewater refers to liquid discharge from residences, business buildings, and institutions. Industrial wastewater is discharge from manufacturing plants. Municipal wastewater is the general term applied to the liquid collected in sanitary sewers and treated in a municipal plant. In addition, interceptor sewers direct dry weather flow from combined sewers to treatment, and unwanted infiltration and inflow enters the collector pipes. A schematic of the system is given in Figure 9-1.

Storm runoff water in most communities is collected in a separate storm sewer system, with no known domestic or industrial connections, and is conveyed to the nearest watercourse for discharge without treatment. Rain water washes contaminants from roofs, streets, and other areas. Although the pollutional load of the first flush may be significant, the total amount from separated storm-water systems is relatively minor compared with other waste discharges. Several large cities have a combined sewer system where both storm water and sanitary wastes are collected in the same piping. Dry weather flow in the combined sewers is intercepted and conveyed to the treatment plant for processing but, during storms, flow in excess of plant capacity is by-passed directly to the receiving watercourse. This can constitute significant pollution and a health hazard in cases where the receiving body is used for a drinking water supply. One solution would be to replace the combined sewers with separate pipes, but the cost in large cities would be prohibitive,

although this technique can be applied where only a few combined sewers exist in a municipal system.

9-1 DOMESTIC WASTEWATER

The volume of wastewater from a community varies from 50 to 250 gal per capita per day (gpcd) depending on sewer uses. A common value for domestic wastewater flow is 120 gpcd (450 l/person · d), which assumes that the residential dwellings have modern water-using appliances, such as automatic washing machines. The organic matter contributed per person per day in domestic wastewater is approximately 0.24 lb (110 g) of suspended solids and 0.20 lb (90 g) of BOD in communities where a substantial portion of the household kitchen wastes is discharged to the sewer system through garbage grinders. In selection of data for design, the quantity and organic strength of wastewater should be based on actual measurements taken throughout the year to account for variations resulting from seasonal climatic changes and other factors. The average values during the peak month may be used for design. Excluding unusual infiltration and inflow, the average daily sanitary wastewater flow during the maximum month of the year is commonly 20 to 30 percent greater than the average annual daily flow. Excluding seasonal industrial wastes, the average daily BOD load from sanitary wastewater during the maximum month is greater than the annual average by 30 percent or

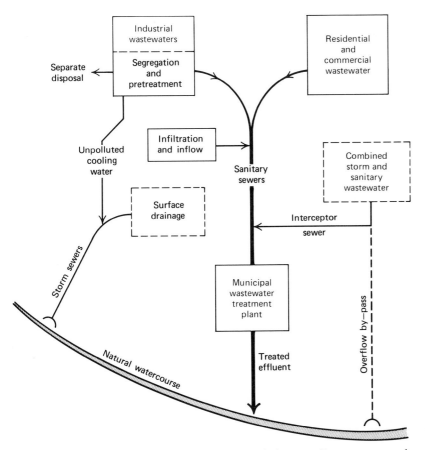

Figure 9-1 Sources of municipal wastewater in relation to collector sewers and treatment.

more in small plants (less than 0.5 mgd) and less than 20 percent in large plants (greater than 50 mgd).

Estimated wastewater flows for residential dwellings and other establishments are listed in Table 9-1. Mobile homes and hotels generate less wastewater than residences, since they have fewer appliances. The quantity and strength of wastewater from schools, offices, factories, and other commercial establishments depend on hours of operation and available eating facilities. Although cafeterias do not provide a great deal of flow, the wastewater strength is increased materially by food preparation and cleanup.

The common value for sanitary wastewater of 120 gpcd includes residential and commercial wastewaters plus reasonable infiltration, but excludes industrial discharges. Characteristics of this wastewater prior to treatment, after settling, and following conventional biological processing are given in Table 9-2. Total solids, residue on evaporation, include both dissolved salts and organic matter; the latter is represented by the volatile fraction. BOD is a measure of the wastewater strength. Sedimentation of a typical domestic wastewater diminishes BOD approximately 35 percent and suspended solids 50 percent. Processing, including secondary biological treatment,

Table 9-1. Approximate Wastewater Flows for Various Kinds of Establishments and Services

Type	Gallons per Person per Day	Pounds of BOD per Person per Day
Domestic wastewater from residential areas		
Large single-family houses	120	0.20
Typical single-family houses	80	0.17
Multiple-family dwellings (apartments)	60 to 75	0.17
Small dwellings or cottages	50	0.17
Domestic wastewater from camps and motels		
Luxury resorts	100 to 150	0.20
Mobile home parks	50	0.17
Tourist camps or trailer parks	35	0.15
Hotels and motels	50	0.10
Schools		
Boarding schools	75	0.17
Day schools with cafeterias	20	0.06
Day schools without cafeterias	15	0.04
Restaurants		
Each employee	30	0.10
Each patron	7 to 10	0.04
Each meal served	4	0.03
Transportation terminals		
Each employee	15	0.05
Each passenger	5	0.02
Hospitals	150 to 300	0.30
Offices	15	0.05
Drive-in theaters, per stall	5	0.02
Movie theaters, per seat	3 to 5	0.02
Factories, exclusive of industrial and cafeteria wastes	15 to 30	0.05

Table 9-2. Approximate Composition of Average Sanitary Wastewater (mg/l) Based on 120 gpcd (450 l/person · d)

Parameter	Raw	After Settling	Biologically Treated
Total solids	800	680	530
Total volatile solids	440	340	220
Suspended solids	240	120	30
Volatile suspended solids	180	100	20
Biochemical oxygen demand	200	130	30
Inorganic nitrogen as N	15	15	24
Total nitrogen as N	35	30	26
Soluble phosphorus as P	7	7	7
Total phosphorus as P	10	9	8

reduces the suspended solids and BOD content more than 85 percent, volatile solids 50 percent, total nitrogen about 25 percent, and phosphorus only 20 percent.

The surplus of nutrients in the treated effluent indicates that sanitary wastewater has nitrogen and phosphorus in excess of biological needs. The generally accepted BOD/N/P weight ratio required for biological treatment is 100/5/1 (100 mg/l BOD to 5 mg/l nitrogen to 1 mg/l phosphorus). Raw sanitary wastewater has a ratio of 100/17/5 and after settling 100/23/7, and thus contains abundant nitrogen and phosphorus for microbial growth. (The exact BOD/N/P ratio needed for biological treatment depends on the process method and availability of the N and P for growth; 100/6/1.5 is often related to unsettled sanitary wastewater, while 100/3/0.7 is used where the nitrogen and phosphorus are in soluble forms.) Another important wastewater characteristic is that not all of the organic matter is biodegradable. Although a substantial portion of the carbohydrates, fats, and proteins are converted to carbon dioxide by microbial action, a waste sludge equivalent to 20 to 40 percent of the applied BOD is generated in biological treatment.

Loadings on treatment units are often expressed in terms of pounds of BOD per day or pounds of solids per day, as well as quantity of flow per day. The relationship between the parameters of concentration and flow is based on the following conversion factors: 1.0 mg/l, which is the same as 1.0 part per million parts by weight, equals 8.34 lb/mil gal, since 1 gal of water weighs 8.34 lb; and, used less frequently, the value 62.4 lb/mil cu ft, since 1 cu ft of water weighs 62.4 lb. These relationships are defined by the following equations:

$$\text{Pounds of } C = \text{concentration of } C(\text{mg/l})$$
$$\times Q(\text{mil gal}) \times 8.34 \qquad (9\text{-}1)$$

or

$$\text{Pounds of } C = \text{concentration of } C(\text{mg/l})$$
$$\times Q(\text{mil cu ft}) \times 62.4 \qquad (9\text{-}2)$$

where C = BOD, SS, or other constituent, milligrams per liter

Q = volume of wastewater, million gallons or million cubic feet

$$8.34 = \frac{\text{lb/mil gal}}{\text{mg/l}}$$

$$62.4 = \frac{\text{lb/mil cu ft}}{\text{mg/l}}$$

Calculations in Example 9-1 show that 120 gal of the sanitary wastewater as described in Table 9-2 contains 0.20 lb of BOD and 0.24 lb of suspended solids; Examples 9-2 and 9-3 illustrate applications of Eqs. 9-1 and 9-2.

EXAMPLE 9-1

Sanitary wastewater from a residential community is 120 gpcd containing 200 mg/l BOD and 240 mg/l suspended solids. Compute the pounds of BOD per capita and pounds of SS per capita.

Solution

Using Eq. 9-1,

$$\text{BOD} = 200 \text{ mg/l} \times 0.000,\ 120 \text{ mil gal}$$

$$\times 8.34 \frac{\text{lb}}{\text{mil gal} \times \text{mg/l}} = 0.20 \text{ lb}$$

$$\text{SS} = 240 \text{ mg/l} \times 0.000,\ 120 \text{ mil gal}$$

$$\times 8.34 \frac{\text{lb}}{\text{mil gal} \times \text{mg/l}} = 0.24 \text{ lb}$$

EXAMPLE 9-2

Industrial wastewaters (Table 9-4) have a total flow of 2,930,000 gpd, BOD of 21,600 lb/day, and suspended solids of 13,400 lb/day. Calculate the BOD and suspended solids concentrations.

Solution

From the relationship in Eq. 9-1,

$$\text{BOD concentration} = \frac{21,600 \text{ lb/day}}{2.93 \text{ mil gal/day} \times 8.34}$$

$$= 880 \text{ mg/l}$$

$$\text{SS concentration} = \frac{13,400 \text{ lb/day}}{2.93 \text{ mgd} \times 8.34} = 550 \text{ mg/l}$$

EXAMPLE 9-3

An aeration basin with a volume of 300 m³ contains a mixed liquor (aerating activated sludge) with a suspended solids concentration of 2000 mg/l (g/m³). How many kilograms of mixed liquor suspended solids are in the tank?

Solution

$$\text{MLSS} = \frac{2000 \text{ g/m}^3 \times 300 \text{ m}^3}{1000 \text{ g/kg}} = 600 \text{ kg}$$

9-2   INDUSTRIAL WASTEWATERS

Industries within municipal limits ordinarily discharge their wastewater to the city's sewer system after pretreatment. In joint processing of wastewater, the municipality accepts responsibility of final treatment and disposal. The majority of manufacturing wastes are more amenable to biological treatment after dilution with domestic wastewater; however, large volumes of high-strength wastes must be considered in sizing of a municipal treatment plant. Uncontaminated cooling water is directed to the storm sewer.

A sewer code, user fees, and separate contracts between an industry and city can provide adequate control and sound financial planning while they accommodate industry by joint treatment. Pretreatment at the industrial site must be considered for wastewaters having strengths or characteristics significantly different from sanitary wastewater. Consideration should be given to modifications in industrial processes, segregation of wastes, flow equalization, and waste strength reduction. Process changes, equipment modifications, by-product recovery, and in-plant wastewater reuse can result in cost savings for both water supply and wastewater treatment. Modern industrial plant design dictates segregation of separate waste streams for individual pretreatment, controlled mixing, or separate disposal. The latter applies to both uncontaminated cooling water that can be discharged directly to surface watercourses, and toxic wastes that cannot be adequately processed by the municipal plant and must be handled by the industry. Manufacturing plants using a diversity of operations may be required to equalize wastewaters by holding them in a basin for stabilization prior to their discharge to the sewer. Unequalized flows may have dramatic fluctuations in quality that would upset the efficiency of a biological treatment system. Certain industrial discharges, such as dairy wastes, can be more easily reduced in strength by treatment in their concentrated form at the industrial site. Others, like metal-plating wastes, often require pretreatment for the removal of toxic metal ions. If reuse of the municipal wastewater is planned, rather stringent controls on industrial discharges are needed, since many of the substances in manufacturing wastes are refractory to conventional treatment and will interfere with water reuse.

The characteristics of four selected industrial wastewaters are listed in Table 9-3 for comparison with sanitary wastewater in Table 9-2. BOD concentrations range from 5 to 20 times greater than for domestic wastewater. Total solids are also greater but vary in character from colloidal and dissolved organics in food processing wastewaters to predominantly inorganic salts, such as the chlorophenolic waste. Suspended solids concentration relative to BOD is important when considering conventional primary sedimentation and secondary biological treatment. Settling of the synthetic textile wastewater with a suspended solids to BOD ratio of 2000 mg/l to 1500 mg/l would be as effective as clarifying a sanitary wastewater with a ratio of 240/200, but settling a milk-processing wastewater with a suspended solids to BOD ratio of 300 mg/l to 1000 mg/l would remove very little organic matter. In addition to high strength and settleability, particular consideration must be given to nutrient content, grease, and toxicity. Food-processing wastes generally contain sufficient nitrogen and phosphorus for biological treatment while discharges from chemical and materials industries are deficient in growth nutrients. Handling animal fats, plant oils, and petroleum products may result in a wastewater too high in grease content for admission to a municipal system without pretreatment. The chlorophenolic waste in Table 9-3 could not be discharged to a sewer without extensive reduction in phenol; the limit applied by sewer ordinances is in the range of 0.5 to 1.0 mg/l. Metal finishing wastes are pretreated to remove oil, cyanide, chromium, and other heavy metals such that the pretreated discharge has fewer contaminants than

Table 9-3. Average Characteristics of Selected Industrial Wastewaters

	Milk Processing	Meat Packing	Synthetic Textile	Chlorophenolic Manufacture
BOD, mg/l	1,000	1,400	1,500	4,300
COD, mg/l	1,900	2,100	3,300	5,400
Total solids, mg/l	1,600	3,300	8,000	53,000
Suspended solids, mg/l	300	1,000	2,000	1,200
Nitrogen, mg N/l	50	150	30	0
Phosphorus, mg P/l	12	16	0	0
pH	7	7	5	7
Temperature, °C	29	28	—	17
Grease, mg/l	—	500	—	—
Chloride, mg/l	—	—	—	27,000
Phenols, mg/l	—	—	—	140

domestic wastewater. Each municipality should have an inventory of industrial wastewaters discharged to the sanitary sewer system as is illustrated in Table 9-4. In this city the major wastewater contributors are food-processing industries. The manufacturing wastewaters from rubber products, metal working, and carpet weaving have strengths comparable to, or less than, domestic wastewater.

Table 9-4. Results from a Municipal Industrial Wastewater Survey Listing Discharges to the Sanitary Sewer in a City with a Population of 145,000

	Flow (gpd)	BOD (mg/l)	BOD (lb/day)	Suspended Solids (mg/l)	Suspended Solids (lb/day)	COD (mg/l)	Grease (mg/l)
Meat processing	1,200,000	1,300	13,000	960	9,600	2,500	460
Soybean oil extraction	478,000	220	880	140	560	440	—
Rubber products	189,000	200	310	250	390	300	—
Ice cream	138,000	910	1,050	260	300	1,830	—
Cheese	110,000	3,160	2,900	970	890	5,600	—
Metal plating	108,000	8	7	27	24	36	—
Carpet mill	103,000	140	120	60	51	490	—
Candy	97,700	1,560	1,270	260	210	2,960	200
Motor scooters	93,500	30	23	26	20	70	—
Potato chips	90,400	600	450	680	510	1,260	—
Flour	83,100	330	230	330	250	570	—
Milk processing	65,100	1,400	760	310	170	3,290	—
Industrial laundry	50,000	700	290	450	190	2,400	520
Pharmaceuticals	40,700	270	91	150	50	390	160
Chicken hatchery	35,300	200	59	310	90	450	—
Luncheon meats	20,900	270	47	60	10	420	—
Soft drinks	16,000	480	64	480	64	1,000	—
Milk bottling	12,700	230	24	110	12	420	—
Totals	2,930,000		21,600		13,400		

Industrial wastewaters expressed in terms of quantity of flow and pounds of BOD are relatively meaningless to the general public. Therefore, the quantity and strength can be related to the number of persons that would be required to contribute an equivalent quantity of wastewater. Hydraulic and BOD population equivalents, based on average sanitary wastewater, are 120 gpcd and 0.20 lb BOD per person per day, respectively. In addition to equivalent populations, it is desirable to express the quantity of wastewater produced per unit of raw material processed or finished product manufactured. Examples 9-4 and 9-5 illustrate wastewater production and equivalent population calculations.

EXAMPLE 9-4

A dairy processing about 250,000 lb of milk daily produces an average of 65,100 gpd of wastewater with a BOD of 1400 mg/l. The principal operations are bottling of milk and making ice cream, with limited production of cottage cheese. Compute the flow and BOD per 1000 lb of milk received, and the equivalent populations of the daily wastewater discharge.

Solution

Flow per 1000 lb of milk

$$= \frac{1000 \text{ lb}}{250,000 \text{ lb/day}} \times 65,100 \text{ gpd} = 260 \text{ gal}$$

BOD per 1000 lb of milk

$$= \frac{0.0651 \text{ mil gal/day} \times 1400 \text{ mg/l} \times 8.34}{250 \text{ thousands of lb/day}}$$

$$= 3.0 \text{ lb}$$

BOD equivalent population

$$= \frac{0.0651 \text{ mil gal/day} \times 1400 \text{ mg/l} \times 8.34}{0.20 \text{ lb BOD/person/day}}$$

$$= 3800 \text{ persons}$$

Hydraulic equivalent population

$$= \frac{65,000 \text{ gal/day}}{120 \text{ gal/person/day}} = 540 \text{ persons}$$

EXAMPLE 9-5

A meat-processing plant slaughters an average 500,000 kg of live beef per day. The majority is shipped as dressed halves with some production of packaged meats. Blood is recovered for a salable by-product, paunch manure (undigested stomach contents) is removed by screening and hauled to land burial, and process wastewater is settled and skimmed to recover heavy solids and some grease for inedible rendering with other meat trimmings. After this pretreatment, the waste discharged to the municipal sewer is 4500 m³/d containing 1300 mg/l BOD. Calculate the BOD waste per 1000 kg LWK (live weight kill) and the equivalent populations of the daily wastewater flow.

Solution

BOD per 1000 kg LWK

$$= \frac{4500 \text{ m}^3/\text{d} \times 1300 \text{ mg/l}}{500 \text{ thousands of kg/d} \times 1000 \text{ g/kg}} = 11.7 \text{ kg}$$

BOD equivalent population

$$= \frac{4500 \text{ m}^3/\text{d} \times 1300 \text{ mg/l}}{90 \text{ g BOD/person} \cdot \text{d}} = 65,000 \text{ persons}$$

Hydraulic equivalent population

$$= \frac{4500 \text{ m}^3/\text{d} \times 1000 \text{ l/m}^3}{450 \text{ l/person} \cdot \text{d}} = 10,000 \text{ persons}$$

9-3   INFILTRATION AND INFLOW

Infiltration is groundwater entering sewers and building connections through defective joints, and broken or cracked pipe and manholes. Inflow is water discharged into sewer pipes or service connections from such sources as foundation drains, roof leaders, cellar and yard area drains, cooling water from air conditioners, and other clean-water discharges from commercial and industrial establishments. In comparison to storm sewers, sanitary lines are small, being sized to handle only domestic and industrial wastewaters plus reasonable infiltration. Excessive infiltration and inflow can create several serious problems including surcharging of sewer lines with backup of sanitary wastewaters into house basements, flooding of street and road areas, overloading of treatment facilities, and by-passing of pumping stations and treatment works.

The quantity of infiltration water entering a sewer depends on condition of pipe and pipe joints, groundwater levels, and permeability of the soil. Seepage into new lines is controlled by proper design, selection of sewer pipe, close supervision of construction, and limiting infiltration allowances. Construction specifications usually permit a maximum infiltration rate of 500 gpd per mile of sewer length and inch of pipe diameter (46 l/d per kilometer of length and millimeter of pipe diameter). The quantity of this seepage flow is equal to 3 to 5 percent of the peak hourly domestic flow rate, or approximately 10 percent of the average flow. With development of better pipe jointing materials and tighter control of construction methods, infiltration allowances as low as 200 gpd/mile/in. (19 l/d·km·mm) of pipe diameter are being considered. Correction of infiltration conditions in existing sewer systems involves evaluation and interpretation of wastewater flow conditions in determining the source and rate of excessive infiltration, followed by consideration of corrective measures. Present techniques to reduce infiltration are grouting or sealing of soils surrounding the sewer pipe, pipe relining, and sewer replacement; all of them are costly.

Inflow is the result of deliberately planned, or expediently devised, connections of extraneous water sources to sanitary sewer systems. Although unwanted storm water or drainage should be disposed of in storm sewers, the sanitary system is often a more convenient conduit because of greater depth of burial and more convenient location. Excess inflow can be prevented by establishing and enforcing a sewer use regulation that excludes storm and surface waters from separate sanitary collectors. The ordinance should be explicit in directing surface runoff from roofs and other areas, foundation drainage, unpolluted water from air conditioning systems, industrial cooling operations, swimming pools, and the like, to storm lines leading to natural drainage outlets. A few ordinances allow cellar drainage into sanitary sewers; however, this is no longer considered proper under present-day conditions. This permit was probably derived from the days when basements were built with stone walls

and unpaved floors. Where inflow problems already exist, surveys can be conducted to locate connections and to institute corrective measures.

EXAMPLE 9-6

Calculate the infiltration and compare this quantity to the average daily and peak hourly domestic wastewater flows for the following:

Sewered population = 24,000 persons
Average domestic flow = 100 gpcd
Peak hourly domestic flow = 240 gpcd
Infiltration rate = 500 gpd/mile/in. of pipe diameter
Sanitary sewer system
4-in. building sewers = 36 miles
8-in. street laterals = 24 miles
10-in. submains = 6 miles
12-in. trunk sewers = 6 miles

Solution

Infiltration (gpd)

$$= \text{rate}\left(\frac{\text{gal}}{\text{day} \times \text{miles} \times \text{in.}}\right) \times \text{dia(in.)} \times \text{length (miles)}$$

$$= 500(4 \times 36 + 8 \times 24 + 10 \times 6 + 12 \times 6)$$

$$= 234,000 \text{ gpd}$$

Average domestic flow = $24,000 \times 100 = 2,400,000$ gpd

$$\frac{\text{Infiltration}}{\text{Average domestic flow}} = \frac{234,000}{2,400,000} \times 100 = 9.8 \text{ percent}$$

Peak hourly domestic flow

$$= 24,000 \times 240 = 5,760,000 \text{ gpd}$$

$$\frac{\text{Infiltration}}{\text{Peak hourly flow}} = \frac{234,000}{5,760,000} \times 100 = 4.1 \text{ percent}$$

9-4 MUNICIPAL WASTEWATER

As shown in Figure 9-1, the flow in sanitary sewers is a composite of domestic and industrial wastewaters, infiltration and inflow, and intercepted flow from combined sewers. Collector sewers must have hydraulic capacities to handle maximum hourly flow including domestic and infiltration, plus any additional discharge from industrial plants. New sewer systems are usually designed on the basis of an average daily per capita flow of 100 gal (400 liters), which includes normal infiltration. However, pipes must be sized to carry peak flows that are often assumed to be 400 gpcd (1500 l/person·d)

for laterals and submains when flowing full, 250 gpcd (950 l/person·d) for main, trunk, and outfall sewers; and in the case of interceptors, collecting from combined sewer systems, 350 percent of the average dry weather flow. Peak hourly discharges in main and trunk sewers are less than the maximum flows in laterals and sub-

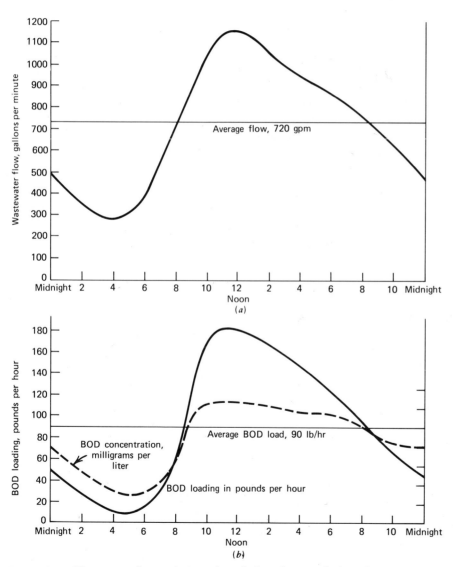

Figure 9-2 Wastewater flow and strength variations for a typical medium-sized city. (*a*) Typical municipal wastewater flow pattern. (*b*) Variation in concentration of BOD in the wastewater and resulting BOD loading pattern.

mains, since hydraulic peaks tend to level out as the wastewater flows through a pipe network picking up an increasing number of connections.

A typical discharge pattern from a separate sanitary sewer system is illustrated in Figure 9-2a. Hourly flow rates range from a minimum to a maximum of 20 to 250 percent of the average daily rate for small communities, and from 50 to 200 percent for larger cities. Lowest flows occur in early morning about 5 A.M. and peak discharge takes place near midday. The BOD concentration in wastewater also varies with time of day in a path that follows the flow variation (Figure 9-2b). Waste strength is greatest during the workday when household and industrial activities are contributing a large amount of organic matter, and is reduced during the night when entering flow is less contaminated and slow velocities in pipes permit settling of solids. If both flow and BOD concentration variations are known, the time-BOD loading on a treatment plant can be calculated and plotted as shown in Figure 9-2b. Knowledge of influent hydraulic and BOD loadings are essential in evaluating the operation of a treatment plant.

The quantity and characteristics of wastewater fluctuate with season of the year and between weekdays and holidays. Summer discharges frequently exceed winter flows by 10 to 20 percent, and industrial contributions are reduced on Sundays. Hourly fluctuations in large cities are modified in comparison with small towns because of the diversity of activities and operations that take place throughout the 24-hr day. Large volumes of high-strength industrial waste contributions can distort typical flow and BOD patterns by accentuating the peak hydraulic and BOD loadings during operational hours. Excessive infiltration and inflow, while diluting wastewater strength, can have considerable impact on a treatment facility by increasing both the average and peak flows during periods of high rainfall. All of these factors must be considered in assessing the wastewater flow and strength variations for a particular community.

EXAMPLE 9-7

The sanitary and industrial waste from a community consists of domestic wastewater from a sewered population of 7500 persons; potato processing waste of 30,000 gpd containing 550 lb BOD; and creamery wastewater flow of 120,000 gpd with a BOD concentration of 1000 mg/l. Estimate the combined wastewater flow in gallons per day and BOD concentration in milligrams per liter.

Solution

Source	Flow in Gallons per Day	BOD in Pounds per Day
Domestic	$7,500 \times 100 = 750,000$	$0.17 \times 7,500 = 1,270$
Potato	$= 30,000$	$= 550$
Creamery	$= 120,000$	$0.120 \times 1,000$
		$\times 8.34 \quad = 1,000$
Total	$\overline{900,000}$	$\overline{2,820}$

$$\text{BOD Concentration} = \frac{2820 \text{ lb/day}}{0.90 \text{ mil gal/day} \times 8.34}$$

$$= 380 \text{ mg/l}$$

EXAMPLE 9-8

A city with a sewered population of 145,000 has an average wastewater flow of 18.9 mgd with an average BOD of 320 mg/l. An inventory of the industrial wastewaters entering the sanitary sewer system is given in Table 9-4. (a) Compute the equivalent populations for this municipal wastewater flow that includes both sanitary and industrial wastewaters. (b) Determine the per capita contribution of sanitary wastewater flow and BOD based on the city's population excluding the industrial wastewaters.

Solution

(a) For the municipal wastewater,

$$\frac{\text{Hydraulic equivalent}}{\text{population}} = \frac{18,900,000 \text{ gpd}}{120 \text{ gpcd}} = 158,000$$

$$\frac{\text{BOD equivalent}}{\text{population}} = \frac{18.9 \text{ mgd} \times 320 \text{ mg/l} \times 8.34}{0.20 \text{ lb/person/day}}$$

$$= 252,000$$

(b) Per capita contributions excluding industrial wastewaters are

$$\text{Sanitary flow} = \frac{18,900,000 - 2,930,000}{145,000}$$

$$= 110 \text{ gpcd}$$

$$\text{Sanitary BOD} = \frac{18.9 \times 320 \times 8.34 - 21,600}{145,000}$$

$$= 0.20 \text{ lb/person/day}$$

9-5   EVALUATION OF WASTEWATER

Proper sampling techniques are vital for accurate testing in evaluation studies. To be representative of the entire flow, samples should be taken where the wastewater is well mixed. An instantaneous grab sample represents conditions at the time of sampling only, and cannot be considered to represent a longer time period, since the character of a wastewater discharge is not stable. A composite sample is a mixture of individual grabs proportioned according to the wastewater flow pattern. Compositing is commonly accomplished by collecting individual samples at regular time intervals, for example, every hour on the hour, and by storing them in a refrigerator or ice chest; coincident flow rates are read from an installed flow meter or are determined from some other flow recording device. A representative sample is then integrated by mixing together portions of individual samples relative to flow rates at sampling times. Example 9-9 illustrates this procedure.

Composite samples representing specified time periods are tested to appraise plant performance and loadings. Weekday specimens collected over a 24-hr period are most common. Average daily BOD and suspended solids data are used to calculate plant loadings while mean influent and effluent concentrations yield treatment efficiencies. Integrated samples during the period of peak flow, usually 8 to 12 hr depending on influent variation, allow determination of maximum loadings on treatment units.

The most common laboratory analyses for defining the characteristics of a municipal wastewater are BOD and suspended solids. BOD and flow data are basic to the design of biological treatment units, while the concentration of suspended solids relative to BOD indicates to what degree organic matter is removable by primary

settling. Occasionally, comprehensive testing to yield the type of information given in Table 9-2 should be performed. Treatability as well as wastewater characteristics can be defined from these data. In addition, it may be advantageous to record wastewater temperature, pH, COD, alkalinity, color, grease, and presence of specific heavy metals. Selection of these tests depends on the industrial wastes contributed to the sewer system. Importance of comprehensive BOD testing, as described in Section 3-11, is extremely vital, since the BOD test relates more closely than any other to the operation of a biological treatment system. Besides the five-day value and k-rate, the ratios of BOD to COD and BOD to volatile solids indicate biodegradability of waste organics. Presence of inhibiting substances or toxins, resulting from industrial wastewaters, are often indicated by increasing BOD values with increasing dilution, lag periods at the beginning of properly seeded tests, and erratic test results. To be significant, laboratory analyses must be correlated with plant operations that prevailed at the time of sampling. Wastewater flow, day of the week, weather and rainfall, and abnormal wastewater discharges, for example, breakdown of pretreatment facilities at a major industry, are essential for future interpretation of recorded testing data. Also, it may be advantageous to record sludge production, equipment performance, chemical and power usage, operational problems, and information on industrial wastewater flows.

Occasionally, a biological process treating municipal wastewater is afflicted by unknown substances that reduce efficiency by inhibiting microbial activity, for example, bulking of activated sludge in an aeration system. If the plant operates satisfactorily under the same loadings at other times, the problem is most likely related to an industrial wastewater being discharged to the sewer system. Often laboratory testing of the industrial wastewater, or comprehensive tests on the municipal wastewater, can pinpoint the difficulty. However, if results do not prove adequate, a treatability study using a small, laboratory, biological treatment unit is warranted.

Wastewater treatability studies are particularly appropriate for evaluating proposed industrial wastewater discharges from new industries that plan to use the municipal plant for treatment. Although most wastewaters are more amenable to biological treatment after dilution with domestic wastewater, they should be subjected to laboratory analyses prior to permitting their discharge to a municipal sewer. This is especially true if the projected volume is substantial in comparison to the sanitary flow, and if the type involved has a history of creating treatment problems. Management of the proposed industry should be required to furnish precise information as to the quantity of wastewater flow, including anticipated flow patterns both daily and weekly, wastewater characteristics with anticipated hourly and daily variations from the norm, and proposed pretreatment facilities.

A laboratory apparatus for studying aerobic treatability is pictured in Figure 9-3. Wastewater is pumped from a refrigerated supply bottle into the aeration chamber; air is supplied through a porous stone diffuser at the bottom. Aerated mixed liquor flows through a connecting tube to the clarifier for gravity separation. Clear supernatant is the effluent while the settled solids are returned to the aeration cylinder by air lifting. Aeration period and BOD loading should be similar to those used in the full-scale treatment system being simulated. Treatability of the applied wastewater is measured in terms of BOD or COD removal efficiency, settleability of the mixed liquor, and microscopic examination of the activated sludge biota. The unit should be operated both on a feed of pure industrial wastewater and in a mixture with domestic wastewater. When treated alone, the industrial wastewater may require neutralization and/or addition of inorganic nitrogen and phosphate for nutrient balance. Joint treatment should be conducted at several different ratios of industrial to domestic wastewater to include the dilution anticipated in the actual municipal system.

EXAMPLE 9-9

Hourly samples were taken of the wastewater entering a treatment plant. The recorded flow pattern is given in Figure 9-2a. Tabulate the portions to be used from the hourly grabs to provide composite samples for the 24-hr duration and during the period of maximum 8-hr loading, between 9 A.M. and 5 P.M. The composite sample volumes needed for laboratory testing are approximately 2500 ml.

Solution

The following relationship determines the portion per unit of flow needed from individual samples to provide a desired composite volume.

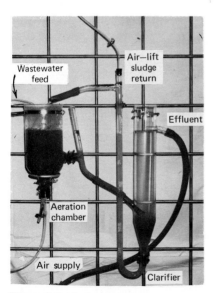

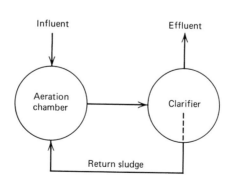

Figure 9-3 Laboratory apparatus for aerobic treatability studies on industrial and municipal wastewaters.

Portion of sample needed per unit of flow

$$= \frac{\text{total volume of sample desired}}{\text{average flow rate} \times \text{number of portions}} \quad (9\text{-}3)$$

Portion for the 24-hr period $= \dfrac{2500 \text{ ml}}{720 \text{ gpm} \times 24}$

$$= 0.15 \text{ ml/gpm}$$

Portion for the 8-hr period $= \dfrac{2500 \text{ ml}}{1000 \text{ gpm} \times 8}$

$$= 0.3 \text{ ml/gpm}$$

Calculations for the portions of hourly samples to be used in compositing are tabulated as follows:

Time	Flow (gpm)	Portions of Hourly Samples in Milliliters for: 24-hr Composite	8-hr Composite
Midnight	490	$0.15 \times 490 = 74$	
1 A.M.	420	$0.15 \times 420 = 63$	
2 A.M.	360	$0.15 \times 360 = 54$	
3 A.M.	310	$0.15 \times 310 = 47$	
4 A.M.	290	$0.15 \times 290 = 43$	
5 A.M.	310	$0.15 \times 310 = 46$	
6 A.M.	390	$0.15 \times 390 = 58$	
7 A.M.	560	$0.15 \times 560 = 84$	
8 A.M.	620	$0.15 \times 620 = 93$	
9 A.M.	900	$0.15 \times 900 = 135$	$0.3 \times 900 = 270$
10 A.M.	1040	$0.15 \times 1040 = 156$	$0.3 \times 1040 = 310$
11 A.M.	1130	$0.15 \times 1130 = 170$	$0.3 \times 1130 = 340$
Noon	1160	$0.15 \times 1160 = 174$	$0.3 \times 1160 = 350$
1 P.M.	1120	$0.15 \times 1120 = 168$	$0.3 \times 1120 = 340$
2 P.M.	1060	$0.15 \times 1060 = 159$	$0.3 \times 1060 = 320$
3 P.M.	1000	$0.15 \times 1000 = 150$	$0.3 \times 1000 = 300$
4 P.M.	950	$0.15 \times 950 = 143$	$0.3 \times 950 = 290$
5 P.M.	910	$0.15 \times 910 = 136$	
6 P.M.	870	$0.15 \times 870 = 130$	
7 P.M.	810	$0.15 \times 810 = 121$	
8 P.M.	760	$0.15 \times 760 = 114$	
9 P.M.	690	$0.15 \times 690 = 103$	
10 P.M.	630	$0.15 \times 630 = 94$	
11 P.M.	540	$0.15 \times 540 = 81$	
Total composite sample volumes		2596 ml	2520 ml

PROBLEMS

9-1 Using values from Table 9-1, estimate the daily quantity of wastewater and pounds of BOD produced by a transportation terminal based on the following: 30 employees, 500 passengers passing through the terminal each day, and a restaurant that serves 600 patrons daily with a staff of 10 employees. Calculate the average BOD concentration in the wastewater by Eq. 9-1. (*Answers* 9300 gpd, 37 lb/day, 470 mg/l)

9-2 Calculate the removal efficiencies for biologically treated sanitary wastewater for suspended solids, BOD, total nitrogen, and total phosphorus from the values listed in Table 9-2. (*Answer* suspended solids 88 percent)

9-3 Data given in Table 9-4 can be used to practice converting pollutant concentrations in milligrams per liter to pounds per day, and vice versa, employing Eq. 9-1. For example, calculate the pounds per day of BOD and SS in 110,000 gpd of cheese-processing wastewater, or compute the concentration of BOD in the potato-chip–manufacturing wastewater of 90,400 gpd containing 450 lb of BOD.

9-4 Compute the hydraulic and BOD equivalent populations for the meat-processing wastewater listed in Table 9-4. (*Answers* 10,000 and 65,000 persons, respectively)

9-5 A community with a population of 4500 has one major industry that produces an average of 430 m³/d of wastewater containing 950 kg of BOD. What is the estimated BOD concentration in milligrams per liter of the composite wastewater?

9-6 The combined wastewater in a municipality with a sewered population of 2000 persons includes wastes from a dairy and a poultry plant. The milk-processing wastewater is 20,000 gpd with a BOD concentration of 900 mg/l. The poultry plant processes 5000 chickens per day, discharging 16,000 gpd containing 150 lb of BOD. Estimate the total combined wastewater flow from the community in gallons per day and the average BOD concentration in milligrams per liter. Compute the BOD and hydraulic equivalent populations of the combined wastewater. (*Answers* 276,000 gpd, 300 mg/l of BOD, 3500 persons, 2300 persons)

9-7 A poultry-processing plant kills 10,000 chickens per day, producing a wastewater flow of 52,000 gpd containing mean BOD and SS concentrations of 920 mg/l and 420 mg/l, respectively. Compute the wastewater flow, BOD and SS produced per 1000 chickens processed, and the hydraulic and BOD equivalent populations per 1000 chickens.

9-8 A small city has a sanitary flow of 2300 m³/d with characteristics as given in Table 9-2 plus the following industrial wastewaters:

	Flow (m³/d)	BOD (mg/l)	SS (mg/l)
Soft-drink bottling	57	430	220
Bakery	23	3000	2000
Brewery	455	390	72

Calculate the composite BOD and SS concentrations of the municipal wastewater in milligrams per liter.

9-9 A town with a sewered population of 4500 has a daily wastewater flow of 420,000 gpd with an average BOD of 280 mg/l including industrial wastes. The industrial discharges to the municipal sewers are 15,000 gpd at 1800 mg/l BOD from a slaughterhouse and 20,000 gpd containing 70 lb of BOD from a soup-canning plant. Calculate the per capita contribution of domestic flow and BOD excluding the food-processing wastewaters.

9-10 The wastewater flow applied to a biological aeration process is a composite of 75 percent domestic wastewater with 130 mg/l of BOD, 30 mg/l of nitrogen, and 9 mg/l of phosphorus and 25 percent industrial wastewater with 1200 mg/l of BOD, 40 mg/l of nitrogen, and no phosphorus. Calculate the BOD/N/P ratio of the wastewater. Does the composite wastewater contain adequate nutrients? (*Answers* 100/8.1/1.7, yes)

9-11 What is the BOD/N/P ratio of the synthetic textile wastewater described in Table 9-3? If a BOD/N/P ratio of 100/3.0/0.7 is considered the minimum nutrient level for completely mixed activated-sludge treatment, calculate the additions of NH_4OH and H_3PO_4 required in pounds per million gallons of textile wastewater. (*Answers* 100/2.0/0, 280 lb of H_3PO_4, 310 lb of NH_4OH)

9-12 A pharmaceutical wastewater needs a minimum BOD/N/P of 100/3.0/0.7 for biological treatment. The wastewater characteristics are BOD = 740 mg/l, SS = 250 mg/l, soluble nitrogen = 24 mg/l, and soluble phosphorus = 12 mg/l. What ammonium nitrate and phosphoric acid additions would you recommend to ensure adequate nutrients for a daily flow of 28 m³/d?

9-13 Unsettled sanitary wastewater, with characteristics as given in Table 9-2 under the heading "Raw," is to be blended with the synthetic textile wastewater described in Table 9-3 to provide a proper nutrient balance for biological treatment. The appropriate BOD/N/P ratio is 100/6/1.5. If the blended wastewater is one third textile waste and two thirds raw sanitary wastewater by volume, are the concentrations of nitrogen and phosphorus adequate?

9-14 Hourly samples were taken of the wastewater entering a treatment plant with the magnitudes of flow at the times of sampling listed below. Tabulate the portions to be used from the hourly grabs to provide a 24-hr composite volume of approximately 2500 ml for laboratory testing.

Time	Flow (mgd)	Time	Flow (mgd)	Time	Flow (mgd)
0600	9.7	1400	25.6	2200	16.7
0700	12.1	1500	24.0	2300	15.5
0800	14.8	1600	23.5	2400	14.4
0900	18.6	1700	22.2	0100	13.8
1000	22.5	1800	21.8	0200	13.1
1100	24.8	1900	20.3	0300	11.9
1200	26.6	2000	18.9	0400	10.7
1300	28.5	2100	17.8	0500	10.1

9-15 Samples were taken at one-half hour intervals of the industrial wastewater discharge plotted in Figure 10-12. Tabulate the portions from half-hour grabs to provide a composite sample for the 12-hr period from 7 A.M. to 7 P.M. The sample volume needed for laboratory testing is approximately 2500 ml.

chapter 10
Wastewater Collection Systems

Sewers are underground, watertight conduits for conveying wastewaters by gravity flow from urban areas to points of disposal. The earliest drainage systems, constructed in the 16th- and 17th-century cities, were to carry storm runoff from built-up areas protecting them against inundation. Privies and cesspools were used for disposal of human excreta, and household wastes were often thrown on the streets. Although this created deplorable sanitary conditions, such cities as London and Philadelphia prohibited discharge of household wastes to storm drains as late as 1850. The development of steam-driven pumps and cast-iron pipes for pressure water distribution, however, led to indoor plumbing and flush toilets. Soon cesspools were banned and waterborne wastes were piped to the storm drains, converting them to combined sewers. While this water carriage system improved conditions within the city, untreated wastes were now being directed to surface watercourses.

Combined sewers were constructed in many cities of the United States prior to 1900 without recognizing the need for segregation and treatment of domestic and industrial wastewaters. Although these systems still exist in older municipalities, separate sewers have dominated construction during the 20th century. Storm sewers carry only surface runoff and other uncontaminated waters to natural channels directly, while sanitary sewers convey domestic and industrial wastewaters to treatment works for processing prior to their disposal. Where combined drain pipes exist, intercepting sewers have been con-structed to collect the dry-weather flow from a number of transverse outfalls and to transport it to a treatment plant; wet-weather flow in excess of treatment plant capacity is still routed directly to disposal in most cases.

10-1 STORM SEWER SYSTEM

Surface waters enter a storm drainage system through inlets located in street gutters or depressed areas that collect natural drainage. Cooling water from industries and groundwater seepage that enters footing drains are pumped to the storm sewer, since the pipes are usually set too shallow for gravity flow; furthermore, direct connections would be subject to backflow when the pipe surcharges. Figure 10-1 illustrates two common types of storm-water inlets for streets. The curb inlet has a vertical opening to catch gutter flow. Although the gutter may be depressed slightly in front of the inlet, this type offers no obstruction to traffic. The gutter inlet is an opening covered by a grate through which the drainage falls. The disadvantage is that debris collecting on the grate may result in plugging of the gutter inlet. Combination inlets composed of both curb and gutter openings are also common. Street grade, curb design, and gutter depression define the best type of inlet to select; nevertheless, minimizing traffic interference and eliminating plugging often take precedence over hydraulic efficiency.

Catch basins under street inlets are connected to the main storm sewer located in the street right-of-way, often

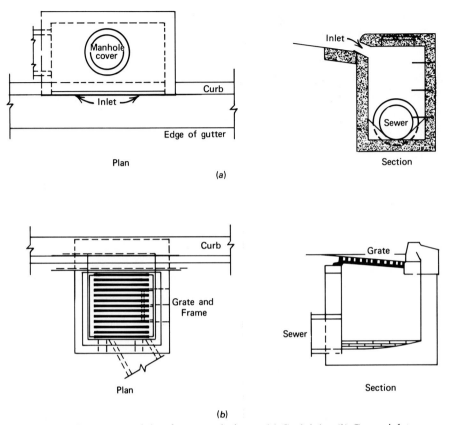

Figure 10-1 Storm-water inlets for street drainage. (*a*) Curb inlet. (*b*) Gutter inlet.

along the center line, by short pipelines. Manholes are placed at curb inlets, intersections of sewer lines, and regular intervals to facilitate inspection and cleaning. Pipeline gradients follow the general slope of the ground surface such that water entering can flow downhill to a convenient point for discharge. Sewer pipes are set as shallow as possible to minimize excavation while providing 2 to 4 ft of cover above the pipe to reduce the effect of wheel loadings. Sewer outlets that terminate in natural channels subject to tides or high water levels are equipped with flap gates to prevent backflooding into the sewer system. Backwater gates are also used on combined sewer outfalls and effluent lines from treatment plants where needed.

The rational method, described in Section 4-11, is used to calculate the quantity of runoff for sizing storm sewers. Climatic conditions are incorporated by using local rainfall intensity-duration formulas or curves. In dry regions, sewers may be placed only in high-value districts, while streets and roadside ditches serve as surface drains in sparsely populated areas. On the other hand, in regions of the country having intense thunderstorm weather, lined open channels are often found to be more economical compared with large buried conduits; sewers leading from small drainage areas terminate in grassed or concrete-lined ditches that discharge to surface watercourses.

Flowing full velocities used in design of storm sewers are a minimum of 3.0 ft/sec (0.9 m/s) and a maximum of about 10 ft/sec (3.0 m/s). The lower limit is set so that the lines are self-cleansing to avoid deposition of solids, and the upper limit is fixed to prevent erosion of the pipe by grit transported in the water.

A major difference in design philosophy between sanitary and storm sewers is that the latter are assumed to surcharge and overflow periodically. For example, a storm drain sized on the basis of a ten-year rainfall frequency presumes that one storm every ten years will exceed the capacity of the sewer. Sanitary sewers are designed and constructed to prevent surcharging. Where backup of sanitary sewers does occur, it is more frequently attributable to excess infiltration of groundwater through open pipe joints and unauthorized drain connections. A second easily recognizable difference between sanitary and storm sewers is the pipe sizes that are needed to serve a given area. As illustrated in Example 10-1, storm drains are many times larger than the pipes collecting domestic wastewater. Consequently, only a small amount of infiltrating rain water results in overloading domestic sewers.

Circular concrete pipe is commonly used for storm sewers. Nonreinforced concrete pipe is available in sizes up to 24-in. diameter in lengths of 3 or 4 ft. Reinforced pipe has steel imbedded in the concrete to give added structural strength. Circular pipe is manufactured in diameters from 12 to 144 in., and in laying lengths that range from 4 to 12 ft. Elliptically shaped and arch-type concrete pipe are also manufactured for special applications. Various types of joints are available for connecting pipe sections. The one selected depends on construction conditions, pipe size, manufacturer, and whether the pipe is reinforced or nonreinforced. The single-rubber-ring gasket joint has largely superseded all other types for use in concrete sewer lines because of economy, ease of construction, and satisfactory performance. In manufacture, the pipe ends are carefully cast with a space left for the rubber ring. With the gasket in place and lubricated, the spigot end is pushed into the bell to seat the joint properly.

EXAMPLE 10-1

(a) What is the maximum population that can be served by an 8-in. sanitary sewer laid at minimum grade using a design flow of 400 gpcd and a flowing full velocity of 2.0 ft/sec? (b) Compute the diameter of storm drain to serve the same population based on population density = 30 persons per acre, coefficient of runoff = 0.40, 10-yr rainfall frequency curve in Figure 4-34, a duration (time of concentration) = 20 min, and a velocity of flow = 5.0 ft/sec.

Solution

(a) Based on Figure 4-25, Q flowing full at a velocity of 2.0 ft/sec for an 8-in.-diameter pipe is 0.67 cu ft/sec. The maximum population that can be served

$$= \frac{0.67 \text{ cu ft/sec} \times 7.48 \text{ gal/cu ft} \times 86,400 \text{ sec/day}}{400 \text{ gal/day}}$$

$$= 1080 \text{ persons}$$

(b) Drainage area $= \dfrac{1080 \text{ persons}}{30 \text{ persons per acre}} = 36 \text{ acres}$

From Figure 4-34, for duration = 20 min., I = 4.2 in./hr. Using Eq. 4-31, Q = 0.40 × 4.2 × 36 = 60 cu ft/sec. Based on Figure 4-25, for Q = 60 cu ft/sec and V = 5.0 ft/sec, pipe diameter = 42 in.

Therefore, a 42-in.-diameter storm sewer is needed to drain the housing area of 1200 persons that can be served by an 8-in.-diameter sanitary sewer.

10-2   SANITARY SEWER SYSTEM

Sanitary sewers transport domestic and industrial wastewaters by gravity flow to treatment facilities. A lateral sewer collects discharges from houses and carries them to another branch sewer, and has no tributary sewer lines. Branch or submain lines receive wastewater from laterals and convey it to large mains. A main sewer, also called trunk or outfall sewer, carries the discharge from large areas to the treatment plant. A force main is a sewer through which wastewater is pumped under pressure rather than by gravity flow.

Design flows for sewer systems are based on population served, using the following per capita quantities: laterals and submains 400 gpcd (1500 l/person · d), main and trunk 250 gpcd (950 l/person · d), and interceptors 350 percent of the estimated average dry-weather flow. These figures include normal infiltration and are based on flowing full capacity. Excluded are industrial wastewaters and excessive infiltration. Sewer slopes should be sufficient to maintain self-cleansing velocities; this is normally interpreted to be 2.0 ft/sec (0.60 m/s) when flowing full. Table 10-1 lists sewer size, minimum slope for 2 ft/sec, and the corresponding quantity of flow. Slopes slightly less than those listed may be permitted in lines where the design average flow provides a depth of flow greater than one third the diameter of the pipe. Where velocities are greater than 10 ft/sec (3.0 m/s), special provision must be made to protect the pipe and manholes against displacement by erosion and shock hydraulic loadings.

Sanitary sewers are placed at sufficient depth to prevent freezing and to receive wastewater from basements. As a general rule, laterals placed in the street right-of-way are set at a depth of not less than 11 ft (3.3 m) below the top of the house foundation. To provide economical access to the sewer after street construction, service connections are generally extended from laterals to outside the curb line at the time of sewer placement. An alternative is to place the sanitary sewer behind the curb on one side of the street, making it readily accessible for service connections on

that side. Pipe connections from the opposite side of the street are accomplished by excavating working pits on each side and by jacking the house sewer into position for connection to the lateral on the opposite side. In jacking, pipe sections are pushed beneath the roadway by hydraulic jacks. For hard soils, boring machines are used to cut an opening through which the pipeline is pushed. The cutter head operates in front of the first pipe section and an auger pulls the excavated material out through the pipe to the jacking pit.

Ease of maintenance dictates many of the design criteria for wastewater collection systems. The minimum recommended size for laterals is an 8-in. (200 mm) diameter. Manholes located at regular intervals allow access to the pipe for inspection and cleaning. Pipes laid on too flat a grade require periodic flushing and cleaning to remove deposited solids and to prevent pipe plugging. Sewers less than 24 in. should be laid on a straight line between manholes, although in recent years curves have been permitted to follow the curvature of the streets. The degree of curvature is determined by length of the pipe sections and flexibility of the joints. Curved sewer lines can be cleaned with a high-pressure jet without damaging the pipe interior. Pumping stations are used in a collection system only when continuing flow by gravity is impractical because of their high cost and potential maintenance problems.

Sewer Profile

The drawing for a section of sanitary submain is given in Figure 10-2. To illustrate more detail, the vertical scale is ten times greater than the horizontal. The profile shows the ground surface, sewer line giving slope and diameter, manholes, connecting sewers, all other underground utilities, elevation of the sewer invert at critical points, and special features, such as an inverted siphon. In design drawings the profile is always accompanied by a plan view diagraming the location of the sewer in the street right-of-way.

Layout of new sewer lines and water mains must ensure adequate separation to prevent any possible passage of polluted water into the potable supply. Generally sewers are laid at least 10 ft (3 m) horizontally from a water main.

Table 10-1. Minimum Slopes for Various Sized Sewers at a Flowing Full Velocity of 2.0 ft/sec and Corresponding Discharges[a]

Sewer Diameter (in.)	Minimum Slope (ft/100 ft)	Flowing Full Discharge	
		(cu ft/sec)	(gpm)
8	0.33	0.7	310
10	0.25	1.1	490
12	0.19	1.6	700
15	0.14	2.4	1080
18	0.11	3.5	1570
21	0.092	4.8	2160
24	0.077	6.3	2820
27	0.066	8.0	3570
30	0.057	9.8	4410
36	0.045	14.1	6330

[a] Based on Manning's formula with $n = 0.013$

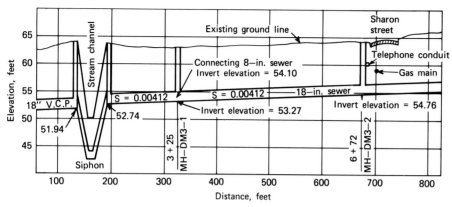

Figure 10-2   Profile of a sanitary sewer.

If local conditions prevent this lateral separation, a sewer may be laid closer if the crown of the sewer is at least 18 in. (46 cm) below the bottom of the water main, and the pipes are placed in separate trenches. The minimum vertical separation for sewers crossing under water mains is 18 in. between the bottom of the water main and top of the sewer. When this separation cannot be provided the water main should be relocated, or the sewer should be constructed of materials and with joints that are equivalent to water main standards of construction and are tested to insure watertightness. If a sewer must pass over a water main, the main should be constructed with compression-type or mechanical joint pipe and the sewer should be made with mechanical joint iron pipe; both services are then pressure tested for leakage. Sewer plan and profile drawings must show existing or proposed water mains and special sewer pipe construction where minimum distances between facilities cannot be maintained.

Inverted Siphons

A siphon is a depressed sewer that drops below the hydraulic gradient to avoid an obstruction, such as a stream, railway cut, or depressed highway. The design incorporates provisions for maintenance-free operation and minimum hydraulic head loss. Since a depressed sewer acts as a trap, the velocity of flow in the pipes should be greater than 3 ft/sec (0.9 m/s) to prevent deposition of

solids. This is accomplished by constructing an inlet splitter box that directs flow to two or more siphon pipes placed in parallel. For example, in Figure 10-3, flow in the 18-in.-diameter sewer is carried by two depressed pipes. During low flow in the main sewer, all of the wastewater is directed to the 8-in. pipe. When the depth of flow in the main sewer exceeds 6 in., the wastewater stream is split in the inlet box and is directed to both siphon pipes; with the main sewer flowing half full, the velocity in both is greater than 3 ft/sec. In addition to control of flow, the inlet and outlet structures provide access for cleaning the sewer lines.

Manholes

Most manholes are circular in shape with an inside diameter of 4 ft, which is considered sufficient to perform sewer inspection and cleaning (Figure 10-4). For small-diameter pipes, the manhole is usually constructed directly over the center line of the sewer. For very large sewers, access may be provided on one side with a landing platform for the convenience of introducing cleaning equipment. Manhole frames and covers are usually cast iron with a minimum clear opening of 21 in. (54 cm). Solid covers are used on sanitary sewers, while open-type covers are common on storm sewers. Steps or ladder rungs are placed for access. Walls may be constructed of precast concrete rings, concrete block, brick, or poured concrete.

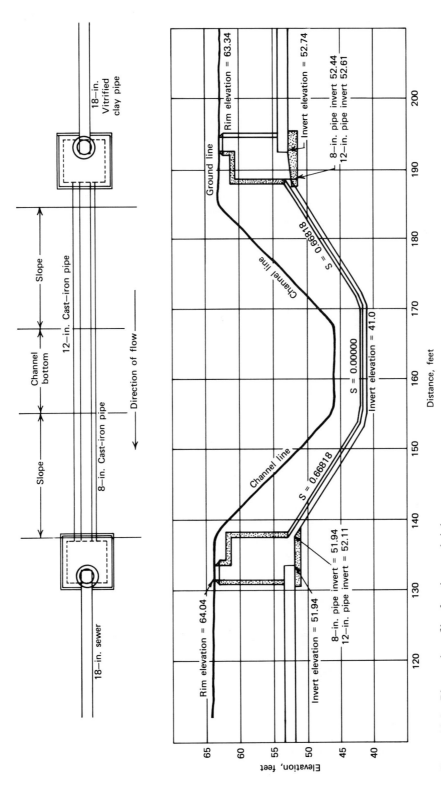

Figure 10-3 Plan and profile of an inverted siphon.

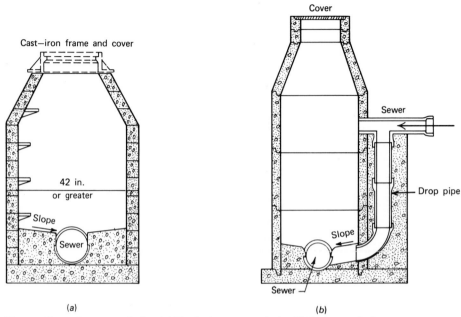

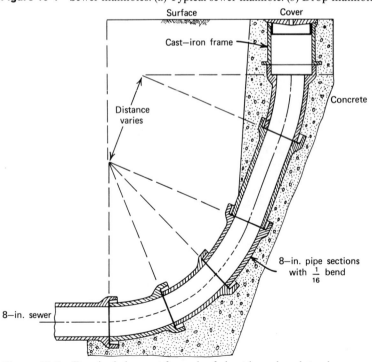

Figure 10-4   Sewer manholes. (*a*) Typical sewer manhole. (*b*) Drop manhole.

Figure 10-5   Terminal cleanout for ends of short branch or lateral sewers.

Wastewater flow is conveyed through the manhole in a smooth U-shaped channel formed in the concrete base. Where more than one sewer enters a manhole, the flowing-through channels should be curved to merge the flow streams. If a sewer changes direction in a manhole without change of size, a drop of 0.05 to 0.10 ft (1.5 to 3.0 cm) is provided in the manhole channel to account for head loss. When a smaller sewer joins a larger one, the bottom of the larger pipe should be lowered sufficiently to maintain uniform flow transition. An approximate method for achieving this is to place the 0.8 depth point of both sewers at the same elevation; an alternate technique is to have the pipe crowns at the same height.

A drop manhole is used when it is necessary to lower the elevation of a sewer in a manhole more than 24 in. (Figure 10-4*b*). This construction is needed to protect a man entering the structure, and to eliminate nuisance created by solids splashed onto the walls. The tee pipe section extending past the drop into the manhole allows access to the sewer line for cleaning tools.

Manholes should be placed at all changes in sewer grade, pipe size, or alignment; at all intersections; at the end of each line; and at distances not greater than 400 ft (120 m) for sewers 15 in. or less, and 500 ft (150 m) for sewers 18 to 30 in. In some municipalities, a spacing of 300 ft (90 m) is considered the maximum for smaller sized sewers. Distances of greater than 500 ft may be used where the pipe is large enough to permit walking in the sewer. A terminal cleanout sometimes substitutes for a manhole at the end of a short lateral sewer. The cleanout structure shown in Figure 10-5 is an upturned pipe coming to the surface of the ground for flushing or inserting cleaning tools. This type of unit should not be installed at the end of a lateral greater than 150 ft from a manhole.

Service Connections

House sewers are laid on a straight line and grade using 4-in. or 6-in. (100- or 150-mm) pipe. The preferred minimum slope is 2 percent, or $\frac{1}{4}$ in./ft, although slopes as shallow as $\frac{1}{8}$ in./ft are occasionally used. In some housing developments, the setback of dwellings from the street dictates the slope of the connection. Pipe trenches for laying the service line should be straight and should be ex-cavated to the required slope. Where the soil is suitable for pipe support, the natural floor of the trench can be shaped to support the barrel of the pipe. Crushed stone or coarse sand may be used for bedding when fill is required to provide uniform support. High-quality pipe, watertight joints, and good workmanship are required to minimize infiltration and root penetration.

The service connection to a sanitary sewer is made through a tee branch turned upward 45° or more from the horizontal, as shown in Figure 10-6, so that backflooding does not occur when the collecting sewer is flowing full. For a deep sewer, the tee connection and riser pipe are often vertical and may be encased in concrete to prevent damage during backfilling. Although a section of pipe with a tee branch can be installed during sewer construction, the location of future service connections is difficult to predict. Therefore, the tee connection is commonly made with a sewer pipe saddle installed after the sewer is in use. When the hole is excavated to uncover the buried sewer, the soil adjacent to the pipe is removed using a hand shovel to prevent damaging the sewer with the power excavator. Before workers enter the excavation, the soil walls must be braced for safety.

The first step in installing a pipe saddle is to cut a circular hole through the existing sewer pipe. The coring machine can be powered by water pressure, compressed air, or a gasoline engine. The rotary bit is cooled and lubricated by water as the hole is cut. Two different kinds of saddles are manufactured; both have connecting ends contoured for proper fit against the curved exterior of the sewer pipe. The clay-pipe saddle with a flange (shown in Figure 10-7*a*) is sealed into the hole using an epoxy glue. This tee, manufactured in 4-, 6-, and 8-in. (100-, 150-, and 200-mm) diameters, has a bell end for inserting the service pipe. The clamped cast-iron saddle in Figure 10-7*b* uses a rubber gasket for a watertight seal and a stainless steel strap to hold the saddle in position. Clamped saddles with 4- to 8-in. tee diameters are available with contoured ends to fit against sewer pipes 6 to 60 in. in diameter.

Building vents attached to sewer drains provide a connection between the air in sewer pipes and the atmosphere. Plumbing traps on toilet and sink drains prevent backup of sewer gases into the building interior, while the fall of wastewater down the stack draws air into

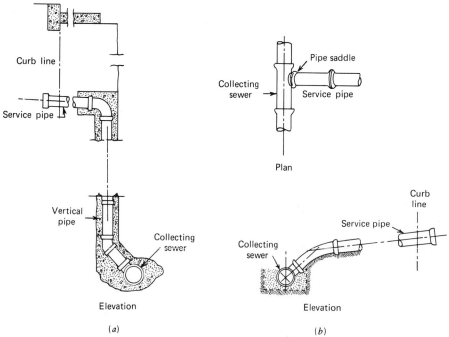

Figure 10-6 Typical sanitary sewer service connections. (*a*) Connection to a deep sewer. (*b*) Connection to a shallow sewer.

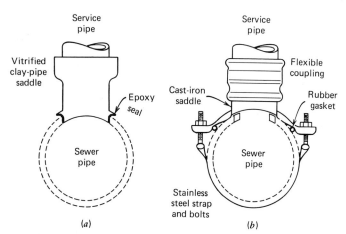

Figure 10-7 Pipe saddles for service connections to a sanitary sewer pipe. (*a*) Clay pipe saddle with an epoxy seal. (*b*) Clamped pipe saddle with a rubber gasket seal.

the pipe. Oxygen in the sewer atmosphere reduces production of hydrogen sulfide, and ventilation carries off volatile gases that may originate from illegal disposal of flammable liquids. In some instances where natural aeration is not sufficient because of long sewer lines with few service connections, forced ventilation is installed to draw air out of the sewer, exhausting it to a high stack or some deodorizing process.

10-3 MEASURING AND SAMPLING OF FLOW IN SEWERS

Monitoring sewer use requires measuring of flow and composite sampling at critical points in the collection system. Evaluation of peak flow periods defines the unused capacity of pipes and indicates infiltration-inflow problems. Sampling stations on industrial waste discharges are necessary to regulate sewer use and to determine flow and strength data to establish user fees.

The estimated quantity of flow in a sewer can be

computed from the measured depth of flow using the Manning formula if the slope of the pipe is known (Section 4-8). Another approximate method is to multiply the measured wetted cross-sectional area of flow times the velocity obtained by a current meter or the use of dyes or floats.

Weirs are employed where accurate flow measurement is desired. For installation in a manhole, a watertight bulkhead, containing the triangular or rectangular weir

opening on the top, is constructed at right angles to the wastewater flow. Water elevation in the stilling pool behind the bulkhead relates to the depth of flow over the weir for calculating discharge. Periodic observations may be performed with a vertical rule or hook gauge. However, for flow variation and total quantity of discharge, a continuous recording device is convenient and more accurate. The unit shown in Figure 10-8 has a float linked mechanically to a clock-driven, drum-type,

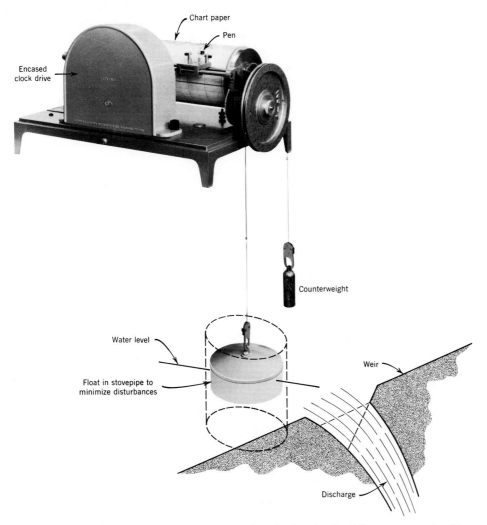

Figure 10-8 Continuous water-level recorder for plotting depth of flow over a weir with respect to time. (Courtesy of Leupold & Stevens, Inc.)

head-recording device. Rise and fall of the float with changing water levels turns the recording drum proportionally as the clock-controlled pen moves across the chart at constant speed. The resulting graph traced on removable chart paper is a record of water level with respect to time (Figure 10-11). With the proper weir forula, this head-time plot can be converted into a flow diagram (Figure 10-12) as demonstrated in Example 10-2.

Sampling stations for industrial wastewaters vary from a manhole to a separate building that is large enough to house flow recording and automatic sampling equipment. A manhole on the service sewer to a commercial business is generally adequate, provided that neither flow measurements nor composite sampling are necessary. Frequently for small enterprises, the quantity of wastewater flow can be determined from water consumption, and grab sampling is adequate for monitoring wastewater quality.

For moderate-sized industries and commercial businesses, a sampling station as shown in Figure 10-9 is appropriate. The sewer passes through a reinforced concrete chamber of sufficient size to conveniently set up sampling equipment. The flow measuring device is a flume inserted in the sewer line; these devices are available commercially for various sized pipes. A slightly elevated bench and stilling well are furnished to accommodate a water-level recorder of the type shown in Figure 10-8. A portable sampler is used to collect wastewater samples automatically at set time intervals. The intake hose is placed in the flow downstream from the flume, and the sampler is set on the bench above. Composite samples can then be integrated from the individual portions based on the recorded flow pattern.

The portable wastewater sampler illustrated in Figure 10-10 can be powered by either a battery or line current to

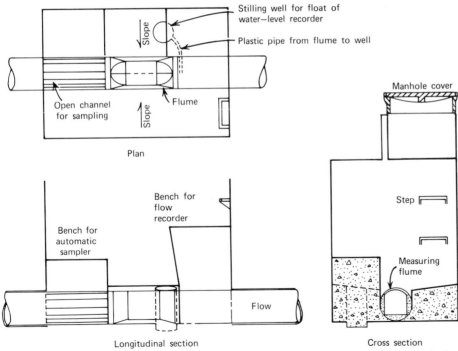

Figure 10-9 Sampling chamber to monitor industrial wastewater discharge to a municipal sewer where only intermittent analyses are required. Measuring flume is permanently installed but portable water-level recorder and automatic sampler are needed to perform composite sampling.

Cover

Sampling
pump

Rotating
funnel

Distributor
plate

Base with
individual
bottles

(b)

(c)

Suction tubing
with strainer
attached to end

(a)

Figure 10-10 Automatic wastewater sampler that collects sample portions at regular intervals of time. (*a*) Expanded sampler showing individual components. (*b*) The peristaltic pump draws wastewater through a flexible tube by squeezing action of moving rollers. Pump motor, gears, and electronic components are in a watertight case. (*c*) Assembled sampler. (Courtesy of ISCO.)

collect 28 samples at preset intervals of time. A peristaltic pump draws a volume of up to 500 ml through a flexible suction tube with a strainer on the end. Both before and after sampling, the pump reverses to air purge the suction line to minimize cross-contamination between samples. After each withdrawal, a rotating funnel is repositioned over the next sample bottle under the distributor plate. The base has an ice compartment in the center to provide cold storage. For a composite sample, the base with individual bottles can be replaced by a base with one large bottle. Sample collection in proportion to flow is controlled by an electronic attachment that receives signals from a flow meter. For large industries, completely automated systems are essential for monitoring wastewater discharges economically. The flow measuring system consists of a Parshall flume with flow recorder and totalizer, while the sampler performs continuous, automatic compositing of wastewater samples. This may be done by connecting a positive displacement sampling pump electrically to the flow recorder such that the amount pumped to a refrigerated collection bottle is proportional to the flow. Although costly, this type of station eliminates the time-consuming task of installing and removing portable equipment, and of compositing samples manually.

EXAMPLE 10-2

Wastewater flow entering a municipal sewer from an industry was monitored by using a continuous flow recorder installed in a stilling chamber behind a 90° V-notch weir (Figure 10-8). The recorded water-level varia-

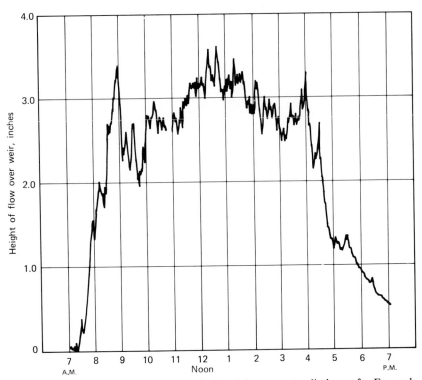

Figure 10-11 Water-level chart of an industrial wastewater discharge for Example 10-2. Flow recorder and 90° V-notch weir were arranged as illustrated in Figure 10-8.

tion with time is given in Figure 10-11; flow was negligible prior to 7 A.M. Plot the flow pattern in gallons per minute versus time, and compute the total discharge between 7 A.M. and 7 P.M.

Solution

Height at zero flow was located by marking the pen position on the chart paper when the wastewater level in the stilling chamber reached the weir crest. After the paper was removed from the recorder drum at 7 P.M., the horizontal time scale and vertical discharge-head scale were written on the chart.

To develop a flow pattern from the water-level graph, the heights of flow at various breaks in the curve in Figure 10-11 were converted to flow using the Eq. 4-29. For example, at 9:00 A.M. the height of flow over the weir

equaled 3.4 in., which is a discharge of 48 gpm.

Figure 10-12 is a plot of the wastewater flow diagram from the water-level chart in Figure 10-11. The total quantity of flow from 7 A.M. to 7 P.M. was 15,000 gal and the average flow rate was 20.8 gpm. Total discharge is determined graphically by counting the square spaces under the curve, or by measuring the area using a planimeter. Average flow equals total discharge divided by time.

EXAMPLE 10-3

A survey of the sewer line shown in Figure 10-2 was conducted to determine the approximate unused capacity. The maximum depth of flow in the pipe entering

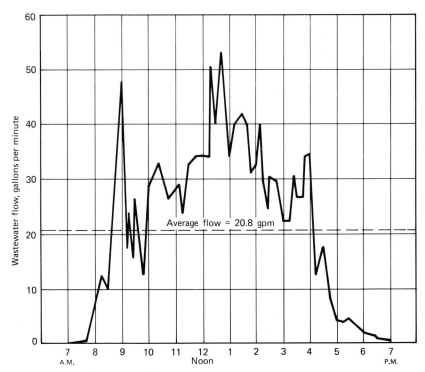

Figure 10-12   Wastewater flow pattern calculated from weir overflow recording in Figure 10-11 as the solution to Example 10-2.

manhole DM3-1 was 12 in. Calculate the quantity of flow at this depth, and the unused capacity of the sewer based on the Manning formula (Eq. 4-25) using $n = 0.013$.

Solution

From Figure 4-24, for an 18-in. diameter and slope of 0.00412 ft/ft, flowing full $Q = 6.7$ cu ft/sec.
Entering Figure 4-26 with the depth to diameter ratio of $12/18 = 0.67$, the partial flow to full flow is $q/Q = 0.80$. Therefore,

Quantity of flow at 12 in. $= 0.80 \times 6.7 = 5.3$ cu ft/sec

Unused capacity $= 6.7 - 5.3 = 1.4$ cu ft/sec

10-4   SEWER PIPES AND JOINTING

Circular sewer pipe is manufactured with inside diameters from 4 to 144 in., at intervals of 2 in. from 4 to 12 in., increases of 3 in. between 12 and 36, and 6-in. diameter intervals from 36 to 144. Maximum sizes vary with materials, for example, the maximum clay pipe is 42-in. diameter. The physical characteristics essential for sewer pipe are durability for long life; abrasion-resistant interior to withstand scouring action of wastewater carrying gritty materials; impervious walls to prevent leakage of water; and adequate strength to resist failure or de-

formation under backfill and traffic loads. Joints should be durable, easy to install, and watertight to prevent leakage or entrance of roots.

The most important chemical characteristics of a pipe material are resistance to dissolution in water and corrosion. Pipe surfaces must be able to withstand both electrochemical and chemical reactions from the surrounding soil and wastewater conveyed in the pipe. Figure 10-13 illustrates the process of crown corrosion in sanitary sewers. Bacterial activity in anaerobic wastewater produces hydrogen sulfide gas, particularly in warm climates when sewers are laid on flat grades. Hydrogen sulfide absorbed in the water condensed on the crown is converted to sulfuric acid by aerobic bacterial action. If the pipe is not chemically resistant, the acid deteriorates it and eventually results in collapse of the crown. The most effective preventative measure is to select a pipe material resistant to corrosion, such as vitrified clay or plastic. When reinforced concrete pipe is needed for larger sizes, an interior protective coating of coal tar, vinyl, or epoxy should be considered. Generation of hydrogen sulfide in a sewer can be reduced by placing the pipe on as steep a gradient as possible, and by ventilating if necessary. Corrosion of the pipe bottom is caused by disposal of acidic industrial wastewaters. This problem is best solved by limiting the discharge of acid wastes to the municipal

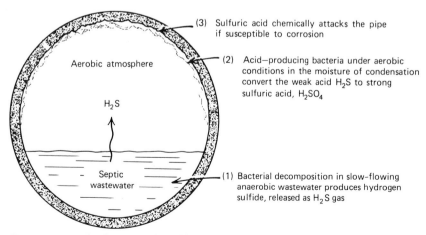

Figure 10-13   Environmental conditions leading to crown corrosion occur in sanitary sewers as a result of flat grades and warm wastewaters.

sewer system. In concrete pipe, corrosion-resistant liners, such as vitrified clay plates, may be placed in the invert of the sewer for protection.

Vitrified clay is the most common material used in sanitary sewer pipe. It is manufactured of clay, shale, or combinations of these materials that are pulverized and mixed with a small amount of water. The moistened clay is extruded through dies under high pressure to form the barrel and socket, dried, and then fired in a kiln for vitrification. Vitrified clay pipe (VCP) is manufactured in standard strength to 36-in. (900-mm) diameter and in extra strength to 42-in. (1050-mm) diameter. Laying lengths range from 2 to 7 ft depending on diameter. Curves, elbows, and branching fittings, including single and double tee and wye shapes, are available in most pipe sizes.

Bell-and-spigot vitrified clay pipes are connected by compression joints with plastic or rubber sealing ele-ments. The joints allow limited deflection without leaking; for example, pipes with diameters up to 12 in. (300 mm) are leak tested at a deflection of 0.5 in. per linear ft of length (42 mm/linear m). One kind of joint is a plastic liner bonded in the bell of the pipe. When they are joined, the ring and liner lock together with the ring compressed for a watertight seal. A second kind consists of bonded plastic liners on both the bell and spigot ends and a separate compression ring. The adjoining ends are designed to compress the O-ring in the annu-lar space between the liners to form a seal as detailed in Figure 10-14b.

Asbestos cement pipe is manufactured for service connections, gravity sewers, and pressure pipe for force mains. The pipe material is composed of cement, silica, and asbestos fibers. It is formed under high pressure and is heat treated to develop strength. Asbestos cement is

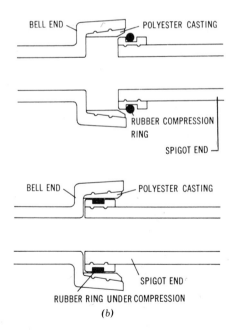

Figure 10-14 Picture of vitrified clay sewer pipe and details of an O-ring compression joint. (a) Vitrified clay sewer pipes packaged for delivery to job site. (b) Details of an O-ring compression joint. Both bell and spigot ends have bonded plastic liners with an annular space for sealing with a rubber ring. (Courtesy of the Logan Clay Products Co.)

durable and has a smooth surface without lining, but epoxy-lined pipe is available for special applications. Joints are made with an outside coupling and flexible rubber rings (see Figure 6-15). Nonpressure sewer pipe is available in the size range from 4- to 42-in. diameter in 13-ft lengths; shorter lengths are available in smaller sizes, and $6\frac{1}{2}$-ft sections are available in 10- to 21-in. sizes. Pipe is available in class designations based on supporting strength.

Plastic pipe used most frequently in sewer systems is of PVC (polyvinyl chloride) and PE (polyethylene). PVC pipe is produced in two strength classifications in sizes from 4 to 12 in. A few manufacturers make pipe up to 30-in. diameter. The standard length is 20 ft with others available. PVC pipe sections have a deep-socket, bell end to accommodate a chemical weld joint. Solvent is spread both on the inside of the bell and on the plain end before they are pushed together. PE pipe is joined by softening the aligned faces of the ends in a suitable apparatus and pressing them together under controlled pressure. PVC pipe is used for building connections and branch sewers, while the popular application of PE pipe has been for long pipelines, often laid under adverse conditions, for example, in swamp or underwater crossings.

Precast concrete sewer pipe may be obtained in many sizes with several kinds of joints. Choice of a particular type of pipe and joint depends on application, location, and conditions for installation. Nonreinforced concrete pipe is available in 4- to 24-in. diameters in 3- or 4-ft lengths. It is manufactured in standard strength and extra strength grades. The bell and spigot ends are normally joined by using a rubber ring gasket. Circular reinforced concrete pipe, produced in a size range from 12 to 108 in., is available in five classes based on strength. Precast conduits are also manufactured in elliptical and arch-type shapes. Tongue and groove joints, which are common in reinforced pipe, use a mastic compound or rubber gasket to form a watertight seal.

Concrete pipe is used extensively in storm sewer systems for its abrasion resistance, availability in large sizes, high crushing strength, and, generally, lower cost relative to pipes of other materials. Since concrete is an alkaline material subject to attack by acids, it should not be used for small-sized sanitary sewers where hydrogen sulfide production can cause internal corrosion. However, reinforced concrete pipe is installed for sanitary trunk sewers where the diameter exceeds the available sizes of vitrified clay pipe. Here the installation must be protected by control of acidic, high-temperature, and high-sulfate wastewaters by adding chemicals to control biological growth, maintaining flushing velocities and adequate ventilation, and installation of pipe lining if necessary. Epoxy or plastic lining may be cast into the concrete pipe during manufacture; or bitumastic or coal-tar epoxy may be painted on the concrete pipe surfaces after installation.

Ductile-iron pipe is used for force mains, inverted siphons, inside pumping stations and treatment plants, when proper separation from water mains cannot be maintained, and where poor sewer foundation conditions exist. Other pipe materials for special applications in wastewater collection systems are smooth wall and corrugated steel pipe, bituminized fiber, and reinforced resin pipe.

10-5 LOADS ON BURIED PIPES

Sewer design requires prior knowledge of soil and site conditions to determine overburden loads that will be placed on the buried pipe. Crushing strength of the pipe and type of bedding define the load that can be safely supported by a sewer. The paragraphs that follow are an overview of considerations in sewer design.

Backfill load on a pipe depends on trench width, depth of burial, unit weight of the fill material, and frictional characteristics of the backfill. These parameters are formulated in Eq. 10-1 and are defined in Figure 10-15. Soil exploration data, size of pipe, depth of burial, and trench excavation technique by the contractor are needed to compute backfill load. Unit weight of the fill and trench width are applied directly in Eq. 10-1. Coefficient C_d depends on the ratio of trench depth to width and the backfill material as graphed in Figure 10-16.

$$W_d = C_d w B_d^2 \qquad (10\text{-}1)$$

where W_d = load on buried pipe as a result of backfill, pounds per linear foot

C_d = coefficient based on the backfill and ratio of trench depth to width, Figure 10-16

w = unit weight of backfill, pounds per cubic foot

B_d = width of trench at top of pipe, feet

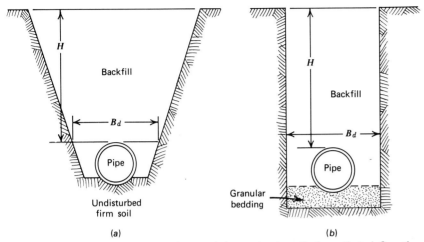

Figure 10-15 Common trench cuts used for sewer installations that define the dimensions B_d and H as applied in Eq. 10-1 and Figure 10-16. The minimum trench width for working space to install pipe is commonly taken as $\frac{4}{3}$ the pipe diameter plus 8 in. (a) Trench with sloping sides in cohesionless soil. (b) Vertical trench walls in stiff clayey soil.

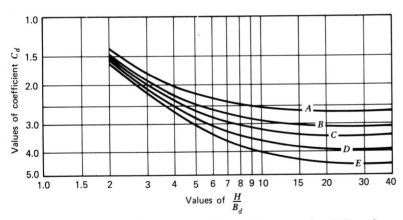

Figure 10-16 Diagram for the coefficient C_d in Eq. 10-1 based on H/B_d and type of trench backfill: A, granular material without cohesion; B, sand and gravel; C, maximum for saturated top soil; D, ordinary clay; E, maximum for saturated clay.

Design manuals published by the American Concrete Pipe Association, National Clay Pipe Institute, and others provide tables listing backfill loads for various conditions based on Eq. 10-1. An example is Table 10-2 for 8-in. circular pipe and ordinary clay fill with a unit weight of 100 lb/cu ft. Tables such as this have been developed for other pipe diameters and backfill materials. From the proper table, height of fill above the pipe and trench width are used to select backfill load. This number is then adjusted for the proper unit weight; for backfill

Table 10-2.   Backfill Loads in Pounds per Linear Foot on 8-in. Circular Pipe in a Trench Installation Based on 100 lb/cu ft Ordinary Clay Fill

Height of Backfill H Above Top of Pipe, Feet	Trench Width at Top of Pipe										Transition Width[a]
	1′-6″	1′-9″	2′-0″	2′-3″	2′-6″	2′-9″	3′-0″	3′-3″	3′-6″	4′-0″	
5	501	603[b]									1′- 9″
6	559	694	**724**								1′-10″
7	608	761	**847**								1′-11″
8	649	819	**969**								2′- 0″
9	683	868	**1088**								2′- 0″
10	712	911	1119	**1213**							2′- 1″
11	736	948	1170	**1332**							2′- 2″
12	757	979	1215	**1458**							2′- 3″
13	774	1007	1254	1513	**1575**						2′- 4″
14	788	1030	1289	1560	**1698**						2′- 4″
15	801	1051	1319	1603	**1818**						2′- 5″
16	811	1068	1346	1640	**1942**						2′- 6″
17	819	1083	1369	1674	1993	**2065**					2′- 7″
18	827	1096	1390	1703	2034	**2182**					2′- 7″
19	833	1107	1408	1730	2070	**2308**					2′- 8″
20	838	1117	1424	1754	2103	**2429**					2′- 9″
21	842	1125	1438	1775	2133	**2553**					2′- 9″
22	846	1133	1450	1793	2159	2545	**2673**				2′-10″
23	849	1139	1461	1810	2184	2578	**2788**				2′-11″
24	851	1144	1470	1825	2205	2607	**2910**				2′-11″
25	854	1149	1478	1838	2225	2635	**3041**				3′- 0″
26	855	1153	1486	1850	2242	2658	**3154**				3′- 0″
27	857	1156	1492	1861	2258	2682	3128	**3278**			3′- 1″
28	858	1159	1498	1870	2273	2702	3155	**3398**			3′- 2″
29	859	1162	1502	1878	2286	2721	3181	**3514**			3′- 2″
30	860	1164	1507	1886	2297	2738	3204	**3646**			3′- 3″
31	861	1166	1511	1892	2308	2753	3225	**3776**			3′- 3″
32	862	1167	1514	1898	2317	2767	3245	3748	**3880**		3′- 4″
33	862	1169	1517	1904	2326	2780	3263	3772	**4004**		3′- 4″
34	862	1170	1519	1908	2333	2791	3279	3794	**4124**		3′- 5″
35	863	1171	1522	1913	2340	2802	3294	3815	**4241**		3′- 5″
36	863	1172	1524	1916	2346	2811	3308	3834	**4384**		3′- 6″
37	863	1173	1525	1920	2352	2820	3321	3851	**4495**		3′- 6″
38	864	1173	1527	1922	2357	2828	3333	3868	4431	**4603**	3′- 7″
39	864	1174	1528	1925	2362	2835	3343	3883	4451	**4740**	3′- 7″
40	864	1174	1529	1927	2366	2842	3353	3896	4470	**4877**	3′- 8″

Source: Concrete Pipe Design Manual, American Concrete Pipe Association.

[a] The trench width at which the backfill load on the pipe is a maximum and remains constant regardless of increase in trench width.

[b] The bold printed figures are the maximum load at the transition width for any given height of backfill.

weighing 110 lb/cu ft, increase loads 10 percent; for 120 lb/cu ft, increase 20 percent, and so forth.

Wheel loads from trucks and other vehicles transmit live loads to buried sewer lines. As illustrated in Figure 10-17, the vertical pressure produced in underlying soil dissipates laterally with depth, such that only a portion of the concentrated tire pressure is realized by the underground pipe. Live loads on the surface rarely influence design of sanitary sewers because of their great depth and small size. On the other hand, supporting strength of storm sewers placed at shallower depths must consider loads transmitted from heavy vehicles. The equations for computing live loads are complex, and therefore designers often use precalculated data that are organized

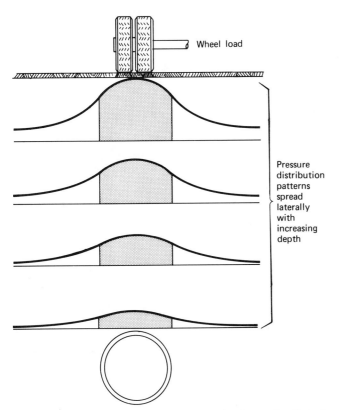

Figure 10-17 Wheel loads and other live loads are distributed with depth such that only a portion of the surface pressure reaches a buried pipe.

Pressure distribution patterns spread laterally with increasing depth

as illustrated in Table 10-3. If pipe diameter and height of fill are known, the load transmitted by a standard wheel load can be selected. If a pipe has less than 3-ft cover, the traffic load is multiplied by an impact factor as recommended in Table 10-4.

The three-edge bearing test is the standard for crushing strength of sewer pipe. The laboratory apparatus (see Figure 10-18) loads full-size pipe sections until failure occurs. Specifications published by the American Society for Testing and Materials list the minimum crushing strength requirements for various types and sizes of pipe. These values can then be reliably used in design when the pipe manufacturer subscribes to ASTM specifications. Crushing strength for clay pipe based on three-edge bearing are given in Table 10-5. Similar tables are available for asbestos cement and nonreinforced concrete pipe.

Strength requirements for reinforced concrete sewer pipe are given in Table 10-6. Since numerous pipe sizes are available, three-edge bearing strengths are classified by D loads to produce both a 0.01-in. crack and ultimate failure of the section. The D-load concept provides a strength classification independent of pipe diameter; consequently, strength is expressed as pounds per linear foot per foot of inside diameter. Three-edge bearing test load in pounds per linear foot for a specific size pipe is calculated by multiplying the D load for the particular class pipe times inside diameter in feet.

If sewers were simply laid by placing the pipe barrels on the flat trench bottom, as illustrated in Figure 10-19a, the pipe would not be able to support a load significantly greater than the three-edge bearing load. However, if bedding bears, at least, the lower quadrant of the barrel and backfill material is carefully placed and tamped around the sides of the pipe, the supporting strength is increased significantly. Load factor expresses this increase numerically as the ratio of the supporting strength in the trench, as determined by bedding conditions, to the three-edge bearing test strength. Supporting strength is 50 percent greater when a bedding of coarse sand or crushed stone carries the lower quadrant of the pipe (Figure 10-19b). Granular bedding carefully placed in the trench bottom and tamped under the pipe haunches up to the center line raises the load factor to 1.9. Concrete

Table 10-3. Highway Truck Loads Transmitted to Buried Circular Pipe in Pounds per Linear Foot[a]

Pipe Diameter, inches	0.5	1.0	1.5	2.0	2.5	3.0	3.5	4.0	5.0	6.0	
12	4849	2730	1542	954	574	361	316	237	168	46	12
15	**5915**	3322	1872	1163	700	441	358	289	205	56	15
18		3926	2196	1369	825	520	454	341	242	66	18
21		**4498**	2525	1580	950	599	522	392	279	77	21
24			2856	1788	1077	679	592	444	316	87	24
27			**3096**	1998	1201	757	660	496	353	97	27
30				2204	1328	837	731	548	390	107	30
33				**2370**	1452	915	798	599	426	117	33
36					1580	996	869	652	463	127	36
42					**1679**	1154	1007	755	537	147	42
48						**1221**	1146	859	611	168	48
54							**1185**	962	684	188	54
60								**998**	758	208	60
66									832	228	66
72									**857**	249	72
78										269	78
84										**275**	84

Source: Concrete Pipe Design Manual, American Concrete Pipe Association.

[a] Values based on a 16,000-lb dual-tire wheel load (32,000-lb axle load) and 80-psi tire pressure.

Table 10-4. Recommended Impact Factors for Calculating Loads on Pipe with Less Than 3-ft Cover Subjected to Highway Truck Loads

Height of Cover H	Impact Factor
0 to 1 ft-0 in.	1.3
1 ft-1 in. to 2 ft-0 in.	1.2
2 ft-1 in. to 2 ft-11 in.	1.1
3 ft-0 in. and greater	1.0

Source: AASHO Standard Specifications for Highway Bridges.

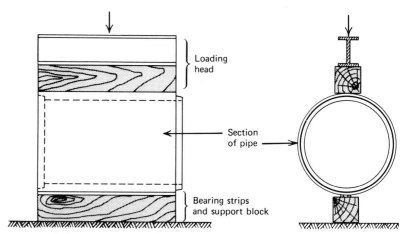

Figure 10-18 The three-edge bearing test is the standard laboratory procedure for determining crushing strength of sewer pipe.

Table 10-5. Crushing Strength Requirements for Vitrified Clay Sewer Pipe Based on the Three-Edge Bearing Test, Pounds per Linear Foot

Nominal Size (in.)	Standard Strength	Extra Strength
4	1200	2000
6	1200	2000
8	1400	2200
10	1600	2400
12	1800	2600
15	2000	2900
18	2200	3300
21	2400	3850
24	2600	4400
27	2800	4700
30	3300	5000
33	3600	5500
36	4000	6000

Source: ASTM Specification C700 for Standard and Extra Strength Clay Pipe.

Table 10-6. Strength Requirements for Reinforced Concrete Sewer Pipe Based on the Three-Edge Bearing Test, Pounds per Linear Foot per Foot of Inside Pipe Diameter

Classification	D Load to Produce 0.01-in. Crack	D Load at Failure	Pipe Size (diameter in in.)
Class I	800	1200	
Concrete strength 4000 psi			60 to 96
Concrete strength 5000 psi			102 to 108
Class II	1000	1500	
Concrete strength 4000 psi			12 to 96
Concrete strength 5000 psi			102 to 108
Class III	1350	2000	
Concrete strength 4000 psi			12 to 72
Concrete strength 5000 psi			78 to 108
Class IV	2000	3000	
Concrete strength 4000 psi			12 to 66
Concrete strength 5000 psi			60 to 84
Class V	3000	3750	
Concrete strength 6000 psi			12 to 72

Source: ASTM Specification C76-66T.

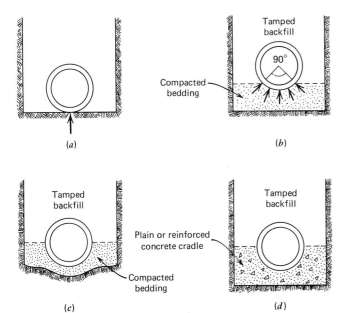

Figure 10-19 Common types of bedding for sewer pipe installations and corresponding load factors. (*a*) Class D bedding, improper support (not to be used). (*b*) Class C bedding, support of lower pipe quadrant (load factor = 1.5). (*c*) Class B bedding, compacted backfill (load factor = 1.9). (*d*) Class A concrete cradle bedding (load factor = 2.3 to 3.2).

bedding or encasement provides even higher load factors (Figure 10-19*d*).

A factor of safety is applied to decrease the calculated field supporting strength to the safe supporting strength as expressed in Eq. 10-2. In effect, this gives a margin of safety against variations in soil conditions, quality of construction, and other deviations not anticipated during design. The factor of safety normally used for clay or plain concrete pipe is 1.2 to 1.5, while for reinforced concrete pipe a factor of 1.0 based on 0.01-in. crack load is generally considered sufficient.

Safe supporting strength, lb/lin ft

$$= \frac{\text{field supporting strength, lb/lin ft}}{\text{factor of safety}} \quad (10\text{-}2)$$

where safe supporting strength must be greater than, or equal to, backfill load (W_d) plus live load (W_l)
For clay or plain concrete pipe,

$$\begin{array}{c}\text{Field supporting}\\\text{strength}\end{array} = \text{crushing strength} \times \text{load factor}$$

$$\text{Factor of safety} = 1.2 \text{ to } 1.5$$

For reinforced concrete pipe,

$$\begin{array}{c}\text{Field supporting}\\\text{strength}\end{array} = \text{D load} \times \text{diameter} \times \text{load factor}$$

$$\begin{array}{c}\text{Factor of}\\\text{safety}\end{array} = \begin{array}{c}1.0 \text{ based on the } 0.01\text{-in.}\\\text{crack strength}\end{array}$$

EXAMPLE 10-4

An 8-in.-diameter, vitrified clay sewer is being placed in a 2.0-ft-wide trench at a depth of 19.2 ft to the pipe invert. The backfill is ordinary clay with a unit weight of 120 lb/cu ft. Determine the strength of pipe and class of bedding that should be used for a factor of safety of 1.5.

Solution

$$H = \text{depth to bottom} - \text{pipe diameter}$$
$$H = 19.2 - 0.7 = 18.5 \text{ ft}$$
$$\frac{H}{B_d} = \frac{18.5}{2.0} = 9.2$$

From Figure 10-16, for $H/B_d = 9.2$ and curve *D* for

ordinary clay,

$$C_d = 3.5$$

By Eq. 10-1,

$$W_d = 3.5 \times 120 \times 2.0 \times 2.0 = 1680 \text{ lb/lin ft}$$

Alternate solution is from Table 10-2, under the 2'-0" column interpolating at $H = 18.5$, load = 1400 lb/lin ft for 100 lb/cu ft backfill.
Therefore, $W_d = 1400 \times 1.2 = 1680$ lb/lin ft for 120 lb/cu ft.
Consider standard strength pipe with a crushing strength of 1400 lb/lin ft (Table 10-5) and Class C bedding (Figure 10-19*b*); substituting into Eq. 10-2,

Safe supporting strength

$$= \frac{1400 \times 1.5}{1.5} = 1400 \text{ lb/lin ft} < 1680 \text{ (No Good)}$$

Try Class B bedding (Figure 10-19*c*).

Safe supporting strength

$$= \frac{1400 \times 1.9}{1.5} = 1770 \text{ lb/lin ft} > 1680 \text{ (Adequate)}$$

Consider extra strength pipe and Class C bedding.

Safe supporting strength

$$= \frac{2200 \times 1.5}{1.5} = 2200 \text{ lb/lin ft} > 1680 \text{ (Adequate)}$$

Either standard strength VCP with Class B bedding, or extra strength with Class C bedding, is satisfactory.

EXAMPLE 10-5

What class of reinforced concrete pipe should be placed for a storm sewer based on the following: pipe diameter = 33 in., $H = 6.0$ ft, $B_d = 4.5$ ft, sand and gravel backfill with a unit weight of 110 lb/cu ft, Class C bedding, highway truck loads, and a factor of safety = 1.0 based on the 0.01-in. cracking load?

Solution

From Figure 10-16, for sand and gravel backfill,

$$\frac{H}{B_d} = \frac{6.0}{4.5} = 1.3, \; C_d = 1.1$$

Using Eq. 10-1,

$$W_d = 1.1 \times 110 \times 4.5 \times 4.5 = 2450 \text{ lb/lin ft}$$

From Table 10-3 for $H = 6.0$ ft and diameter = 33 in., $W_1 = 117$ lb/lin ft.

Therefore, total load = 2450 + 117 = 2570 lb/lin ft.

Substituting into Eq. 10-2 to find required D load,

$$2570 = \frac{\text{D load} \times 33 \times 1.5}{12 \times 1.0}$$

$$\text{D load} = 620 \text{ lb/lin ft/ft of diameter}$$

Based on D loads in Table 10-6, Class I at 800 lb/lin ft/ft of diameter would be satisfactory, but 33-in. pipe is not precast in this classification. Therefore, use a Class II pipe and 4000-psi concrete with a specified D load of 1000 lb/lin ft/ft of diameter.

10-6   SEWER INSTALLATION

A backhoe is used commonly for trench excavation since it can serve also for backfilling and lifting pipe sections. Trenches can be excavated with vertical walls in stiff clayey soils, but sloping sides are often necessary in less cohesive soils. Most excavations require bracing to protect workers during pipe placement. The most common shores are special steel planks positioned vertically against the trench walls and held apart by hydraulic jacks. If the soil walls are self-supporting, this shoring is adequate without horizontal members. In the case of loose soils, however, sheeting is placed behind the vertical braces to form either a criss-cross pattern or solid wall to prevent chunks of soil from falling into the excavation. Digging below the water table in permeable soils requires a well-point system for dewatering to keep the water level below the trench bottom. A trench is overexcavated and backfilled with a layer of granular bedding so that the thickness of bedding under the pipe is at least one quarter of the pipe diameter, but never less than 4 in.

Construction methods for a deep sanitary sewer in unstable soils are shown in Figure 10-20. The 54-in.-diameter pipe sections are reinforced concrete precast in 12-ft lengths. On the right side of the top picture is a pipeline connected to dewatering wells spaced alongside the trench to lower the water table below the bottom of the excavation. A backhoe is being used to excavate the trench ahead of a steel trench box in which the pipe

Figure 10-20   Construction of a deep sanitary sewer. (a) Backhoe excavating the trench ahead of a safety shield (trench box) in which pipe sections are placed. (b) Workers in the trench box leveling bedding on the bottom of the excavation. (c) Lowering a pipe section into the trench box.

sections are placed. This bottomless steel safety shield consists of side walls connected together by heavy braces so that the distance between the walls is the trench width. The purpose of the box is to provide a safe working space for laying the pipe without having to set shoring. In the top picture, a section of pipe has already been placed in the space protected by the safety shield, and as the trench excavation is extended, the backhoe is used to pull the box forward. With the trench box in the forward position, granular bedding is poured into the trench using a hopper box while the workers climb into the completed pipeline for protection. Figure 10-20*b* shows the workers leveling the bedding on the trench bottom with hand shovels. In

the bottom picture, a pipe section is being lowered into the box by a crane. The workers in the trench place the rubber gasket on the spigot of the pipe and guide the section into position to join the buried pipe. The hopper box for pouring bedding into the trench is standing on the ground surface in the upper center of the picture. The front-end loader on the right fills the hopper from a stockpile of bedding gravel. The front-end loader is used also to backfill the trench as the safety shield is pulled forward. In this method of construction, the trench is open only where a pipe section is being placed. Each pipe joint is tested for tightness immediately after placement with a special circular expander that presses against the

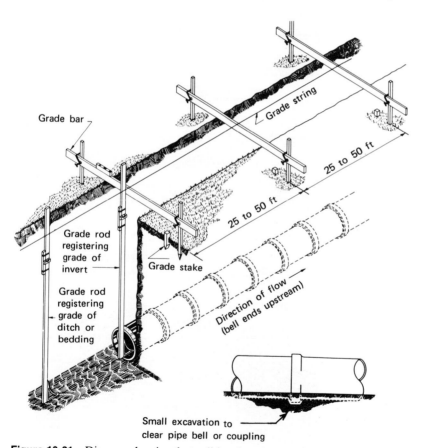

Figure 10-21   Diagram showing the traditional method of laying sewer pipe to line and grade. The grade bars carry both vertical and horizontal levels. (*Clay Pipe Engineering Manual*, National Clay Pipe Institute.)

walls of the pipe on both sides of the joint, forming an enclosure that is filled with compressed air to test for leakage. Alignment of pipe sections is done using a laser beam.

In the traditional method of pipe placement, an offset line paralleling the proposed sewer is located with transit and tape. The trench location is then staked out on the ground surface by measuring from the offset line. After excavation at the proper width, grade bars (batter boards) are placed across the trench at 25- or 50-ft intervals and are supported by stakes as illustrated in Figure 10-21. The center line of the sewer is located on the grade bars by a nail or edge of an upright cleat. Elevations are then run and the grade bars are set, or a mark is placed on each cleat, at some even foot distance above the invert of the sewer. A cord is stretched from grade bar to grade bar along the center line. Sewer alignment is transferred from this stringline to the trench bottom by means of a plumb bob. Sewer slope is transferred by using a wooden rod, marked in even foot marks, that has a short piece fastened at right angles to the lower end. After bedding and joining each pipe section, grade is checked by placing the foot of the grade rod on the sewer pipe invert and noting whether the proper foot mark touches the cord (Figure 10-21).

Special laser systems are used commonly for aligning sewers. Such a unit eliminates the stringline and the need to transfer line and grade into the trench every time a pipe section is laid. It also serves as a guide in the excavation of the trench to proper slope. The laser shown in Figure 10-22a projects a pencil-thin ray of red light through the pipe onto a target for position reference. After the instrument is leveled, desired grade is set by a dial.

Figure 10-22 A laser beam for laying sewer pipe eliminates the need for grade bars and a stringline. (*a*) Laser with a dial for setting the desired sewer grade. (*b*) Laser installed in a manhole. (*c*) A transit is used to set the laser beam on line. (Courtesy of Spectra-Physics.)

A laser installed in a manhole is illustrated in Figure 10-22b. The base plate is set over the starting point by using a center hole for location, and a vertical pole, with a bullseye level on top, is placed in position on the base and held plumb with extension braces. The laser is then attached to the vertical pole at the proper height to put the beam over the pipe center line. The beam is aligned in the proper horizontal direction by using an above-ground tilting telescope assembly, or transit, to transfer a distant reference point into the trench just ahead of the laser. The procedure is to swing the transit so that the vertical hairline centers on the next manhole, and then to dip it to focus on the trench bottom in front of the laser. Next, the laser beam is swung into view and is set so that it coincides with the vertical hairline on the transit. The laser instrument is then leveled, and the proper grade is set on the dial. Each pipe section is placed with a removable target set in the bell so that the pipe may be shifted into position by bringing the laser spot on target center.

10-7   SEWER TESTING

New sewers are tested for watertightness by monitoring infiltration, measuring exfiltration with the pipeline full of water, and air testing. Infiltration testing consists simply of measuring, through use of a weir, the amount of flow in a sewer before any building connections are made. This technique, of course, is not applicable if the sewer line is above the water table. Preferably, the observed groundwater level should be at least 4 ft above the top of the pipe. Even when sewers are laid below the groundwater level, interpretation of results are questionable, since the head of water above the pipe greatly influences the quantity entering through pipe cracks and joint defects. Another problem with the infiltration procedure is that long sections of sewers must be tested to get measurable flow rates; for example, infiltration at a rate of 500 gal/in. of sewer diameter per mile per day (46 l/d·km·mm) would yield a flow rate of only 0.2 gpm for a 400-ft manhole-to-manhole section of 8-in. pipe (0.8 l/min for 120 m of 200-mm pipe). Testing of long sewer lines, while having the advantage of higher flow yields, may not identify the precise location of broken pipes and poor joints.

Exfiltration testing, the reverse of measuring actual infiltration, is used mainly in dry areas where the groundwater table is below the pipe crown. Filling the pipe with water and observing the loss during a specified time period is a positive method, since it subjects the sewer and manholes to a known water pressure. Excessive pressures can produce destructive results in lower sections of a sewer; however, testing of sections between manholes has few hazards. The maximum hydrostatic head applied is usually 10 ft, and the water is allowed to stand for at least 4 hr before measuring exfiltration. This allows the pipe and joint material to become saturated with water and permits the removal of entrapped air. The maximum allowable exfiltration specified varies from 100 to 500 gal/in. of diameter/mile/day. For example, a typical allowable limit is 290 gal/in. of diameter/mile/day (27 l/d·km·mm) under a 10-ft head of water, while another specifies a maximum rate of 200 gal/in. of diameter/mile/day plus 10 percent increase for each 2 ft of head over the initial 2 ft. Table 10-7 gives infiltration and exfiltration rates for various sized sewer pipes.

Table 10-7. Conversion of Infiltration or Exfiltration Rates from Gallons per Inch of Diameter per Mile per Day to Gallons per Hour per 100 ft for Various Sized Sewer Pipes

Diameter of Sewer, (in.)	Filtration Rates in gal/hr/100 ft for the Following Rates in gal/in. diameter/mile/day				
	100	200	300	400	500
8	0.63	1.3	1.9	2.5	3.2
10	0.79	1.6	2.4	3.2	4.0
12	0.95	1.9	2.8	3.8	4.7
15	1.2	2.4	3.5	4.7	5.9
18	1.4	2.8	4.3	5.7	7.1
21	1.7	3.3	5.0	6.6	8.3
24	1.9	3.8	5.7	7.6	9.5
27	2.1	4.3	6.4	8.5	10.7
30	2.4	4.7	7.1	9.5	11.8
36	2.8	5.7	8.5	11.4	14.2
42	3.3	6.6	10.0	13.3	16.6
48	3.8	7.6	11.4	15.2	19.0

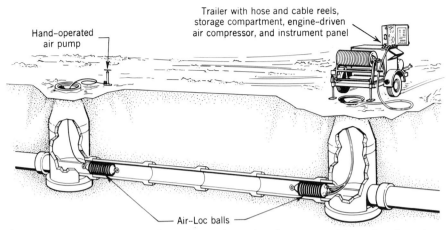

Trailer with hose and cable reels, storage compartment, engine–driven air compressor, and instrument panel

Hand–operated air pump

Air–Loc balls

Figure 10-23 Low-pressure air testing to determine the structural integrity of newly constructed sewer lines. (Courtesy of Cherne Industrial, Inc.)

Low-pressure air testing is applicable to sewer lines constructed of vitrified clay pipe or combinations of clay and other pipe materials. Although air leakage cannot be correlated to water leakage, the air test does demonstrate the structural integrity of a sewer line. A well-constructed sewer will exhibit some loss of air pressure even if watertight; however, significant air loss is caused by damaged pipe or defective joints. To reduce the passage of air through the walls, the interior of the sewer line is wetted before conducting the air test.

For the air test, a section of sewer between manholes is isolated as shown in Figure 10-23 by plugging both ends and tightly capping any service connections. Air pressure is applied, raised to 4.0 psi, and held for 2 to 5 min while all plugs are checked for leakage and the temperature of the air stabilizes. The air supply is then disconnected and the time required for the pressure to drop from 3.5 psi to 2.5 psi is measured with a stopwatch. If the 1.0 psi drop does not occur within the calculated test time, the sewer line has passed the test. If the pressure drops more than 1.0 psi during the test time, the sewer has failed the test. The test times required per 100 ft of sewer for pipes of various diameters are given in Table 10-8. If the section of sewer line tested includes more than one size pipe, the test time is calculated for each diameter and the total of these times is the test time for the section.

Table 10-8. Minimum Test Times per 100 Feet of Sewer Line for Low-Pressure Air Testing

Pipe Diameter (in.)	Test Time (min)	Pipe Diameter (in.)	Test Time (min)
3	0.2	21	3.0
4	0.3	24	3.6
6	0.7	27	4.2
8	1.2	30	4.8
10	1.5	33	5.4
12	1.8	36	6.0
15	2.1	39	6.6
18	2.4	42	7.3

Source: ASTM C828-80 Low-Pressure Air Test of Vitrified Clay Pipe Lines.

EXAMPLE 10-6

A 400-ft length of 8-in. vitrified clay sewer was air tested for structural integrity. The time measured for the air pressure to drop from 3.5 to 2.5 psi was 6.5 min. Did the sewer line pass this low-pressure air test?

Solution

From Table 10-8, the minimum test time per 100 ft of 8-in. pipe is 1.2 min. Therefore,

$$\text{Test time} = \frac{400 \text{ ft} \times 1.2 \text{ min}}{100 \text{ ft}} = 4.8 \text{ min}$$

The measured time of 6.5 min is greater than this, so the sewer line passes the test.

10-8 LIFT STATIONS IN WASTEWATER COLLECTION

Lift stations are required to elevate and transport wastewater in collection systems when continuation of gravity flow is no longer feasible. In flat terrain, sewers en route to a treatment plant may increase in depth to the point where it is impractical to continue gravity flow. Here, a pumping station can be installed to lift the wastewater to an intercepting sewer at a higher level. Or the pumps can discharge to a force main conveying wastewater under pressure to the treatment plant. New subdivisions or industries locating on the edge of an existing sewer system may find that the pipe is too shallow to receive sewer connections from basement level. In this case, a small lift station can pump wastewaters from the area to the main sewer.

Pumping stations are expensive to install, require periodic inspection and maintenance, and strain public relations if they malfunction and allow backup of domestic wastewaters into residences or businesses. For these reasons, they are avoided wherever possible by recognizing sewer planning and treatment plant location in municipal land use zoning and comprehensive planning.

Pneumatic ejectors are commercially available for pumping flows not exceeding about 100 gpm, for example, for several homes or a commercial building. They have the important advantage of being able to handle unscreened domestic wastewater without clogging. (Although low capacity pumps can be employed, the smallest centrifugal pump designed to pass 3-in. solids and operate with reasonable freedom from clogging has a rated capacity greater than 100 gpm.) The essential components of a pneumatic ejector are an air compressor,

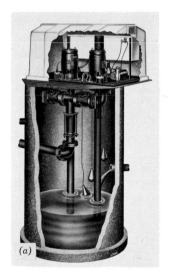
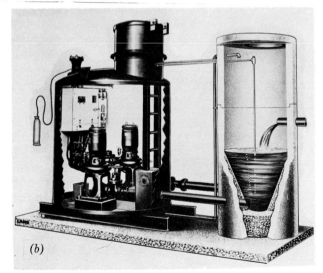

Figure 10-24   Typical collection system pumping stations for lifting wastewater from a deep sewer to an intercepting sewer or for conveying to a treatment plant by a force main. (*a*) Wet well–mounted lift station for moderate wastewater flows. (*b*) Standard pump station with separate wet- and dry-well structures. (Courtesy of Smith & Loveless, Inc.)

wastewater receiving chamber, piping with appropriate valves, and automatic system for operation. Wastewater flows by gravity into an enclosed receiving chamber displacing air through a vent. When the receiver fills, an electrical control system shuts off the vent connection and supplies compressed air to the chamber. The pressure closes the inlet check valve and forces the wastewater through a discharge check valve in the outlet pipe. After it is emptied, the chamber is again vented, allowing wastewater to flow in while the outlet check valve shuts, preventing the backflow of ejected wastewater. Except in the smallest installations, it is recommended that at least two ejector units be provided, each having sufficient capacity to handle the expected maximum flow.

Wet-well pumping stations are often installed for peak flows between 100 and 600 gpm. They can be mounted directly over an enlarged manhole providing a compact low-cost station. The pumps may be submersible pumps or turbine pumps with the motors mounted above. The pump control devices hanging in the wet well shown in Figure 10-24a are mercury switches enclosed in teardrop-shaped plastic casings. These units tip when floated and activate the enclosed switch.

Pumping stations with separate wet and dry wells are often preferred to wet-well–mounted lift stations. The unit illustrated in Figure 10-24b is a factory-built pumping station that is available for maximum flows in the range of 100 to 1500 gpm. The pumps are automatically controlled using an air-bubbler pipe to sense the depth of wastewater in the wet well. Back pressure on the air supply to the bubbler tube actuates a mercury pressure switch that controls the pump starter. Wastewater flows from the wet well through a suction pipe, gate valve, pump housing, check valve, and discharge gate valve to the force main. The dry well is ventilated and dehumidified to protect the pump control equipment and to make the chamber safe for entry by workers.

FURTHER READINGS

1. *Clay Pipe Engineering Manual*, National Clay Pipe Institute, Washington, D.C., 1982.
2. *Concrete Pipe Design Manual*, American Concrete Pipe Association, Arlington, Va., 1970.
3. *Gravity Sanitary Sewer Design and Construction*, Water Pollution Control Federation, Manual FD-5, Washington, D.C., 1982.

PROBLEMS

10-1 If a storm sewer surcharges during a heavy rainstorm, was it improperly sized? Explain.

10-2 How do combined sewers differ from storm sewers? Why are separate sewer collection systems preferred in current design?

10-3 Why is the hydraulic design for a sanitary sewer serving a residential area based on a flow of 400 gpcd when the average wastewater contribution per person is commonly about 100 gal/day?

10-4 What is the maximum population that can be served by a 12-in. submain laid on minimum slope? (*Answer* 2500 persons)

10-5 The trunk sewer illustrated in Figure 10-2 collects wastewater from a large residential area. What is the maximum design population that can be served by this main sewer?

10-6 For a peak flow of 1000 gpm, what size sewer is needed for construction at minimum slope? At a slope of 0.40 ft/100 ft? (*Answers* 15 in., 12 in.)

10-7 A submain sewer with a 450-mm diameter is constructed on the minimum allowable slope for a sanitary sewer. What is the flowing full quantity of flow? At what depths of flow in the pipe are the velocities of flow equal to 0.60 m/s? At what slope must the pipe be placed for a velocity of 4.5 m/s when flowing full?

10-8 For a peak flow of 500 gpm, what size sewer is needed for construction at minimum slope? At a slope with an average flowing full velocity of 4.0 ft/sec?

10-9 Explain the hydraulic operation of the inverted siphon illustrated in Figure 10-3. How is an inverted siphon designed to prevent plugging by deposition of solids in the depressed pipes?

10-10 What is the primary reason for installing manholes spaced along a sewer line? What factors dictate their placement? Where are drop manholes used and why?

10-11 How are lateral sewers ventilated?

10-12 The depth of flow in a 24-in.-diameter sewer placed on a 0.0030 ft/ft grade is 16 in. What is the quantity of flow? (*Answer* 9.6 cu ft/sec)

10-13 A 90° V-notch weir was installed in a manhole to measure flow in a sewer. If the maximum height of flow over the weir crest was 6.5 in., what was the maximum flow? (*Answer* 0.35 mgd)

10-14 Additional housing to accommodate 2000 persons is proposed for a residential area served by an 18-in.-diameter vitrified clay sanitary sewer constructed on a slope of 0.00412 ft/ft. To determine the feasibility of this extension, a 90° V-notch weir was placed in a sewer manhole to determine the existing peak flow. The maximum height of flow over the weir was measured at 16.5 in. Estimate the additional residential population that can be served by the main. Is the sewer capacity adequate for the proposed housing development?

10-15 If hydrogen sulfide is too weak an acid to cause corrosion of concrete, why does the release of hydrogen sulfide from septic wastewater result in crown corrosion in concrete sanitary sewers?

10-16 Why are compression joints the most popular method of joining sewer pipes?

10-17 A vitrified clay sewer with a diameter of 18 in. is to be placed in a trench excavated 3.0-ft wide with vertical walls. Depth to the crown of the pipe is 24 ft, and the backfill is saturated clay with a unit weight of 125 lb/cu ft. What type of bedding and strength of pipe do you recommend?

10-18 Standard strength 12-in.-diameter VCP is placed as shown in Figure 10-15a with $H = 5.0$ ft, $B_d = 2.0$ ft, Class C bedding, and 110 lb/cu ft sand and gravel backfill; it is also subjected to highway truckloads. What is the calculated factor of safety for the pipe installation? (*Answer* 2.9)

10-19 Vitrified clay pipe with an 8-in. diameter is being placed as illustrated in Figure 10-15b with Class B bedding, 2-ft 3-in. trench width, and ordinary clay backfill

with a unit weight of 120 lb/cu ft. What is the maximum allowable depth of burial for standard strength pipe maintaining a factor of safety of 1.5? For extra strength pipe? (*Answers* $H = 12$ ft, $H > 40$ ft)

10-20 Vitrified clay pipe with a 12-in. diameter is being placed in an excavation that has a width of 2 ft 6 in. at the top of the pipe. The trench bottom provides Class B bedding, and the backfill is ordinary clay with a unit weight of 120 lb/cu ft. What is the maximum allowable depth of burial to the top of the sewer for standard strength pipe maintaining a factor of safety of 1.5?

10-21 The soil profile at the site of a proposed sewer is a 6.5-ft layer of silty sand underlain by stiff silty clay that extends to below the depth of the proposed sewer invert 13.0 ft below the ground surface. The anticipated trench excavation is 3.0 ft wide with vertical walls in the silty clay and sloping walls on a 1 to 1 slope in the overlying silty sand. The pipe is 12-in.-diameter vitrified clay, and the trench is to be backfilled with the excavated silty sand at a density of 100 lb/cu ft. What type of bedding and strength of pipe are recommended?

10-22 The cross section of a proposed sewer construction is diagramed in Figure 10-15a. The soil profile is medium-dense sand that can be trenched with side slopes of 45° by lowering the water table during construction using well points. The 15-in.-diameter, standard strength VCP is to be placed with the invert at a depth of 10.0 ft below the original ground surface. The pipe is to be bedded with the lower quadrant supported by compacted sand on a trench bottom measuring 24 in. in width. Excavated sand is to be compacted on both sides of the installed pipe and used for trench backfill at a density of 130 lb/cu ft. Review this proposed design.

10-23 A proposed sewer is placed in a trench with a bottom width of 4.0 ft, depth of 16.0 ft, and 45° sloping side walls. The compacted granular bedding is placed under the pipe and along the sides to mid-height of the pipe. The backfill is sand with a density of 110 lb/cu ft. The buried pipe is subject to highway truckloads. The pipe is 36-in. diameter, Class II, reinforced concrete. Calculate the factor of safety based on a supporting strength to produce a 0.01-in. crack.

10-24 What is the field supporting strength of a 33-in.-diameter, Class II, reinforced concrete pipe based on the 0.01-in. crack strength and a load factor of 1.5? (*Answer* 4100 lb/lin ft)

10-25 Explain how the beam of a laser is aligned in the proper direction and set on the desired slope after the unit is placed in a manhole as shown in Figure 10-22.

10-26 What is the maximum allowable infiltration flow for a 400-ft-long, 10-in.-diameter sewer if the specification permits 400 gal/in. diameter/mile/day? (*Answer* 12.8 gal/hr)

10-27 An exfiltration test is to be specified for a new sewer with a length of 350 ft. The pipe is vitrified clay with a diameter of 15 in. What maximum allowable water loss do you recommend for this pipe section? State the test criteria and express the water loss in gallons per day.

10-28 During exfiltration testing of 400 ft of 18-in. sewer, the water loss was 8.0 gal/hr. Does this exceed 200 gal/in. diameter/mile/day?

10-29 A 300-ft length of 10-in. vitrified clay sewer was tested for structural integrity by low-pressure air testing. If the pressure drop of 1.0 psi occurred in 290 sec, was the test satisfactory?

chapter 11
Wastewater Processing

Conventional wastewater treatment is a combination of physical and biological processes designed to remove organic matter from solution. The earliest method was plain sedimentation in septic tanks. Imhoff tanks used by municipalities were two-story septic tanks that separated the upper sedimentation zone from the lower sludge digestion chamber by a sloping bottom with a slot opening. Solids settling in the upper portion of the tank passed through the slot into the bottom compartment from which the digested sludge was periodically withdrawn for disposal. The final step in development of primary treatment was complete separation of sedimentation and sludge processing units. Currently, raw sludge is handled independently by mechanical dewatering processes or biological digestion.

Primary sedimentation of municipal wastewater has limited effectiveness, since less than one half of the waste organics are settleable. The initial attempt at secondary treatment involved chemical coagulation to improve settleability of the wastes. Although this provided considerable improvement, the heavy chemical dosages resulted in high cost and dissolved organics were still not removed. The first major breakthrough in secondary treatment occurred when it was observed that the slow movement of wastewater through a gravel bed resulted in rapid reduction of organic matter. This process, referred to as trickling filtration, was developed for municipal installations starting in about 1910. A more accurate term for a trickling filter is biological bed, since the process is microbial oxidation of organic matter by slimes attached to the stone, rather than a straining action.

A second major advancement in biological treatment took place when it was observed that biological solids, developed in polluted water, flocculated organic colloids. These masses of microorganisms, referred to as activated sludge, rapidly metabolized pollutants from solution and could be subsequently removed by gravity settling. In the 1920s, the first continuous-flow treatment plants were constructed using activated sludge to remove BOD from wastewaters. (Biological systems are also discussed in Section 3-12.)

The pictorial diagram in Figure 11-1 summarizes processes applied in conventional municipal wastewater treatment. Preliminary steps include screening to remove large solids, grit removal to protect mechanical equipment against abrasive wear, flow measuring, and pumping to lift the wastewater above ground. Primary treatment is to remove settleable organic matter, amounting to 30 to 50 percent of the suspended solids, and scum that floats to the surface. Secondary treatment is by aeration in open basins with return biological solids, or fixed-media (trickling) filters, followed by final settling. Excess microbial growth settled out in the final clarifier is wasted while the clear supernatant may be disinfected with chlorine prior to discharge to a receiving watercourse. Waste sludges from primary settling and secondary biological flocculation are thickened and dewatered in preparation for disposal. Anaerobic bacterial digestion may be used to

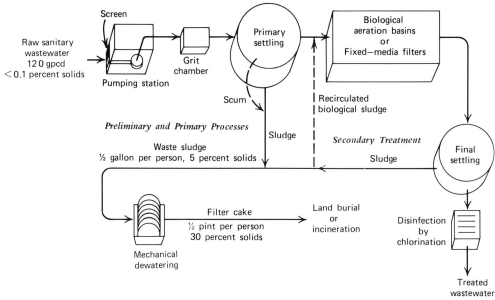

Raw sanitary
wastewater
120 gpcd
<0.1 percent solids

Screen

Pumping station

Grit
chamber

Primary
settling

Biological
aeration basins
or
Fixed—media filters

Scum

Recirculated
biological sludge

Preliminary and Primary Processes

Sludge

Secondary Treatment

Final
settling

Waste sludge
½ gallon per person, 5 percent solids

Sludge

Mechanical
dewatering

Filter cake
½ pint per person
30 percent solids

Land burial
or
incineration

Disinfection
by
chlorination

Treated
wastewater

Figure 11-1 Schematic of a conventional, municipal, wastewater treatment plant. Floating, settleable, and biologically flocculated waste solids are removed from the water and thickened for ease of disposal.

stabilize the sludge prior to dewatering, or a mechanical process may be used to extract water directly from raw sludge after chemical conditioning. Ultimate disposal of dewatered solids may be by landfill, incineration, or land application if biologically stabilized.

The overall process of conventional treatment can be viewed as thickening; pollutants removed from solution are concentrated in a small volume convenient for ultimate disposal. The contribution of raw sanitary wastewater is about 120 gal/person (400 l) with a total solids content of less than 0.1 percent, 240 mg/l suspended solids, and 200 mg/l BOD. Liquid waste sludge withdrawn from primary and secondary processing amounts to approximately 0.5 gal/person (2 l) with a solids content of 5 percent by weight (Figure 11-1). This is further concentrated to a handleable material by mechanical dewatering; the extracted water is returned for reprocessing. Cake from dewatering amounts to about 0.5 pt/person (0.25 l) with a 30 percent solids concentration. This type of physical-biological scheme is effective in reducing the organic content of wastewater, and accomplishes the major objective of BOD and sus-

pended solids removal. Dissolved salts and other refractory pollutants are removed to a lesser extent. Referring to Table 9-2, 50 percent of total volatile solids, 60 percent of total nitrogen, and 70 percent of total phosphorus remain in the effluent after secondary biological treatment. Advanced wastewater treatment processes are needed to remove these refractory contaminants. Orthophosphate can be precipitated by chemical coagulation, nitrogen may be reduced by biological nitrification and denitrification, and activated carbon takes out refractory soluble organics.

Completely mixed aeration without primary sedimentation (Figure 11-2a) is popular for treatment of small wastewater flows, for example, from subdivisions, villages, and towns. The size of aeration basins ranges from factory-built metal tanks with diffused aeration to accommodate the waste flow from a few hundred persons to mechanically aerated concrete-lined basins, such as an oxidation ditch, to serve a town with a population of several thousand. Elimination of primary settling dramatically affects the character of waste sludge. Instead of a septic sludge with relatively high solids content, the

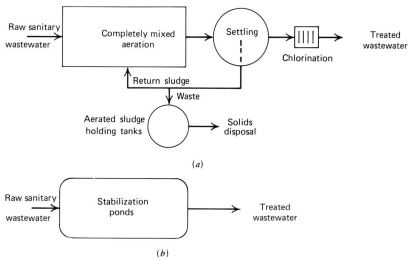

Figure 11-2 Processing diagrams of systems for treatment of small wastewater flows. (*a*) Biological processing without primary sedimentation. (*b*) Natural biological stabilization in lagoons.

waste is aerobic and much more voluminous with solids in the range of 0.5 to 2 percent. Therefore, the solids handling system for the flow scheme in Figure 11-2*a* frequently consists of aerobic digesters (aerated sludge holding tanks) and hauling of the stabilized liquid sludge by tank wagon to land burial or surface spreading on farmland. Larger plants may reduce the volume of sludge by gravity thickening in hopper-bottomed holding tanks, or mechanical flotation thickening, to consolidate the waste solids.

Hundreds of villages and commercial establishments in rural areas use stabilization ponds for wastewater treatment (Figure 11-2*b*). The organic loading applied to lagoons is very low and the liquid retention time is long, normally more than 90 days. The natural biological processes that stabilize the waste are sketched in Figure 3-24. In dry climates the evaporation rate may equal or exceed the liquid loading so that the ponds provide complete retention of the wastewater. In other cases, stabilized wastewater is used for irrigation. Where lagoons overflow to streams, discharge of the treated wastewater may be limited to only warm seasons of the year when the natural purification in the lagoons is most efficient.

11-1 TREATMENT PLANT DESIGN STANDARDS

Municipal facilities are conventionally designed on the basis of quantity of flow and organic content of the raw wastewater. Required degree of treatment is designated by effluent standards including BOD, suspended solids, fecal coliforms, pH, phosphorus, ammonia nitrogen, heavy metals, and toxic organic compounds.

Design Loading

The wastewater quantity for sizing basins is the average weekday flow during that season of the year when discharge is greatest. For example, in warm climates summer flows are often 20 to 30 percent greater than the annual daily average. Saturdays and Sundays generally have lower flows, since industries are not in operation. If sewer pipes in the collection system are in poor condition, the selected design flow includes normal infiltration that occurs during the wet season. Storm runoff collected in combined sewers draining to the plant must also be considered. Where few data are available, the engineer may increase the quantity of flow from areas with older sewers by about 25 percent and refers to this value as the wet-weather design flow.

Maximum and minimum hourly rates, in addition to the average daily, are applicable for sizing certain units, for example, lift pumps. Normally, low flows range from 20 to 50 percent of the average daily, while maximum rates are 200 to 250 percent. Sometimes these daily extremes are confusing, since they may be expressed in units of million gallons per day while referring to flows that occur only over a 1-hr period. For example, a specification may state that an aeration system is planned for an average daily flow of 5 mgd and a maximum (or peak) flow of 10 mgd. This means that the unit has a capability of treating 5 mil gal during one day with a flow variation such that the peak hourly rate does not exceed 10 mgd, and should not be interpreted that the system is able to process 10 mil gal in a 24-hr time span.

Organic content of a municipal wastewater is defined by the concentrations of BOD and suspended solids. Design organic loadings on a treatment plant are expressed in terms of average pounds per weekday. Values given in the units of milligrams per liter can be converted to pounds per day, using the design flow.

Effluent Quality

Rules and regulations of the Environmental Protection Agency describe the minimum level of effluent quality that must be attained by secondary treatment under the National Pollutant Discharge Elimination System (NPDES). Acceptable secondary effluent is commonly defined in terms of BOD, suspended solids, fecal coliform bacteria, and pH. For BOD, the arithmetic mean of the values for effluent samples collected in a period of 30 consecutive days must not exceed 30 mg/l, and the arithmetic mean of values in any period of 7 consecutive days must not exceed 45 mg/l. Therefore, those days when the BOD exceeds 30 mg/l have to be compensated for by improved quality on other days during a 30-day period so that the average is not greater than 30 mg/l. Furthermore, the mean of the effluent values should not exceed 15 percent of the mean of the influent BOD concentrations, which means a minimum of 85 percent removal. The effluent standard for suspended solids is the same as the

BOD limitation—a maximum 30-day mean of 30 mg/l, a maximum 7-day mean of 45 mg/l, and a minimum removal of 85 percent. The quality standard for fecal coliform bacteria is not universally applied. Disinfection of a wastewater effluent is required only where necessary to protect public health. In this case, the geometric mean of the values for effluent samples collected in a period of 30 consecutive days must not exceed 200 per 100 ml, and the geometric mean in a period of 7 consecutive days must not exceed 400 per 100 ml. The effluent values for pH should remain within limits of 6.0 to 9.0.

Design Parameters

Standards for design of treatment units are expressed by a variety of terms. The following common definitions are inserted here as examples. Sedimentation basins are sized on overflow rate, detention time, and tank depth. Overflow rate is the average daily effluent divided by the surface area of the tank, expressed in units of million gallons per day per square foot. Detention time in hours is calculated by dividing the tank volume by influent flow. Tank depth is the water depth at the side of the tank measured from the bottom to top of the overflow weir.

Organic loadings on biological treatment units are stated in terms of pounds of applied five-day BOD. Loading on an aeration basin is commonly expressed as pounds BOD applied per 1000 cu ft of tank volume per day. A biological filter loading uses the same units except the volume refers to the quantity of media rather than liquid volume. Aeration period in an activated-sludge process is equal to volume of the aeration basin divided by flow of raw wastewater. Since biological filters do not contain a liquid volume, hydraulic loading is presented as the amount of wastewater applied per unit of surface area, for example, million gallons per acre per day (mgad).

The specific numerical values of design parameters used by an engineer in sizing treatment units are frequently recommended in published standards. For example, the *Recommended Standards for Sewage Works, Great Lakes Upper Mississippi River Board of State Sanitary Engineers* (*GLUMRB*) are used by regulatory agencies in the 10 states surrounding the Great Lakes in reviewing plans for sanitary facilities. However, many design specifications are not listed and must be selected by the engineer based on experience, or on recommendations of equipment manufacturers. These may not be readily available in published literature, and for a particular treatment plant the criteria applied must be obtained from the designing engineer or equipment supplier.

Construction drawings show plant layout, flow schemes, and tank dimensions, but do not list design loadings or show proprietary equipment. Contract specifications describe quality of workmanship, materials, and equipment to be installed and include performance requirements for mechanical units, such as pumps and aerators. The preliminary study report, written prior to preparation of plans and specifications, does discuss conditions relative to plant design and includes loadings, flows, and alternate treatment processes. Despite this, values in this document or amendments to it must be confirmed by comparing the proposed scheme of the system selected against the as-built drawings, since designs may be modified between the time of preliminary report writing and plant construction. The designing engineer is required to write operational manuals that should include design loadings and effluent quality requirements. The latter should be confirmed by referring to the most recent NPDES permit issued by the state regulatory agency.

11-2 PRELIMINARY TREATMENT

Screening, pumping, flow measuring, and grit removal are normally the first steps in processing a municipal wastewater. Flotation is sometimes included to handle greasy industrial wastes, but these are usually more effectively handled by pretreatment at the industrial site. Chemical coagulation is sometimes incorporated to increase removal in primary settling. However, this is costly and generally is applied only when a treatment plant is overloaded. Chlorination of raw wastewater may be employed for odor control and to improve settling characteristics of the solids. Arrangement of preliminary units varies but the following general rules always apply. Screens protect pumps and prevent solids from fouling subsequent units; therefore, they are always placed first. A Parshall flume is located ahead of constant speed lift pumps because the turning on and off of the pumps produces a pulsing flow that cannot be graphed by a flow recorder. With variable-speed pumps, the flume may be placed on either side, since the pumping rate is paced to be identical to the influent flow. Grit removal reduces abrasive wear on mechanical equipment and prevents accumulation of sand in tanks and piping. Although ideally it should be taken out ahead of the lift pumps, grit chambers located above ground are far more economical and offset the cost of pump maintenance. Two possible

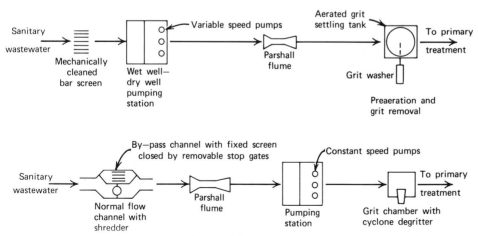

Figure 11-3 Typical arrangements of preliminary treatment units in municipal wastewater processing. The lower sequence is common to smaller plants.

arrangements for preliminary units are illustrated in Figure 11-3. The upper scheme is typical of a large plant while the lower sequence is common in small systems.

Screens and Shredders

Mechanically cleaned screens have clear bar openings of approximately 1 in. As illustrated in Figure 11-4, collected solids are removed from the bars by a traveling rake that lifts them to the top of the unit. Here the screenings may be fed to a shredder and returned to the wastewater flow, or they may be deposited in a portable receptacle for hauling to land burial. Inclined bar screens with front rakes are also manufactured. Mechanically cleaned

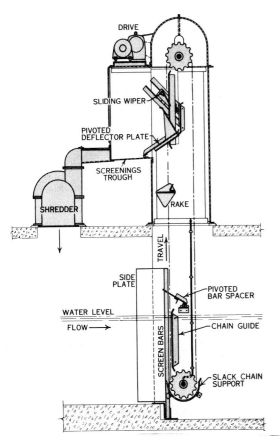

Figure 11-4 Wastewater screen with vertical bars cleaned by a traveling rake. (Courtesy of FMC Corp., Material Handling Systems Div.)

screens are preceded by hand-cleaned bar racks for protection in large combined sewer systems.

A shredder or comminutor cuts solids in the wastewater passing through the device to about 0.25 in. Such devices are installed directly in a flow channel and are provided with a by-pass so that the section containing the shredder can be isolated and drained for machine maintenance. A comminutor is not often used with a mechanically cleaned bar screen; nevertheless, if installed, it is placed at some point after the screen. In a small municipal plant, a shredder can be placed in the flow stream. For this type of arrangement (Figure 11-3), the by-pass channel contains a hand-cleaned bar screen for emergency use. The channels are equipped with stop gates so that wastewater can be directed through the fixed screen while maintenance is performed on the shredder.

Two common types of shredding devices are pictured in Figure 11-5. The comminutor consists of a vertical revolving slotted drum screen submerged in a flow channel such that the wastewater passes into the drum through slots and is discharged from the bottom of the unit. Solids too large to enter the slots are cut into pieces small enough to pass through while stringy material that enters partway is cut by a shear bar. The *Barminutor*, combination screening and comminuting machine, has a high-speed rotating cutter that travels up and down on the front of the bar screen. Accumulated solids are shredded and flushed through with the wastewater. Cutter action can be automatically initiated when solids have accumulated on the screen face and can be stopped when they are shredded and passed through.

Grit Chambers

Grit includes sand and other heavy particulate matter, such as seeds and coffee grounds, which settle from wastewater when the velocity of flow is reduced. If not removed in preliminary treatment, grit in primary settling tanks can cause abnormal abrasive wear on mechanical equipment and sludge pumps, can clog pipes by deposition, and can accumulate in sludge holding tanks and digesters. Grit chambers are designed to remove particles equivalent to a fine sand, defined as 0.2-mm-diameter particles with a specific gravity of 2.7, with a minimum of organic material included. A variety of systems are employed depending on the quantity of grit in the wastewater, the size of the treatment plant, and the expenditures allocated to installation and operation. Standard chambers include channel-shaped settling tanks, aerated units with hopper bottoms, and clarifiers with mechanical scraper arms. Separated grit may be further processed in a screw-type grit washer or cyclone separator.

The earliest grit chamber consisted of two or more long narrow channels constructed in parallel with space for grit storage. While one was in service, the idle channel could be cleaned by shoveling out the accumulated grit. A proportional weir placed at the discharge end controlled the flow such that the horizontal velocity was maintained at about 1.0 ft/sec independent of the quantity of flow. This allowed the grit to settle while providing a scouring velocity to flush through organic suspended solids. Currently, channel-type grit removal units are equipped with mechanical grit collectors. Buckets on a continuous chain scrape material from the bottom into a receptable, thus avoiding hand cleaning and the need for more than one channel.

The most popular type of grit chamber in smaller plants is a hopper-bottomed tank with the influent pipe entering on one side and an effluent weir on the opposite. The chamber is small with a detention time of approximately 1 min at peak hourly flow, and is often mixed by diffused aeration to keep the organics in suspension while grit settles out. Solids are removed from the hopper bottom by an air-lift pump, screw conveyer, bucket elevator, or gravity flow, if possible. Settled solids from this type of unit are sometimes relatively high in organic content, thus creating a nuisance. A cyclone separator or grit washer may be used to wash and dewater the grit slurry, returning the overflow to the raw wastewater. A common application for a cyclone separator is in processing grit slurry drawn from the hopper of a grit chamber. Feed entering tangentially creates a spiral flow, thereby throwing heavy solids toward the inner wall for discharge from the apex while the liquid vortex moves in the opposite direction for discharge.

Grit clarifiers, sometimes called detritus tanks, are generally square with influent and effluent weirs on opposite sides. A centrally driven scraper arm pushes the

(a)   (b)

Figure 11-5 Two common types of shredders. (*a*) The revolving slotted drum screen of the comminuter has cutting teeth that pass through a stationary comb mounted on the main casing. As wastewater passes through the drum screen, large solids are captured and shredded to pass through. (*b*) The *barminutor* has a high-speed rotating cutter that travels up and down the vertical bar screen shredding accumulated solids. (Courtesy of FMC Corp., Material Handling Systems Div.)

grit into a hopper on the bottom for removal by a mechanical conveyer or gravity flow. In most installations, the slurry withdrawn is washed and dewatered by a unit such as that shown in Figure 11-6. The slurry is diluted with clean wash water to flush out suspended matter. Grit is moved out of the wash water and up an inclined screw conveyer for drainage prior to discharge. The clarifier may be a shallow tank with a short detention time or a deeper aerated chamber to improve grit separation while freshening the raw wastewater.

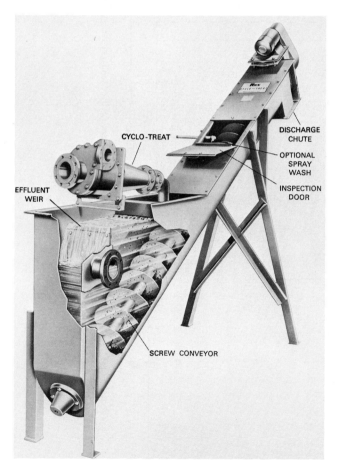

Figure 11-6 The slurry containing grit enters the washer in a liquid chamber above the screw conveyor. Heavy settled solids are moved up the inclined conveyor by the screw while a water spray washes the grit before discharge. (Courtesy of Envirex Inc., a Rexnord Company.)

Preaeration of wastewaters prior to primary settling is practiced to increase dissolved oxygen and scrub out entrained gases for the purpose of improving settleability. The detention period in a combined grit removal and preaeration basin is generally 15 to 20 min.

Flow Measuring

All treatment plants should have a means of monitoring wastewater flow. The best system is a Parshall flume that is equipped with an automatic flow recorder and totalizer (refer to Section 4-10). Advantages of a flume are low head loss and smooth hydraulic flow, preventing deposition of solids. Some existing treatment plants have venturi meters in the discharge piping from the wet well, but these are not recommended because of potential operation and maintenance problems.

Small wastewater plants without installed flow recording equipment usually have provisions for placing a weir in either the plant influent or effluent channel. Recording of flow then involves inserting the weir and installing a portable water-level recorder as is illustrated in Figure 10-8. Measuring flow in this fashion is time consuming and is not recommended for plants where regular wastewater compositing is an anticipated requirement. The measured flow for several study periods can be correlated with pump operating time. Then, the quantity of influent in the future can be estimated by reading the total hours of pump operation from timers attached to the electrical controls of the pumps.

11-3 PUMPING STATIONS

Pumping stations at large treatment plants have wet and dry chambers separated by a common wall as shown in Figure 11-7. Raw wastewater after screening flows into the hopper-bottomed wet well where the pipe intakes are positioned near the bottom of the chamber to prevent deposition of solids. Automatic controls governing pump operation maintain the water surface between preset levels. At the high mark all pumps are running, except standby units, while lowering to the minimum level shuts off the last pump before it can draw in air. In addition to the influent wastewater, drain and recirculation lines return in-plant wastewaters to the wet well. These include

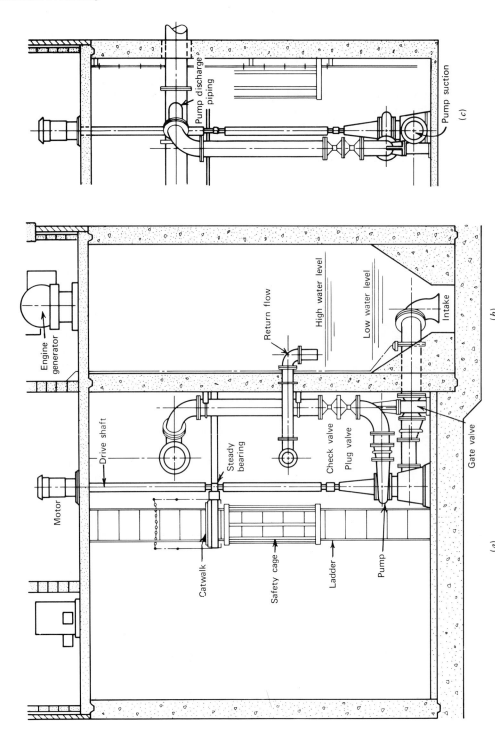

Figure 11-7 Cross section of a typical municipal wastewater pumping station. (*a*) Dry well. (*b*) Wet well. (*c*) Side view of end pump in dry well.

supernatant from digesters, filtrate from sludge dewatering, underflow from final clarifiers in filtration plants, and waste activated sludge in aeration systems.

Effective capacity of a wet well should provide a detention time of 10 to 20 min to permit reasonably long periods of pump operation without allowing septic conditions in the wastewater. A small wet well keeps the wastewater mixed and moving through at sufficient velocity to reduce the accumulation of settled and floating solids. Yet, if the detention time is too short, on-and-off cycling increases mechanical wear of the pumps and temperature of the drive motors. For best performance, with any combination of inflow and pumping, the cycle of operation for each pump is not less than 5 min and the maximum detention time of wastewater in the wet well is 30 min. To meet these conditions, the selection of individual pumps and operating water levels are coordinated in the design of a wet well.

Centrifugal wastewater pumps located in the dry well can be mounted with either horizontal or vertical drive shafts. Placing motors in the housing above the dry well has the advantages of accessibility for maintenance and protection in case the dry well is accidentally flooded. A separate sump pump is installed in a dry well but its capacity is limited to removal of leakage or drainage that enters the chamber. Pump housings are set below the low-water level in the wet well for automatic priming. Suitable valves are placed on the suction and discharge lines of each pump so that it can be removed for maintenance. A check valve is essential in the discharge line, between shutoff valve and pump, to prevent wastewater from flowing back through the pump when not in operation.

Independent mechanical ventilation is required for the dry and wet chambers, even if the latter are not covered. Switches for operation are marked and located conveniently, and if the ventilation system is intermittent rather than continuous, the electrical switches are interconnected with the pit lighting system. A safe means of access by either a stairway or caged ladder, with rest landings at least every 10 ft, is needed for dry wells and wet wells containing mechanical equipment. Alarm systems are installed to notify plant personnel in cases of power or pump failure. Provision of an emergency power supply is accomplished by using two incoming lines with automatic switching equipment to transfer the load from one utility system to a standby source in the event of failure of the former. The ideal situation is two independent public utilities; if they cannot be provided, standby engine-driven generator sets should be installed.

Centrifugal pumps for lifting raw wastewater are designed for ease of cleaning or repair, and are provided with open impellers to reduce the risk of clogging. Most horizontal pumps are made with a split casing so that one half can be removed for access to the interior. For removing obstructions, the casing may also be provided with a hand-sized cleanout. Freedom from clogging depends on type of impeller, shape of casing, and operating clearance between the two. Impellers are generally made with only two or three vanes, set between two parallel plates, so that any object entering the pump passes through. Pumps preceded by a bar screen with clear openings less than $2\frac{1}{2}$ in. should be capable of passing spheres of at least 3-in. diameter.

The number of pumps installed depends on the station capacity and range of flow. The total pumping capacity is equal to or greater than the maximum rate of inflow with at least one of the largest pumps out of service. To meet the variations of flow, the minimum number of pumps is three or more of the same or varying capacities. In small stations with a peak flow of less than 1.0 mgd, two units may be installed with each being able to meet the maximum inflow. The pumps may be either constant speed or variable speed. The characteristics of centrifugal pumps are discussed in Section 4-4, and system head curves are explained in Section 4-5.

Constant-Speed Pumps

A typical pumping station lifts wastewater from the wet well for discharge into an open tank, which is usually the grit chamber or preaeration basin. The pumping head is determined by the water level in the wet well and the friction head loss in the discharge piping. The hydraulic characteristics of a station represent its entire useful life; therefore, the system head-discharge curves are drawn for minimum and maximum pipe roughness at both high and low water levels in the wet well. Since the system curves include only the head losses in the piping from the common header to the discharge tank, the pump characteristic curves are modified to account for losses in the individual pump suction and discharge lines. These modified curves are then considered the pump head-discharge curves. The curves sketched in Figure 11-8 are for a station with three constant-speed pumps (two of equal size and one larger). The pump characteristic curves have been modified, and the system head-discharge curves bracket the minimum-to-maximum operating range.

Pumps should operate near their optimum efficiency along the average system head curve, which is the line traced midway between the lower and upper system curves established by water levels in the wet well. In Figure 11-8, two average system head curves could be traced, one between the two $C = 140$ lines applicable

when the discharge piping is new and one between the $C = 100$ lines for 20-year-old ductile-iron pipe. In practice, however, the selection of pumps is based on maximum and minimum operating conditions rather than the average; this ensures satisfactory performance over the entire operating range. When pumping at maximum capacity, the wet-well water level is at the maximum elevation and all three pumps are operating. The maximum resistance to flow is along the lower system head-discharge curve labeled $C = 100$. Therefore, the operating point is at the intersection of the combined pump curve for pumps 1, 2, and 3 and the $C = 100$ system curve. A horizontal line drawn through this intersection locates the operating points on the modified pump curves. Then, by projecting vertically to the pump characteristic curves, the pumping rates at maximum anticipated head are located

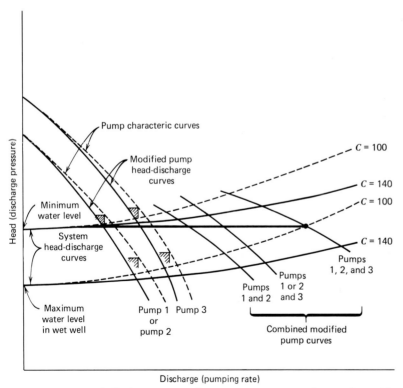

Figure 11-8 Head-discharge curves for a wastewater pumping station with three constant-speed pumps. (From *Design of Wastewater and Stormwater Pumping Stations*, Manual FD-4, Water Pollution Control Federation.)

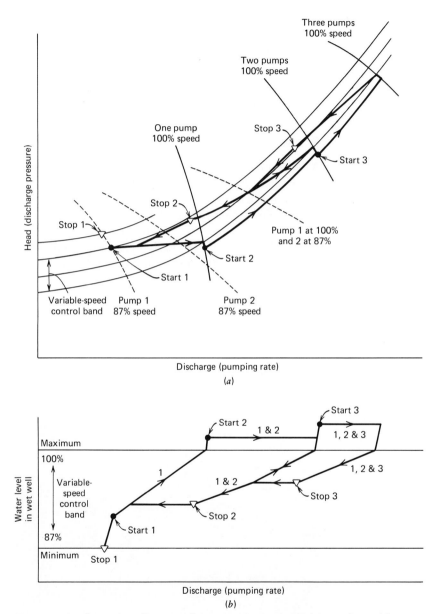

Figure 11-9 Operating diagrams for a wastewater pumping station with two variable-speed pumps, numbers 1 and 2, and one constant-speed pump, number 3, controlled by water level in the wet well. (*a*) System head-discharge curves with superimposed diagram of pump operation. (*b*) Straightline chart relating station discharge with pump speeds and water levels in the wet well. (From *Design of Wastewater and Stormwater Pumping Stations*, Manual FD-4, Water Pollution Control Federation.)

as indicated by the shaded angles. At minimum pumping capacity, the wet-well water level is at the maximum elevation and the pumps are operating individually against the lower system head curve at a $C = 140$. Pump operating points are where the modified pump curves cross the $C = 140$ curve. These are projected vertically to locate the minimum pumping rates indicated by the shaded angles on the pump characteristic curves.

Variable-Speed Pumps

The graphical analysis of a typical pumping station with a combination of two variable-speed pumps and one constant-speed pump is shown in Figure 11-9 (on page 387). All three pumps, plus a third variable-speed standby pump, have identical characteristic curves so they can be operated in parallel in alternating combinations. The speed reduction to discharge at the desired minimum capacity is 87 percent of full speed. The first variable-speed pump starts when the water level in the wet well reaches START 1 (Figure 11-9a). If the pumping rate exceeds inflow, the water level drops to the minimum level and the pump stops, STOP 1. On the other hand, if the inflow continues to rise, the pump speed increases to 100 percent. As pumping discharge continues to rise, pump number 2 starts (START 2) and numbers 1 and 2 operate in parallel at the same speed. When the constant-speed pump number 3 starts, the discharge of the two variable-speed pumps share in pumping the remaining discharge. Finally, at maximum design flow all three pumps are operating at full speed. The straight-line chart in Figure 11-9b relates pump discharge to pump speeds and water levels in the wet well. Because the most efficient pump operation is at the highest speed, keeping a minimum number of pumps in operation is an economic advantage. Thus, with diminishing flow rates, pumps are stopped as soon as possible. To prevent pumps from cycling on and off during inflow reduction, pump speeds are slowed to less than 100 percent at a discharge slightly lower than the inflow rate. The slight rise in water level is then handled by an increase in speed of the operating pumps rather than by restarting the stopped pump. For example, when pump number 3 is stopped, the variable-speed pumps numbers 1 and 2 are operating at about 94 percent of full speed. Then

as the level rises, these two pumps increase in speed to a value somewhat less than maximum to match pump number 3 stop capacity. The objective is to have a sufficient capacity overlap between stopping and starting of pumps to prevent fluctuation between the starting pumping rate and the stopping rate.

Screw Pumps

A screw pump is a rotating helix conveyor that pushes wastewater up an inclined trough to a higher elevation as shown in Figure 11-10. The main advantage is the wide range of flows that can be efficiently pumped at a constant speed of rotation. Also, when used to pump return activated sludge, the reduced turbulence compared to centrifugal pumps prevents breakup of the biological floc.

(a)

Figure 11-10   A screw pump to lift wastewater. (a) Installation of three-flight (triple helix) screw pumps lifting wastewater a height of 8 ft. (b) Diagram showing screw pump operation. (Courtesy of Lakeside Equipment Corp.)

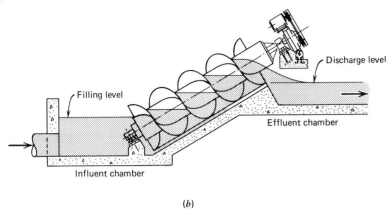

Discharge level

Filling level

Effluent chamber

Influent chamber

(b)

Figure 11-10 (*continued*)

The filling level (Figure 11-10b) is the depth in the influent chamber when a screw pump operates at full capacity, best efficiency, and highest power consumption. If the level rises, the capacity remains the same but efficiency decreases. If the level drops below the filling point, capacity, efficiency, and power consumption are reduced.

Practical lift heights of screw pumps range from 6 to 30 ft (2 to 10 m) with screw diameters increasing from 12 to 144 in. (0.3 to 3.7 m) in proportion to height. Screw inclinations are 22°, 30°, and 38° with decreased discharge at steeper slopes. Therefore, a screw installed at a 30° angle pumps more than one at a 38° angle; however, decreased inclination takes more space and may be a disadvantage in some installations. The capacity of a pump also varies with diameter, speed of rotation, and number of flights mounted on the screw shaft. The diameter is related directly to pumping capacity. The best speed is the highest rate of rotation possible without having wastewater overflow the screw shaft into a lower chamber. Screws are manufactured with one, two, and three flights (single, double, and triple helices). A single flight is a 360° helix in a length of one diameter along the shaft. Discharge capacity of a screw increases with the number of flights. For the same diameter, the capacity of a two-flight screw is approximately 80 percent of that of a three-flight screw.

11-4   SEDIMENTATION

Clarification is performed in rectangular or circular basins where the wastewater is held quiescent to permit particulate solids to settle out of suspension. To prevent short-circuiting and hydraulic disturbances in the basin, flow enters behind a baffle to dissipate inlet velocity. Overflow weirs, placed near the effluent channel, are arranged to provide a uniform effluent flow. Floating materials are prevented from discharge with the liquid overflow by placing a baffle in front of the weir. A mechanical skimmer collects and deposits the scum in a pit outside of the basin. Settled sludge is slowly moved toward a hopper in the tank bottom by a collector arm. Clarifiers following activated-sludge aeration may be equipped with hydraulic pickup pipes for rapid sludge return.

Criteria for sizing settling basins are overflow rate (surface settling rate), tank depth at the side wall, and detention time. Surface settling rate is defined as the average daily overflow divided by the surface area of the tank, expressed in terms of gallons per day per square foot (cubic meters per square meter per day), Eq. 11-1. Area is calculated by using inside tank dimensions, disregarding the central stilling well or inboard weir troughs. The quantity of overflow from a primary clarifier is equal to the wastewater influent, since the volume of sludge withdrawn from the tank bottom is negligible. However, secondary settling tanks may have recirculation drawing liquid from the tank bottom, for example, recirculation of activated sludge, in which case the influent flow is equal to the effluent plus returned flow. For these, the flow used for design is the effluent, or overflow, and not the influent, which includes recirculation.

$$V_0 = \frac{Q}{A} \qquad (11\text{-}1)$$

where V_0 = overflow rate (surface settling rate),
gallons per day per square foot
(cubic meters per square meter per day)
Q = average daily flow, gallons per day
(cubic meters per day)
A = total surface area of tank, square feet
(square meters)

Detention time is computed by dividing tank volume by influent flow expressed in hours, Eq. 11-2. Numerically, it is the time that would be required to fill the tank at a uniform rate equivalent to the design average daily flow. Depth of a tank is taken as the water depth at the side wall measuring from the tank bottom to the top of the overflow weir; this excludes the additional depth resulting from the slightly sloping bottom that is provided in both circular and rectangular clarifiers. Effluent weir loading is equal to the average daily quantity of overflow divided by

the total weir length, expressed in gallons per day per linear foot (cubic meters per meter per day).

$$t = 24\frac{V}{Q} \qquad (11\text{-}2)$$

where t = detention time, hours
V = tank volume, million gallons (cubic meters)
Q = average daily flow, million gallons per day
(cubic meters per day)
24 = number of hours per day

Primary Clarifiers

Settling basins that receive raw wastewater prior to biological treatment are called primary tanks. Figure 11-11 is a sectional view of a rectangular tank. Raw wastewater enters through a series of ports near the surface along one end of the tank. A short baffle dissipates the influent velocity directing the flow downward. Water moves through at a very slow rate and discharges from the opposite end by flowing over multiple effluent weirs. Settled solids are scraped to a sludge hopper at the inlet end by redwood flights that are attached to endless chains riding on sprocket wheels. Sludge is withdrawn periodically from the sludge hopper for disposal. The upper run of flights protrudes through the water surface pushing floating matter to a skimmer placed in front of the effluent weir. The scum trough is a cylindrical tube with a slit opening along the top. When the skimmer is manually or

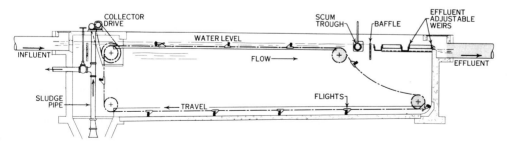

Figure 11-11   Longitudinal section of a rectangular, primary settling tank. Sludge collector scrapes settled solids to hopper at inlet end. The upper run of flights pushes floating matter to a scum trough. (Courtesy of FMC Corp., Material Handling Systems Div.)

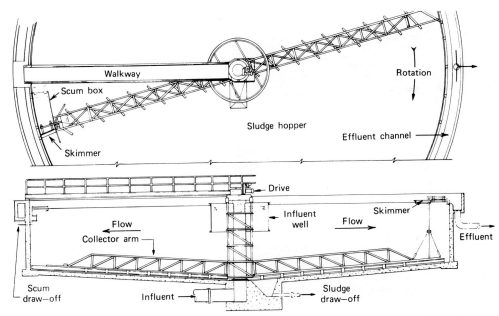

Figure 11-12 Circular primary clarifier with a pier-supported center drive and peripheral effluent weir. (Courtesy of Welles Products Corp.)

mechanically rotated, scum collected on the surface flows through the slot into the tube that slopes toward a scum pit. Length to width ratio of rectangular tanks varies from about 3:1 to 5:1 with liquid depths of 7 or 8 ft (2 to 2.5 m). The bottom has a gentle slope toward the sludge hopper.

Views of a circular primary clarifier are shown in Figure 11-12. Raw wastewater enters through ports in the top of a central vertical pipe, and flows radially to a peripheral effluent weir. The influent well directs flow downward to reduce short-circuiting across the top. A very slowly rotating collector arm plows settled solids to the sludge drawoff at the center of the tank. Floating solids migrating toward the edge of the tank are prevented from discharge by a baffle set in front of the weir. A skimmer attached to the arm collects scum from the surface and drops it into a scum box that drains outside the tank wall. Circular tanks are from 30 to 150 ft (9 to 46 m) in diameter, although some are as large as 200 ft (60 m). Side water depths range from 7 to 12 ft (2.1 to 3.7 m), and bottom slopes are about 8 percent.

Circular basins are generally preferred to rectangular tanks in new construction because of lower installation and maintenance costs. Bridge or pier supported, center-driven collector arms have fewer moving parts than the chain-and-sprocket scraper mechanisms in rectangular tanks. Although inlet turbulence is greater behind the small influent well of a circular clarifier, as the flow radiates toward the effluent weir the wastewater movement slows, thus reducing the exit velocity. Greater weir lengths can be more easily achieved around the periphery of a circular tank than across the end of a rectangular one. A typical installation of two primary settling basins located on opposite sides of a sludge pumping station is pictured in Figure 11-13. Pumps housed inside draw sludge from the bottom of the clarifiers at preset time intervals and discharge it to holding tanks for further processing. (Where holding tanks are not employed, sludge is allowed to accumulate in the clarifiers and is pumped directly to digesters or dewatering equipment.) Scum boxes drain to pits adjacent to the pump building so

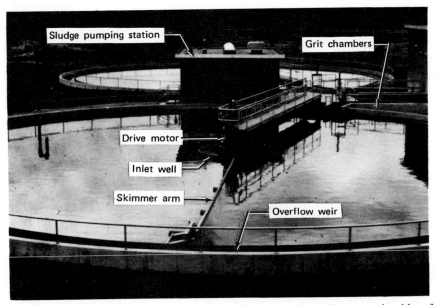

Figure 11-13 Two 90-ft-diameter, primary settling basins located on opposite sides of a sludge pumping station. Waste sludge drawn from the clarifiers is pumped to sludge holding tanks at preset time intervals. Floating solids are collected by skimmers and are deposited in scum boxes that drain to a pit for disposal with the waste sludge. (Grand Island, Nebraska.)

that the scum can be disposed of with the waste sludge. On the right side of the picture are two aerated grit chambers with grit washers housed in an attached building.

Overflow rates for sizing primary clarifiers fall in a range between 400 and 800 gpd/sq ft (16 to 33 m³/m²·d), with 600 being a common design value. These hydraulic loadings realize 30 to 40 percent BOD removal in settling raw domestic wastewater. Effectiveness of plain sedimentation, of course, depends largely on the character of the wastewater. If a municipal waste contains a large amount of soluble organic matter, BOD removal may drop to less than 20 percent while, on the other hand, industrial wastes that contribute settleable solids may increase BOD removal as high as 60 percent. Hydraulic loading also influences the density of the accumulated sludge. With overflow rates less than 600 gpd/sq ft (24 m³/m²·d), the settled solids tend to thicken in the bottom of the tank as the collector arm

slowly moves through the accumulated sludge. Rates in excess of 800 gpd/sq ft may create hydraulic movements in the tank that inhibit consolidation of the sludge. In addition to a more dilute waste sludge, a hydraulically overloaded clarifier is identified by increased turbidity in the effluent during the period of maximum wastewater flow. Storing sludge in a basin for too long a time can also upset clarification, particularly if waste activated sludge is returned to the head of the plant for settling with the raw wastewater. Microorganisms decomposing the waste organics produce gas that makes the solids more buoyant, thus expanding the sludge blanket and reducing the solids concentration. A severe case of detrimental biological activity is characterized by foul odors, floating sludge, and a darkening of the wastewater color.

The *Recommended Standards for Sewage Works, Great Lakes-Upper Mississippi River Board of State Sanitary Engineers* (*GLUMRB*) suggest that the liquid depth of mechanically cleaned primary tanks be as shallow as practicable but not less than 7 ft (2.1 m). Designers sometimes increase the side water depth to 7.5 or 8 ft to provide additional volume for accumulated sludge. Detention time is generally not a specified criterion for sizing primary clarifiers, since it is already defined by overflow rate and depth as related in Eq. 11-3. For example, an overflow rate of 600 gpd/sq ft and depth of 7 ft yield a detention time of 2.1 hr.

$$t = \frac{180 \times H}{V_0} \qquad (11\text{-}3)$$

where t = detention time, hours
  H = depth of water in tank, feet
  V_0 = overflow rate, gallons per day per square foot
  180 = 7.48 gal/cu ft × 24 hr/day

Weir loading is the hydraulic flow over an effluent weir. For primary tanks, weir loadings are not to exceed 10,000 gpd/lin ft (125 m^3/m·d) for plants of 1 mgd or smaller, and preferably not more than 20,000 gpd/ft (250 m^3/m·d) for design flows above 1 mgd. These values limit the water velocity approaching the effluent weir to minimize carryover of suspended solids.

EXAMPLE 11-1

Two primary settling basins are 90 ft in diameter with a 7-ft side water depth. Single effluent weirs are located on the peripheries of the tanks. For a wastewater flow of 7.0 mgd, calculate the overflow rate, detention time, and weir loading.

Solution

Calculating surface area and volume,

  Surface area = $2\pi r^2$ = 2 × 3.14(45)^2 = 12,700 sq ft

  Volume = 12,700 × 7 = 89,000 cu ft = 0.665 mil gal

From Eq. 11-1,

  V_0 = 7,000,000/12,700 = 550 gpd/sq ft

By Eq. 11-2,

  t = (24 × 0.665)/7.0 = 2.3 hr

Alternate calculation for t using Eq. 11-3

$$t = \frac{180 \times 7}{550} = 2.3 \text{ hr}$$

Weir length = circumference of both tanks = $2\pi d$

$$\text{Weir loading} = \frac{7,000,000}{2 \times 3.14 \times 90} = 12,400 \text{ gpd/ft}$$

Intermediate Clarifiers

Sedimentation tanks between trickling filters, or between a filter and subsequent biological aeration, in two-stage secondary treatment are called intermediate clarifiers. *GLUMRB Standards* recommend the following criteria for sizing intermediate settling tanks: Overflow rate should not surpass 1000 gpd/sq ft (41 m^3/m^2·d), minimum side water depth of 7 ft, and weir loadings less than 10,000 gpd/lin ft for plants of 1 mgd or smaller and should not be over 20,000 for larger plants.

Final Clarifiers

Settling tanks following biological filters are similar to the one illustrated in Figure 11-12. Sometimes the effluent channel is an inboard weir trough that allows overflow to enter the channel from both sides to reduce weir loading. Common criteria for final clarifiers of trickling filter plants are overflow rate not exceeding 800 gpd/sq ft (33 m^3/m^2·d), minimum side water depth of 7 ft, and maximum weir loadings the same as for intermediate tanks, although lower values are preferred.

The purpose of gravity settling following filtration is to collect biological growth, or humus, flushed from filter media. These sloughed solids are generally well-oxidized particles that settle readily. Therefore, a collector arm that slowly scrapes the accumulated solids toward a hopper for continuous or periodic discharge gives satisfactory performance. Gravity separation of biological growths suspended in the mixed liquor of aeration systems is more difficult. Greater viability of activated sludge results in lighter, more buoyant flocs with reduced settling velocities. In part, this is the result of microbial production of gas bubbles that buoy up the tiny biological clusters. Depth of accumulated sludge in a trickling filter final is normally a few inches if recirculation flow is drawn from the tank bottom. Even if sludge is drained only twice a day, the blanket of settled solids rarely exceeds 1 ft. In contrast, the accumulated microbial floc in a final basin for separating activated sludge may be 1 to 2 ft thick in a well-operating plant. During peak loading periods, the sludge blanket may expand further to incorporate one third to one half of the tank volume; this is particularly true in high-rate aeration systems.

The final clarifier in Figure 11-14 is specially designed for activated-sludge systems. The wastewater flow pattern is the same as that of other circular clarifiers, but the sludge collection system is unique. Suction pipes are attached to and spaced along a V-plow-scraper mechanism rotated by a turntable above the liquid surface. The discharge elevation of the suction riser pipes in the sight well is lower than the water surface in the tank so that sludge is forced up and out of the uptake pipes by water pressure. The flow from each suction pipe is controlled by elevation adjustment of a slip tube over the riser pipe. The sludge collected in the sight well flows out through a vertical pipe centrally located in the influent wastewater pipe. A seal between the sight well and the wastewater riser allows the well to rotate around with the collector arm without leaking sludge into the tank. While the unit illustrated is referred to as a sight well clarifier, other manufacturers produce similar units using such descriptive titles as a rapid-sludge-removal clarifier. Each

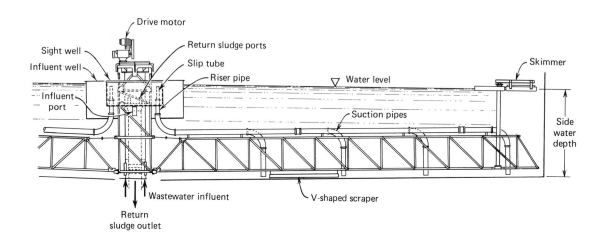

Figure 11-14 Final clarifier designed for use with biological aeration. Activated sludge is withdrawn through suction (uptake) pipes that are located along the collector arm for rapid return to the aeration basin. Sludge flowing from each pipe can be observed in the sight well. (Courtesy of Walker Process Corp.)

proprietary unit has features that may be of importance in the operation of a specific biological aeration system. For example, one mechanism separates the sludge collected by the uptake pipes from the heavier solids that are plowed to a sludge hopper. The collected sludge is returned as activated sludge while the solids scraped from the bottom are wasted.

Rapid uniform withdrawal of sludge across the entire bottom of an activated-sludge final clarifier has two distinct advantages. The retention time of solids that settle near the tank's periphery is not greater than those that land near the center; thus, aging of the biological floc and subsequent floating solids due to gas production are eliminated. With a scraper collector, the residence time of settled solids depends directly on the radial distance from the sludge hopper. The second advantage is that the direction of activated-sludge return flow is essentially perpendicular to the tank bottom, rather than horizontal toward a centrally located sludge hopper. Downward flow through a sludge blanket enhances gravity settling of the floc and increases sludge density. This is an important factor when one considers that the return flow may be as great as one half of the influent flow.

Design parameters for final clarifiers in activated-sludge processes take into account the reduced settleability of a flocculent biological suspension. Compared to other wastewater sedimentation tanks, activated-sludge clarifiers are deeper to accommodate the greater depth of settled solids, have a lower overflow rate to reduce carryover of light biological floc, and have longer weir lengths by the installation of an inboard wier channel to reduce the approach velocity of the effluent. Typical overflow rates are 600 gpd/sq ft (24 m^3/m^2·d) for plants smaller than 1 mgd and 800 gpd/sq ft (33 m^3/m^2·d) for larger plants. During the peak hydraulic flow of the day, the overflow rate should not exceed 1200 gpd/sq ft and 1600 gpd/sq ft (33 m^3/m^2·d) for small and larger plants, respectively. The recommended minimum side water depth is 10 ft (3.1 m) with greater depths for larger diameter tanks, for example, 11 ft for 50-ft diameter and 12 ft for 100-ft diameter. Depending on the values selected for overflow rate and depth, detention time is in the range of 2.0 to 3.0 hr. The maximum recommended weir loading is 10,000 to 20,000 gpd/ft (125 to 250 m^3/m·d).

EXAMPLE 11-2

Determine the size of two identical circular final clarifiers for an activated-sludge system with a design flow of 20,000 m^3/d with a peak hourly flow of 32,000 m^3/d. Use maximum overflow rates of 33 m^3/m^2·d at design flow and 66 m^3/m^2·d at peak hourly flow.

Solution

At design flow, the surface area required for each tank is

$$\text{Area} = \frac{20,000 \text{ m}^3/\text{d}}{2 \times 33 \text{ m}^3/\text{m}^2 \cdot \text{d}} = 303 \text{ m}^2$$

$$\text{Peak overflow rate} = \frac{32,000}{2 \times 303}$$

$$= 53 \text{ m}^3/\text{m}^2 \cdot \text{d} < 66 \text{ m}^3/\text{m}^2 \cdot \text{d} \quad (\text{OK})$$

$$\text{Tank diameter} = \left(\frac{4 \times 303}{\pi} \right)^{0.5} = 19.6 \text{ m} = 64 \text{ ft}$$

Recommended side water depth for a tank diameter greater than 50 ft is 11 ft. Therefore,

$$\text{Tank side water depth} = 11 \text{ ft} = 3.4 \text{ m}$$

Use an inboard weir channel set on a diameter of 18 m.

$$\text{Weir loading} = \frac{10,000 \text{ m}^3/\text{d}}{2\pi \times 18 \text{ m}}$$

$$= 88 \text{ m}^3/\text{m} \cdot \text{d} < 125 \text{ m}^3/\text{m} \cdot \text{d} \quad (\text{OK})$$

11-5 BIOLOGICAL FILTRATION

Fixed-growth biological systems are those that contact wastewater with microbial growths attached to the surfaces of supporting media. Where the wastewater is sprayed over a bed of crushed rock, the unit is commonly referred to as a trickling filter. With the development of synthetic media to replace the use of stone, the term biological tower was introduced, since these installations are often about 20 ft in depth rather than the traditional 6-ft stone-media filter. Another type of fixed-growth system is a rotating biological contactor, where a series of circular plates on a common shaft are slowly rotated while partly submerged in wastewater. Microbes attached to the disks extract waste organics. Although the physical structures differ, the biological process is essentially the same in all of these fixed-growth systems.

Biological Process

Domestic wastewater sprinkled over fixed media produces biological slimes that coat the surface. The films consist primarily of bacteria, protozoa, and fungi that feed on waste organics. Sludge worms, fly larvae, rotifers, and other biota are also found, and during warm weather sunlight promotes algal growth on the surface of a filter bed. The scheme in Figure 11-15 describes the biological activity. As the wastewater flows over the slime layer, organic matter and dissolved oxygen are extracted, and metabolic end products such as carbon dioxide are released. Dissolved oxygen in the liquid is replenished by absorption from the air in the voids surrounding the filter media. Although very thin, the biological layer is an-

aerobic at the bottom. Therefore, although biological filtration is commonly referred to as aerobic treatment, it is in fact a facultative system incorporating both aerobic and anaerobic activity.

Organisms attached to the media in the upper layer of a bed grow rapidly, feeding on the abundant food supply. As the wastewater trickles downward, the organic content decreases to the point where microorganisms in the lower zone are in a state of starvation. Thus, the majority of BOD is extracted in the upper 2 or 3 ft of a 6-ft filter. Excess microbial growth sloughing off of the media is removed from the filter effluent by a final clarifier. Purging of a bed is necessary to maintain voids for passage of wastewater and air. Organic overload of a stone-media filter, in combination with insufficient hydraulic flow, can result in plugging of passages with biological growth causing ponding of wastewater on the bed, reduced treatment efficiency, and foul odors from anaerobic conditions.

Stone–Media Trickling Filters

A cutaway view of a trickling filter is shown in Figure 11-16. The major components are a rotary distributor, underdrain system, and filter media. Influent wastewater is pumped up a vertical riser to a rotary distributor for spreading uniformly over the filter surface. Rotary arms are driven by reaction of the wastewater flowing out of the distributor nozzles. Bed underdrains carry away the effluent and permit circulation of air. Ventilation risers and the effluent channel are designed to permit free passage of air. In some installations, the underdrain blocks empty into a channel between double exterior walls to allow improved aeration and access for flushing of underdrains.

The most common media in existing filters are crushed rock, slag, or field stone that are durable, insoluble, and resistant to spalling. The size range preferred for stone media is 3- to 5-in. diameter. Although smaller stone provides greater surface area for biological growth, the voids tend to plug and limit passage of liquid and air. Bed depths range from 5 to 7 ft; greater depths do not materially improve BOD removal efficiency. Stone-media filters in the treatment of municipal wastewaters are

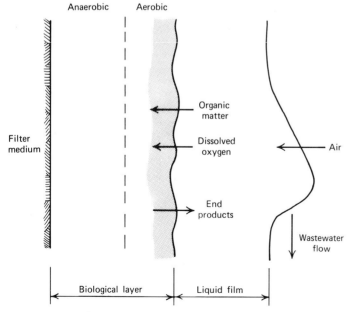

Figure 11-15   Sketch illustrating the biological process in a filter bed.

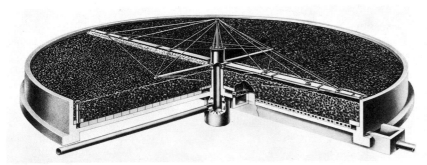

Figure 11-16 Cutaway view of a stone-media trickling filter. (Courtesy of Dorr-Oliver, Inc.)

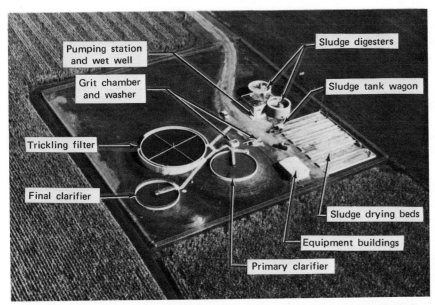

Figure 11-17 Aerial view of a single-stage trickling filter plant. (Shenandoah, Iowa.)

always preceded by primary settling to remove larger suspended solids.

An aerial view of a single-stage trickling filter plant is shown in Figure 11-17. The sequence of treatment units is a hopper-bottomed grit chamber with separate grit washer, primary settling tank, trickling filter, final clarifier with a gravity-flow recirculation line back to the wet well, waste sludge treatment by anaerobic digestion, and

digested sludge disposal by drying beds or spreading on adjacent farmland using a tank wagon. The following waste flows return to the raw wastewater wet well: overflow from grit washer, drainage from drying beds, supernatant from digesters, and underflow recirculation (humus return) from the final clarifier. Waste sludge is pumped from the bottom of the primary clarifier into floating-cover digestion tanks.

BOD load on a trickling filter is calculated using the raw BOD in the primary effluent applied to the filter, without regard to any BOD contribution in the recirculated flow from the final clarifier, Eq. 11-4. Hydraulic loading is the amount of liquid applied to the filter surface including both untreated wastewater and recirculation flows, Eq. 11-5. This surface loading is commonly expressed in units of million gallons per acre of surface area per day (mgad) or gallons per minute per square foot (cubic meters per square meter per day). Recirculation ratio, calculated by Eq. 11-6, is the ratio of recirculated flow to the wastewater entering the treatment plant.

$$\text{BOD loading} = \frac{\text{settled wastewater BOD}}{\text{volume of filter media}} \quad (11\text{-}4)$$

where   BOD loading = pounds of BOD applied per 1000 cu ft per day (grams per cubic meter per day)

Settled BOD = raw wastewater BOD remaining after primary sedimentation, pounds per day (grams per day)

Volume of media = volume of stone in the filters, thousands of cubic feet (cubic meters)

$$\text{Hydraulic loading} = \frac{Q + Q_R}{A} \quad (11\text{-}5)$$

where   Hydraulic loading = million gallons per acre per day (cubic meters per square meter per day)

Q = raw wastewater flow, million gallons per day (cubic meters per day)

Q_R = recirculation flow, million gallons per day (cubic meters per day)

A = surface area of filters, acres (square meters)

$$R = \frac{Q_R}{Q} \quad (11\text{-}6)$$

where R = recirculation ratio
Q_R and Q = same as above

Table 11-1. Typical Loadings for Trickling Filters with a 5- to 7-ft Depth of Stone or Slag Media

	High Rate	Two Stage
BOD loading		
lb/1000 cu ft/day[a]	30 to 90	45 to 70
lb/acre-ft/day	1300 to 3900	2000 to 3000
Hydraulic loading		
mil gal/acre/day[b]	10 to 30	10 to 30
gpm/sq ft	0.16 to 0.48	0.16 to 0.48
Recirculation ratio	0.5 to 3.0	0.5 to 4.0

[a] 1.0 lb/1000 cu ft/day = 16.0 g/m^3·d
[b] 1.0 mil gal/acre/day = 0.935 m^3/m^2·d

Typical loadings for stone-media filters are listed in Table 11-1. Filter plants return sufficient flow from the final clarifier hopper to the wet well for removal of accumulated settled solids and to prevent stalling of the distributor arm during low wastewater flow. Also, in-plant recirculation of wastewater increases liquid flow through the filter bed to allow greater organic loading without filling the bed voids with biological growths that would inhibit aeration. Experience has shown that BOD loadings in excess of about 25 lb/1000 cu ft/day (400 g/m^3·d) require a minimum hydraulic flushing of 10 mil gal/acre/day to keep a stone-filled bed open. In addition, BOD removal efficiency is enhanced by passing wastewater through a filter more than once.

Numerous recirculation patterns are applied in existing filter plants. The most popular in new design is gravity return of underflow from the final clarifier to the wet well during periods of low wastewater flow (Figure 11-18); and direct recirculation by pumping filter discharge back to the influent as shown by the dotted line in Figure 11-18a. Recycling of final clarifier underflow is generally limited to a recirculation ratio of about 0.5, since this is adequate to return settled solids for removal in the primary clarifier and to maintain adequate flow for turning the distributor arm. Yet, by limiting this return flow, the peak outflow rate of the primary clarifier is not increased. Direct recirculation has the advantage of influencing neither the primary nor secondary settling tank; the disadvantage is that a separate pumping station is required.

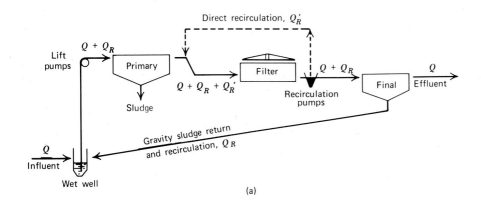

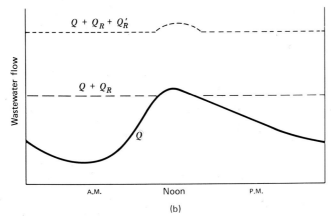

Figure 11-18 Profile of a filter plant and related wastewater flow diagrams including in-plant recirculation. (*a*) Flow diagram for a single-stage trickling filter plant. (*b*) General flow patterns: Q = wastewater influent and final clarifier overflow, $Q + Q_R$ = flow through primary tank and applied to the final based on a gravity return of about $0.5\,Q$, and $Q + Q_R + Q_R'$ = flow through the filter with both indirect and direct recirculation.

A two-stage trickling filter consists of two filter-clarifier units in series (Figure 11-19); sometimes the intermediate settling tank is omitted. This type of system is needed to achieve an effluent BOD of 30 mg/l when processing a wastewater that is stronger than average domestic wastewater. Both filters are normally constructed the same size for economy and optimum operation. Generally, several options for recirculation are incorporated in design. For example, in Figure 11-19, underflow from the inter-mediate and final clarifiers is returned to the wet well for solids disposal, while direct recirculation around each filter stage is possible. Direct recirculation may be pumped from the bottom of a settling tank, or directly from the influent well (filter discharge).

Several mathematical equations have been developed for calculating BOD removal efficiency of biological filters based on such factors as depth of bed, kind of media, temperature, recirculation, and organic loading.

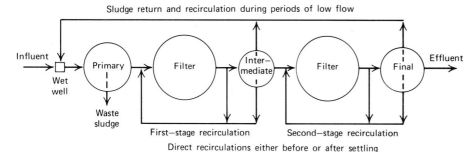

Figure 11-19 depiction with labels: Sludge return and recirculation during periods of low flow; Influent; Wet well; Primary; Waste sludge; Filter; Inter-mediate; Filter; Final; Effluent; First-stage recirculation; Second-stage recirculation; Direct recirculations either before or after settling

Figure 11-19 Typical flow diagram for a two-stage trickling filter plant.

The exactness of these formulas depends on the filter media having a uniform biological layer, and an evenly distributed hydraulic load. These conditions rarely occur in stone-media filters that may develop unequal biological growths resulting in short-circuiting of wastewater through the bed. Therefore, general practice has been to use empirical relationships based on operational data collected from existing treatment plants. One of the most popular formulations evolved from National Research Council (NRC) data that were collected from filter plants at military installations in the United States during the early 1940s. The results, graphed in Figure 11-20, are considered applicable to single-stage stone-media trickling filters followed by a final settling tank and treating settled domestic wastewater with a temperature of 20° C. Efficiency is read along the bottom scale below the point where the horizontal load line intersects the curved line with the proper recirculation ratio. For example, at a load of 40 lb BOD/1000 cu ft/day and a recirculation ratio of 0 the efficiency is 74 percent, with the same BOD load and $R = 3$, efficiency is 81 percent.

A second-stage filter is less efficient than the first stage because of the decreased treatability of the waste fraction applied to the second bed. In other words, the most available biological food is taken out first, passing the organics that are more difficult to remove through to the second-stage filter. Based on NRC observations, this effect can be incorporated by increasing the actual load to the second-stage filter as shown in the following relationship:

Second-stage BOD load adjusted for treatability

$$= \frac{\text{Actual second-stage BOD load}}{\left[\dfrac{(100 - \text{percentage of first-stage efficiency})}{100}\right]^2} \quad (11\text{-}7)$$

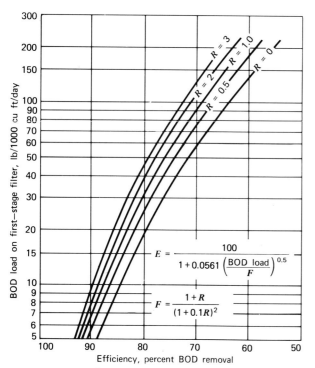

Figure 11-20 showing curves labeled $R = 3$, $R = 2$, $R = 1.0$, $R = 0.5$, $R = 0$; with equations:

$$E = \frac{100}{1 + 0.0561 \left(\dfrac{\text{BOD load}}{F}\right)^{0.5}}$$

$$F = \frac{1 + R}{(1 + 0.1R)^2}$$

Axis: BOD load on first-stage filter, lb/1000 cu ft/day; Efficiency, percent BOD removal

Figure 11-20 Efficiency curves for single-stage stone-media trickling filters treating domestic wastewater at 20° C, based on the National Research Council data.

The second-stage BOD load is calculated by Eq. 11-4 where the settled wastewater is taken as the overflow from the intermediate clarifier. After adjusting for treatability

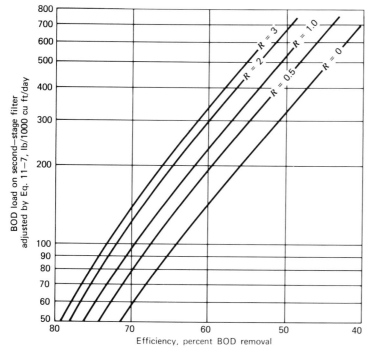

Figure 11-21 Efficiency curves for second-stage stone-media trickling filters treating domestic wastewater at 20°C based on the National Research Council data. Adjusted BOD load for entering diagram is calculated using Eq. 11-7.

by Eq. 11-7, efficiency of the second stage can be determined from Figure 11-21.

Overall treatment plant efficiency of a two-stage filter system can be calculated by Eq. 11-8.

$$E = 100 - 100\left[\left(1 - \frac{35}{100}\right)\left(1 - \frac{E_1}{100}\right)\left(1 - \frac{E_2}{100}\right)\right] \quad (11\text{-}8)$$

where E = treatment plant efficiency, percent
   35 = percentage of BOD removed in primary settling
   E_1 = BOD efficiency of first-stage filter and intermediate clarifier corrected for temperature, percent
   E_2 = BOD efficiency of second-stage filter and final clarifier corrected for temperature, percent

BOD removal in biological filtration is influenced

significantly by wastewater temperature. Filters in northern climates operate at efficiencies about 5 percent or more below the yearly average during winter months. The plot in Figure 11-22 can be used to adjust efficiencies from Figures 11-20 and 11-21 for temperatures above or below 20°C. Filter covers may be required in cold climates to achieve the effluent standard of 30 mg/l BOD or less. An installation of the type shown in Figure 11-23 protects the filters from wind and snow, thus preventing excess cooling of wastewater sprayed onto the beds. Positive ventilation is provided to maintain passage of air through the bed and to dissipate corrosive gases, namely, hydrogen sulfide.

EXAMPLE 11-3

A trickling filter plant as illustrated in Figure 11-17 has the following: a primary clarifier with a 55-ft diameter, 7.0-ft side water depth, and single peripheral weir; an 85-ft

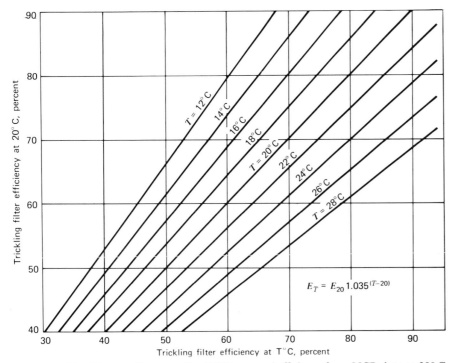

Figure 11-22   Diagram for correcting BOD removal efficiency from NCR data at 20° C to efficiency at other temperatures between 12 and 28° C.

Figure 11-23   Trickling filters that are domed with fiberglass covers to reduce cooling of the wastewater and prevent loss of treatment efficiency during winter months. (Kearney, Nebraska.)

diameter trickling filter with a 7.0-ft deep stone-filled bed; and a final settling tank with a 50-ft diameter, 7.0-ft side water depth, and single peripheral weir. The normal operating recirculation ratio is 0.5 with return to the wet well from the bottom of the final during periods of low influent flow (Figure 11-18). The daily wastewater flow is 1.38 mgd with an average BOD of 180 mg/l, essentially all domestic waste. Calculate the loadings on all of the units, and the anticipated effluent BOD at 20° C and 16° C.

Primary Clarifier

$$A = \frac{3.14 \times (55)^2}{4} = 2370 \text{ sq ft}$$

$$V = 2370 \times 7 = 16,600 \text{ cu ft} = 0.124 \text{ mil gal}$$

From Eq. 11-1,

$$V_0 \text{ without recirculation flow} = \frac{1,380,000 \text{ gpd}}{2370 \text{ sq ft}}$$

$$= 589 \text{ gpd/sq ft}$$

Using Eq. 11-2,

$$t = 24\frac{\text{hr}}{\text{day}} \times \frac{0.124 \text{ mil gal}}{1.38 \text{ mgd}} = 2.2 \text{ hr}$$

$$\text{Weir loading} = 1,380,000/(3.14 \times 55)$$

$$= 8000 \text{ gpd/lin ft}$$

Trickling Filter

$$A = \frac{3.14 \times (85)^2}{4} = 5660 \text{ sq ft,}$$

$$A = \frac{5660 \text{ sq ft}}{43,560 \dfrac{\text{sq ft}}{\text{acre}}} = 0.130 \text{ acres}$$

$$V = 5660 \times 7 = 39,600 \text{ cu ft}$$

Substituting Eq. 11-6 in Eq. 11-5,

$$\text{Hydraulic load} = \frac{Q + R \times Q}{A}$$

$$= \frac{1.38 \text{ mgd} + 0.5 \times 1.38 \text{ mgd}}{0.130 \text{ acres}}$$

$$= 15.9 \text{ mgad}$$

Since the wastewater is primarily domestic and the primary overflow rate is reasonable, assume a BOD removal of 35 percent by settling in the primary clarifier.

Therefore, settled wastewater BOD = 0.65 × 180 = 117 mg/l. By Eq. 11-4,

$$\text{BOD load} = \frac{1.38 \text{ mgd} \times 117 \text{ mg/l} \times 8.34}{39.6 \text{ thou cu ft}}$$

$$= 34 \text{ lb BOD/1000 cu ft/day}$$

Final Settling Tank

$$A = \frac{3.14 \times (50)^2}{4} = 1960 \text{ sq ft,}$$

$$V = \frac{1960 \times 7 \times 7.48}{1,000,000} = 0.103 \text{ mil gal}$$

$$V_0 = \frac{1,380,000}{1960} = 705 \text{ gpd/sq ft,}$$

$$t = 24\frac{0.103}{1.38} = 1.8 \text{ hr}$$

$$\text{Weir loading} = \frac{1,380,000}{3.14 \times 50} = 8800 \text{ gpd/lin ft}$$

Treatment Plant Efficiency and Effluent BOD

The trickling filter loading is in the normal range and the final settling tank overflow rate is reasonable. Therefore, Figure 11-20 can be used to estimate filter efficiency. Entering with a BOD load of 34 lb/1000 cu ft/day and R = 0.5, effciency is 78 percent at a wastewater temperature of 20° C.

$$\text{Effluent BOD} = \frac{100 - 78}{100} 117 \text{ mg/l} = 26 \text{ mg/l}$$

The trickling filter efficiency at 16° C, entering Figure 11-22 with 78 percent at 20° C, is 68 percent.

$$\text{Effluent BOD at 16° C} = \frac{100 - 68}{100} 117 \text{ mg/l} = 37 \text{ mg/l}$$

Overall treatment plant efficiencies are

$$\text{at 20° C, } E = \frac{180 - 26}{180} 100 = 86 \text{ percent}$$

$$\text{at 16° C, } E = \frac{180 - 37}{180} 100 = 80 \text{ percent}$$

EXAMPLE 11-4

The design flow for a two-stage trickling filter plant is 1.2 mgd with an average BOD concentration of 280 mg/l. Calculate the unit loadings, and treatment plant efficiency at a wastewater temperature of 17° C.

> Primary tank surface area = 2400 sq ft
> First-stage filter: area = 4880 sq ft = 0.112 acres
> volume = 29,300 cu ft
> Intermediate clarifier area = 1500 sq ft
> Second-stage filter is identical to first stage
> Final clarifier area = 1500 sq ft

Recirculation pattern is as shown in Figure 11-19: return to wet well is 0.3 mgd underflow from each clarifier for total $Q_R = 0.6$ mgd; and direct recirculation around each filter Q'_R is 0.8 mgd.

Solution

Primary Settling Tank

$$V_0 \text{ without recirculation flow} = 1,200,000/2400$$
$$= 500 \text{ gpd/sq ft}$$
$$V_0 \text{ with } Q_R \text{ of 0.6 mgd} = 1,800,000/2400$$
$$= 750 \text{ gpd/sq ft}$$

Both of these values are satisfactory; therefore, assume a BOD removal of 35 percent.

$$\text{BOD of settled wastewater} = 0.65 \times 280 = 182 \text{ mg/l}$$

First-stage Filter and Intermediate Clarifier

$$\text{BOD load} = \frac{1.2 \text{ mgd} \times 182 \text{ mg/l} \times 8.34}{29.3 \text{ thou cu ft}}$$
$$= 62 \text{ lb/1000 cu ft/day}$$
$$\text{Hydraulic load} = \frac{1.2 + 0.6 + 0.8}{0.112} = 23 \text{ mgad (OK)}$$

From Eq. 11-6,

$$R = (0.6 + 0.8)/1.2 = 1.2$$

Efficiency from Figure 11-20 at 62 lb/1000 cu ft/day and $R = 1.2$ is 75 percent at 20° C.
Efficiency corrected for 17° C from Figure 11-22 is 67 percent.

V_0 of intermediate clarifier

$$= (1,200,000 + 300,000)/1500 = 1000 \text{ gpd/sq ft (OK)}$$

Effluent BOD from intermediate clarifier

$$= 0.33 \times 182 = 60 \text{ mg/l}$$

Second-stage Filter and Final Clarifier

$$\text{BOD load} = \frac{1.2 \text{ mgd} \times 60 \text{ mg/l} \times 8.34}{29.3 \text{ thou cu ft}}$$
$$= 20 \text{ lb/1000 cu ft/day}$$

From Eq. 11-7,

$$\text{Adjusted BOD load} = \frac{20}{(1 - 0.67)^2}$$
$$= 180 \text{ lb/1000 cu ft/day}$$
$$\text{Hydraulic load} = \frac{1.2 + 0.3 + 0.8}{0.112}$$
$$= 21 \text{ mgad (OK)}$$
$$R = \frac{0.3 + 0.8}{1.2} = 0.9 \qquad (Note: \text{0.3 mgd was returned to wet well in first stage.})$$

Efficiency from Figure 11-21 at 180 lb/1000 cu ft/day and $R = 0.9$ is 63 percent. Efficiency corrected for 17° C from Figure 11-22 is 56 percent.

$$V_0 \text{ of final clarifier} = \frac{1,200,000}{1500}$$
$$= 800 \text{ gpd/sq ft (OK)}$$
$$\text{Plant effluent BOD} = 0.44 \times 60 = 26 \text{ mg/l}$$

Treatment Plant Efficiency at 17° C

$$E = \frac{(280 - 26)}{280} 100 = 91 \text{ percent}$$

Biological Towers

Several forms of manufactured media are marketed for trickling filters. The main advantage, relative to crushed stone, is the high specific surface (sq ft/cu ft) with a corresponding high percentage of void volume that permits substantial biological slime growth without inhibiting passage of air supplying oxygen. Other advantages include uniform media for better liquid distribution, light weight facilitating construction of deeper beds, chemical resistance, and the ability to handle high-strength wastewaters. The two common kinds of plastic media are vertical-sheet packing and random packing. Another medium consists of racks of redwood slates.

Vertical-sheet packing is manufactured of polyvinyl chloride (PVC) in modules approximately 2 ft wide, 4 ft long, and 2 ft high. As illustrated in Figure 11-24a, the corrugated sheets are bonded between flat sheets to prevent clear vertical openings; therefore, the wastewater flowing down through the packing is distributed over the surfaces of the media. The specific surface varies with the manufacturer from 26 to 43 ft^2/ft^3 (85 to 140 m^2/m^3), and

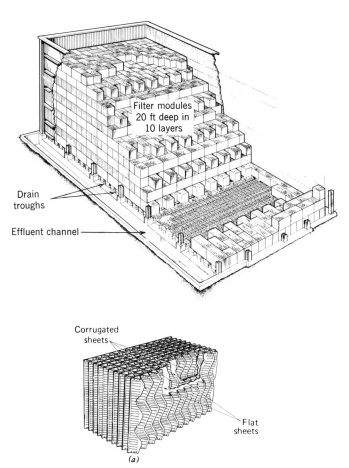

Figure 11-24 Cutaway view showing construction of a rectangular biological tower. Wastewater is piped to the top of the tower, is spread over the packing, and trickles down through the bed. The underflow is conveyed by drains to an effluent channel. (a) Detail of a vertical-sheet packing module 2 ft wide, 4 ft long, and 2 ft deep. (b) Photograph of fixed-nozzle distributors. (Courtesy of Imperial Chemical Industries Limited, Pollution Control Systems.)

the void space is about 95 percent. The strength of the modules is adequate to support the packing with attached biological growths in towers of about 20 ft (6 m) in height. The media can be cut to fit in a circular tower equipped with a rotary distributor, or the modules can be stacked in a rectangular tower and the wastewater sprayed on top with fixed-nozzle distributors.

Construction features of a rectangular biological tower are illustrated in Figure 11-24. Modules of self-supporting media are stacked so that they interlock for structural stability. The underdrain system consists of supporting block with drain troughs leading to an effluent channel. Wastewater is pumped to the top of the bed and is spread by fixed distributors. Piping is supported by horizontal beams that rest on vertical columns. The siding may be corrugated metal, plastic, wood, or block; all of these permit architectural design to enhance the appearance of a treatment plant.

The typical random packing is composed of small (3 to 4 in.) cylinders with perforated walls and internal ribs made of plastic placed unarranged in a trickling filter or biological tower. Because of the random placement, the wastewater can be spread effectively over the surface of the bed in a circular tower using a rotary distributor. The specific surface of packing is 30 to 40 ft^2/ft^3 (100 to 130 m^2/m^3), and the void space is 91 to 94 percent.

Efficiency equations have been developed for plastic-media filters since the uniformity of the packing can be defined by the specific surface. Yet, no universal formulas exist that can precisely describe removal of organic matter, partially because of the different geometric shapes of the media. Packing configuration influences residence time of the liquid in the bed, which is in turn related to hydraulic loading and filter depth. Furthermore, removal of organic matter depends directly on its solubility. In theoretical equations, the BOD is the filtered (soluble) BOD, not the total BOD as normally measured.

The efficiency equation based on soluble BOD removal by first-order kinetics is

$$E = \left(1 - \frac{e^{-KD/Q^n}}{(1 + R) - Re^{-KD/Q^n}} \right) 100 \qquad (11\text{-}9)$$

where E = efficiency, percent
K = reaction-rate constant, per minute (per hour)
D = depth of packing, feet (meters)
Q = hydraulic loading of raw wastewater without recirculation, gallons per minute per square foot (cubic meters per square meter per hour)
n = constant related to the specific surface and configuration of the packing
R = recirculation ratio, Q_R/Q
e = base of Napierian logarithms, 2.718

If the recirculation is zero, the efficiency is computed as the term in the numerator of Eq. 11-9. The reaction-rate constant K is corrected for temperature by the equation

$$K = K_{20}(1.035)^{T-20} \qquad (11\text{-}10)$$

where K = reaction-rate constant at temperature T in degrees Celsius, per minute (per hour)
K_{20} = reaction-rate constant at 20° C, per minute (per hour)

The values for n and K are determined by pilot-plant studies treating the wastewater being evaluated in a column filled with the specific packing. After a biological growth on the media is established, experiments are conducted at several recirculation rates to find out the influence of hydraulic loading on removal efficiency. Normally, wastewater samples are collected and analyzed at various depths in the column, as well as the influent and effluent. In the case of a typical domestic wastewater, most manufacturers can recommend values for n and K based on the performance of full-scale installations.

EXAMPLE 11-5

A strong municipal wastewater with a soluble BOD of 200 mg/l at a temperature of 20° C is treated in a biological tower with an 18-ft depth of plastic packing. The hydraulic loading of the raw wastewater is 0.35 gpm/sq ft, and the recirculation ratio is 3.0. The constants based on a pilot-plant study are $K = 0.075$ per min and $n = 0.39$. Calculate the effluent soluble BOD.

Solution

The value of the exponential term in Eq. 11-9 is

$$e^{-0.075 \times 18/0.35^{0.39}} = 0.131$$

Substituting into Eq. 11-9,

$$E = \left(1 - \frac{0.131}{(1+3) - 3 \times 0.131} \right) 100 = 96 \text{ percent}$$

Hence, soluble effluent BOD is $0.04 \times 200 = 8$ mg/l.

Combined Filtration and Aeration Process

Redwood packing is made of rough-sawn redwood slats ($\frac{3}{8}$ in. $\times 1\frac{1}{2}$ in.) stapled to rails ($1\frac{3}{4}$ in. deep) to form racks 4 ft by 4 ft. For installation, the racks are laid flat in stacks with slats and rails aligned vertically. The packing is commonly placed in a rectangular tower to a depth of 14 ft. Fixed nozzle distributors, positioned about 3.5 ft above the surface of the packing, are used to spray wastewater on the bed.

The most popular application of redwood-media filtration is in series with aeration to develop suspended biological growth before final sedimentation as diagramed in Figure 11-25. The biological tower is a filtration process with direct recirculation of underflow. The second-stage aeration develops an activated sludge that is settled in the final clarifier and recirculated to the tower influent for mixing with the raw wastewater. By returning settled biological solids, the wastewater flowing through the system develops an activated sludge. Combining the processes of fixed growth with suspended growth results in a stable biological system that can

tolerate shock loads. Therefore, activated biological filtration is best suited to treatment of strong municipal or industrial wastewaters with variations of flow and organic content.

Typical design criteria for redwood-media filtration with supplemental aeration treating a clarified municipal wastewater are a BOD loading on the tower of 75 to 150 lb/1000 cu ft/day (1.2 to 2.4 kg/m³·d), a surface hydraulic loading from 1 to 5 gpm/sq ft (60 to 300 m³/m²·d), and a second-stage aeration period of 1 to 2 hr. To achieve an adequate hydraulic loading, the recommended recirculation flows are $1.5Q$ direct recycle of tower underflow and $0.5Q$ indirect recycle to return the settled activated sludge from the final clarifer. The mixed liquor suspended solids concentrated in the aeration tank is maintained in the range of 1000 to 3000 mg/l.

Operational Problems

Two major problems of stone-media trickling filters are effluent quality and odors; both are associated with organic loading, industrial wastes, and cold weather operation. The average BOD removal efficiency of a single-stage filter plant is about 85 percent. Therefore, to achieve an effluent BOD of 30 mg/l, the raw wastewater must be essentially a domestic waste with a BOD not greater than 200 mg/l. If a municipal wastewater contains significant industrial waste contributions, a two-stage filtration system is necessary to meet the required effluent standards. In northern climates, the temperature of wastewater passing through the bed may be considerably lower in the winter and may adversely influence BOD removal. Covers can be placed over trickling filters to help maintain temperature during cold weather.

The microbial zone immediately adjacent to the surface of the media is anaerobic and capable of producing metabolic end products that have offensive odors. Reduced compounds formed in treating domestic wastewaters, such as hydrogen sulfide, appear to be oxidized as they move through the aerobic zone with adequate aeration. However, if voids fill with excess biological growth, foul odors can be emitted during spring and fall when air temperatures reduce natural air circulation through the bed. Industrial wastes, particularly from

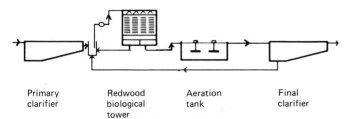

Primary   Redwood   Aeration   Final
clarifier   biological   tank   clarifier
             tower

Figure 11-25 Flow diagram for a combined filtration and aeration process. (Courtesy of Neptune Microfloc Inc.)

food-processing industries, have characteristic odors that are not easily oxidized in a trickling filter, and they can cause problems even when design loadings are not exceeded. Filter covers have been used in some locations to reduce the spread of nuisance odors, but the installation must be carefully planned. For example, forced air ventilation with an air scrubbing tower to remove the odors from exhausted air may be required to maintain adequate air passage through the bed and to prevent a corrosive atmosphere under the dome.

Biological towers using synthetic media do not appear to be as susceptible to operational difficulties of quality control and odors as are stone-media beds. This is primarily attributed to improved aeration and hydraulic distribution of the wastewater. Nevertheless, potential odor problems and the influence of cold weather must be considered in design and operation of treatment systems that employ biological towers.

Filter flies, *Psychoda*, are a nuisance problem near filters during warm weather. They breed in sheltered zones of the media, and on the inside surfaces of the retaining walls. Although wind can carry these small flies considerable distances, their greatest irritation is to operating personnel. Periodic spraying of the peripheral area and walls of the filter with an insecticide is a common method of fly control.

11-6   ROTATING BIOLOGICAL CONTACTORS

A rotating biological contactor (RBC) is constructed of bundles of plastic packing attached radially to a shaft, forming a cylinder of media. The shaft is placed over a contour-bottomed tank so that the media are submerged approximately 40 percent. The contactor surfaces are spaced so that during submergence wastewater can enter the voids in the packing. When rotated out of the tank, the liquid trickles out of the voids between the surfaces and is replaced by air. A fixed-film biological growth, similar to that on a trickling filter packing, adheres to the media surfaces. Alternating exposure to organics in the wastewater and oxygen in the air during rotation is like the dosing of a trickling filter with a rotating distributor.

Excess biomass sloughs from the media and is carried out in the process effluent for gravity separation.

The RBC unit pictured in Figure 11-26 consists of bundles of polyethylene media in the shape of truncated pyramids fitted into structural supports mounted on a central shaft. By using bundles rather than full-size circular disks, the RBC unit can be shipped as components for assembly on site. Also, media can be installed or replaced with the shaft mounted in position. In order to reduce the probability of biological growth filling the voids, the plastic sheets in the bundle are separated by conical spacers to maximize the open area between layers. The spacing between sheets in the media used for BOD removal is $\frac{3}{4}$ in.; the spacing used for nitrification is $\frac{1}{2}$ in. The disks are 12 ft (3.7 m) in diameter and operate at 40 percent submergence. At the operating speed of 1.5 rpm, the peripheral velocity is 57 ft/min for a 12-ft-diameter cylinder.

A treatment system consists of primary sedimentation preceding and final sedimentation following the rotating biological contactors. Since recirculation through RBC units is not normally practiced, only sufficient underflow from the final clarifier is returned to allow removal of excess biological solids in primary sedimentation as sketched in Figure 11-27a. Waste sludge, similar in character to that from a trickling filter plant, is withdrawn from the primary clarifiers for disposal. In large plants, the RBC shafts are placed in a baffled tank perpendicular to the direction of wastewater flow. In small plants, a common shaft is placed over a contoured tank with the wastewater flow parallel to the shaft. In this arrangement, the bundles of media are spaced so that baffles can be placed between the stages (Figure 11-27c). A series of four stages are normally installed in the treatment of domestic wastewater for BOD reduction; additional stages may be added to initiate nitrification. Each stage acts as a completely mixed chamber and the movement of the wastewater through the series of tanks simulates plug flow. Biological solids washed off of the media are transported hydraulically under the baffles to be carried out with the effluent. RBC units are protected by installation either in a building with adequate ventilation or under separate plastic covers lined with insulation.

(a)

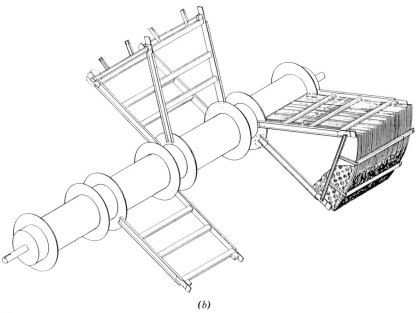

(b)

Figure 11-26   Rotating biological contactor media. (*a*) Four stages of 12-ft-diameter media assembled on a 27-ft shaft for a total surface area of 100,000 sq ft. (*b*) Shaft construction showing a structural support and a bundle of media mounted on the shaft. (Courtesy of CLOW, Waste Treatment Division.)

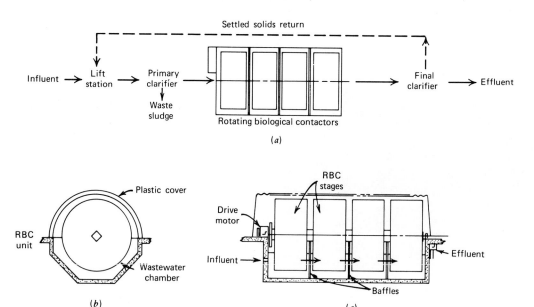

Figure 11-27 Schematic diagrams of a rotating biological contactor plant. (*a*) Flow scheme. (*b*) Cross section of a covered RBC unit. (*c*) Four-stage system with the wastewater flow parallel to the rotating shaft.

The efficiency of BOD removal in processing domestic wastewater is based on empirical data from operating RBC plants. Mathematical equations have been developed, but their prediction of performance is not consistently reliable. For a municipal wastewater containing a significant quantity of industrial wastewater, pilot-plant studies are recommended to determine design parameters. Preferably, the studies should be performed with a full-scale package plant, nevertheless, smaller RBC units with a diameter of about 2 ft are often used. Typical recommendations for secondary treatment of domestic wastewater to produce an effluent of less than 30 mg/l of BOD and 30 mg/l of suspended solids are: (1) an average organic loading based on the total RBC surface of 1.5 lb/1000 sq ft/day (7.5 g/m²·d) of soluble BOD, or 3.0 lb/1000 sq ft/day (15 g/m²·d) of total BOD; (2) a maximum loading on the first stage of 6.0 lb/1000 sq ft/day (30 g/m²·d) of soluble BOD, or 12 lb/1000 sq ft/day (60 g/m²·d) of total BOD; and (3) a temperature correction for additional RBC surface

area of 15 percent for each 5° F below a design wastewater temperature of 55° F (2.8° C below 13° C).

EXAMPLE 11-6

Determine the size of an RBC unit needed for treatment of a settled domestic wastewater having a BOD of 130 mg/l. The design flow is 0.25 mgd at a temperature of 50°F. The required effluent quality is 30 mg/l BOD.

Solution

Using the recommended design criteria given in the text,

$$\text{RBC area at } 55° \text{F} = \frac{0.25 \text{ mgd} \times 130 \text{ mg/l} \times 8.34}{3.0 \text{ lb}/1000 \text{ sq ft/day}}$$
$$= 90{,}400 \text{ sq ft}$$

Correcting the surface area for a temperature of 50° F by a 15 percent area increase per 5° F decrease,

$$\text{RBC area at } 50° \text{ F} = 1.15 \times 90{,}400 = 104{,}000 \text{ sq ft}$$

Use an RBC shaft with a nominal surface area of 100,000 sq ft mounted in a baffled tank as illustrated in Figure 11-27.

11-7 BIOLOGICAL AERATION

Understanding of aeration processes in wastewater treatment requires a greater knowledge of biology than that required for biological filtration. Therefore, the reader is encouraged to review the discussions on bacteria and protozoa presented in Sections 3-1 and 3-4, and population dynamics in Section 3-12.

The generalized biological process that takes place in an aeration system is sketched in Figure 11-28. Raw wastewater flowing into the aeration basin contains organic matter (BOD) as a food supply. Bacteria metabolize the waste solids, producing new growth while taking in dissolved oxygen and releasing carbon dioxide. Protozoa graze on bacteria for energy to reproduce. Some of the new microbial growth dies, releasing cell contents to solution for resynthesis. After the addition of a large population of microorganisms, aerating raw wastewater for a few hours removes organic matter from solution by synthesis into microbial cells. Mixed liquor is continuously transferred to a clarifier for gravity separation of the biological floc and discharge of the clarified effluent. Settled floc is returned continuously to the aeration basin for mixing with entering raw wastewater.

The liquid suspension of microorganisms in an aeration basin is generally referred to as a mixed liquor, and the biological growths are called mixed liquor suspended solids (MLSS). The name activated sludge was originated in referring to the return biological suspension, since these masses of microorganisms were observed to be very "active" in removing soluble organic matter from solution. This extraction process is a metabolic response of bacteria in a state of endogenous respiration, or starvation. The activated-sludge process is truly aerobic since the biological floc is suspended in mixed liquor containing oxygen.

Aeration Tank Loadings

Aeration period, BOD loading per unit volume, food-to-microorganism ratio, and sludge age define an activated-sludge process. Aeration period is calculated in the same manner as detention time. Using Eq. 11-2, V is the volume of the aeration tank and Q the quantity of wastewater entering, without regard to recirculation flow. BOD load is usually expressed in terms of lb BOD applied per day per 1000 cu ft of liquid volume in the aeration tank (grams per cubic meter per day). This can be calculated by Eq. 11-4 where V is the liquid volume of the aeration basins, rather than volume of the trickling filter media.

The food-to-microorganism ratio (F/M) is a way of expressing BOD loading with regard to the microbial

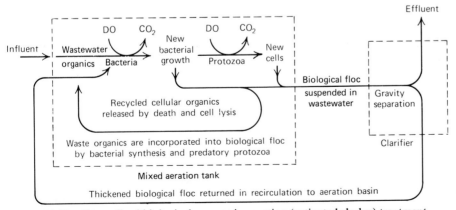

Figure 11-28 Generalized biological process in aeration (activated sludge) treatment.

mass in the system. Equation 11-11 calculates the F/M value as BOD applied/day/unit mass of MLSS in the aeration tank.

$$\frac{F}{M} = \frac{Q \times BOD}{V \times MLSS} \qquad (11\text{-}11)$$

where   F/M = food-to-microorganism ratio, pounds of BOD per day per pound of MLSS (grams of BOD per day per gram of MLSS)

Q = wastewater flow, million gallons per day (cubic meters per day)

BOD = wastewater BOD, milligrams per liter (grams per cubic meter)

V = liquid volume of aeration tank, million gallons (cubic meters)

MLSS = mixed liquor suspended solids in the aeration basin, milligrams per liter (grams per cubic meter)

Some authors express the food-to-microorganism ratio in terms of mass of BOD/day/unit mass of MLVSS (mixed liquor volatile suspended solids).

Sludge age, also referred to as mean cell residence time, is an operational parameter related to the F/M ratio. While liquid retention times (aeration periods) vary from 3 to 30 hr, the residence time of biological solids in a system is much greater and is measured in terms of days. In other words, while wastewater passes through aeration only once and rather quickly, the resultant biological growths and extracted waste organics are repeatedly recycled from the final clarifier back to the aeration tank. Several formulations have been developed to compute a value for sludge age. The most common establishes sludge age on the basis of the mass of MLSS in the aeration tank relative to the mass of suspended solids in the wastewater effluent and waste sludge.

$$\text{Sludge age} = \frac{MLSS \times V}{SS_e \times Q_e + SS_w \times Q_w} \qquad (11\text{-}12)$$

where   Sludge age = mean cell residence time, days

MLSS = mixed liquor suspended solids, milligrams per liter

V = volume of aeration tank, million gallons (cubic meters)

SS_e = suspended solids in wastewater effluent, milligrams per liter

Q_e = quantity of wastewater effluent, million gallons per day (cubic meters per day)

SS_w = suspended solids in waste sludge, milligrams per liter

Q_w = quantity of waste sludge, million gallons per day (cubic meters r per day)

Sludge age can also be expressed in terms of volatile suspended solids founded on the argument that the volatile portion of suspended solids is more representative of biological mass.

BOD loading per unit volume and aeration period are interrelated parameters dependent on the concentration of BOD in the wastewater entering and the volume of the aeration tank. (For example, if a 200 mg/l wastewater flows into a tank with an aeration period of 24 hr, the resulting BOD load is 12.5 lb/1000 cu ft/day. If the aeration period is reduced to 8 hr, by increasing the waste flow, the BOD load would be tripled or 37.5 lb/1000 cu ft/day.) But the F/M ratio as an expression of BOD loading relates to the metabolic state of the biological system rather than to the volume of aeration tank. The advantage of this expression is that lb BOD/day/lb MLSS defines an activated-sludge process without reference to aeration period or strength of applied wastewater. Two systems quite different may operate at the same F/M ratio. For example, an extended aeration process with a 24-hr aeration period and MLSS concentration of 1000 mg/l would have an F/M of 0.2 for an influent wastewater with a BOD of 200 mg/l. A conventional activated-sludge system can treat this same wastewater at the same F/M ratio by operating with a higher MLSS of 3000 mg/l to compensate for reducing the aeration period from 24 to 8 hr.

A summary of volumetric BOD loadings, F/M ratios, sludge ages, aeration periods, return sludge rates, and typical BOD removal efficiencies are listed in Table 11-2. The concentration of MLSS maintained in an aeration tank varies with the kind of process and the method of operation. In general, less than 1000 mg/l does not

Table 11-2. Summary of Loading and Operational Parameters for Aeration Processes

Process	BOD Loading $\left(\dfrac{\text{lb BOD/day}}{1000 \text{ cu ft}}\right)^a$	F/M Ratio $\left(\dfrac{\text{lb BOD/day}}{\text{lb MLSS}}\right)^b$	Sludge Age (days)	Aeration Period (hr)	Return Sludge Rates (percent)	BOD Removal Efficiency (percent)
Conventional	30–40	0.2–0.5	5–15	6.0–7.5	20–40	80–90
Step aeration	30–50	0.2–0.5	5–15	5.0–7.0	30–50	80–90
Contact stabilization	30–50	0.2–0.5	5–15	6.0–9.0	50–100	75–90
High rate	80 up	0.5–1.0	3–10	2.5–3.5	50–100	70–85
High-purity oxygen	120 up	0.6–1.5	3–10	1.0–3.0	30–50	80–90
Extended aeration	10–20	0.05–0.2	20 up	20–30	50–100	85–95

^{a} 1.0 lb/1000 cu ft/day = 16.0 g/m³·d
^{b} 1.0 lb/day/lb = 1.0 g/d·g

provide a low enough F/M for good sludge settleability and over 4000 mg/l results in loss of suspended solids in the clarifier overflow. The common range is 2500 to 3500 mg/l. The rate of return sludge from the final clarifier to the aeration basin is expressed as percentage of the raw wastewater influent. For example, if the return activated sludge rate is 30 percent and the raw wastewater flow into the plant is 10 mgd, the recirculated flow equals 3.0 mgd. BOD efficiency is calculated by dividing the quantity of BOD removed in aeration and subsequent settling by the raw BOD entering.

Sludge Settleability

Degree of treatment achieved in an aeration process depends directly on settleability of the activated sludge in the final clarifier. A biological floc that agglomerates and settles by gravity leaves a clear supernatant for discharge. Conversely, poorly flocculated particles (pin floc), or buoyant filamentous growths, that do not separate by gravity contribute to BOD and suspended solids in the process effluent. Excessive carryover of floc resulting in inefficient operation is referred to as sludge bulking. This may be caused by adverse environmental conditions that are created by insufficient aeration, lack of nutrients, presence of toxic substances, or overloading.

Settleability of a biological sludge, under a normal operating environment, depends on the food-to-microorganism ratio and sludge age. Figure 11-29 graphically relates F/M values to sludge settleability. Extended aeration systems with long aeration periods and relatively high MLSS concentrations operate in the endogenous growth phase. This yields high BOD efficiency since starving microorganisms effectively scavenge the organic matter, and flocculate readily under quiescent conditions. At the opposite extreme, high-rate aeration employs a high F/M ratio, which allows a greater volumetric BOD loading and reduced aeration period. The resulting activated sludge has reduced settleability that can be partially overcome by using a rapid sludge return clarifier that helps to thicken the biological floc hydraulically (Figure 11-14). Nevertheless, high-rate systems are not as efficient in BOD removal primarily because of carryover of microscopic floc, for instance, groups of bacteria or protozoa not captured by the settleable matter. This can result in suspended solids greater than 30 mg/l in the effluent. The F/M range between 0.5 and 0.2 appears to be optimum for achieving the efficiency needed for treating municipal wastewater. Volumetric loadings in this range are great enough to allow aeration periods of 5 to 7 hr, thus permitting economical construction and operating costs.

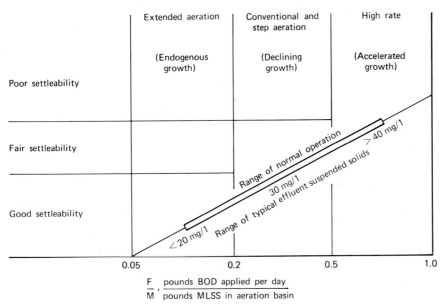

Figure 11-29 Approximate relationship between activated-sludge settleability and operating food-to-microorganism ratio.

Mathematical Relationships

One of the most common tests for monitoring operation of an aeration system is the sludge volume index (SVI). The procedure involves determining the mixed liquor suspended solids concentration (MLSS), and sludge settleability using a standard laboratory graduated cylinder with a capacity of 1 liter. Mixed liquor for testing is drawn from the aeration basin near the discharge end. Sludge volume is measured by filling the graduated cylinder to the 1-liter mark, allowing undisturbed settling for 30 min, and then reading the volume occupied by the settled solids in milliliters. A second portion of the mixed liquor, or contents of the cylinder after remixing, is tested for MLSS by the procedure for suspended solids presented in Section 2-11. Sludge volume index is then calculated by Eq. 11-13. SVI is the volume in milliliters occupied by 1 gram of settled suspended solids. In general, the range of 50 to 150 ml/g indicates a good settling sludge. The MLSS in conventional and step aeration basins is normally held between 2000 and 3000 mg/l; high-rate processes carry about 4000 mg/l.

$$SVI = \frac{V \times 1000}{MLSS} \qquad (11\text{-}13)$$

where   SVI = sludge volume index, milliliters per gram
   V = volume of settled solids in a 1-liter graduated cylinder after 30 min, milliliters per liter
   MLSS = mixed liquor suspended solids, milligrams per liter
   1000 = milligrams per gram

Sludge volume index can be hypothetically related to the quantity and solids concentration in return activated sludge as depicted in Figure 11-30. In the following mathematical relationships, the final clarifier of the treatment system is assumed to respond identically to the graduated cylinder used in the SVI test. This assumption appears reasonable with respect to extended aeration and conventional processes, while in high-rate systems there may be considerable deviation in sludge settleability between an actual clarifier and that measured in a laboratory container. In either case, the illustration gives

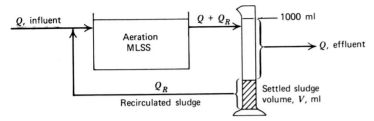

Figure 11-30 Hypothetical relationship between the settled sludge volume from the SVI test and required recirculation flow of activated sludge in an aeration process, Eqs. 11-13 and 11-15.

a pictorial representation of process operation. Quantity of return sludge flow is keyed to settled sludge volume by the relationship in Eq. 11-14. These ratios were derived by noting in Figure 11-30 that recirculated sludge flow is to settled sludge volume as the flow entering the clarifier relates to clarifier volume. This assumes that sludge recirculation is exactly at the required rate. A greater quantity returns clarified wastewater unnecessarily, while a lesser flow leaves settled solids in the clarifier and results in eventual loss in the process effluent. By rearranging Eq. 11-14 to the form in Eq. 11-15, the theoretical required flow of recirculated activated sludge can be calculated based on settled sludge volume. Concentration of suspended solids in the recirculated sludge can be computed from the SVI by Eq. 11-16. This formula is useful in estimating the solids concentration in excess activated sludge that is usually wasted from the return sludge line.

$$\frac{Q_R}{Q + Q_R} = \frac{V}{1000} \tag{11-14}$$

$$Q_R = \frac{V \times Q}{1000 - V} \tag{11-15}$$

where Q_R = flow of recirculated activated sludge, million gallons per day (cubic meters per day)

Q = average flow of wastewater to aeration basin, million gallons per day (cubic meters per day)

V = volume of settled solids in 1-liter graduated cylinder, milliliters per liter

1000 = milliliters per liter

$$\frac{\text{SS in recirculated}}{\text{activated sludge}} = \frac{1,000,000}{\text{SVI}} \tag{11-16}$$

where SS = suspended solids, milligrams per liter
SVI = sludge volume index, milliliters per gram

For a recirculation rate greater than the minimum, resulting in some clarified wastewater being returned with the settled sludge, the concentration of suspended solids in the recirculated sludge is

$$\frac{\text{SS in recirculated}}{\text{activated sludge}} = \frac{\text{MLSS} (Q + Q_R)}{Q_R} \tag{11-17}$$

Both this equation and Eq. 11-16 assume no loss of suspended solids in the effluent.

EXAMPLE 11-7

The mixed liquor suspended solids concentration in the basin of a step aeration system is 2400 mg/l, and the sludge volume after settling in a 1-liter graduated cylinder is 220 ml. Compute the SVI. Is this value within the proper range? Based on the hypothetical relationships in Figure 11-30, what is the required return sludge ratio and suspended solids concentration in the recirculated sludge?

Solution

Applying Eq. 11-13,

$$\text{SVI} = \frac{220 \text{ ml/l} \times 1000}{2400 \text{ mg/l}} = 92 \text{ ml/g} (<150 \text{ OK})$$

By Eq. 11-15,

$$\frac{Q_R}{Q} = \frac{220}{1000 - 220} = 0.28 \qquad \text{or 28 percent}$$

From Eq. 11-16,

$$SS = \frac{1,000,000}{92} = 11,000 \text{ mg/l} = 1.1 \text{ percent}$$

Conventional and Step Aeration

These processes are similar to the earliest activated-sludge systems that were constructed for secondary treatment of municipal wastewater. As diagramed in Figure 11-31, the aeration basin is a long rectangular tank with air diffusers along the bottom for oxygenation and mixing. In a conventional basin, the air supply is tapered along the length of the tank to provide greatest aeration at the head end where raw wastewater and return activated sludge are introduced. Air is provided uniformly in step aeration while wastewater is introduced at intervals, or steps, along the first portion of the tank.

Air diffusers are set on the bottom of the tank, or attached to pipe headers along one side at a depth of 8 ft or more, to provide deep mixing and adequate oxygen transfer. Fine-bubble ceramic diffusers, in the form of aeration domes or plates mounted over channels in the floor, are distributed over the entire bottom of the tank to provide uniform vertical mixing. In spiral-flow aeration, a large number of diffuser tubes or nozzles are attached to air headers along one side of the tank. A number of different kinds are manufactured including perforated tubes with or without removable sleeves, jet nozzles, and a variety of air spargers. These produce coarser bubbles than a ceramic diffuser. The air headers are connected to a jointed arm so that the diffusers can be swung out of the tank for cleaning and maintenance (Figure 11-32).

Plug-flow pattern of long rectangular tanks produces an oscillating biological growth pattern. The relatively

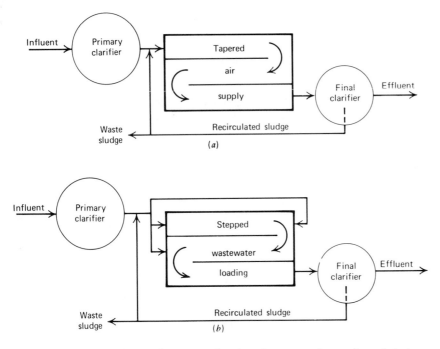

Figure 11-31 Flow schemes for conventional and step-aeration activated-sludge processes. (*a*) Conventional tapered aeration. (*b*) Step aeration.

Figure 11-32   Picture of a conventional diffused aeration tank. Air headers on the right side with diffusers attached are retracted for inspection.

high food-to-microorganism ratio at the head of the tank decreases as mixed liquor flows through the aeration basin. Since the aeration period is 5 to 8 hr, and can be considerably greater during low flow, the microorganisms move into the endogenous growth phase before their return to the head of the aeration basin. This starving microbial population must quickly adapt to a renewed supply of waste organics. The process has few problems of instability where wastewater flows are greater than 0.5 mgd; however, because of wide hourly variations in waste loads from small cities, the conventional plug-flow system can experience serious problems of biological instability. This phenomenon was a major factor contributing to the development of completely mixed aeration for handling small flows.

EXAMPLE 11-8

The following are average operating data from a conventional activated-sludge secondary:

Wastewater flow = 7.7 mgd
Volume of aeration basins = 300,000 cu ft = 2.24 mil gal
Influent total solids = 599 mg/l
Influent suspended solids = 120 mg/l
Influent BOD = 173 mg/l

Effluent total solids = 497 mg/l
Effluent suspended solids = 22 mg/l
Effluent BOD = 20 mg/l
Mixed liquor suspended solids = 2500 mg/l
Recirculated sludge flow = 2.7 mgd
Waste sludge quantity = 54,000 gal/day
Suspended solids in water sludge = 9800 mg/l

Based on these values, calculate the following: aeration period; BOD load in lb BOD/1000 cu ft/day; F/M ratio in lb BOD/day/lb MLSS; total solids, suspended solids, and BOD removal efficiencies; sludge age; and return sludge rate.

Solution

$$t = \frac{2.24 \text{ mil gal}}{7.7 \text{ mgd}} \times 24 = 7.0 \text{ hr}$$

$$\text{BOD load} = \frac{7.7 \text{ mgd} \times 173 \text{ mg/l} \times 8.34}{300 \text{ thou cu ft}}$$

$$= 37.1 \frac{\text{lb BOD/day}}{1000 \text{ cu ft}}$$

$$\frac{F}{M} = \frac{7.7 \text{ mgd} \times 173 \text{ mg/l} \times 8.34}{2.24 \text{ mil gal} \times 2500 \text{ mg/l} \times 8.34}$$

$$= 0.24 \frac{\text{lb BOD/day}}{\text{lb MLSS}}$$

$$\text{Total solids efficiency} = \frac{599 - 497}{599} \times 100 = 17 \text{ percent}$$

$$\text{Suspended solids efficiency} = \frac{120 - 22}{120} \times 100$$

$$= 82 \text{ percent}$$

$$\text{BOD efficiency} = \frac{173 - 20}{173} \times 100 = 88 \text{ percent}$$

Suspended solids in effluent

$$= 7.7 \text{ mgd} \times 22 \text{ mg/l} \times 8.34 = 1400 \text{ lb/day}$$

Suspended solids in waste sludge

$$= 0.054 \text{ mgd} \times 9800 \text{ mg/l} \times 8.34 = 4400 \text{ lb/day}$$

$$\text{Sludge age} = \frac{2.24 \text{ mil gal} \times 2500 \text{ mg/l} \times 8.34}{1400 \text{ lb/day} + 4400 \text{ lb/day}} = 8.1 \text{ days}$$

$$\text{Return sludge rate} = \frac{2.7 \text{ mgd}}{7.7 \text{ mgd}} \times 100 = 35 \text{ percent}$$

Contact Stabilization

The sequence of operations in this process is aeration of raw wastewater with return activated sludge, sedimentation to yield a clarified effluent, and reaeration of the clarifier underflow with a portion wasted to an aerobic digester. Supernatant drawn from the digester is returned to the process influent. The raw wastewater aeration chamber, also referred to as the contact zone, is approximately one third of the total aeration volume. Therefore, the aeration chamber has a detention time of 2 to 3 hr while reaeration is a 4- to 6-hr period. Normal operating sludge recirculation is 100 percent. Aerated compartments may be rectangular to simulate plug flow, or completely mixed.

The sequence of aeration-clarification-reaeration in processing large waste flows is neither as economical nor as efficient in BOD reduction as conventional or step aeration. Hence, current use is largely limited to factory-built field-erected plants capable of handling 0.05 to 0.5 mgd. A typical unit (Figure 11-33) consists of two concentric circular tanks about 14 ft deep with the inner shell 15 to 30 ft in diameter and the outer tank 30 to 70 ft

across. The doughnut-shaped space between the two walls is segmented into three chambers for aeration, reaeration, and aerobic digestion. The central tank serves as a clarifier.

High-Rate Aeration

The primary motivation in developing the high-rate process was to reduce the cost of construction by increasing BOD load per unit volume of aeration tank, with a corresponding decrease in aeration period. This was achieved by operating at a higher food-to-microorganism ratio and reduced sludge age, while increasing the mixed liquor suspended solids carried in the basin to 4000 mg/l. Problems in oxygen transfer and sludge settleability were rectified by modifying processing equipment. Complete mixing was necessary to distribute shock loads throughout the tank contents and maintain a homogeneous mixed liquor for efficient oxygen transfer. The unit illustrated in Figure 11-34 uses a combination of compressed air aeration and mechanical mixing. Air is introduced through holes in a sparge ring located at the base of the mixer shaft, and turbines churn the tank contents shearing the coarse bubbles. Hydraulic thickening action of a rapid sludge return clarifier is mandatory to offset decreased settleability of the biological floc. If a high-rate process is not pushed to excessive BOD loads, exceeding 80 lb BOD/1000 cu ft/day (1.3 kg/m^3·d), or extremely shortened aeration periods, the features of completely mixed aeration and rapid sludge return clarification can provide an efficient treatment process. Overloading of a high-rate system is generally reflected by insufficient aeration capacity to maintain adequate dissolved oxygen during peak loading periods and carryover of suspended floc in the clarifier effluent.

High-Purity Oxygen

The major components of a system using high-purity oxygen in lieu of air are a gas generator, a specially compartmented aeration tank, final clarifier, pumps for recirculating activated sludge, and sludge disposal facilities. Oxygen supply is generated either by manu-

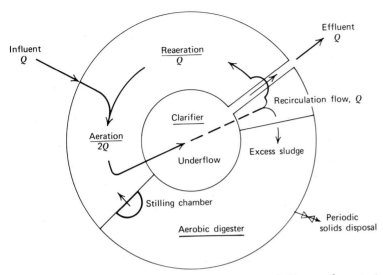

Figure 11-33 Diagram of a typical factory-built field-erected contact stabilization plant.

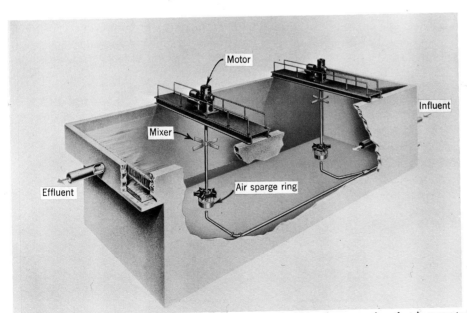

Figure 11-34 High-rate aeration tank with a large-bubble air sparge ring that is mounted under dual mixing impellers. (Courtesy of Dorr Oliver, Inc.)

facturing liquid oxygen or by producing high-purity gas by adsorption separation from air. Standard cryogenic air separation, involving liquefaction of air followed by fractional distillation to separate the major components of nitrogen and oxygen, is employed at large installations. For the majority of treatment plants, the simpler pressure-swing adsorption system is more efficient. Air is compressed in a vessel filled with a granular adsorbent that takes in carbon dioxide, water, and nitrogen gas under high pressure, leaving a gas relatively high in oxygen. The adsorber is regenerated of depressurizing and purging. Three vessels are installed for continuous flow of high-purity oxgyen, with two units alternating between operation and regeneration and the third as a standby or backup unit.

The aeration tank is divided into stages by means of baffles and covered with a gas-tight enclosure; thus, the liquid and gas phases flow concurrently through the sections (Figure 11-35). Raw wastewater, return activated sludge, and oxygen gas under a slight pressure are introduced to the first stage. Oxygen may be mixed with the tank contents by recirculation through a hollow shaft to a rotating sparger device as illustrated in Figure 11-35, or a surface aerator may be installed at the top of the mixer turbine shaft to contact oxygen gas with the mixed liquor. Successive aeration chambers are connected to each other so that liquid flows through submerged ports,

and head gases pass freely from stage to stage with only a slight pressure drop. Exhausted waste gas is a mixture of carbon dioxide, nitrogen, and about 10 to 20 percent of the oxygen applied. Effluent mixed liquor is settled in a rapid sludge return clarifier, and activated sludge is recirculated to the aeration tank.

Several advantages are attributed to high-purity oxygen aeration compared with air systems. As is shown in Table 11-2, high efficiency is possible at increased BOD loads and reduced aeration periods. Disposal of waste sludge is easier because of the higher solids content and smaller quantity produced; the settled sludge generally contains 1 to 2 percent solids. Effective odor control is easier to achieve with oxygen aeration, covered tanks, and the reduced volume of exhaust gases.

EXAMPLE 11-9

A high-purity oxygen aeration system, as diagramed in Figure 11-35, is being considered for treatment of a combined domestic and industrial wastewater. Since the combined wastewater is high in soluble BOD and low in suspended solids, primary clarification is not included in the processing scheme. The design flow is 3000 m^3/d with an average BOD of 300 mg/l. From previous experience in the aerobic treatment of similar wastewaters, the design F/M is 0.60 g/d of BOD per g of MLSS and the operating

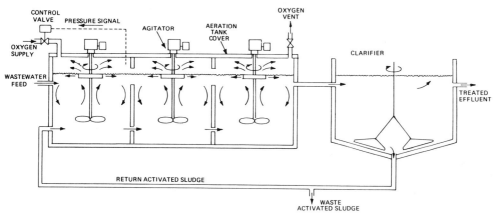

Figure 11-35 Schematic longitudinal section of a high-purity oxygen aeration system. (Courtesy of UNOX System, Lotepro Corp.)

MLSS is 4000 mg/l. The maximum volumetric BOD loading is 2.5 kg/m³·d, minimum aeration period is 2.0 h, and minimum recommended sludge age is 3.0 d. Based on these criteria, determine the volume of the aeration tank.

Solution

Rearranging Eq. 11-11,

$$V = \frac{Q \times BOD}{F/M \times MLSS} = \frac{3000 \text{ m}^3/\text{d} \times 300 \text{ mg/l}}{0.6 \text{ g/d·g} \times 4000 \text{ mg/l}} = 375 \text{ m}^3$$

$$\text{BOD loading} = \frac{3000 \text{ m}^3/\text{d} \times 300 \text{ g/m}^3}{375 \text{ m}^3 \times 1000 \text{ g/kg}}$$

$$= 2.4 \text{ kg/m}^3\text{·d} < 2.5 \text{ kg/m}^3\text{·d} \quad (\text{OK})$$

$$\text{Aeration period} = \frac{375 \text{ m}^3 \times 24 \text{ h/d}}{3000 \text{ m}^3/\text{d}}$$

$$= 3.0 \text{ h} > 2.0 \text{ h} \quad (\text{OK})$$

Assuming excess suspending solids production of 0.4 g of MLSS per g of BOD applied,

$$\text{Estimated sludge age} = \frac{375 \text{ m}^3 \times 4000 \text{ g/m}^3}{0.4 \times 3000 \text{ m}^3/\text{d} \times 300 \text{ g/m}^3}$$

$$= 4.2 \text{ d} > 3.0 \text{ d} \quad (\text{OK})$$

Extended Aeration

The most popular application of this process is in treating small flows from schools, subdivisions, trailer parks, and villages. The cross-sectional diagram of a common design is given in Figure 11-36. Aeration basins may be cast-in-place concrete or steel tanks fabricated in a factory. Continuous complete mixing is either by diffused air or by mechanical aerators, and aeration periods are 24 to 36 hr. Because of these conditions, as well as low BOD loading, the biological process is very stable and can accept intermittent loads without upset. For example, a unit serving a school may receive wastewater during a 10-hr period each day, for only five days a week.

Clarifiers for small plants are conservatively sized with low overflow rates, ranging from 200 to 600 gpd/sq ft (8 to 24 m³/m²·d) and long detention times. Sludge may be returned to the aeration chamber through a slot opening or by employing an air-lift pump. A slot return requires periodic cleaning to prevent plugging by settled solids. Although satisfactory performance can be achieved, returning settled solids by pumping provides a more positive process control (Figure 11-36). Sludge that floats to the surface of the sedimentation chamber is returned to the aeration tank by a skimming device attached to the air-lift pump return.

Usually no provision is made for wasting of excess activated sludge from small extended aeration plants. Instead, the mixed liquor is allowed to increase in solids concentration over a period of several months and then is removed directly from the aeration basin. This is performed by allowing the suspended solids to settle in the tank with the aerators off, and then pumping the concentrated sludge from the bottom into a vehicle for hauling away. The MLSS operating range varies from a minimum of 1000 mg/l to a maximum of about 10,000 mg/l. In treating domestic wastewater under normal loading,

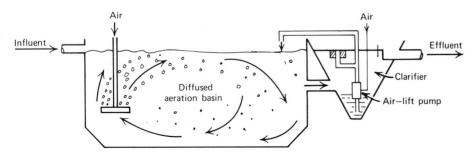

Figure 11-36 Cross-sectional diagram illustrating the operation of a typical small extended aeration plant with diffused aeration and an air-lift pump for return of settled solids and scum to the aeration tank.

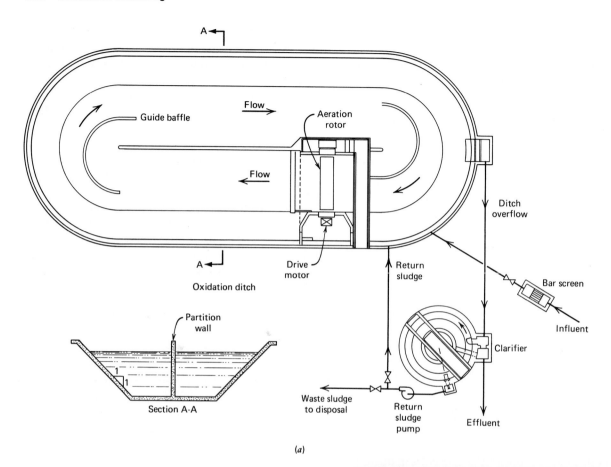

Oxidation ditch

Section A-A

(a)

the mixed liquor concentration increases at the rate of approximately 30 to 50 mg/l suspended solids per day.

Larger extended aeration plants consisting of an aeration tank, clarifier, and aerobic sludge digester are used by small municipalities. The aeration unit may be a concrete tank with diffused aeration, a lined earthen basin with mechanical aerators, or a channel with horizontal rotor aeration. Final clarifiers are usually separate circular concrete tanks with mechanical sludge collectors. Aerated holding tanks are used to digest and store waste activated sludge prior to disposal. The oxidation ditch plant illustrated in Figure 11-37 is popular because of its high efficiency and easy operation. The aeration tank is usually an elongated oval with a liquid depth of 4 to 6 ft. Depending on the size of of the plant, the outside channel walls may be sloping or vertical, and the median may be

Figure 11-37 Extended aeration process employing a horizontal mechanical aerator mounted over an aeration channel. (*a*) Flow diagram for an oxidation-ditch plant. (*b*) Picture of a horizontal rotor aerator and drive motor. (Courtesy of Lakeside Equipment Corp.)

either a wide strip with sloping walls or a dividing wall with flow guide baffles in the channel. The rotor aerator oxygenates the mixed liquor while propelling it around in the channel. The aerator pictured in Figure 11-37b is a 42-in.-diameter, high-capacity rotor with blades mounted on a 14-in.-diameter shaft. In most cases, the BOD loading on larger extended aeration systems is less than 20 lb/1000 cu ft/day (320 g/m³·d), and the aeration period is greater than 12 hr.

EXAMPLE 11-10

A small extended aeration plant without sludge wasting facilities is loaded at a rate of 10.5 lb BOD/1000 cu ft/day with an aeration period of 24 hr. The measured suspended solids buildup rate in the aeration tank is 30 mg/l/day. What percentage of the raw influent BOD is converted and retained as MLSS? If the MLSS concentration is allowed to increase from 1000 mg/l to 7000 mg/l before wasting solids, how long would this buildup take?

Solution

Since the aeration period is 24 hr,

$$\frac{\text{BOD load/day}}{\text{liter of tank}} = 10.5 \frac{\text{lb BOD/day}}{1000 \text{ cu ft}}$$

$$\times \frac{\text{mg/l}}{62.4 \text{ lb/1,000,000 cu ft}} = 168 \text{ mg/l}$$

$$\frac{\text{MLSS buildup}}{\text{BOD applied}} = \frac{30 \text{ mg/l/day}}{168 \text{ mg/l/day}} \times 100 = 18 \text{ percent}$$

$$\text{Buildup time} = \frac{7000 \text{ mg/l} - 1000 \text{ mg/l}}{30 \text{ mg/l}} = 200 \text{ days}$$

Aeration and Oxygen Transfer

Oxygen is supplied to the mixed liquor in an aeration tank by dispersing air bubbles through submerged diffusers or by entraining air into the liquid by mechanical means. Air diffusers are porous plates, tubes, or nozzles attached either to air piping on the bottom of the tank or to pipe headers that can be lifted out of the tank. Ceramic diffusers are manufactured as tubes, domes, or plates. Compressed air forced through the porous material is released as fine bubbles. Medium-bubble tube diffusers are often constructed as perforated tubes covered by a deflector plate, flexible tubing, or bags made of synthetic cloth. Coarse-bubble devices are orifices or nozzles designed so that the discharged air is broken up into bubbles and dispersed in the surrounding liquid. Although each kind of diffuser has individual features, coarse-bubble nozzles are noted for maintenance-free operation and fine-bubble diffusers have the advantage of higher oxygen-transfer efficiency. Mechanical aerators are horizontal paddle, vertical turbine, and vertical-turbine draft tube. A horizontal rotor rotates partially submerged in an aeration channel. A vertical turbine may be a surface unit or completely submerged with compressed air supplied under the rotating blade. For deep mixing, a vertical turbine may be in a draft tube so that liquid from the bottom of the tank is drawn up through the tube and discharged at the surface.

Oxygen transfer is a two-phase process as diagramed in Figure 11-38. First, gaseous oxygen is dissolved in the wastewater by diffused or mechanical aeration. Then, the dissolved oxygen is taken up by the microorganisms in metabolism of the waste organic matter. If the rate of oxygen utilization exceeds the rate of dissolution, the dissolved oxygen in the mixed liquor is depleted.

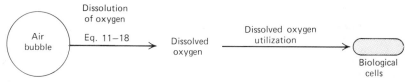

Figure 11-38 Two-phase oxygen transfer in activated-sludge aeration. Gaseous oxygen is first converted to dissolved oxygen and is then taken up by the biological floc.

The rate of oxygen transfer from the air bubbles into solution is expressed by Eq. 11-18. The K-factor depends on wastewater characteristics and, more importantly, on the physical features of the aeration system, such as kind of diffuser or aerator, liquid depth of the tank, mixing turbulence, and basin configuration.

$$R = K(\beta C_s - C_t) \tag{11-18}$$

where R = rate of oxygen transfer from air to dissolved oxygen, milligrams per liter per hour

K = transfer coefficient depending on aeration equipment and wastewater characteristics, per hour

β = oxygen saturation coefficient of the wastewater, usually 0.8 to 0.9

C_s = dissolved oxygen concentration at saturation for pure water, milligrams per liter (Table 2-5)

C_t = dissolved oxygen concentration existing in the mixed liquor, milligrams per liter

$(\beta C_s - C_t)$ = dissolved oxygen deficit, milligrams per liter

The rate of dissolved oxygen utilization is essentially a function of the food-to-microorganism ratio (BOD loading and aeration period) and temperature. This biological uptake is generally less than 10 mg/l/hr for extended aeration processes, about 30 mg/l/hr for conventional, and as great as 100 mg/l/hr for high-rate aeration. Aerobic biological activity is independent of dissolved oxygen above a minimum critical value. Below this concentration, the metabolism of microorganisms is limited by reduced oxygen supply. Critical concentrations reported for various systems range from 0.2 to 2.0 mg/l, the most common being 0.5 mg/l.

Knowledge of basic concepts of oxygen transfer and uptake, which are sketched in Figure 11-38, is helpful in understanding operational problems generally associated with aeration processes. An oxygen deficiency in an aerating basin is possible if the rate of biological utilization exceeds the capability of the equipment. For example, organic overloading of an extended aeration system that is equipped with coarse-bubble diffusers set at a shallow depth can result in a dissolved oxygen level less than 0.5 mg/l even though the tank contents are being vigorously mixed by air bubbles emitting from diffusers. Perhaps a situation that occurs more frequently in practice is uneconomical operation from overaeration producing a dissolved oxygen level greater than is necessary in the mixed liquor. Since biological activity is just as great at low levels and the transfer rate from air to dissolved oxygen increases with decreasing concentration, it is logical to operate a system as close to critical minimum dissolved oxygen as possible. Operation of the air compressors at reduced capacity, or even turning off one blower on weekends, may be feasible to conserve electrical energy with no adverse effects to the biological process. The best method for determining suitable operation is to make dissolved oxygen measurements at various times, particularly during periods of maximum loading, and then to adjust the air supply accordingly.

GLUMRB Standards provide the following guidelines as minimum design air requirements for diffused air systems: conventional, step aeration, and contact stabilization 1500 cu ft of air applied per lb BOD aeration tank load; modified or high-rate 400 to 1500 cu ft per lb BOD load; and extended aeration 2000 cu ft per pound of BOD load. They further recommend that the aeration system produce thorough mixing of the basic contents, and be capable of maintaining a minimum of 2.0 mg/l of dissolved oxygen at all times. For mechanical aeration units, as well as diffused air, the equipment should be capable of transferring at least 1.0 lb of oxygen into the mixed liquor per lb of BOD aeration tank loading.

EXAMPLE 11-11

An aeration system has a transfer coefficient K of 3.0 per hour for a wastewater at 20° C with a β of 0.9. What is the rate of oxygen transfer when the dissolved oxygen is 4.0 mg/l? When the DO is 1.0 mg/l?

Solution

From Table 2-5, $C_s = 9.2$ mg/l.
Using Eq. 11-18,

R at DO of 4.0 mg/l = $3.0(0.9 \times 9.2 - 4.0) = 13$ mg/l/hr

R at DO of 1.0 mg/l = $3.0(0.9 \times 9.2 - 1.0) = 22$ mg/l/hr

These calculations illustrate that the rate of oxygen transfer is greater, and the aeration system more efficient, at an increased dissolved oxygen deficit.

EXAMPLE 11-12

A conventional diffused aeration process supplies 1000 cu ft of air per pound of BOD applied to the tank. The installed equipment is capable of transferring 1.0 lb of atmospheric oxygen to dissolved oxygen per pound of BOD applied, as specified by *GLUMRB*. Calculate the oxygen transfer efficiency of the aeration equipment. (One cubic foot of air at standard temperature and pressure, 20° C and 760 mm, contains 0.0174 lb oxygen.)

Solution

$$\frac{\text{Oxygen supplied}}{\text{lb BOD applied}} = 1000 \text{ cu ft} \times 0.0174 \frac{\text{lb O}_2}{\text{cu ft}} = 17.4 \text{ lb}$$

$$\frac{\text{Oxygen transferred}}{\text{lb BOD applied}} = 1.0 \text{ lb (specified rate)}$$

$$\text{Oxygen transfer efficiency} = \frac{1.0}{17.4} \times 100 = 6 \text{ percent}$$

This means that about 1 percent of the air diffused into the mixed liquor is absorbed, since air is only 21 percent oxygen by volume.

Empirical Design of Processes

Activated-sludge systems treating municipal wastewaters are commonly designed on the basis of operating experience of existing plants using the same aeration processes. Local conditions, such as sewer infiltration and the affect of industrial wastewaters, are taken into account by the designer from experience. Climatic conditions, particularly wastewater temperature and seasonal variations in flow, also have an influence on the selection of loading parameters used in sizing aeration tanks. For instance, to produce an effluent of less than 30 mg/l BOD in a moderate climate, the typical design values for conventional secondary aeration on essentially domestic wastewater are a volumetric BOD loading of 40 lb/1000 cu ft/day (640 g/m^3·d) and a minimum aeration period of 6 hr. Based on this aeration tank sizing, the process can be operated at an F/M of 0.2 to 0.3 lb BOD/day/lb MLSS with a mixed liquor suspended solids concentration of approximately 2500 to 3000 mg/l.

Operation and Control

Aeration processes are regulated by the quantity of air supplied, the rate of activated-sludge recirculation, and the amount of sludge wasted, which, in turn, controls the MLSS, F/M ratio, and sludge age. A properly designed system, treating domestic wastewater, experiences few problems if the system is operated in a steady-state condition. Namely, the quantity of wastewater applied each day is approximately the same, aeration mixing and dissolved oxygen concentration are relatively steady, and excess sludge is wasted continuously in quantities necessary to maintain a proper food-to-microorganism ratio. The latter may require reduced sludge wasting on weekends when the organic load on the plant is less.

Perhaps the most common problem associated with domestic wastewater treatment is overaeration resulting in a bulking activated sludge. When a plant is operated considerably under the design BOD load, nitrifying bacteria can convert ammonia to nitrate in the aeration basin. During subsequent detention in the final clarifier, the nitrate can be used as an oxygen source under anaerobic conditions, releasing nitrogen gas that buoys up biological floc. The best solution is to increase sludge wasting to deplete the nitrifying bacterial populations and to decrease the air supply to reduce the dissolved oxygen concentration, provided that these control measures do not reduce the carbonaceous BOD removal efficiency.

Laboratory tests for monitoring activated sludge treatment are dissolved oxygen, MLSS concentration, SVI, and effluent BOD and suspended solids. Influent BOD and flow are needed to calculate organic loading, F/M ratio, and aeration period. Concentration of solids in the return activated sludge, effluent quality from the final clarifier, probing to determine depth of the sludge blanket in the clarifier, and SVI data yield information to establish the proper recirculation rate for optimum process efficiency and maximum solids concentration in the waste sludge.

Industrial wastes are frequently the source of operational instability of aeration processes. Shock loads of high-strength wastewater can deplete dissolved oxygen and unbalance the biological system. Toxic wastes in sufficient amounts interfere with microbial metabolism, and high-carbohydrate wastes may cause nutrient deficiency. All of these result in loss of MLSS in the clarifier overflow, thus decreasing treatment efficiency and shifting the F/M ratio by this unintentional wasting of microorganisms. Section 9-5 discusses techniques to evaluate the treatability of wastewater. Hydraulic shock loads resulting from excessive inflow and infiltration to the sewer system can be just as detrimental as toxins or organic overload. High rates of overflow from a final clarifier, even for short time periods, can carry over a substantial portion of the viable activated sludge needed in the system. Several days may be required to rebuild the density of mixed liquor and to return to a proper operating F/M ratio.

Determining the cause of poor settleability of an activated sludge involves investigating biological, chemical, and physical factors. First, the operating parameters of volumetric and F/M BOD loadings, MLSS concentration, and sludge age should be calculated and compared to the recommended values for the particular aeration process. Microscopic examination and visual observation of the settling characteristics of the MLSS can reveal filamentous growths aand poorly agglomerated floc. Nitrification-denitrification is evidenced by sludge bulking in the final clarifier and testing of the mixed liquor for nitrate concentration. Chemical factors that can cause poor biological growth are insufficient dissolved oxygen, lack of nutrients, presence of toxins, and low temperature. Laboratory tests can be used to establish the adequacy of dissolved oxygen and availability of nitrogen and phosphorus nutrients. Since lowering the temperature of the mixed liquor decreases the rate of biological activity, cold temperature has an effect similar to increasing the BOD loading. Finally, the physical characteristics of the aeration system and final clarifier should be considered. Excessive agitation due to underloading or overly aggressive mechanical mixing can cause shearing of biological floc into tiny particles with reduced settleability. Ineffective final clarification can result from an inadequate rate of returning sludge, excessive overflow rate, hydraulic turbulence, or a faulty collector mechanism.

11-8   MATHEMATICAL MODEL FOR COMPLETELY MIXED AERATION

The principles of biological kinetics as defined by Monod in the batch growth of pure cultures (Section 3-13) have been applied in developing a mathematical model for biological aeration. The two key assumptions are that the kinetics reactions of pure cultures are the same as those of mixed cultures and that growth and substrate metabolism in a batch culture simulates BOD or COD removal in a completely mixed, continuous-flow, aeration tank.

The basic mathematical relationship between rate of substrate utilization and substrate concentration in the declining growth phase is depicted in Figure 11-39 as a hyperbolic function. The shape of this curve is identical to Monod's equation graphed in Figure 3-26. For good activated-sludge settleability, aeration processes operate at a low substrate concentration (low F/M ratio) in the declining and endogenous growth phases. The unique

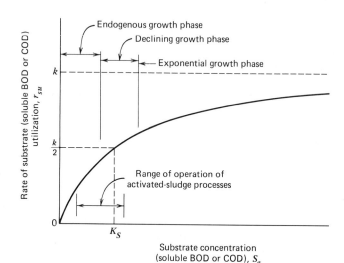

Figure 11-39   Rate of substrate utilization versus substrate concentration in the aeration tank (Eq. 11-19). The constant K_S is equal to the concentration of substrate S when the rate of substrate utilization r_{su} equals one half the maximum rate k.

mathematical feature of a hyperbolic function is that the constant K_S is equal to the substrate concentration when the substrate utilization is equal to one half of the maximum rate of substrate utilization. The equation for the curve in Figure 11-39 is

$$r_{su} = k \frac{XS_e}{K_S + S_e} \qquad (11\text{-}19)$$

where r_{su} = rate of substrate utilization, milligrams per liter of soluble BOD or COD per day

k = maximum rate of substrate utilization, milligrams per liter of soluble BOD or COD per milligram per liter of volatile MLSS per day

X = concentration of biomass, milligrams per liter of volatile MLSS

S_e = concentration of substrate surrounding the microorganisms, milligrams per liter of soluble BOD or COD

K_S = saturation constant, equal to substrate concentration when K_S equals $k/2$, milligrams per liter of soluble BOD or COD

The flow scheme shown in Figure 11-40 is used in deriving the mathematical equations for the completely mixed, activated-sludge process. The symbols of the diagram are

Q = rate of influent wastewater flow, cubic meters per day

Q_w = rate of excess sludge wasting from the aeration tank, cubic meters per day

$Q - Q_w$ = rate of effluent flow, cubic meters per day

R = recirculation ratio (Q_R/Q)

RQ = rate of activated-sludge recirculation

$Q(1 + R)$ = rate of flow from aeration tank

V = volume of aeration tank, cubic meters

X = concentration of biomass in aeration tank, milligrams per liter of volatile MLSS

X_R = concentration of biomass in recirculated activated sludge, milligrams per liter of volatile SS

X_e = concentration of biomass in effluent

S_0 = concentration of substrate in influent wastewater, milligrams per liter of soluble BOD or COD

S_e = concentration of substrate in effluent flow, recirculating sludge, and aeration tank

The following relationships use the symbols shown in the flow scheme of Figure 11-40.

$$E = \frac{(S_0 - S_e)\,100}{S_0} \qquad (11\text{-}20)$$

where E = efficiency of soluble BOD or COD removal, percent.

$$\theta = \frac{V}{Q} \qquad (11\text{-}21)$$

where θ = aeration period, days.

$$\text{F/M} = \frac{QS_0}{VX} = \frac{S_0}{\theta X} \qquad (11\text{-}22)$$

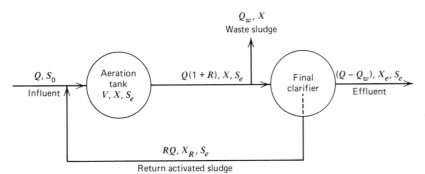

Figure 11-40   Flow scheme for a completely mixed activated-sludge process.

where F/M = food-to-microorganism ratio, gram of soluble BOD or COD applied per day per gram of volatile MLSS in the aeration tank.

$$\theta_c = \frac{VX}{Q_wX + (Q - Q_w)X_e} \qquad (11\text{-}23)$$

where θ_c = sludge age (mean cell residence time), days.

When the activated-sludge process in Figure 11-40 is operated under steady-state conditions, the rate of biomass growth equals the rate of biomass losses in the effluent and waste sludge.

$$r_gV = Q_wX + (Q - Q_w)X_e \qquad (11\text{-}24)$$

Substituting the endogenous rate of growth from Eq. 3-25 in Section 3-13 for r_g, Eq. 11-24 becomes

$$\frac{1}{\theta_c} = Y\frac{r_{su}}{X} - k_d \qquad (11\text{-}25)$$

where Y = growth yield, milligrams per liter of biomass increase per milligram per liter of substrate utilized. Inserting U as the symbol for specific substrate utilization rate r_{su}/X,

$$\frac{1}{\theta_c} = YU - k_d \qquad (11\text{-}26)$$

Also, at steady state, the substrate utilization rate r_{su} and specific substrate utilization rate U are constant values defined by the following equations:

$$r_{su} = \frac{S_0 - S_e}{\theta} \qquad (11\text{-}27)$$

$$U = \frac{r_{su}}{X} = \frac{S_0 - S_e}{\theta X} \qquad (11\text{-}28)$$

where U = specific substrate utilization rate per day. Substituting the last term of Eq. 11-28 into Eq. 11-25 for r_{su}/X and V/Q for θ, solving for aeration tank volume yields

$$V = \frac{\theta_c YQ(S_0 - S_e)}{X(1 + k_d\theta_c)} \qquad (11\text{-}29)$$

The biological kinetics relationship given by Eq. 11-19 is now introduced to derive an equation incorporating the saturation constant K_S. By setting Eq. 11-27 equal to Eq. 11-19, dividing by X, inverting and linearizing, and

substituting in U as defined in Eq. 11-28, the resulting relationship is

$$\frac{1}{U} = \left(\frac{K_S}{k}\right)\left(\frac{1}{S_e}\right) + \frac{1}{k} \qquad (11\text{-}30)$$

Based on observed growth yield Y_{obs} (Eq. 3-26), the production of excess biomass in the waste activated sludge is calculated from Eq. 11-31.

$$P_x = \frac{YQ(S_0 - S_e)}{1 + \theta_c k_d} \qquad (11\text{-}31)$$

where P_x = volatile solids in the waste sludge, grams per day.

The assumed operating conditions of the aeration process in the derivation of these equations are as follows:

1. Complete mixing in the aeration tank.
2. Steady-state flows, biomass concentrations, and substrate concentrations.
3. Soluble substrates, filtered BOD or COD.
4. Excess activated sludge is wasted from the aeration tank, rather than the sludge recirculation line.
5. Substrate concentration in the aeration tank equals substrate concentration in the effluent.
6. Biological activity takes place only in the aeration tank.

Kinetic Constants

In order to apply the mathematical equations for completely mixed aeration, numerical values for the following constants are required:

Y = growth yield, milligrams of volatile suspended solids per milligram of soluble BOD or COD

k_d = biomass decay coefficient, per day

K_S = saturation constant, milligrams per liter of soluble BOD or COD

k = maximum rate of substrate utilization, (milligrams of soluble BOD or COD per milligram of volatile suspended solids) per day

A wastewater treatability study using a bench-scale activated-sludge unit is required to determine these constants, unless full-scale study data are available. The operating conditions in the laboratory are the same as those assumed in deriving the mathematical equations. To collect sufficient data, the experimental runs are conducted at several sludge ages in the range of 3 to 20 days under steady-state conditions including continuous uniform flow rates, complete mixing in the aeration tank, and constant substrate loading and sludge wasting. The resulting operating conditions should be constant θ, volatile MLSS, F/M, and θ_c. Temperature, pH, and dissolved oxygen concentration are held constant throughout the series of tests.

The operating time for each test period should be at least twice the sludge age to ensure a steady-state analysis. From the average of daily measurements of flow rates, soluble BOD or COD, and volatile suspended solids concentrations, the specific substrate utilization rate U is calculated by Eq. 11-28 and the sludge age θ_c is calculated by Eq. 11-23. Values for $1/\theta_c$ and U for each test are plotted as shown in Figure 11-41a. Equation 11-26 is the equation of a straight line drawn through these plotted points. The slope of the line is the growth yield Y, and the intercept with the vertical axis is the biomass decay coefficient k_d. Values of $1/U$ are then plotted against $1/S_e$ as shown in Figure 11-41b. Based on Eq. 11-30, the slope of a straight line drawn through the data is K_S/k and the intercept with the vertical axis is $1/k$.

Table 11-3. Typical Ranges for Kinetic Constants for Completely Mixed Activated-sludge Aeration Treating Municipal Wastewater at a Temperature of Approximately 20° C

Constant	Units of Expression	Range
Y	mg VSS/mg BOD	0.4–0.8
Y	mg VSS/mg COD	0.3–0.4
k_d	per day	0.04–0.08
K_S	mg/l of BOD	25–100
K_S	mg/l of COD	25–100
k	per day	4–8

The general ranges of magnitude for kinetic constants for completely mixed activated-sludge aeration are listed in Table 11-3.

Process Design

The aeration process designed using the kinetics model must be completely mixed activated sludge, since the application of Monod's growth kinetics in a batch culture to those of a continuous-flow system is valid only with complete mixing. Other ideal conditions assumed during derivation of the mathematical model must be compensated for in selecting design parameters. Steady-state

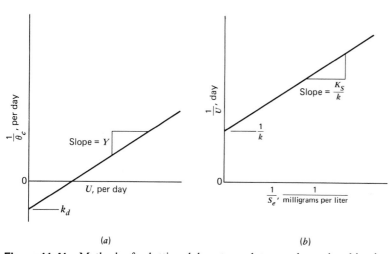

(a)   (b)

Figure 11-41 Method of plotting laboratory data to determine kinetic constants from a wastewater treatability study. (a) Graph of Eq. 11-26. (b) Graph of Eq. 11-30.

conditions do not exist in operation of an actual treatment plant; therefore, allowance must be made for diurnal and random variations in wastewater flow and loading. Since most wastewaters contain suspended solids, the oxygen demand of the wastewater is from both suspended and soluble organic matter, while the kinetics equations are founded on soluble BOD or COD. In order to correlate total to soluble oxygen demand, BOD or COD tests can be conducted on both unfiltered and filtered samples during the treatability analyses. Finally, sludge settleability must be considered in the selection of design parameters to ensure good separation of the biological suspended solids. Design criteria for the final clarifiers are very important.

The first step in aeration design is to select the desired concentration of effluent soluble BOD, based on the allowable total BOD. The criterion for aeration loading can be either selection of a sludge age (Eq. 11-23) or food-to-microorganism ratio (Eq. 11-22). Sizing an aeration tank based on either of these parameters requires selection of a design flow and operating volatile MLSS. Knowing these data, the aeration tank volume can be calculated by Eq. 11-29.

The design sludge age for a conventional loading rate, to produce an effluent of less than 30 mg/l of total BOD, is in the range of 5 to 15 days. The selection of an operating volatile MLSS concentration is based on settleability of the activated sludge as well as food-to-microorganism loading. A low design value results in an extended aeration period, while an excessive MLSS concentration results in poor gravity separation of the activated sludge in the final clarifier. In general, a conventional activated-sludge process operates in the MLSS range of 1500 to 3000 mg/l with 70 to 80 percent of the suspended solids being volatile. Therefore, a typical design value in a volatile MLSS concentration is between 1200 and 2400 mg/l.

EXAMPLE 11-13

The aeration tank for a completely mixed aeration process is being sized for a design wastewater flow of 2.0 mgd. The influent total BOD is 130 mg/l with a soluble BOD of 90 mg/l. The design effluent total BOD is 20 mg/l with a soluble BOD of 7.0 mg/l. Recommended design parameters are a sludge age of 10 days and volatile MLSS of 1400 mg/l. Selection of these values takes into account the anticipated variations in wastewater flows and strengths. The kinetic constants from a bench-scale treatability study are $Y = 0.60$ mg VSS/mg soluble BOD and $k_d = 0.06$ per day.

Solution

Using Eq. 11-29, the required aeration tank volume is

$$V = \frac{10 \times 0.60 \times 2.0(90 - 7)}{1400(1 + 0.06 \times 10)}$$

$$= 0.44 \text{ mil gal}$$

From Eq. 11-21,

$$\theta = \frac{0.44 \times 24}{2.0} = 5.3 \text{ hr}$$

From Eq. 11-22,

$$\text{F/M} = \frac{90}{(5.3/24)\,1400}$$

$$= 0.29 \frac{\text{lb soluble BOD/day}}{\text{lb volatile MLSS}}$$

Using Eq. 11-31, the excess biomass per day is

$$P_x = \frac{0.60 \times 2.0(90 - 7)\,8.34}{1 + 10 \times 0.06}$$

$$= 519 \text{ lb VSS/day}$$

11-9   STABILIZATION PONDS

Stabilization ponds, also called lagoons or oxidation ponds, are generally employed as secondary treatment in rural areas. Although they serve only about 7 percent of the population, thousands of lagoon installations exist in the United States with about 90 percent in communities of less than 10,000 population. Ponds are classified as facultative, tertiary, aerated, and anaerobic according to the type of biological activity that takes place in them. Each of these kinds of ponds is discussed in the following paragraphs.

Facultative Ponds

These are the most common lagoons employed for stabilizing municipal wastewater. The bacterial reactions include both aerobic and anaerobic decomposition and, hence, the term facultative pond. Figure 11-42 illustrates the basic biological activity. Waste organics in suspension are broken down by bacteria releasing nitrogen and phosphorus nutrients, and carbon dioxide. Algae use these inorganic compounds for growth, along with energy from sunlight, releasing oxygen to solution. Dissolved oxygen is in turn taken up by the bacteria, thus closing the symbiotic cycle. Oxygen is also introduced by reaeration through wind action. Settleable solids decomposed under anaerobic conditions on the bottom yield inorganic nutrients and odorous compounds, for instance, hydrogen sulfide and organic acids. The latter are generally oxidized in the aerobic surface water, thus preventing their emission to the atmosphere.

Bacterial decomposition and algal growth are both severely retarded by cold temperature. During winter when pond water is only a few degrees above freezing, the entering waste organics accumulate in the frigid water. Microbial activity is further reduced by ice and snow cover that prevents sunlight penetration and wind re-aeration. Under this environment, the water can become anaerobic causing odorous conditions during the spring thaw, until algae become reestablished. This may take several weeks depending on climatic conditions and the amount of waste organics accumulated during the cold weather.

Operating water depths range from 2 to 5 ft, with 3 ft of dike freeboard above the high water level (Figure 11-42). The minimum 2-ft depth is needed to prevent growth of rooted aquatic weeds, but exceeding a depth of 5 ft may create excessive odors because of anaerobiosis on the bottom.

Typical construction is two or more shallow pools with flat bottoms enclosed by earth dikes (Figure 11-43). Wastewater enters through an inlet division box, flows between cells by valved cross-connecting lines, and overflows through an outlet structure. Inlet lines, controlled by stop gates in the division box, discharge near the pond centers. Operating water depth is managed by a valving arrangement in the discharge structure. Connections between cells permit either parallel or series operation. Dikes are constructed with relatively flat side slopes to facilitate grass mowing and reduce slumping of the earth wall into the lagoon. Often the inside slopes along the waterline are protected by stone riprap to prevent erosion by wave action. If the soil is pervious, the pond bottoms should be sealed with bentonite clay, or lined with plastic, to prevent groundwater pollution. The area should be fenced to keep out livestock and discourage trespassing.

BOD loadings on stabilization ponds are expressed in terms of lb BOD applied per day per acre of water surface area (grams BOD per day per square meter), or sometimes as BOD equivalent population per acre. The maximum

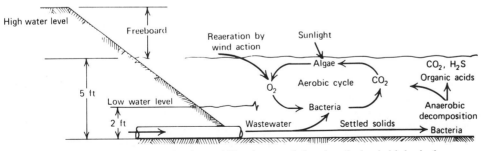

Figure 11-42 Schematic of a facultative stabilization pond showing the basic biological reactions of bacteria and algae. (Also refer to Figure 3-24.)

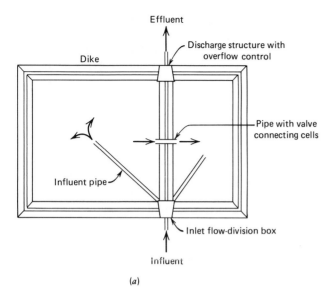

(a)

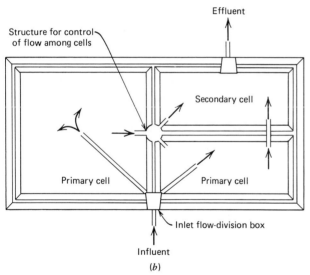

(b)

Figure 11-43 Typical stabilization pond arrangements. (a) Two-cell lagoon that can be operated in either parallel or series. (b) Three-cell lagoon with primary ponds, operated in parallel or series, followed by a secondary pond to provide additional storage capacity and controlled discharge.

allowable loading is about 20 lb BOD per acre per day ($2.2 \text{ g/m}^2 \cdot \text{d}$) in northern states to minimize odor nuisance in the spring of the year. In those climates where ice coverage does not prevail, higher organic loadings may be used; for example, in the South and Southwest a loading of 50 lb BOD per day per acre ($5.6 \text{ g/m}^2 \cdot \text{d}$) is practical. These loadings are 0.9 to 1.3 lb BOD per 1000 cu ft per day for a 5-ft water depth, which are substantially less than the volumetric loads applied to aeration and filtration units. Retention time of the wastewater in lagoons is 3 to 6 months depending on applied load, depth of wastewater, evaporation rate, and loss by seepage.

Lagoon treatment is regulated by the sequence of cell operation and water-level control. Series operation prevents short-circuiting and increases BOD reduction, but the load on the first cell is increased and may cause odor problems. Functioning in parallel distributes the raw BOD load but allows short-circuiting. Regulating lagoon discharge to a receiving stream can minimize water pollution. Pond water level may be slowly lowered in the fall and early winter when adequate dilution flow is in the receiving stream. Discharge in the winter is then minimized or stopped, and influent wastewater is stored until spring. Overflow is then permitted after the weather warms and the biological processes reduce the BOD concentration in the impounded water to an acceptable effluent level.

The two-cell lagoon in Figure 11-43a can be operated in either parallel or series. The flow pattern for series operation is shown by the arrows. In larger installations, a minimum of three cells is recommended with a secondary pond that cannot receive raw wastewater. The arrangement illustrated in Figure 11-43b allows parallel or series operation of the primary cells, which operate in series with the secondary cell. This flow pattern reduces short-circuiting. The secondary cell also provides additional storage capacity and permits isolation of the effluent wastewater before discharge. In sizing of ponds, the secondary cell is not included in BOD loading calculations; however, the volume is considered in calculating retention time of the wastewater. The water depth in a secondary pond is often increased to 8 ft, since odors are not as likely to be emitted from a cell that does not receive raw wastewater.

Facultative ponds treating only domestic wastewater normally operate odor free except for a short period of time in the spring of the year. On the other hand, lagoons treating municipal wastewaters that include industrial wastes can produce persistent obnoxious odors. Often this is the result of organic overload from food-processing industries or a result of the odorous nature of the industrial waste itself or both. The best solution is to require pretreatment of the offending wastewaters prior to discharge to the sewer system.

Stabilization ponds treating domestic wastewater produce an effluent with a BOD less than 30 mg/l during warm weather operation. Nevertheless, meeting an effluent standard of 30 mg/l of suspended solids is unlikely since algae suspended in the water generally contribute 50 to 70 mg/l. In some instances, the algal population can be reduced in lightly loaded ponds by series operation. If an effluent is unacceptable for discharge to a watercourse, an alternate may be land disposal by irrigation of agricultural crops near the lagoon site.

In summary, facultative ponds are best suited for small towns that do not anticipate industrial expansion and where extensive land area is available for construction and effluent disposal. The advantages are low initial cost and ease of operation as compared to a mechanical plant. The potential problems are poor assimilation of industrial wastewaters, emission of odors, and meeting the minimum effluent standards for discharge to surface waters.

EXAMPLE 11-14

Stabilization ponds for a town of 3000 population are constructed as is shown in Figure 11-43b with the larger cell having 14 acres and the two smaller cells each having 7 acres. The average daily wastewater flow is 0.24 mgd containing 450 lb of BOD; this is equivalent to 80 gpcd and 0.15 lb BOD per person per day, or 225 mg/l BOD. (a) Calculate the BOD loadings based on the total area of primary cells. (b) Estimate the number of days of winter storage available between the 2-ft and 5-ft water levels assuming an evaporation and seepage loss of 0.10 in. of water per day.

Solution

(a) BOD load on primary pond area is $450/21 = 21.4$ lb per day per acre. Equivalent population load is $21.4/0.17 = 125$ persons per acre.

(b) Storage volume:

$(5 \text{ ft} - 2 \text{ ft}) \times 28 \text{ ac} \times 43{,}560 \text{ sq ft per acre}$

$$= 3{,}660{,}000 \text{ cu ft} = 27.4 \text{ mil gal}$$

Water loss:

$$\frac{0.10 \text{ in./day}}{12 \text{ in./ft}} \times 28 \text{ ac} \times 43{,}560 \frac{\text{sq ft}}{\text{acre}} \times \frac{7.48 \text{ gal}}{\text{cu ft}}$$

$$= 76{,}000 \text{ gpd} = 0.076 \text{ mgd}$$

Storage time available:

$$\frac{27.4 \text{ mil gal}}{0.24 \text{ mgd} - 0.076 \text{ mgd}} = 170 \text{ days}$$

Tertiary Ponds

These units, also referred to as maturation or polishing ponds, serve as third-stage processing of effluent from activated sludge or trickling filter secondary treatment. Stabilization by retention and surface aeration reduces suspended solids, BOD, fecal microorganisms, and ammonia. The water depth is generally limited to 2 or 3 ft for mixing and sunlight penetration. BOD loads are less than 15 lb BOD per acre per day ($1.7 \text{ g/m}^2\cdot\text{d}$), and detention times are relatively short at 10 to 15 days.

Aerated Lagoons

Completely mixed aerated ponds, usually followed by facultative ponds, are used for first-stage treatment of high-strength municipal wastewaters and for pretreatment of industrial wastewaters. The basins are 10 to 12 ft deep and are aerated with pier-mounted or floating mechanical units (Figure 11-44). The aerators are designed to provide mixing for suspension of microbial floc and to supply dissolved oxygen. The biological process does not include algae, and organic stabilization depends on the mixed liquor that develops within the

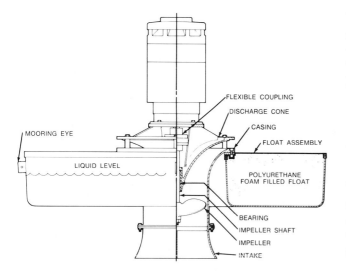

basin, since no provision is made for settling and returning activated sludge. BOD removal is a function of aeration period, temperature, and nature of the wastewater. The common mathematical relationship for reduction of BOD in a completely mixed aerated pond is

$$\frac{\text{Effluent BOD}}{\text{Influent BOD}} = \frac{1}{1 + kt} \qquad (11\text{-}32)$$

where k = reaction-rate constant to base e, per day
t = aeration period, days

The value of k depends primarily on the biodegradability of the waste organics and temperature of the wastewater. At $20°$ C, k values for different wastewaters occur in a wide range from 0.3 to over 1.0 per day; the value for a particular wastewater must be experimentally deter-

Figure 11-44   Cutaway section of a floating aerator and picture of a unit in operation. (Courtesy of Welles Products Corp.)

mined. Equation 3-17 in Section 3-12 relates the value of k to temperature. Design aeration periods are normally in the range of three to eight days depending on the degree of treatment desired and wastewater temperature during the cold season of the year. For example, aerating a typical municipal wastewater for five days at $20°C$ provides about 85 percent BOD reduction while lowering the temperature to $10°C$ reduces the efficiency to approximately 65 percent.

Problems of odors and low efficiency result when aerated lagoons are improperly designed or poorly operated. Thorough mixing and adequate dissolved oxygen ensure odor-free operation. If the aeration equipment is inadequate, deposition of solids and reduced oxygenation can result in anaerobic decomposition that leads to foul odors. Pretreatment and control of industrial wastes are required, since large inputs of either biodegradable or toxic wastes can cause process upset. Infiltration entering the sewer collection system during wet weather can have a detrimental effect on an aerated lagoon by reducing the aeration period and flushing microbial floc out of the basin. Where infiltration is a problem, diverting a portion of wet-weather flow around the aerated lagoon to the second-stage facultative ponds can prevent undesirable hydraulic loading on the aerated basins. In the winter, the aerators should be adjusted and windbreaks should be set up to reduce cooling of the lagoon water.

Anaerobic Lagoons

Bacteria anaerobically decompose organic matter to gaseous end products of carbon dioxide and methane. In addition, intermediate odorous compounds, such as organic acids and hydrogen sulfide, are formed. The two primary advantages of anaerobic treatment compared with an aerobic process are the low production of waste biological sludge and no need for aeration equipment. The disadvantages are that incomplete stabilization requires a second-stage aerobic process and a relatively high temperature for anaerobic decomposition. The important characteristics for a wastewater to be amenable to anaerobic treatment are high organic strength particularly in proteins and fats, relatively high temperature, freedom from toxic materials, and sufficient biological nutrients. Typical meat-processing wastewater with a BOD of 1400 mg/l, grease content of 500 mg/l, temperature of $82°F$, and neutral pH exhibits these characteristics.

A schematic diagram of a first-stage anaerobic lagoon for treatment of slaughterhouse wastewater in rural locations is given in Figure 11-45. Minimum pretreatment of the raw wastewater includes blood recovery for a salable by-product, screening to remove coarse solids (paunch manure), and skimming to reduce the grease content. The basins are constructed with steep side walls and a depth of 15 ft to minimize the surface area relative to total volume. This construction allows several inches of grease accumulation to form a natural cover for retaining heat, suppressing odors, and maintaining anaerobic conditions. Influent wastewater enters near the bottom so that it mixes with the active microbial solids in the sludge blanket. The discharge pipe is located on the opposite end and is submerged below the grease cover. Upward flow of the discharge allows settling of the bacterial floc so that the anaerobic mixed liquor is retained in the lagoon. Sludge recirculation is not necessary, since gasification and the inlet-outlet flow pattern provide adequate mixing. Series operation is not recommended because it is difficult

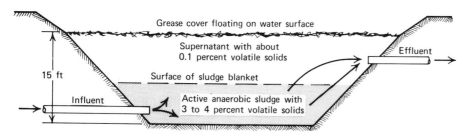

Figure 11-45 Schematic of an anaerobic lagoon for treating meat-processing wastewater.

to maintain an adequate grease cover on a second-stage lagoon.

The normal operating standards to achieve a BOD removal efficiency of at least 75 percent are a loading of 20 lb BOD per 1000 cu ft per day, a minimum detention time of four days, and a minimum operating temperature of 75° F. The most common operating problems result from reduced temperature in the liquid caused by an insufficient cover of grease for thermal insulation and protection from wind mixing. Too low a BOD loading and overly efficient pretreatment can result in inadequate grease input to build up and maintain a cover layer. Anaerobic lagoons do not create serious odor problems when operating properly, that is, when there is complete anaerobiosis and adequate grease cover. One exception is when the water supply to the industry is high in sulfate ion, which is reduced in the anaerobic environment and emitted as hydrogen sulfide.

Success of anaerobic lagoons in handling meat-processing wastes should not be translated to municipal applications. Domestic wastewater with its relatively low BOD and grease content and cool temperature is simply not suitable to develop the system in Figure 11-45. Even in instances where municipal wastewater has a high industrial waste content, application of first-stage anaerobic treatment has led to serious odor problems. This can generally be attributed to the fact that the lagoons are facultative cells instead of being strictly anaerobic.

11-10   DISINFECTION

Although other bactericidal agents may be used to disinfect wastewater, chlorine is the only one that has widespread application. The purpose for chlorinating wastewater effluents is to protect public health by inactivating pathogenic organisms including enteric bacteria, viruses, and protozoans. Satisfactory disinfection of a secondary effluent is defined by an average fecal coliform count of less than 200/100 ml. The number of coliform organisms in effluents from trickling filters, activated-sludge systems, and stabilization ponds is in the range of 100,000 to 10,000,000/100 ml. Safety of effluent disposal by dilution in surface waters after chlorination is based on the argument that reduction of fecal coliforms from $10^6/100$ ml to 200/100 ml eliminates the great majority of bacterial pathogens and inactivates large numbers of enteric viruses.

A properly designed system provides rapid initial mixing of chlorine solution with the wastewater, contact time in a plug-flow basin for at least 30 min at peak flow, and automatic chlorine residual control. Chlorine solution should be applied either in a pressure conduit under conditions of highly turbulent flow, or in a channel immediately upstream from a mechanical mixer. Adding the solution to an open channel results in very little dispersion, and in poor chlorination efficiency because the flow is usually stratified. Following initial blending, the most efficient tank for disinfection is a long narrow chamber that approaches plug-flow conditions. Circular and rectangular tanks that allow backmixing are not as effective, and the design criterion of detention time has little meaning.

Precise measurements and control of chlorine residual are extremely important for efficient and economical operation. An automatic residual monitoring and feedback control is essential to prevent low concentrations and inadequate disinfection, as well as excessive chlorination resulting in discharge of an effluent toxic to aquatic life. Existing installations may require structural modifications to ensure proper disinfection without discharging wastewater with harmful chlorine residuals. Some regulatory agencies have specified maximum chlorine residuals in undiluted effluent of 0.1 to 0.5 mg/l to prevent potential toxicity in the receiving stream. Dechlorination may be required to detoxify the discharge after disinfection has been achieved. The least expensive and most effective method is by adding sulfur dioxide; for small flows, sodium metabisulfite may be considered as an alternate.

Chlorine dosage needed for disinfection depends on wastewater pH, presence of interfering substances, temperature, and contact time. Applications of 8 to 15 mg/l provide adequate disinfection in well-designed units with a minimum contact time of 20 to 30 min. (Chlorine chemistry and chlorination equipment are discussed in Section 7-6.)

11-11   INDIVIDUAL HOUSEHOLD DISPOSAL SYSTEMS

Approximately one fourth of the homes in the United States are located in unsewered areas and must rely on individual treatment. The quantity of wastewater is from 60 to 100 gpcd with BOD concentrations varying from 200 to 400 mg/l. The number of bathrooms, automatic washing machines, and garbage grinders, as well as the number of residents, define the magnitude and character of the wastewater.

The most popular system is the septic tank and absorption field because of its low cost and the desirable characteristic of underground disposal of effluent (Figure 11-46). A septic tank is an underground concrete box sized for a detention time of approximately two days. With garbage grinders and automatic washers, the recommended minimum capacity is 750 gal (2.8 m^3) for a two-bedroom house, 900 gal for three bedrooms, 1000 for four bedrooms, and 250 for each additional bedroom. Inspection and cleaning ports must be accessible for maintenance, usually by removing about 1 ft of earth cover. Inlet and outlet pipe tees prevent clogging of the drains with scum that accumulates on the liquid surface. The functions of a septic tank are settling of solids, flotation of grease, anaerobic decomposition of accumulated organic matter, and storage of sludge. Retaining large solids is essential to prevent plugging of the percolation field.

An absorption field, where the majority of the biological stabilization takes place, consists of looped or lateral trenches 18 to 24 in. wide and at least 18 in. deep. Drain tile or perforated pipe in an envelope of gravel are used to distribute the wastewater uniformly over the trench bottom. Organics are decomposed in the aerobic-facultative environment of the bed, and water seeps down into the soil profile. Air enters the drain tile through the backfill covering of the trenches and by ventilation through the house plumbing stack. The percolation area required depends directly on soil permeability, and for a four-bedroom dwelling the area needed ranges from 300 to 1300 sq ft. Trench area for a particular location is determined by subsurface soil exploration and percolation tests. Most state environmental control agencies and county health departments have guidelines for installations based on local conditions.

The most frequent complaints in operation of septic tank-absorption field systems concern plumbing stoppages and odorous seepage through the ground surface. Tanks fill with nonbiodegradable solids and must be pumped out every few years. When cleaning them, leaving a small amount of the black-colored digesting sludge in the tank ensures adequate bacterial seeding to continue solids digestion. Introduction of chemical or enzymatic conditioners have not been shown to be of any significant value in reviving or increasing bacterial activity in a septic tank. Attempting to delay cleaning by adding a large quantity of caustic chemical can be

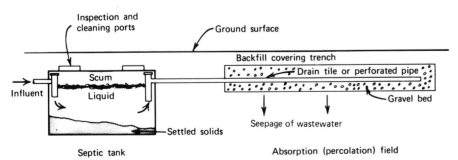

Figure 11-46   A typical septic tank–absorption field system for disposal of household wastewater. The tank provides for sedimentation, scum accumulation, anaerobic destruction of organic matter, and sludge storage. The absorption field permits aerobic-facultative decomposition of organics and seepage of the wastewater.

harmful to the disposal system. It disrupts bacterial decomposition in the tank and "cleans" the unit by discharging suspended solids into the drain tile, which promotes clogging of the percolation field.

Proper functioning of the absorption field relies on leaching of the wastewater into the soil profile and adequate aeration of the bed. Obstruction of percolation may result for any of the following reasons: construction in clayey soils of low permeability that are simply inadequate for placement of a seepage system, high water table conditions that saturate the soil profile during the wet-weather season, inadequate design resulting in hydraulic and organic overloading, and improper operation of the septic tank. In attempting to solve the problem of

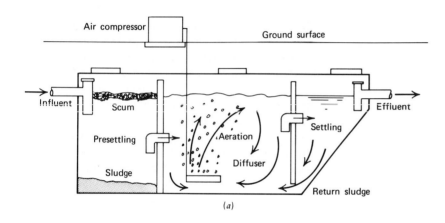

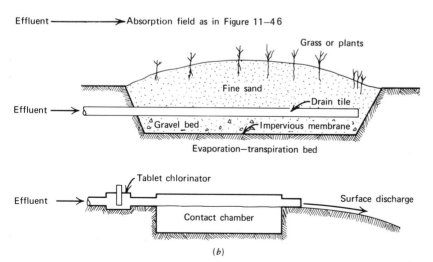

Figure 11-47 Schematics of an aeration system for treating wastewater from an individual household and alternatives suggested for disposal of the process effluent. (*a*) Compartmented aeration tank. (*b*) Alternative methods proposed for disposal of effluent from the aeration unit above.

leaching fields in unsuitable soils, small aeration treatment units have been developed in recent years.

Individual household aeration units are manufactured in many different shapes and sizes using concrete, steel, or fiber glass tanks. The cross section of a typical compartmented aeration tank is shown in Figure 11-47a. Some designs have a presedimentation chamber that performs similarly to a septic tank. The aeration portion is sized for a 24- to 48-hr retention period, and is mixed with either compressed air or some type of mechanical aerator. A final settling chamber is for separation and return of activated sludge. Although these systems have the appearance of an extended aeration process, in reality, they seem to function more like an aerated stabilization pond. The main reason is that gravity return of solids is not effective in maintaining an adequate mixed liquor solids concentration, particularly with the surging influent caused by high rates of discharge from household appliances. Consequently, the aeration unit is not efficient in solids capture or retention, which leads to inefficient treatment and a poor-quality effluent.

Several methods have been proposed for the disposal of aeration effluent (Figure 11-47b). An underground absorption field is acceptable, but then there is no purpose in installing a more expensive aeration tank when an ordinary septic tank produces an equally suitable effluent for leaching. Evaporation-transpiration beds may be satisfactory with proper climatic conditions; however, the bed must not be subjected to freezing. Furthermore, evapotranspiration must be predictable, and salt buildup is always a potential problem.

The main motivation for development of a miniature aeration system was an attempt to permit surface effluent discharge and thus to avoid the problems associated with absorption fields. But this raises the question of disinfection to meet the required microbiological quality. The simplest system is a chlorine contact chamber equipped with a tablet-type chlorinator. One such unit consists of a slotted tube containing solid tablets of calcium hypochlorite suspended in the wastewater flow such that a suitable amount of chlorine is dissolved for disinfection. Control problems include potential overchlorination during low flow periods and underchlorinating at peak flows, and the possibility that homeowners may not be conscientious in replacing hypochlorite tablets because of the inconvenience and cost. Also, a poorly operating aeration tank discharges excessive suspended solids that prevent effective disinfection. Little evidence exists to substantiate the idea that effluent standards of 30 mg/l BOD and 30 mg/l suspended solids can be met by small compartmented aeration tanks.

In summary, septic tanks are economical and practical for the disposal of household wastewaters in areas with acceptable soils for percolation. Where leaching fields are impossible, rather extensive treatment is required to meet established quality standards for surface discharge. Few of the proposed aeration-chlorination systems appear to provide realistic performance. The labor and cost of operating, maintaining, and monitoring to ensure satisfactory performance may approach a level of economic infeasibility.

11-12   CHARACTERISTICS AND QUANTITIES OF WASTE SLUDGES

The purpose of wastewater settling and biological aeration is to remove organic matter and to concentrate it in a much smaller volume for ease of handling and disposal. As illustrated in Figure 11-1, the solids in 100 gal of wastewater are consolidated into approximately 0.5 gal of liquid sludge; this amounts to about 5000 gal of slurry per 1.0 mil gal of wastewater treated. The cost of facilities for stabilizing, dewatering, and disposal of this concentrate is about one third of the total investment in a treatment plant. Operating expenses in sludge handling may amount to an even larger fraction of the total plant operating costs depending on the system employed. For these reasons, a properly designed and efficiently run sludge disposal system is essential.

The quantity and nature of sludge generated relate to the character of the raw wastewater and processing units employed. Primary settling produces an anaerobic sludge of raw organics that are being actively decomposed by bacteria. Therefore, these solids must be handled properly to prevent emission of obnoxious odors. In comparison with secondary biological waste, primary sludges thicken and dewater readily because of their fiberous and coarse nature. The following formula can be used to estimate the raw solids that are removed by plain sedimentation.

$$W_p = f \times SS \qquad (11\text{-}33)$$

where W_p = raw primary sludge solids, pounds of dry weight per day (grams per day)

f = fraction of suspended solids removed in primary settling (f is about 0.5 for domestic wastewater)

SS = suspended solids in unsettled wastewater, pounds per day (grams per day)

= concentration of suspended solids in unsettled wastewater in mg/l × wastewater flow in mil gal/day × 8.34 lb/mil gal per mg/l (concentration of suspended solids in mg/l × flow in m³/d)

Waste from aeration is flocculated microbial growths with entrained nonbiodegradable suspended and col- loidal solids. It is relatively odor-free because of biological oxidation, but the finely divided and dispersed particles make it difficult to dewater. Excess activated-sludge solids from aeration processes and humus from biological filtration can be estimated by Eq. 11-34, which relates solids production to BOD load. Coefficient K depends on the process food-to-microorganism ratio shown in Figure 11-48. Although this formulation is reasonable for domestic wastewater, calculated values may differ considerably from real sludge yields when treating municipal discharge that contains a substantial portion of industrial waste.

$$W_s = K \times BOD \qquad (11\text{-}34)$$

where W_s = biological sludge solids, pounds of dry weight per day (grams per day)

K = fraction of applied BOD that appears as excess biological solids from Figure 11-48 assuming about 30 mg/l of suspended solids in the process effluent

BOD = BOD in applied wastewater after primary sedimentation, pounds per day (grams per day)

= concentration of BOD in applied wastewater in mg/l × wastewater flow in mil gal/day × 8.34 lb/mil gal per mg/l (concentration of BOD in mg/l × flow in m³/d)

The total sludge solids production in a conventional treatment plant with primary sedimentation and secondary aeration is equal to the sum of the values calculated by Eqs. 11-33 and 11-34.

$$W_{ps} = W_p + W_s \qquad (11\text{-}35)$$

where W_{ps} = total sludge solids from primary sedimentation and secondary biological aeration.

The sludge solids production for an activated-sludge system treating unsettled wastewater, for example, extended aeration without primary sedimentation, will be less than W_{ps} from Eq. 11-35 but more than W_s from Eq. 11-34. The quantity of solids produced can be estimated based on the influent BOD of the unsettled waste-

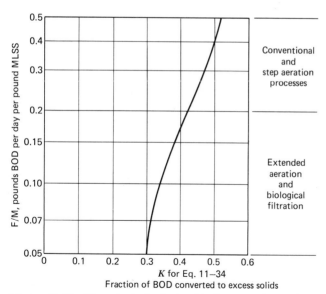

Figure 11-48 Hypothetical relationship between the food-to-microorganism ratio and coefficient K in Eq. 11-34 for calculating excess activated-sludge production in biological aeration of wastewater, based on an effluent suspended solids of approximately 30 mg/l.

water, disregarding the suspended solids, by increasing the value determined from Eq. 11-48 by 100 percent. Thus, for aeration systems without primary clarifiers,

$$W_{as} = 2.0K \times \text{BOD} \qquad (11\text{-}36)$$

where W_{as} = biological sludge solids from activated-sludge processing without primary sedimentation

K = value from Figure 11-48

BOD = BOD in the applied unsettled wastewater, pounds per day (grams per day)

Design and operation of a sludge disposal system are based on volume of the wet sludge as well as the dry solids content. Once the dry weight of solids has been determined, the volume of wet sludge can be calculated using Eq. 11-37 by knowing the percentage of solids, or water content. This formula assumes a specific gravity for the wet sludge of 1.0, which is sufficiently accurate for normal computations. For example, a slurry with 10 percent organics has a specific gravity of about 1.02.

$$V = \frac{W}{\left(\dfrac{s}{100}\right)8.34} = \frac{W}{\left(\dfrac{100-p}{100}\right)8.34} \qquad (11\text{-}37)$$

where V = volume of wet sludge, gallons

W = weight of dry solids, pounds

s = solids content, percent

p = water content, percent

Equation 11-37 in metric units is

$$V = \frac{W}{\left(\dfrac{s}{100}\right)} = \frac{W}{\left(\dfrac{100-p}{100}\right)} \qquad \text{(SI units)} \quad (11\text{-}38)$$

where V = volume of wet sludge, liters

W = weight of dry solids, kilograms

Typical solids concentrations in waste sludges are raw primary sludge only, 6 to 8 percent; primary plus humus from secondary filtration, 4 to 6 percent; primary plus thickened waste activated sludge from secondary aeration, 4 to 5 percent; and unthickened waste activated sludge only, 0.5 to 2.0 percent. In all of these wastes, the

percentage of volatile solids is about 70 percent of the total dry solids. For a given quantity of dry solids, the percentage concentration dramatically influences the sludge volume. Doubling the solids concentration reduces the wet volume by one half; conversely, the amount would double if the slurry were diluted to one half the original percentage of solids. For example, using Eq. 11-37, the volume of wet sludge containing 10 lb of dry solids at 2 percent concentration is 60 gal. If thickened to 4 percent solids, the volume becomes 30 gal, and a further concentration to 8 percent reduces the quantity to 15 gal. With this concept in mind, it is easier to understand why dilute secondary sludges are more difficult to process and handle than primary wastes.

EXAMPLE 11-15

A domestic wastewater with 200 mg/l BOD and 220 mg/l of suspended solids is processed in a trickling filter plant. (a) Using Eqs. 11-33 and 11-34, estimate the quantities of sludge solids from primary settling and secondary biological filtration per million gallons of wastewater treated. (b) How many pounds of dry solids are produced relative to the BOD equivalent population load on the plant? (c) What is the volume of wet sludge per million gallons of wastewater if underflow from the final clarifiers returns filter humus to the plant influent, and the combined sludge drawn from the primaries has a solids content of 5.0 percent?

Solution

(a) Assuming 50 percent suspended solids reduction from settling the raw wastewater, and substituting into Eq. 11-33,

$$W_p = 0.50 \times 220 \times 1.0 \times 8.34 = 916 \ \frac{\text{lb dry solids}}{\text{mil gal}}$$

From Eq. 11-34, assuming 35 percent BOD removal in the primary and an estimated K of 0.34 from Figure 11-48,

$$W_s = 0.34 \times 0.65 \times 200 \times 1.0 \times 8.34$$

$$= 369 \ \frac{\text{lb dry solids}}{\text{mil gal}}$$

(b) For 1.0 mgd with a BOD of 200 mg/l,

$$\frac{\text{BOD equivalent}}{\text{population}} = \frac{1.0 \times 200 \times 8.34}{0.20} = 8340 \text{ persons}$$

$$\frac{\text{Sludge production}}{\text{per capita}} = \frac{W_p + W_s}{\text{population}} = \frac{916 + 369}{8340}$$

$$= 0.15 \frac{\text{lb dry solids}}{\text{person}}$$

(*Note.* The value of 0.20 lb of dry solids per person per day has often been used as a conservative estimate of solids yield in trickling filter plants treating domestic wastewater).

(c) Substituting into Eq. 11-37 with $W_p + W_s$,

$$V = \frac{916 + 369}{\left(\dfrac{5.0}{100}\right)8.34} = 3080 \frac{\text{gal of sludge}}{\text{mil gal of wastewater}}$$

11-13 SLUDGE PUMPS

High-viscosity sludges are primary and mixtures of primary and waste activated sludges, scum from clarifiers, and raw and digested sludges that have been thickened by gravity, flotation, or centrifugation. Because of their high suction capacity, positive-displacement pumps are used to move these dense sludges. High-viscosity slurries will not flow to a pump. Another important feature of positive-displacement pumps is the absolute control of pumping rate. Positive displacement allows metering of the quantity of sludge pumped with respect to time. Also, heavy solids can be moved at a slow rate without clogging the pump chambers.

Reciprocating plunger pumps are commonly installed to pump dense sanitary sludges. Plunger pumps can handle heavy solids with reduced probability of internal damage because of the large clearances through the pump combined with the relatively slow operating speed. As with other positive-displacement pumps, the main advantages are self-priming and positive metering. As shown in Figure 11-49a, a rotating eccentric drive moves the plunger up and down in a cyclindrical body. The eccentric drive bearing is turned by an electric motor through a belt

drive and reduction gearing. A shear pin is located in the eccentric drive to disengage the plunger and prevent damage in case of a stoppage in the pump. Ball valve chambers are installed in the inlet and outlet pipes to

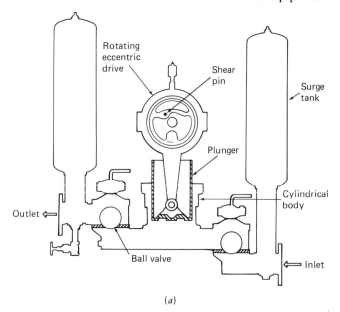

(*a*)

(*b*)

Figure 11-49 Positive-displacement plunger pump for pumping heavy, viscous sludges. (*a*) Cross section of a plunger pump. (*b*) Picture of a triplex pump. (Courtesy of ITT Marlow Pumps.)

prevent backflow, and air chambers are provided for surge control.

The pumping capacity of a plunger pump is varied by adjusting stroke length and speed of rotation. Where continuous operation is not necessary, the capacity can also be varied by intermittent operation. In general, for least wear on a pump, it is operated at the slowest practical speed and a long stroke. For most applications, the pulsating discharge of a reciprocating pump is not a problem. The exception is when the sludge must be fed at a uniform rate for in-line chemical conditioning prior to mechanical dewatering. Plunger pumps are available as simplex, duplex, triplex (Figure 11-49*b*), and quadruplex units with different volume plunger chambers and discharge heads. Pumping capacity increases with the number of chambers.

The progressing cavity pump is a positive-displacement pump that discharges at a uniform rate of flow, Figure 11-50*a*. The chamber is an abrasion-resistant rubber stator with a double-helix opening, and the moving element is a steel single-helix rotor driven by an electric motor. When the single-helix rotor revolves within the double-helix stator, cavities are formed that fill with sludge and progress toward the outlet end. The progressing cavity concept is illustrated in Figure 11-50*b*. Pumping capacity is approximately proportional to the speed of rotation.

Low-viscosity sludges have hydraulic characteristics similar to wastewater. These include waste activated sludge and unthickened digested sludge. Centrifugal pumps are preferred for low-viscosity slurries because of their greater efficiency, higher capacity, ease of maintenance, and low cost. Screw pumps may be used to lift return activated sludge, and air-lift pumps can serve the same purpose in small treatment plants.

11-14   SELECTION AND ARRANGEMENT OF PROCESSES

Techniques for processing waste sludges depend on the type, size, and location of the wastewater plant, unit operations employed in treatment, and the method of ultimate solids disposal. The system selected must be able to receive the sludge produced and economically to

Table 11-4. Common Methods for Handling, Processing, and Disposing of Waste Sludge

Storage Prior to Processing
    In the primary clarifiers
    Separate holding tanks
Thickening Prior to Dewatering or Digestion
    Gravity settling
    Dissolved air flotation
Conditioning Prior to Dewatering
    Chemical treatment
    Stabilization by anaerobic digestion
    Stabilization by aeration
Dewatering
    Vacuum filtration
    Pressure filtration
    Centrifugation
    Composting
    Drying beds or lagoons
Solids Disposal
    Burial in landfill
    Incineration
    Spreading on farmland
    Production of soil conditioner

convert it to a product that is environmentally acceptable for disposal. Popular methods are listed in Table 11-4.

Cost of chemicals for sludge conditioning and optimum use of biological digesters is directly related to the concentration of solids in waste sludge. As a general rule, the solids content must be at least 4 percent for feasible dewatering. Secondary biological sludge from aeration processes is thin and dilutes the more concentrated primary waste, and blending the two is likely to produce a sludge with less than 4 percent solids. Therefore, gravity thickening may be applied to mixtures of primary and secondary sludges, and dissolved air flotation may be employed for concentrating waste activated sludge.

Typical unit operations selected for processing sludge from filtration plants are diagramed in Figure 11-51. Humus from intermediate or final clarifiers is returned with recirculated flow to the head of the plant for separation in the primary. Underflow from the clarifiers may be gravity thickened, although this is not common,

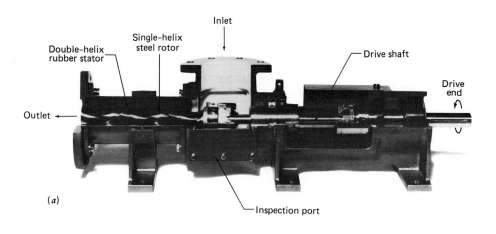

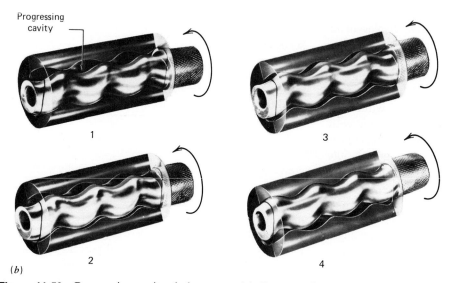

Figure 11-50 Progressing cavity sludge pump. (*a*) Cutaway view of pump. (*b*) Forward movement of progressing cavity with rotation of rotor. (Courtesy of Moyno Products, Robbins & Myers, Inc.)

since the solids concentration averages an acceptable 5 percent. Anaerobic digestion is a popular method for stabilizing raw sludge. Liquid digested sludge may be applied to farmland or may be dried on sand beds from which the solids are later removed for burial. In larger installations, raw solids may be trucked to landfill after

chemical treatment and dewatering by vacuum filtration or belt filter pressing.

The scheme for filter plants is not optimum for activated-sludge systems. Most aeration plants designed for returning secondary waste sludge to the primary tank have operating problems that relate both to upset of

Waste-Water Processing

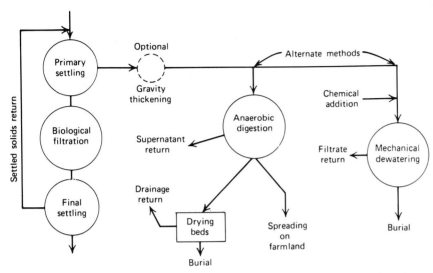

Figure 11-51 Typical unit operations for processing waste sludge from filtration plants.

primary clarification, as a result of biological activity in the settled sludge, and high chemical costs resulting from dewatering thin sludge. The waste activated sludge is often thickened separately by air flotation and then blended with the primary sludge, as shown in Figure 11-52, before processing. Alternately, some systems blend the waste activated and primary sludges for gravity thickening.

EXAMPLE 11-16

A conventional aeration plant treats 7.7 mgd of municipal wastewater with a BOD of 240 mg/l and suspended solids of 200 mg/l. The sludge flow pattern is shown in Figure 11-52 with a flotation thickener to concentrate the excess activated sludge, and a holding tank to blend primary and secondary wastes. Estimate the quantities and solids contents of the primary, secondary, and mixed sludges. Assume the following: 50 percent SS removal and 35 percent BOD reduction in the primary, a raw sludge with 94.0 percent water content, an operating F/M ratio of one third in the aeration basin, a solids concentration of 15,000 mg/l (1.5%) in the waste activated, and 3.5 percent solids in flotation thickener discharge.

Solution

Applying Eqs. 11-33 and 11-37 for the primary sludge,

$$W_p = 0.50 \times 200 \text{ mg/l} \times 7.7 \text{ mgd} \times 8.34$$
$$= 6410 \text{ lb/day}$$

$$V_p = \frac{6410}{\left(\dfrac{100 - 94}{100}\right)8.34} = 12,800 \text{ gal/day}$$

From Figure 11-48, for an F/M ratio of 1/3, the K is 0.48. Substituting in Eqs. 11-34 and 11-37,

$$W_s = 0.48 \times 0.65 \times 240 \text{ mg/l} \times 7.7 \text{ mgd} \times 8.34$$
$$= 4800 \text{ lb/day}$$

$$V_s = \frac{4800}{(1.5/100)8.34} = 38,400 \text{ gal/day}$$

After flotation thickening, assuming 100 percent solids capture, the waste activated sludge volume is

$$V_s = \frac{4800}{(3.5/100)8.34} = 16,400 \text{ gal/day}$$

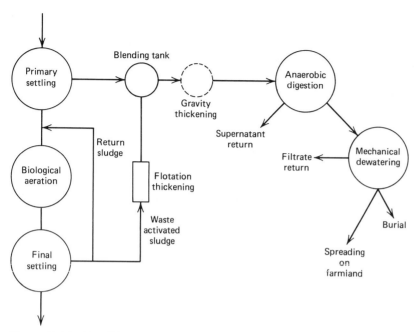

Figure 11-52 Possible alternatives to handling waste sludges from activated-sludge treatment.

The blended sludge, after mixing primary and thickened secondary, has the following characteristics.

$$W_{ps} = W_p + W_s = 6410 + 4800 = 11,200 \text{ lb/day}$$

$$V_{ps} = V_p + V_s = 12,800 + 16,400 = 29,200 \text{ gal/day}$$

$$s = \frac{100 \times W_{ps}}{8.34 \times V_{ps}} = \frac{100 \times 11,200}{8.34 \times 29,200} = 4.6 \text{ percent}$$

11-15 THICKENING OF WASTE SLUDGES

Gravity thickening is performed in circular settling tanks that are equipped with scraper arms having vertical pickets. The process is described in Section 7-17 and a typical unit is illustrated in Figure 7-35. Sludge withdrawn from primary clarifiers or sludge blending tanks is applied to the gravity thickener through a central inlet well. Overflow containing the nonsettleable fraction is returned to the wet well for reprocessing, while the concentrate is drawn from the tank bottom for processing and disposal. Unit loadings are normally in the range of 6 to

12 lb of solids per square foot of tank bottom per day (30 to 60 kg/m^2·d). Treated wastewater is mixed with the sludge entering the thickener to enhance settling by washing out fine suspended solids and to reduce odors by maintaining a moving water layer over the top of the blanket of consolidating solids in the tank. With the added wastewater, tank overflow rates are in the range of 400 to 800 gpd/sq ft. In handling domestic wastes, the underflow solids concentration is often double that of the applied sludge. For example, expected underflow solids content would be 6 percent for a mixture of primary and activated sludge, and 8 percent for primary plus filter humus. Precise performance of a thickener is difficult to predict because of the varying character of waste sludges. Depending on the characteristics of the sludge, solids capture varies from about 80 to 95 percent. Coagulant chemicals may be used to improve capture of suspended solids and density of the thickened waste.

Dissolved air flotation is achieved by releasing fine air bubbles that attach to sludge particles and cause them to

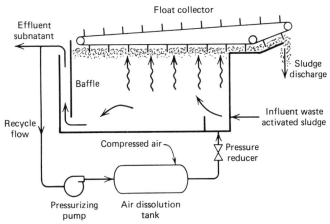

Figure 11-53 Schematic diagram of a dissolved air flotation tank for thickening of waste activated sludge.

float. Figure 11-53 is a flow diagram for a typical unit. Waste activated sludge enters the bottom of the flotation tank where it is merged with recirculated flow that contains compressed air. The latter is a portion of the clarified effluent that has been held in a separate retention tank under an air pressure of approximately 60 psi. On pressure release, air dissolved in the recirculated-pressurized flow forms fine bubbles that agglomerate with the suspended solids. Process underflow is returned to wastewater treatment, and the overflow, discharged by a mechanical skimming device, is the thickened sludge.

Flotation thickeners are used for waste activated sludge and normally produce a concentrate of approximately 4 percent solids with a solids recovery of 85 percent. Polyelectrolytes or other flocculents may be added to increase capture of solids to 95 percent or more (Figure 11-54). Chemical treatment, however, may not increase the concentration of the thickened sludge. Practical loading rates for waste activated sludge are in the range of 2 to 4 lb solids per square foot (10 to 20 kg/m²·d) of flotation area per hour. Since the characteristics of biological floc are not consistent, even in the same treatment plant, the response to chemical flotation aids and solids loadings must be determined in each individual case.

EXAMPLE 11-17

Primary sludge containing trickling filter humus is gravity thickened in a circular tank with a diameter of 12.0 ft and side water depth of 10.0 ft. The applied sludge is 2600 gpd with 4.5 percent solids, and the thickened sludge withdrawn is 1300 gpd at 7.5 percent solids. The blanket of consolidating sludge in the tank has a depth of 3.0 ft. For odor control and to enhance thickening, 45,000 gpd of treated wastewater is pumped to the tank along with the sludge to increase the overflow rate. Calculate the following: the solids loading on the tank bottom, the solids capture, the overflow rate, and the estimated time of solids retention in the tank.

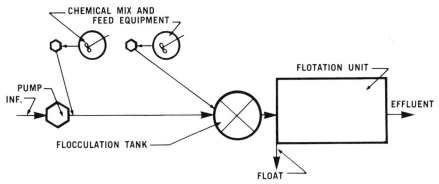

Figure 11-54 Flow diagram for chemical pretreatment of waste sludge prior to flotation thickening. (Courtesy of Komline-Sanderson, Peapack, New Jersey.)

Solution

$$\text{Tank area} = \pi(12.0 \text{ ft})^2/4 = 113 \text{ sq ft}$$

$$\text{Solids loading} = \frac{2600 \text{ gpd} \times 0.045 \times 8.34}{113 \text{ sq ft}}$$

$$= 8.6 \text{ lb/sq ft}$$

$$\text{Solids capture} = \frac{1300 \text{ gpd} \times 0.075 \times 8.34}{2600 \text{ gpd} \times 0.045 \times 8.34}$$

$$= \frac{813 \text{ lb/day}}{976 \text{ lb/day}} = 0.83 = 83\%$$

$$\text{Overflow rate} = \frac{(45{,}000 + 2600 - 1300) \text{ gpd}}{113 \text{ sq ft}}$$

$$= 410 \text{ gpd/sq ft}$$

Estimated solids in tank

$$= 3.0 \text{ ft} \times 113 \text{ sq ft} \times 0.075 \times 62.4$$

$$= 1590 \text{ lb}$$

$$\text{Solids retention time} = \frac{1590 \text{ lb} \times 24 \text{ hr/day}}{813 \text{ lb/day}}$$

$$= 47 \text{ hr}$$

11-16 ANAEROBIC DIGESTION

The purpose of sludge digestion is to convert bulky, odorous, raw sludge to a relatively inert material that can be rapidly dewatered with the absence of obnoxious odors. The bacterial process, as summarized in Eq. 11-39, consists of two successive processes that occur simultaneously in digesting sludge. The first stage consists of breaking down large organic compounds and converting them to organic acids along with gaseous by-products of carbon dioxide, methane, and trace amounts of hydrogen sulfide. This step is performed by a variety of facultative bacteria operating in an environment devoid of oxygen. If the process were to stop there, the accumulated acids would lower the pH and would inhibit further decomposition by "pickling" the remaining raw wastes. In order for digestion to occur, second-stage gasification is needed to convert the organic acids to methane and carbon dioxide. Acid-splitting methane-forming bacteria are strict anaerobes and are very sensitive to environmental conditions of temperature, pH, and anaerobiosis. In addition, methane bacteria have a slower growth rate than the acid formers, and are very specific in food supply requirements. For example, each species is restricted to the metabolism of only a few compounds, mainly alcohols and organic acids, while carbohydrates, fats, and proteins are not available as energy sources.

$$(11\text{-}39)$$

Stability of the digestion process relies on proper balance of the two biological stages. Buildup of organic acids may result from either a sudden increase in organic loading, or a sharp rise in operating temperature. In either case, the supply of organic acids exceeds the assimilative capacity of the methane-forming bacteria. This unbalance results in decreased gas production and eventual drop of pH, unless the organic loading is reduced to allow recovery of the second-stage reaction. Accumulation of toxic substances from industrial wastes, such as heavy metals, may also inhibit the digestion reaction. The exact cause of digester problems is often difficult to determine. Monitoring volatile solids loading, total gas production, volatile acids concentration in the digesting sludge, and percentage of carbon dioxide in the head gases are the methods most frequently employed to give advance warning of pending failure. These measurements can also indicate the most probable cause of difficulties. Gas production should vary in proportion to organic loading. Volatile acids content is normally stable at a given loading rate and operating temperature. The percentage of carbon dioxide should also remain relatively constant. Monitoring digestion by pH measurements is not recommended, since a drop in pH does not herald failure but announces that it has occurred. Table 11-5 lists the general operating and loading conditions for anaerobic digestion.

Table 11-5. General Operating and Loading Conditions for Anaerobic Sludge Digestion

Temperature	
Optimum	98° F (35° C)
General operating range	85 to 95° F
pH	
Optimum	7.0 to 7.1
General limits	6.7 to 7.4
Gas Production	
Per pound of volatile solids added	8 to 12 cu ft
Per pound of volatile solids destroyed	16 to 18 cu ft
Gas Composition	
Methane	65 to 69 percent
Carbon dioxide	31 to 35 percent
Hydrogen sulfide	trace
Volatile Acids Concentration	
General operating range	200 to 800 mg/l
Alkalinity Concentration	
Normal operation	2000 to 3500 mg/l
Volatile Solids Loading	
Conventional single stage	0.02 to 0.05 lb VS/cu ft/day[a]
First-stage high rate	0.05 to 0.15 lb VS/cu ft/day
Volatile Solids Reduction	
Conventional single stage	50 to 70 percent
First-stage high rate	50 percent
Solids Retention Time	
Conventional single stage	30 to 90 days
First-stage high rate	10 to 15 days
Digester Capacity Based on Design	
Equivalent Population	
Conventional single stage	4 to 6 cu ft/PE[b]
First-stage high rate	0.7 to 1.5 cu ft/PE

[a] 1.0 lb/cu ft/day = 16,000 g/m^3·d
[b] 1.0 cu ft = 0.0283 m^3

Single-Stage Digestion

The cross section of a floating-cover digester is illustrated in Figure 11-55. Raw sludge is pumped into the tank through feed pipes that terminate both near the center of the tank and in the gas dome. Contents stratify with a scum layer on top, a middle zone of supernatant (water of separation) underlain by actively digesting sludge, and a bottom layer of digested concentrate. A limited amount of mixing is provided in the upper half of the tank by withdrawing digesting sludge, passing it through a sludge heater, and returning it through the inlet piping. Supernatant is withdrawn from any one of a series of pipes extended through the tank wall. Digested sludge is taken from the tank bottom.

The digester cover floats on the sludge surface, and liquid extending up the sides provides a seal between the tank wall and the side of the cover. Gas rising out of the

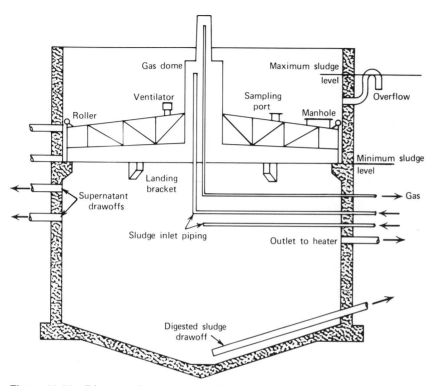

Figure 11-55   Diagram of a single-stage floating-cover anaerobic digester.

digesting sludge is collected in the gas dome and is burned as a fuel in the sludge heater; often the excess is wasted to a gas burner. The cover can rise vertically from the landing brackets to near the top of the tank wall guided by rollers around the circumference to keep it from binding. The volume between the landing brackets and the fully raised cover position is the amount of storage available for digested sludge; this is approximately one third of the total volume.

Digestion in a single-stage floating-cover tank performs the functions of volatile solids digestion, gravity thickening, and storage of digested sludge. When sludge is pumped into the digester from the primary settling tanks, the floating cover rises making room for the sludge. Unmixed operation permits daily drainage of supernatant equal to approximately two thirds of the raw sludge feed. Being high in both BOD and suspended solids, the withdrawn water is returned to the inlet of the treatment plant. Periodically, digested sludge is removed for dewatering and disposal. In large plants, digested sludge may be dewatered mechanically; however, in small installations it is frequently spread in liquid form on farmland or is dried on sand beds and hauled to land burial. Weather often dictates the schedule for land disposal, and, consequently, substantial digester storage volume is required in northern climates. Typical operation lowers the cover to the landing brackets in the fall of the year to provide maximum storage volume for the winter.

A drying bed consists of a 12-in. sand layer underlain by graded gravel that envelopes tile or perforated pipe underdrains. Large beds are partitioned by concrete walls into sections 25 ft wide by 100 to 200 ft long. A pipe header, with gated opening to each cell, is used to apply digested sludge to the bed. Seepage collected in the underdrains is returned to the treatment plant wet well.

The total area of open drying beds at a treatment plant is usually 1 to 2 sq ft/BOD design population equivalent.

Digested sludge applied to a depth of 8 or 10 in. dries to a fibrous layer of 3 or 4 in. in a period of a few weeks. With sufficient bed area and digester storage volume, this process of dewatering can be compatible for most climatic regions. However, the laborious process of removing the digested cake is a major problem for operating personnel. Although some plants have mechanical equipment, the time-honored method is manual removal using a shovel-like fork. Attempting to employ front-end loaders leads to disturbance of the bed and excessive loss of sand. In most instances, an operator will seek an alternate technique, such as spreading the liquid sludge on crop or pastureland or lagooning in shallow trenches that permit the use of a tractor for scraping up dried cake for hauling to landfill.

Two-Stage Digestion

In this process, two digesters in series separate the functions of biological stabilization from gravity thickening and storage as pictured in Figure 11-56. The first-stage high-rate unit is completely mixed and heated for optimum bacterial decomposition. Mixing is accomplished either mechanically by an impeller suspended from the cover, or by recirculation of compressed digestion gases. The latter is done by one of the following three methods: Gas withdrawn from the dome is injected back into the tank through a series of small-diameter pipes hanging

Figure 11-56 Two-stage anaerobic digestion is performed by two tanks in series. The tank on the left is completely mixed for optimum digestion, and the one on the right is for gravity thickening and storage of digested sludge. The right tank has a cover with a gas dome. (Lincoln, Nebraska.)

into the digesting sludge; compressed gas is discharged into a draft tube located in the center of the tank to lift recirculating sludge from the bottom and to spill it out on top; or pressurized gas is forced through diffusers mounted on the bottom of the tank. These systems are available for installation in either fixed- or floating-cover tanks. By using a floating cover, digested sludge does not have to be displaced simultaneously with raw sludge feed as is required with a fixed-cover tank. In either case, however, the sludge cannot be thickened in a high-rate process because continuous mixing does not permit formation of supernatant. Actually, the discharged sludge has a lower solids concentration than the raw feed because of the conversion of volatile solids to gaseous end products.

The second-stage digester must be either a floating-cover or open tank with provisions for withdrawing supernatant. The unit is often unheated, depending on the local climate and degree of stabilization accomplished in the first stage. By minimizing hydraulic disturbances in the tank, density of the digested sludge and clarity of the supernatant are both increased.

Two-stage digestion may be advantageous in some plants while conventional operation may be better in others. The determining factors include size of treatment plant, flexibility of sludge handling processes, method of ultimate solids disposal, storage capacity needed, and the interrelated element of climatic conditions. For large plants with a number of digesters, series operation often provides better utilization of digester capacity, but for small plants with limited supervision the conventional operation is frequently more feasible.

Sizing of Digesters

Historically, conventional single-stage tanks have been sized on the basis of population equivalent load on the treatment plant. Heated digester capacity for a trickling filter plant processing domestic wastewater was established at 4 cu ft (0.11 m³) per capita of design load. For primary plus secondary activated sludge, the total tank volume requirement was increased to 6 cu ft (0.17 m³) per capita. These values are still used as guidelines for sizing conventional digesters for small treatment works.

Total digestion capacity can be calculated for conventional single-stage operation using Eq. 11-40. Application of this formula requires knowing the characteristics of both the raw and digested sludges.

$$V = \frac{V_1 + V_2}{2} T_1 + V_2 \times T_2 \qquad (11\text{-}40)$$

where V = total digester capacity, gallons (cubic meters)
V_1 = volume of daily raw sludge applied, gallons per day (cubic meters per day)
V_2 = volume of accumulated digested sludge, gallons per day (cubic meters per day)
T_1 = period required for digestion, days (approximately 30 days at a temperature of 85 to 95° F or 30 to 35° C)
T_2 = period of digested sludge storage, days

Volume needed for the high-rate unit in a two-stage digestion system is based on a maximum volatile solids loading and minimum detention time. For new design the generally adopted maximum allowable loading is 0.08 lb VS/cu ft/day (1300 g/m^3·d) and the minimum liquid detention time is ten days. At these loadings and a temperature of 95° F, volatile solids reduction should be 50 percent or greater. No specific design criteria are established for second-stage tanks in high-rate systems, since thickening and digested sludge storage requirements depend on local sludge disposal procedures.

Start-up of Digesters

Anaerobic digestion is a difficult process to start because of the slow growth rate and sensitivity of acid-splitting methane-forming bacteria. Furthermore, the number of these microorganisms is very low in raw sludge compared with acid-forming bacteria. The normal procedure for start-up is to fill the tank with wastewater and to apply raw sludge feed at about one tenth of design rate. If several thousand gallons of digesting sludge from an operating digester are used as seed, the new process can be operational in a few weeks. However, if only raw sludge is available, developing the biological process may take months. Careful additions of lime added with raw sludge are helpful in maintaining the pH near 7.0, but erratic dosage can result in sharp pH changes detrimental to the bacteria. After gas production and volatile acids concentration have stabilized, the feed rate is gradually increased by small increments to full loading. Daily monitoring of this process involves plotting the daily gas production per unit of raw sludge fed, percentage of carbon dioxide in the head gases, and concentration of volatile acids in the digesting sludge.

EXAMPLE 11-18

An anaerobic sludge digester operating at a feed rate of 0.06 lb VS/cu ft/day has good gas production and volatile solids reduction. The external sludge heater fails and is not repaired for about 6 weeks. During heater shutdown, the same feed rate is maintained and a noticeable decrease in gas production occurs as the temperature of the digesting sludge drops from 95 to 75° F. When the heater is returned to service, the digester temperature is rapidly elevated back to 95° F with a sudden initial increase in gasification. However, after a few days of renewed operation, gas production decreases sharply, volatile acids are measured at about 2000 mg/l, and pH of the digesting sludge begins to drop. Describe what has happened to the anaerobic biological process.

Solution

Initially, acid-forming and methane-forming populations are in balance and, because they are operating near optimum temperature, the majority of the volatile solids are being converted to gaseous end products. When the heater fails and temperature of the digesting sludge drops, a new equilibrium of populations is established at reduced gasification efficiency. Therefore, raw organic matter accumulates in the digesting sludge. These dormant volatile solids suddenly become available to the bacteria when the temperature is raised back to 95° F. The sharp increase in gas production is related primarily to carbon dioxide produced when first-stage bacteria convert volatile solids to organic acids. The acid-splitting methane formers attempt to respond to the increased organic acid supply. However, their populations have been reduced by the lack of food during low-temperature operation. Furthermore, the methane bacteria, being more sensitive, are inhibited by the acid conditions and, hence, total gas yield decreases. The solution is to drop the temperature of the digesting sludge back to 75° F and to increase it slowly, perhaps one degree per week, to allow the methane formers to adjust gradually to the increasing production of organic acids.

EXAMPLE 11-19

Calculate the capacity required per population equivalent for single-stage floating-cover digester based on the following: 0.24 lb solids contributed per person, 4.0 percent solids content in raw sludge, 7.0 percent solids concentration in digested sludge, 40 percent total solids reduction during digestion, 30-day digestion period, and 90-day storage period for digested sludge.

Solution

Using Eq. 11-37, the daily volumes of raw sludge produced and digested sludge accumulated are

$$V_1 = \frac{0.24 \text{ lb/person/day}}{(4.0/100)8.34}$$

$$= 0.72 \text{ gal/person/day}$$

$$V_2 = \frac{0.60 \times 0.24 \text{ lb/person/day}}{(7.0/100)8.34}$$

$$= 0.25 \text{ gal/person/day}$$

Then, the digester capacity by Eq. 11-40 is

$$V = \frac{0.72 + 0.25}{2} \times 30 + 0.25 \times 90$$

$$= 37 \text{ gal} = 5.0 \text{ cu ft}$$

(*Note.* This calculation verifies the single-stage digester capacity based on design equivalent population listed in Table 11-5.)

EXAMPLE 11-20

A trickling filter plant has two floating-cover digesters each with a total capacity of 30,000 cu ft with 20,000 cu ft below the landing brackets. The remaining 10,000 cu ft in each tank is the volume between the lowered and fully raised cover positions (storage volume). The sludge piping is arranged so that the digesters may be operated in parallel as conventional digesters or in series as a two-stage system. One digester is equipped with gas mixing to serve as a first-stage high-rate process. Daily raw sludge production is 6400 gal containing 2800 lb of dry solids that are 70 percent volatile. Digested sludge is 8.0 percent solids, and the process converts 60 percent of the volatile solids to gas.

(a) For conventional single-stage operation with one half of the waste applied to each tank, calculate the volatile solids loading, supernatant produced, and number of days of digested sludge storage available.

(b) For two-stage operation, compute the volatile solids loading on the first tank, solids content of the digested sludge leaving the high-rate digester, and available sludge storage in the system.

Solution

(a) For conventional single-stage operation

$$\text{VS applied to each tank} = (0.70 \times 2800)/2$$

$$= 980 \text{ lb/day}$$

$$\begin{array}{l}\text{Loading with} \\ \text{covers lowered}\end{array} = \frac{980 \text{ lb VS/day}}{20,000 \text{ cu ft}}$$

$$= 0.049 \text{ lb VS/cu ft/day}$$

$$\begin{array}{l}\text{Loading with} \\ \text{covers raised}\end{array} = \frac{980 \text{ lb VS/day}}{30,000 \text{ cu ft}}$$

$$= 0.033 \text{ lb VS/cu ft/day}$$

$$\begin{array}{l}\text{Supernatant} \\ \text{return per day}\end{array} = \begin{array}{l}\text{raw sludge volume} \\ \text{applied per day } (V_1)\end{array}$$

$$\qquad\quad - \begin{array}{l}\text{digested sludge volume} \\ \text{accumulated/day } (V_2)\end{array}$$

$$V_1 = 6400 \text{ gal}$$

$$V_2 = \frac{\text{inert solids + remaining VS}}{(\text{solids content}/100)8.34}$$

$$= \frac{0.30 \times 2800 + 0.40(0.70 \times 2800)}{(8.0/100)8.34}$$

$$= 2400 \text{ gal}$$

Supernatant return = 6400 − 2400

$$= 4000 \text{ gal/day}$$

Digested sludge storage available

$$= \frac{2 \times 10,000 \text{ cu ft} \times 7.48 \text{ gal/cu ft}}{2400 \text{ gal/day}} = 62 \text{ days}$$

(b) For two-stage operation

VS applied to high-rate tank = 0.7 × 2800

$$= 1960 \text{ lb/day}$$

$$\frac{\text{Loading with}}{\text{cover lowered}} = \frac{1960}{20,000} = 0.098 \text{ lb VS/cu ft/day}$$

$$\frac{\text{Detention time}}{\text{with cover lowered}} = \frac{20,000 \times 7.48}{6400} = 23 \text{ days}$$

With cover raised the volume is 30,000 cu ft.

Loading = 0.065 lb VS/cu ft/day

Detention time = 35 days

Solids content in digested sludge leaving the first-stage high-rate digester (Eq. 11-37)

$$s = \frac{0.30 \times 2800 + 0.40(0.70 \times 2800)}{6400 \times 8.34/100} = 3.0 \text{ percent}$$

(*Note.* The raw sludge is 5.2 percent solids. The percentage of solids reduces to 3.0 percent in high-rate digestion, since volatile solids are converted to gas while no supernatant is withdrawn to thicken the digested sludge.)

Digested sludge storage, available only in the second-stage tank, is equal to 31 days. This reduced storage capacity is a disadvantage of high-rate digestion compared with parallel operation that permits 62 days storage.

11-17 AEROBIC DIGESTION

Waste sludge can be stabilized by long-term aeration that biologically destroys volatile solids. The aerobic digestion process was developed specifically to handle excess activated sludge from aerobic treatment plants without primary clarifiers (Figure 11-2a). Early attempts to anaerobically digest this sludge failed because of the low solids concentration and aerobic nature of the waste. High water content in the range of 98 to 99 percent also prevented economical dewatering by mechanical means without prior thickening. Furthermore, since most of the aeration plants served small communities, large investments for mechanical equipment could not be justified for waste sludge processing. Consequently, aerated holding tanks were introduced to stabilize and store mixed liquor withdrawn from the aeration tank. Aerobic digesters are single or multiple tanks equipped with diffused or mechanical aerators. Gates or pipes are provided to draw of supernatant after gravity settling of the suspended solids with the aerators shut off. Since waste activated sludge contains putrescible solids, it becomes odorous and does not dewater readily an open sand beds without prior aerobic digestion.

Design standards for aerobic digestion vary with character of the waste sludge and method of ultimate disposal. The general guideline for stabilizing waste activated sludge, with a suspended solids concentration of about 1.0 percent, is a maximum volatile solids loading of 0.04 lb/cu ft/day (640 g/m^3·d) and an aeration period in the range of 200 to 300 degree-days, which is computed by multiplying the digesting temperature in °C times the sludge age. This equates to a minimum aeration period of 10 days at 20°C or 20 days at 10°C. Volatile solids and BOD reductions at these loadings are in the range of 30 to 50 percent. In small plants, the aerobic digester volume is often 2 to 3 cu ft per design population equivalent of the treatment plant. This provides a loading in the range of 0.01 to 0.02 lb VS/cu ft/day (160 to 320 g/m^3·d).

Aeration of waste activated sludge creates a digested sludge that resists gravity thickening. By quiescent settling without air mixing, the solids concentration can usually only be gravity thickened to 1.0 to 2.0 percent solids. The maximum concentration with careful withdrawal of supernatant is not likely to exceed 2.5 percent. This poor settleability frequently causes problems in disposing of the large volume of sludge produced. The common methods of disposal are liquid spreading on farmland, lagooning, and drying on sand beds. If drying beds are used, they must be deep enough to allow application of 3 ft or more of digested sludge so that the dried cake will be 1 to 2 in. thick.

An aerobic digester is normally operated by continuously feeding raw sludge with intermittent supernatant and digested sludge withdrawals. The digesting sludge is continuously aerated during filling and for the specified digestion period after the tank is full. Aeration is then discontinued to allow the stabilized solids to settle by gravity. Supernatant is decanted and returned to the head of the treatment plant, and a portion of the gravity-thickened sludge is removed for disposal. In practice, aeration and settlement may be a daily cycle with feed applied early in the day and clarified supernatant decanted later in the day. Digested solids are withdrawn when the sludge in the tank does not gravity thicken to provide a supernatant of adequate clarity.

11-18   VACUUM FILTRATION

Vacuum filtration is used to dewater both raw and digested wastewater sludges. While a reliable process, it is energy intensive and the dryness of the cake is limited by the ability of vacuum to remove water from the chemically conditioned sludge. The competing process is belt filter press dewatering, which has significantly lower energy consumption and is capable of producing a drier cake.

Description of a Vacuum Filter

Rotary-drum vacuum filtration is a mechanical method to dewater sludge. The cylindrical drum covered with a filter medium rotates partially submerged in a vat of chemically treated sludge. The filter medium may be a belt of synthetic cloth, woven metal, or two layers of stainless steel coil springs arranged in corduroy fashion around the drum (Figure 11-57). As the drum slowly rotates, vacuum is applied immediately under the filter medium to attract sludge solids as the drum dips into the vat. Suction continues to draw water from the accumulated solids as the filter slowly rotates. Radial vacuum pipes convey filtrate from collecting channels in the drum surface behind the media. In the discharge sector, suction is broken and the belt or coil springs are drawn over a small-diameter roller for removal of the dried cake. Fork tines help release the solids layer. The filter medium is washed by water sprays, located between wash and return rollers, before it is reapplied to the drum surface for another cycle.

Components of a vacuum filter installation are illustrated in Figure 11-58. Positive-displacement pumps

Figure 11-57 Picture of a vacuum filter with coil-spring media dewatering waste sludge. (Courtesy of Komline-Sanderson, Peapack, New Jersey.)

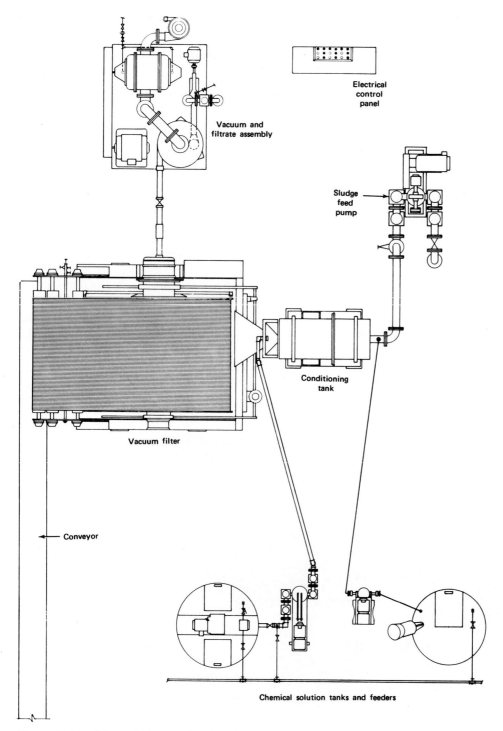

Figure 11-58 Schematic layout showing the major components of a vacuum filtration system. (Courtesy of Komline-Sanderson, Peapack, New Jersey.)

meter sludge to a conditioning tank. Here chemicals are mixed with the liquid to flocculate the solids. Chemical treatment is necessary to capture fines that would otherwise be drawn through the medium, and to prevent cracking of the filter cake during dewatering, which would result in excess air entering the vacuum system. The mixture of filtrate and air drawn by the vacuum pump is separated in a receiving vessel. Air is exhausted through a silencer to reduce noise, and the water is pumped back to the inlet of the treatment plant. Cake peeled from the filter drops onto a belt conveyer for transporting to a truck for hauling to landfill, or directly to an incinerator. Filtrate return accounts for the majority of the process discharge. For example, dewatering 1000 gal of sludge with 5 percent solids to a 30 percent cake results in 830 gal of extracted water and about 140 lb of wet cake; this amounts to a wet sludge weight reduction of 83 percent.

Operation of a Vacuum Filter

Many small plants draw settled sludge directly from the primary tanks without thickening. Although this is often satisfactory for trickling filter installations, vacuum filtration of unthickened sludges from aeration plants often leads to unsatisfactory performance. The generally accepted minimum solids concentration for economical filtration is 4 percent.

Mixtures of raw primary and secondary wastes are gravity thickened, while waste activated may be increased in solids content by flotation. Digested sludge seldom is thickened separately because of its high density when drawn from digesters. However, elutriation may be practiced to remove alkalinity and fine solids to improve filter yields and reduce chemical dosage. This process consists of countercurrent washing of the digested sludge with plant effluent.

Common chemicals in conditioning are ferric chloride and lime, or polyelectrolytes. Ferric chloride is dissolved in water and is applied by a solution feed pump, while lime is fed in a slurry. Polymers available in powder, pellet, or liquid form are dissolved to a stock solution of 0.5 to 5 percent. Most manufacturers recommend further dilution to 0.01 to 0.05 percent before application. This permits uncurling of the long chain molecules. Chemical dosages

for conditioning vary considerably among different sludges. Solids concentration and type of sludge are the most critical factors. With rare exception, digested sludges require considerably greater chemical additions than do fresh raw sludges. Typical inorganic chemical requirements are 2 to 4 percent $FeCl_3$ and 5 to 10 percent CaO, or polyelectrolyte amounts of less than 1 percent. These chemical consumptions are expressed as percentages of the dry solids filtered; for example, 3 percent conditioner means 3 lb of chemical per 100 lb of dry solids in the filter cake. The choice of conditioning chemicals depends on economics, desired filter yield, and method of ultimate disposal. A flexible installation is advantageous so that any of the inorganic metal coagulants, lime, or polyelectrolytes can be handled by the system. The advantages of ferric chloride and lime are that they provide disinfection and stabilization of the waste sludge, thus reducing health hazards and odors. The disadvantage is that they are more difficult to handle and feed, result in increased filter maintenance, and add weight to the filter cake. Polymers do not provide disinfection, but they are easier to feed and are often more economical. The most suitable chemicals for a particular situation can be determined only by laboratory and plant-scale investigations. Final filter cake disposal dictates the extent of dewatering required. Concentration to 20 to 25 percent solids is generally satisfactory for hauling to landfill. Cake with a higher moisture content is troublesome to handle and does not discharge readily from a dump truck. If long distance trucking is involved, it may be more economical to dewater to a lower moisture content and thus reduce the total weight of cake transported. When disposed of by incineration, it is desirable to dewater to as low a moisture content as possible, usually to 30 percent solids or greater to reduce the consumption of auxiliary fuel.

Rate of vacuum filtration is expressed as yield in pounds of dry solids produced per square foot of filter area per hour (kilograms per square meter per hour). Table 11-6 lists the usual filter yields expected in dewatering various types of domestic waste sludges. These values reflect both typical characteristics and solids concentration of the different sludges. As a general rule, the expected filter yield of any chemically conditioned sludge is about 1 lb per square foot per hour for each percentage of

Table 11-6.   Common Vacuum Filter Yields for Chemically Conditioned Waste Sludge

Source of Sludge	Yield, Pounds per Square Foot per Hour[a]	
	Raw	Digested
Primary only	8	7
Primary plus trickling filter humus	7	6
Primary plus waste activated	6	5
Thickened waste activated only	3 to 5	

[a] 1.0 lb/sq ft/hr = 4.88 kg/m^2·h

solids. In addition to yield, solids in the filtrate and final moisture content of the cake are important in judging filter performance. Figure 11-59 shows typical variations of these parameters relative to chemical conditioning. Yield increases continuously with increasing chemical dosage. Filtrate solids are very high when conditioning is not sufficient to flocculate properly the finer particles. It is essential to operate in the region of low filtrate solids to prevent returning an excessive amount of solids back to the treatment plant. Final cake moisture decreases with increasing chemical addition; however, it is more dramatically influenced by solids content of the sludge feed. Where a dry sludge is necessary for incineration, thicken-

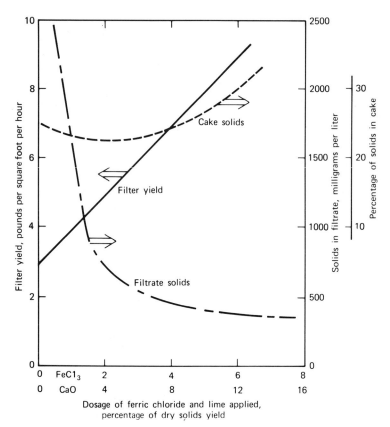

Figure 11-59   Typical variations of filter yield, filtrate solids, and cake moisture relative to dosage of conditioning chemicals.

ing prior to filtration results in chemical savings that usually offset the cost of the thickening operation.

Sizing of Vacuum Filters

Size of vacuum filters for new installations is based on estimated dry solids production, a selected filter yield depending on thickness and character of the sludge, and operating time. For small plants, 35 hr per week is often selected for design purposes. This allows a 7-hr day for five days with 1 hr for start-up and wash-down. For larger plants, filters are often operated for a period of 16 to 20 hr per day.

Where sludge samples are available prior to design and installation, sludge filterability can be tested in the laboratory. The Buchner funnel method consists of pouring a chemically conditioned sludge sample into a funnel lined with a filter paper and dewatered under vacuum. This procedure can predict suitability of a sludge for filtration and applicability of certain chemicals, but it is of little value in sizing or predicting plant-scale operation. A second test procedure uses a filter leaf with 0.10 sq ft of effective area attached to a vacuum apparatus. A medium with characteristics similar to a full-scale filter is placed over the face of the leaf. After suction is applied, the unit is inserted upside down in a representative slurry to simulate cake formation, and then is withdrawn and dewatered for a time period related to drum rotation speed. Cake removed from the leaf face and filtrate drawn into the vacuum flask can then be tested for solids content. If carefully conducted, results can be correlated to indicate expected full-scale vacuum filter operation.

Good filter operation relies on installation of the proper sludge handling equipment, careful monitoring of the process, and a willingness to experiment to determine optimum performance. Uniform sludge and proper chemical dosage are the most essential factors. If the character of the sludge varies, as may be the case if it is drawn directly from primary clarifiers, consistent operation is difficult to maintain. For example, hour to hour fluctuations of solids concentration may lead to poor cake and filtrate quality, or more likely, chemical overdosing to prevent such problems. Laboratory monitoring should include analyses to determine filtrate quality, yield, and cake moisture. The operator should observe sludge level in the vat to ensure proper submergence, clarity of the filtrate, cake thickness, which is normally about $\frac{1}{8}$ in., and drum speed. Slow rotation produces a thicker, drier cake with less yield, while selection of a faster speed produces moister cake and a greater yield. Finally, because of the variety of chemicals marketed for sludge conditioning, particularly polyelectrolytes, special full-scale studies must be conducted periodically to ensure that the selected coagulant is most economical for the performance desired.

EXAMPLE 11-21

The smallest size vacuum filter produced by one of the leading manufacturers has a nominal surface area of 60 sq ft; the drum is 7 ft in diameter and 2 ft 10 in. in width. Estimate the equivalent population that can be served by this unit. Assume 0.20 lb dry solids per capita per day, a filter yield of 5.0 lb per square foot per hour (5 percent solids content in sludge), and 7 hr of operation per day.

Solution

$$\text{Total yield per day} = \frac{5.0 \text{ lb}}{\text{sq ft} \times \text{hr}} \times 7.0 \text{ hr} \times 60 \text{ sq ft}$$

$$= 2000 \text{ lb}$$

$$\text{Equivalent population} = \frac{2000 \text{ lb/day}}{0.20 \text{ lb/capita/day}}$$

$$= 10,000 \text{ persons}$$

EXAMPLE 11-22

A conventional aeration plant produces a daily waste sludge of 25,800 gal containing 10,200 lb of dry solids, which is a 4.7 percent solids concentration. (These values are from Example 11-14.) The installed vacuum filter has a nominal area of 300 sq ft, and the filtration characteristics of the sludge are given in Figure 11-51. For a filter yield of 6.0 lb per square foot per hour, calculate the following: daily consumption of chemicals, amount of cake produced, quantity of filtrate, and hours of filter operation per day.

Solution

From Figure 11-51 for a yield of 6.0 lb per square foot per day, chemical dosages are 3.1 percent ferric chloride and 6.2 percent lime, the filtrate contains 500 mg/l of suspended solids, and the percentage of solids in the cake is 23 percent.

Chemical dosages are

$$FeCl_3 \text{ applied per day} = 0.031 \times 10,200 = 320 \text{ lb}$$

$$CaO \text{ applied per day} = 0.062 \times 10,200 = 630 \text{ lb}$$

After conditioning, the quantity of sludge is increased by the weight and volume of chemical additions. Assume that the $FeCl_3$ and CaO solutions amount to a total of 700 gal/day. Then,

$$\text{Conditioned sludge solids} = 10,200 + 320 + 630$$
$$= 11,200 \text{ lb/day}$$

$$\text{Conditioned sludge volume} = 25,800 + 700$$
$$= 26,500 \text{ gal/day}$$

Weight of filter cake, computed from cake solids content of 23 percent,

$$= \frac{11,200 \text{ lb dry solids}}{0.23 \dfrac{\text{lb dry solids}}{\text{lb wet sludge}}} = 48,700 \text{ lb/day}$$

And the volume of cake using a specific gravity of 1.06

$$= \frac{48,700}{1.06 \times 8.34} = 5500 \text{ gal/day}$$

Quantity of filtrate, volume of conditioned sludge minus the cake volume

$$= 26,500 - 5500 = 21,000 \text{ gal/day}$$

$$\text{Filtrate solids} = 500 \text{ mg/l} \times 0.021 \text{ mgd} \times 8.34$$
$$= 88 \text{ lb/day}$$

Filter operation required per day

$$= \frac{10,200 \text{ lb/day}}{300 \text{ sq ft} \times 6 \text{ lb/sq ft/hr}} = 5.7 \text{ hr/day}$$

11-19   PRESSURE FILTRATION

The belt filter press and the plate-and-frame filter press are pressure filters. Although both are used in dewatering wastewater sludges, the belt filter press is much more popular because it is available in smaller sizes and uses polymer for chemical flocculation of the sludge. In comparison to vacuum filters, belt presses have the advantage of producing a drier cake with much less energy consumption. Plate-and-frame presses, described in Section 7-17, are used primarily to dewater chemical sludges. They are large machines and use ferric chloride and lime for conditioning of organic sludges prior to dewatering. Plant-and-frame presses are noted for producing a very compact and dry cake; hence, they can be economically justified for dewatering wastewater sludges prior to incineration.

Description of a Belt Filter Press

A belt filter press has two continuous porous belts that pass over a series of rollers to squeeze water out of the sludge layer compressed between the belts. Figure 11-60*a* is an operational diagram of a belt press. After addition of polymer to flocculate the solids, the wet sludge is applied at a uniform rate on the upper belt at the beginning of the gravity drainage zone. The belt is supported in this zone on an open framework to allow water that separates by gravity to drain through the porous belt into a collection pan. To enhance this initial dewatering, adjustable plastic vanes are usually installed just above the belt surface to open channels in the sludge for draining of free water through the belt. After about one half of the water is removed in the gravity zone, the sludge drops onto the lower belt and is gradually compressed between the two belts as they come together in the cake-forming wedge zone. The two belts then pass over a series of rollers for high-pressure dewatering by squeezing the water out of the sludge layer through the belts. The dewatering rollers may be plain or perforated cylinders constructed of corrosion-resistant material, while the belt tension, alignment, and drive rollers have a rubber coating to prevent slippage of the belt. The cake is scraped from the belts by doctor blades held against both belts.

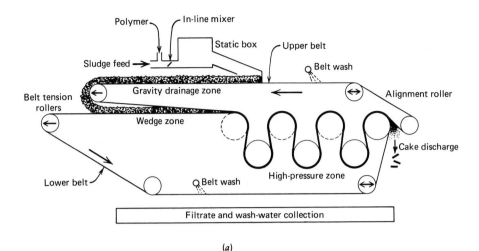

(a)

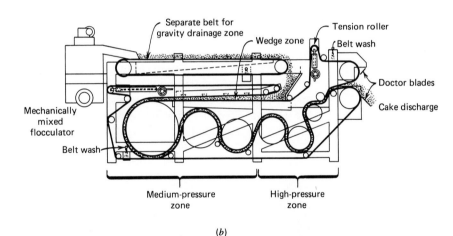

(b)

Figure 11-60 Operational diagrams of belt filter presses. (*a*) Press with two continuous belts for gravity and pressure dewatering with uniform-diameter rollers. (Courtesy of Ashbrook-Simon-Hartley.) (*b*) Press with a separate belt for the gravity drainage zone and rollers with decreasing diameters in the pressure dewatering zone. (Courtesy of Komline-Sanderson, Peapack, New Jersey.)

Features of belt filter presses vary with different manufacturers. The flocculator may consist of an in-line mixer for blending the polymer, followed by a static box that distributes the flocculated sludge onto the belt, or it may be a mechanically mixed tank for blending the polymer and flocculating the sludge, followed by application to the belt. A few manufacturers employ a rotating drum made of filter screen that allows thickening in addition to flocculation. While the aggregated solids are retained on the screen and fed to the press, the separated water is discharged from the enclosure surrounding the drum screen. The sludge must be applied to the press at a uniform rate; therefore, the recommended sludge pump is a progressing cavity pump rather than a plunger pump. As diagramed in Figure 11-60*b*, the gravity drainage zone may be on a separate belt to allow independent control of retention times in the gravity zone and the high-pressure zone. This machine also has rollers that decrease in diameter to gradually increase pressure on the cake. Both compression and shearing in the confined sludge layer increase with decreasing roller size.

Operation of a Belt Filter Press

The main operating parameters of a belt filter press are hydraulic or solids loading and polymer dosage. Hydraulic loading is expressed in gallons per minute of sludge feed per meter of belt width (cubic meters per meter per hour). Solids loading is expressed as the pounds of total dry solids feed per meter per hour (kilograms per meter per hour). Polymer dosage is calculated as the pounds applied per ton of total dry solids in the feed sludge (kilograms per tonne). The key performance parameters are solids recovery, cake dryness, wash-water consumption, and wastewater discharge. Solids recovery is calculated as the total solids in the feed sludge minus the suspended solids in the wastewater (filtrate plus wash water) divided by the total solids in the feed sludge. Cake dryness is expressed as the percentage of dry solids by weight in the cake. For easy comparison with hydraulic sludge loading, wash-water consumption and wastewater discharge are expressed in units of gallons per minute per meter of belt width (cubic meters per meter per hour).

Belt presses can dewater both thickened and dilute sludges that are either raw or digested. The sludge characteristics of greatest importance are solids concentration, nature of the solids, and prior biological or chemical conditioning. When the solids concentration is less than 4 percent, a press is limited to a hydraulic capacity essentially independent of solids concentration. Most manufacturers suggest a maximum hydraulic loading of 50 gpm per meter of belt width (11.4 m³/m·h). With solids concentration greater than about 6 percent, the capacity of a press is restricted by solids loading. The nature of the solids influences both polymer flocculation and the degree of dewatering. Fibrous solids that exist in primary sludge are much easier to dewater than fine biological solids in waste activated sludge.

Polymer is applied to flocculate the sludge solids forming aggregates to allow release of water with dosage depending on the sludge characteristics and sludge feed rate. The sludge feed rate is also related to belt speed and belt tension. Belt speed determines the retention time in the gravity dewatering zone, and tension establishes the pressure on the confined sludge layer. In addition to adequate flocculation, sludge feed rate is governed by solids recovery and cake dryness. If sludge loading is too high, the gravity drainage zone does not release sufficient water, resulting in extrusion of fine solids through the belt fabric and from between the belts at the edges. Excessive belt tension can also cause extrusion of solids. In actual plant operation, the optimum solids loading is based on the economy of operation considering the two major costs of polymer consumption and hours of operation. Adjusting process variables at a given sludge loading is done by selecting initial settings for polymer dosage, belt speed, and belt tension and then readjusting these three settings to achieve the desired cake dryness and solids recovery with minimum polymer dosage.

The belt press shown in Figure 11-61 dewaters a mixture of raw primary and waste activated sludges. The operational diagram for this press is shown in Figure 11-60a. Typical operation in filtering a feed sludge with 3.6 percent solids to a cake density of 21 percent

Figure 11-61 A belt filter press dewatering wastewater sludge. The operational diagram of this press is shown in Figure 11-60a. (Lincoln, Nebraska.)

solids is a hydraulic loading of 44 gpm/m and a solids loading of 790 lb/m/hr with a polymer dosage of 3.3 lb/ton. A cake dryness of 24 percent can be achieved easily by increasing belt tension and polymer dosage; however, a 21 percent cake is adequate for burial with municipal refuse. Sludge loading can also be increased at the expense of higher polymer dosage. The belt presses are in service for two operator shifts of 10 hr each for a total of 19 hr of sludge dewatering and 1 hr for cleanup and routine maintenance. If this daily time schedule is reduced, a substantial increase in emission of odors occurs during dewatering caused by an increase in the anaerobiosis of the sludge. Machine loading is therefore dictated by the dewatering period, and operating parameters optimized for adequate cake density at minimum polymer dosage for greater than 90 percent solids capture. The powdered polymer is applied in a 0.10 percent solution by weight at a rate of 2.6 gpm/m. This concentration provides the most effective flocculation of the sludge. Higher polymer concentrations do not disperse as well in the sludge. The wash-water usage is 40 gpm/m.

Sizing of Belt Filter Presses

Design of a sludge dewatering facility considers size of the treatment plant, desired flexibility of operations, and anticipated conditions of dewatering. The latter depends on the quantity of sludge production, design feed rate, and operating time. Filter presses are manufactured with belt widths in the range of 0.5 to 3.0 m, with the most common sizes between 1.0 and 2.5 m. Unless adequate sludge storage is provided, at least two presses should be installed so that one press can be operated while the other is out of service for maintenance to change belts or repair mechanical components, such as roller bearings. In a small plant, the operating time may be 7 hr per day, whereas in larger plants the presses may be operated for a period of 15 or 23 hr per day. These schedules allow 1 hr for start-up and shutdown.

Typical operating parameters for filter pressing of wastewater sludges are listed in Table 11-7. The solids loadings account for both the feed solids concentration and hydraulic loading. For example, using the average values for anaerobically digested primary sludge, a

6 percent feed at 45 gpm/m yields a solids loading of 1350 lb/m/hr. Also, percentage of cake solids decreases and polymer dosage increases with low feed solids concentration and for those sludges containing waste activated sludge. The data in Table 11-7 should be used only to estimate the capacity of presses required in design. For a given manufacturer's press, operating experience at other installations dewatering a sludge of similar characteristics is considered. In general, press capacity should be prudently selected for new installations to account for the probable inaccuracy in projecting performance.

Sizing of belt filter presses for an existing treatment plant can be determined most reliably by conducting field tests using a narrow belt machine. Most manufacturers have a full-scale 0.5 or 1.0 m press enclosed in a mobile trailer, which is representative of their larger machines. During the preliminary design investigation, a rented trailer unit can be used to determine dewaterability of the sludge and to establish testing criteria for the performance specifications. After the press facility has been sized and designed, selection of the press manufacturer can be based on both competitive bidding and qualification testing using trailer units. The procedure may involve individual testing of different presses, with the manufacturer of lowest bid conducting the first field test. For a larger installation, a group of several different machines can be compared by simultaneous operation in parallel. Field testing reduces the risk in design by demonstrating that the selected manufacturer's press can achieve the results required by the performance specification. Acceptance testing after construction to evaluate installed presses is still conducted to ensure compliance with the specifications.

EXAMPLE 11-23

A belt filter press with an effective belt width of 1.5 m dewaters anaerobically digested sludge at a sludge feed rate of 70 gpm. The polymer dosage is 6.0 gpm containing 0.20 percent polymer by weight, and the wash water usage is 50 gpm. Based on laboratory analyses, total solids in the feed sludge are 4.0 percent, total solids in the cake are 35 percent, and suspended solids in the wastewater (filtrate, polymer feed, and wash water) are 1800 mg/l. Calculate the following: hydraulic feed rate, solids loading rate, polymer dosage, and solids recovery. Compare these values with those listed in Table 11-7.

Solution

Hydraulic loading

$$= \frac{70 \text{ gpm}}{1.5 \text{ m}} = 47 \text{ gpm/m (Table 11-7, 40 to 50 gpm/m)}$$

Solids loading $= 47 \text{ gpm/m} \times 60 \text{ min/hr} \times 0.040 \times 8.34$

$$= 940 \text{ lb/m/hr}$$

The range of solids loadings from Table 11-7 is 1000 to 1600 lb/m/hr.

Polymer dosage $= \dfrac{6.0 \text{ gpm} \times 60 \text{ min/hr} \times 0.0020 \times 8.34}{1.5 \text{ m} \times (940 \text{ lb/m/hr})/(2000 \text{ lb/ton})}$

$$= 8.5 \text{ lb/ton}$$

This polymer dosage is greater than the 3 to 6 lb/ton range given in Table 11-7; however, flocculation effi-ciencies of polymers vary considerably among different brands. Selection of the best polymer is based on cost per ton of cake solids, not the weight per ton.

The quantity of filtrate can be estimated by substracting the calculated cake volume from the sludge feed. Assuming a specific gravity of 1.05, the volumetric flow of sludge cake using Eq. 11-37 with the factor for specific gravity in the denominator is

$$V = \frac{70 \text{ gpm} \times 0.040 \times 8.34}{0.35 \times 8.34 \times 1.05} = 7.6 \text{ gpm} = 8 \text{ gpm}$$

Flow of filtrate $= $ sludge feed $-$ cake volume

$$= 70 - 8 = 62 \text{ gpm}$$

Wastewater flow $=$ filtrate $+$ polymer feed $+$ wash water

$$= 62 + 6 + 50 = 118 \text{ gpm}$$

Solids concentration in wastewater

$$= \frac{1800 \text{ mg/l}}{10,000 \text{ mg/l per percent}} = 0.18 \text{ percent}$$

Table 11-7. Typical Operating Parameters for Belt Filter Press Dewatering of Polymer Flocculated Wastewater Sludges

Type of Sludge	Feed Solids (Percent)	Hydraulic Loading (gpm/m)[a]	Solids Loading (lb/m/hr)[b]	Cake Solids (Percent)	Polymer Dosage (lb/ton)[c]
Anaerobically digested primary only	4 to 6	40 to 50	1000 to 1600	25 to 35	3 to 6
Anaerobically digested primary plus waste activated	2 to 5	40 to 50	500 to 1000	15 to 26	6 to 12
Aerobically digested without primary	1 to 3	30 to 45	200 to 500	11 to 22	8 to 14
Raw primary and waste activated	3 to 6	40 to 50	800 to 1200	16 to 25	4 to 10
Thickened waste activated	3 to 5	40 to 50	800 to 1000	14 to 20	6 to 8
Extended aeration waste activated	1 to 3	30 to 45	200 to 500	11 to 22	8 to 14
Heat-treated primary plus waste activated	4 to 10	35 to 50	1000 to 1800	30 to 40	1 to 2

[a] 1.0 gpm/m = 0.225 m³/m·h
[b] 1.0 lb/m/hr = 0.454 kg/m·h
[c] 1.0 lb/ton = 0.500 kg/tonne

Solids in wastewater

$$= \frac{118 \text{ gpm} \times 60 \text{ min/hr} \times 0.0018 \times 8.34}{1.5 \text{ m}} = 71 \text{ lb/m/hr}$$

$$\text{Solids recovery} = \frac{940 \text{ lb/m/hr} - 71 \text{ lb/m/hr}}{940 \text{ lb/m/hr}}$$

$$= 0.92 = 92 \text{ percent}$$

11-20 CENTRIFUGATION

Centrifuges are employed for dewatering raw, digested, and waste activated sludges, although they are not as popular as vacuum filters and belt presses. A discussion of centrifugation is presented in Section 7-17, and the common types employed are illustrated in Figures 7-36 and 7-37. Typical operation produces a cake with 15 to 30 percent solids concentration, depending on the character of the sludge. Without chemical conditioning the solids capture is in the range of 50 to 80 percent, while proper chemical pretreatment can increase solids recovery to 80 to 95 percent. Disposal of a centrate with high suspended, nonsettleable solids as a result of sludge characteristics and inadequate chemical conditioning can cause problems. For example, high solids in the return to the head of the treatment plant can result in a large load of fine solids recirculating around and around through the centrifuge and the wastewater processing units. Both polyelectrolytes, and ferric chloride and lime may be used as chemical conditioners. The choice between centrifugation and filtration is based on performance and economics. The latter involves both initial cost of installation and operating costs including labor, chemicals, and power.

11-21 COMPOSTING

Composting is the stabilization of moist organic solids by natural biological processes when the organic matter is placed in piles that allow ventilation. In addition to digesting putrescible organic matter, the objectives of sludge composting are to destroy pathogenic organisms and reduce the mass and volume of waste. The optimum moisture content for a compost mixture is 50 to 60 percent. Less than 40 percent limits the rate of decomposition, while over 60 percent is generally too wet for adequate ventilation. The volatile solids reduction and water loss during composting are 50 percent or more. For efficient stabilization and pasteurization, the temperature in a compost pile should rise to a level of 130 to 150° F (55 to 65° C), but not above 176° F (80° C). Composting temperature is influenced by moisture content, degree of aeration, size and shape of the pile, and climatic conditions, particularly air temperature and rainfall. The finished compost is a friable humus with a moisture content less than 40 percent. Although too low in nutrients to be considered a fertilizer, compost is an excellent soil conditioner. For example, when mixed with soil, the added humus content increases the capacity for retention of water.

Sludge cake for composting is usually raw organic solids dewatered using polymer as a conditioning chemical, although partially digested sludge may also be composted for additional stabilization. Dewatered cake alone, commonly having a solids content of 20 to 30 percent, is too wet and too compact to have adequate porosity for aeration of the interior of a pile. If it is not mixed with another substance, the pile of sludge cake will become a dense mass with a wet, anaerobic interior and a dried external crust. Therefore, the dewatered cake is mixed with either an organic amendment like dried manure, straw, or sawdust or a recoverable bulking agent, which is usually wood chips, to reduce the unit weight and increase air voids. Aerobic composting is desired since the end products are carbon dioxide, water, and heat. Anaerobic composting produces intermediate organics, such as organic acids, and gases including hydrogen sulfide and methane. Anaerobic decomposition is more likely to emit odors and releases less heat. Finished compost may also be recycled to help dry the wet cake.

Compost is placed in either windrows, agitated by periodic turning for mixing and aeration, or static piles with forced ventilation. The piles may be either exposed to the atmosphere or sheltered under a roofed structure. The choice of processes and sheltering is based on climate, environmental considerations, availability of a bulking agent, and economics. Compost piles in the windrow system are arranged in long parallel rows. These windrows are remixed and turned at regular intervals to restructure the compost, ensuring uniform biological stabilization. Depending on the equipment used for turning, the shape of the windrows is either triangular or trapezoidal and varies in height and width. The common dimensions are 4 to 8 ft in height and 8 to 12 ft in width. Figure 11-62 pictures a self-propelled windrow composting machine straddling a triangular pile 6 ft high and 12 ft wide that is being restructured. The mixing is done by a rotating agitator mounted near the ground between the side plates located just inside the wheels.

In agricultural regions, manure from confined feeding of cattle can be used for a composting amendment. The manure is aged and dried by stacking in the feedlot, and the wastewater sludge is cake from mechanical dewatering of undigested sludge with polymer conditioning. These are mixed in approximately equal portions using a modified manure spreader with the back beaters reversed so that the mixture is deposited on the ground in a row, rather than being thrown upward by the back beaters for widespread distribution. A windrow composting machine, constructed for heavy-duty mixing (Figure 11-62), is used to stack and periodically turn the piles. Stabilization requires four to six weeks in good weather with weekly turning. In northern climates, the windrows may freeze on the outside preventing turning and slowing the rate of stabilization; however, the process can still be operated in areas where nuisance odors are not a serious problem. The finished compost is stored and applied at appropriate times on grassland and crop land.

Composting by the aerated static pile process requires a structure with special equipment since ventilation is provided by mechanically drawing air through the pile. Porosity of the compost is maintained by wood chips that also reduce the initial water content by absorption. The ratio of dewatered sludge cake to wood chips on a volumetric basis is between 1 to 2 and 2 to 3. After a base of wood chips is prepared over a network of perforated air pipes, the mixture is placed on top in a pile to a depth of 8 to 10 ft. A layer of finished compost is then spread over the pile to form a cover. For a period of about three weeks, air is drawn through the pile by a blower operating

Figure 11-62 Windrow composting machine for mixing and piling compost in triangular shaped rows. (Courtesy of Scarab Mfg., Route 2, White Deer, Texas 79097.)

intermittently to prevent excessive cooling. The blower exhaust is deodorized by venting through a pile of finished compost. After stabilization, the mixture is allowed to cure and dry for several weeks. The wood chips are separated from the compost while the compost is mechanically stockpiled, using a vibrating screen, and stored for reuse.

11-22 LAND DISPOSAL

The majority of wastewater sludges are disposed of on land with about three quarters being used as soil conditioner and the remainder buried in landfill. Burial can be the final disposal for dried digested sludge and dewatered raw sludge cake. Environmental regulations generally require burial of raw sludges and sludges containing excessive heavy metal concentrations or organic toxins. Since these can pose health problems, the landfill sites are designed to prevent groundwater and surface-water pollution. Many cities bury wastewater sludge in their sanitary landfill along with municipal refuse. The basic operations include spreading, compacting, and covering the wastes daily with excavated soil.

Wastewater sludge is spread on agricultural land as a soil conditioner and fertilizer. Liquid disposal of digested sludge is popular where acceptable sites are located within convenient hauling distance. This process is most applicable to small plants that are located in rural areas and are equipped with aerobic or anaerobic digesters to biologically stabilize the sludge before spreading. Even though hauling costs may be high, the alternatives of solid-liquid separation on drying beds or mechanical dewatering are often more expensive. The liquid sludge can be applied on the surface by a vehicle equipped for spreading or can be injected underneath the surface using chisel plows. Spreading from fixed or portable irrigation nozzles may be practiced where odor and insects are not problems.

Subsurface injection of liquid digested sludge is the most environmentally acceptable method since the sludge is incorporated directly with the soil, reducing exposure to the atmosphere. Figure 11-63 pictures an injection unit with seven spring-loaded chisel plow shanks pulled by a tractor. As the blade of each chisel plow passes through

the soil, a cavity is formed approximately 6 in. below the ground surface for injecting the sludge. Sludge is supplied to the injection unit from a remote holding tank through underground piping and a flexible delivery hose pulled behind the tractor. Sludge can also be incorporated into soil using a disc plow with hoses discharging sludge ahead of each disc. This process is less effective in covering the

Figure 11-63 Subsurface application of liquid digested sludge on agricultural land. (a) The sludge is pumped through a flexible hose to the injection unit, which is pulled by a tractor. (b) A hose located behind the shank of each chisel plow blade injects the sludge underground. (Lincoln, Nebraska.)

liquid sludge and requires overturning of the ground surface.

At large plants, digested sludge is dewatered to reduce the weight and cost of hauling. The sludge cake is stored in piles on an open site and placed on crop land at appropriate times using a spreader with back beaters that throw the solids outward from the rear of a wagon or truck-mounted box. Disc cultivators then mix the sludge solids with the surface soils before planting a crop.

The environmental concerns of sludge disposal on agricultural land are pollution of groundwater and surface water, contamination of the soil and crops with toxic substances, and transmission of human and animal diseases. Normally, laboratory analyses of a sludge are conducted for solids concentration; nitrogen, phosphorus, and potassium contents; presence of the heavy metals of cadmium, copper, nickel, lead, and zinc; and selected organic compounds, such as polychlorinated biphenyls. For protection of human health, the preferred crops are fodder grasses, feed grains, and fiber crops that are not in the human food chain. Grasses are considered acceptable provided cattle are restricted from grazing for a specified period after sludge application. Feed grains for animal consumption, like corn, are fertilized by tilling sludge solids into the soil before planting of crops. To ensure adequate environmental control, the sludge, soil, groundwater, and runoff are monitored for fecal coliforms, nutrients, and contaminants.

Cadmium is the heavy metal of greatest concern since it can be taken up by crops and, therefore, has the potential of entering the human food chain. Despite this, plant uptake from soils containing cadmium does not always occur. The pH of the soil and the kind of plant species are important considerations. The highest amounts of cadmium occur in fibrous roots and plant leaves, thus, potential high-risk food plants for humans are leafy vegetables like lettuce and spinach. Fodder crops and cereal grains appear to present the least risk because of their low concentration of cadmium.

Guidelines for restricting the addition of cadmium in sludge to agricultural land are based on loading rates, concentration of cadmium in the sludge, and accumulation in the soil. The maximum permissible level recommended by the Environmental Protection Agency is 2.0 kg/ha·y (1.8 lb/ac/yr) on agricultural land with 500 g/ha·y (0.45 lb/ac/yr) suggested for accumulator crops. Other regulations may limit the concentration in the sludge applied to land to less than 20 mg/kg of dry solids. Based on soil characteristics, the maximum allowable accumulation is in the range of 10 to 50 mg/kg of dry soil.

Nitrogen content of a sludge often determines the allowable rate of application. The concern is pollution of groundwater with nitrate, which is a health-related contaminant in drinking water standards. On porous soils, the recommended rate of nitrogen application is an amount equivalent to the nutrient uptake of the vegetation.

Domestic sludge contains a variety of bacteria, viruses, and parasitic organisms pathogenic to humans and animals. Anaerobic or aerobic digestion of sludge is effective in reducing the numbers of viruses and bacteria; however, roundworm and tapeworm ova or other resistant parasites can still be present in the stabilized sludge. If pathogenic bacteria do survive digestion, they are likely to be inactivated by sunlight, drying, and competition in the soil environment. In contrast, viruses and parasitic ova, being more resistant, can persist in soils or on vegetation for several weeks or even months. Even though the risk of infecting livestock is not great, grazing of animals or harvesting a fodder crop is not advised until six weeks or more after sludge application. Regarding human health, despite the possibility of communicable disease transmission, the lack of epidemiological evidence suggests that the current practice of sludge disposal on land is safe.

11-23 INCINERATION AND DRYING

Mechanical dewatering is used as pretreatment prior to drying or burning of waste solids. Sludge drying involves reducing water content by vaporizing the water to air. Auxiliary fuel is burned to heat the ambient air and to provide the latent heat of evaporation. Incineration is an extension of the drying process and converts solids into an inert ash that can be disposed of easily. If dewatered to approximately 35 percent solids, the process is usually self-sustaining without supplemental fuel, except for initial warm-up and heat control. It is generally preferable

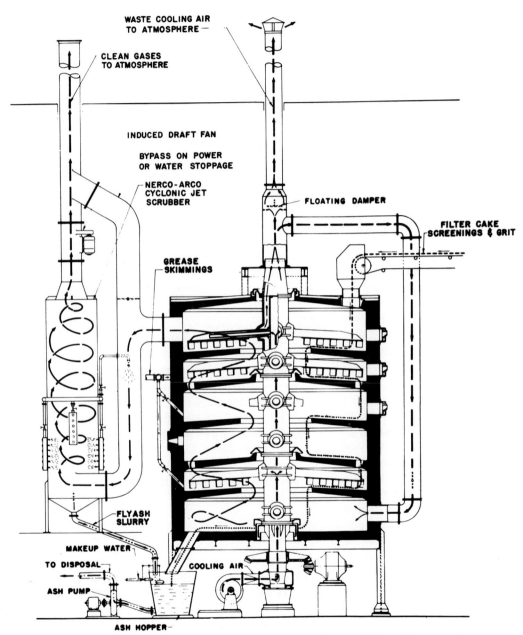

Figure 11-64 Multiple-hearth furnace for burning dewatered waste sludge. (Courtesy of Nichols Engineering and Research Corp., a Signal Co.)

to burn raw rather than digested sludge because of its higher heat value.

The multiple-hearth furnace illustrated in Figure 11-64 has proved to be successful for both sludge drying and incineration. The furnace consists of a circular steel shell containing several hearths arranged in a vertical stack, and a central rotating shaft with rabble arms. Sludge cake is fed onto the top hearth and is raked slowly in a spiral path to the center. Here, it drops to the second level where it is pushed to the periphery and drops, in turn, to the third hearth where it is again raked to the center. The two upper levels allow for evaporation of moisture, the middle hearths burn the solids producing temperatures exceeding 1400° F, and the bottom zone cools the ash prior to discharge.

The hollow central shaft is cooled by forced air vented out the top. A portion of this preheated air from the shaft is piped to the lowest hearth and is further heated by the hot ash and combustion as it passes up into the furnace. The air then cools as it gives up its heat to dry the incoming sludge. Countercurrent flow of air and sludge solids optimize combustion efficiency. After the air passes through the furnace twice, it is discharged through a wet scrubber to remove fly ash. This furnace can also be designed as a dryer only. Hot gases from an external furnace flow parallel with the sludge downward through the multiple hearths to dry the solids without scorching.

The main air pollution of concern in operating a sludge incinerator is emission of particulate matter. The common performance standard specifies that the gas discharged into the atmosphere cannot contain particulate matter in excess of 70 mg/m³, nor exhibit 10 percent or greater opacity (shadiness), except for 2 min in any 1 hr. The presence of uncombined water is the only reason for failure to meet the latter requirement. Performance tests are conducted with the incinerator operating at design load and burning representative sludge. Both the cyclonic water jet scrubber and the perforated plate-type jet scrubber are able to meet the particulate air quality standard, as well as the acceptable stack emissions of nitrogen oxides, sulfur oxides, and odors.

Two other heat-treating processes applied in sludge incineration are the flash-drying incineration system and the fluidized bed reactor. Flash drying involves pulveriz-ing wet sludge cake with recycled dried solids in a cage mill. Hot gases from the incinerating furnace suspend the dispersed sludge particles up into a pipe duct where they are dried. A cyclone separator removes the dried solids from the moisture-laden hot gas, which is returned to the furnace. Part of the dried sludge returns to the mixer for blending with incoming wet cake. The remainder is either withdrawn for fertilizer or incinerated.

A fluidized bed reactor uses a sand bed that is kept in fluid condition by an upflow of air, as a heat reservoir to promote uniform combustion of sludge solids. The bed is preheated to approximately 1200° F by using fuel oil or gas. When sludge cake is introduced, rapid combustion occurs maintaining an operating temperature of about 1500° F. Ash and water vapor are carried out with the combustion gases. A cyclonic wet scrubber is used to remove ash from the exhaust gases. Finally, the ash is separated from the scrubber water in a cyclone separator.

FURTHER READINGS

1. *Recommended Standards for Sewage Works*, Great Lakes–Upper Mississippi River Board of State Sanitary Engineers, 1978 Edition (Health Education Service, Albany, N.Y.).
2. *Wastewater Treatment Plant Design*, WPCF Manual of Practice No. 8, Water Pollution Control Federation, Washington, D.C., 1977.
3. Viessman, W., Jr., and M. J. Hammer. *Water Supply and Pollution Control*, 4th Edition. New York; Harper & Row, 1985.

PROBLEMS

11-1 The design data listed by a sanitary engineer for a wastewater treatment plant are design average flow = 135,000 gpd, design minimum flow = 50,000 gpd, and design maximum flow = 230,000 gpd. Explain the meaning of this information.

11-2 List the standards of effluent quality normally included in an NPDES discharge permit.

11-3 Listed below are effluent BOD concentrations from a municipal treatment plant for 30 consecutive days. Does this plant meet the minimum level of effluent quality for BOD specified by the EPA regulations for secondary treatment?

Day	BOD (mg/1)	Day	BOD (mg/1)	Day	BOD (mg/1)
1	27	11	20	21	32
2	29	12	16	22	34
3	28	13	22	23	30
4	33	14	46	24	26
5	20	15	48	25	22
6	20	16	56	26	16
7	24	17	60	27	18
8	26	18	40	28	22
9	24	19	42	29	24
10	22	20	38	30	26

11-4 List the processes normally included in preliminary treatment, give the primary purpose of each operation, and the typical arrangements of preliminary units.

11-5 The first treatment tanks, following screening and pumping, in the processing of a wastewater are 10 ft by 10 ft with a liquid depth of 8 ft. A flow of 0.55 mgd enters each tank along one side and overflows a weir on the opposite side. The tanks are aerated and equipped with rotary scrapers. Calculate the detention time. What are the purposes of these tanks? What other process unit is normally installed with these tanks?

11-6 A pumping station, similar to the one drawn in Figure 11-7, has a wet-well capacity of 6000 gal below the high water level. The volume between the low and high water levels is 3000 gal with a vertical distance between the low and high levels of 2.5 ft. Four constant-speed pumps are installed with the following design capacities: pump 1 = 400 gpm, pump 2 = 400 gpm, pump 3 = 600 gpm, and pump 4 = 600 gpm. The wastewater inflows for the design period of the pumping station are minimum (night) flows in the range of 90 to 140 gpm, maximum flows during wet weather ranging from 1100 to 1400 gpm, and average daily flows increasing from 350 to 550 gpm during the design period. The outlet from the pumping station is an 8-in. ductile-iron pipe 200 ft in length that discharges into a grit chamber with a water level elevation 32.0 ft above the low water level in the wet well. The pump controls can be programmed to schedule the operation of pumps based on the level of the water in the well. The proposed schedule is all pumps are off at the low water level; pump 1 turns on when the water level rises 1.0 ft; pump 2 turns on when the level rises 2.0 ft above the low water level; and pump 3 turns on when the water rises to the maximum level of 2.5 ft above the low water level. Pumps 1 and 2 and pumps 3 and 4 alternate with each on-off cycle. The volume of the wet well corresponding to a 1.0-ft rise in water level is 1200 gal, and a rise of 0.5 ft increases the storage volume 600 gal. (a) Is the wet-well capacity satisfactory and are the pumps suitably sized? (Criteria are given in Section 11-3.) (b) Draw system head-discharge curves as illustrated in Figure 11-8. Sketch the approximate location of modified pump head-discharge curves for pump 1 or 2, pump 3 or 4, and pumps 1, 2, and 3. Next, sketch estimated pump characteristic curves slightly above the modified pump head-discharge curves and locate the minimum and maximum anticipated pumping rates. Locate their positions with shaded angles as shown in Figure 11-8.

11-7 In Figure 11-9, locate the point where one variable-speed pump is operating at 100 percent speed and the second pump is operating at 87 percent speed. From this point (a) if the water level is rising in the wet well, when does the third pump start, and (b) if the water level is dropping, when does the second pump stop? Why is the water level allowed to drop slightly below the starting water level before a pump is stopped?

11-8 The operating characteristics of a screw pump are the following.

Pump Capacity (percent)	Water Level in Influent Chamber (percent)	Pump Efficiency (percent)
0	0	0
5	25	45
10	34	54
20	48	63
30	58	68
40	66	71
50	73	73
60	79	74
80	90	75
100	100	75

Plot water level in the influent chamber on the vertical axis versus pump capacity on the horizontal axis. On the same graph, also plot efficiency versus capacity. (a) When the water level is 60 percent of the filling level, at what percentage of the maximum capacity is the pump operating? (b) Over what range of pump capacity is the efficiency greater than 70 percent? (c) What is the advantage of a screw pump exemplified by these graphs?

11-9   Two final clarifiers following activated-sludge processing have diameters of 60 ft and side water depths of 11 ft. The effluent weirs are inboard channels set on a mean diameter of 55 ft. For a total flow of 3.4 mgd, calculate the overflow rate, detention time, and weir loading. (*Answers* 600 gpd/sq ft, 3.3 hr, 4900 gpd/ft)

11-10   Calculate the overflow rate and detention time in a primary clarifier with a diameter of 24 m and water depth of 2.3 m for a flow of 10,000 m^3/d. (*Answers* 22 $m^3/m^2 \cdot d$, 2.5 h)

11-11   The processing sequence for treatment of a strong municipal wastewater is primary sedimentation, filtration through biological towers, intermediate clarification, biological step aeration, and final clarification. All processing steps have duplicate units of identical dimensions. The total plant flow is 10.0 mgd. By selecting appropriate design criteria, determine the surface areas and water depths for primary, intermediate, and final clarifiers.

11-12   An extended aeration (activated-sludge) plant has two final clarifiers, each 30 ft in diameter with an 10-ft side water depth. Estimate the design hydraulic loading for the treatment plant.

11-13   List the four main factors that influence the BOD removal efficiency of the second stage in a two-stage stone-media trickling filter plant.

11-14   A trickling filter plant as diagramed in Figure 11-18 has a primary clarifier with a diameter of 45 ft, a filter with a diameter of 70 ft and 7.0-ft depth of stone media, and final clarifier with a diameter of 36 ft. The wastewater influent is 0.80 mgd with a BOD of 200 mg/l. The plant is operated with indirect recirculation during low-flow equal to 0.40 mgd and constant direct recirculation at 600 gpm. Calculate overflow rates of the clarifiers, BOD and hydraulic loadings on the

trickling filter, and effluent BOD and BOD removal efficiency of the plant at a wastewater temperature of 16° C. (*Answers* 500 gpd/sq ft without recirculation, 790 gpd/sq ft, 32 lb/1000 cu ft/day, 23 mgad, 38 mg/l, 71 percent)

11-15   A single-stage filter plant as diagramed in Figure 11-18 treats a domestic wastewater. The clarifiers are sized in accordance with recommended practice. The trickling filter, containing stone media to a depth of 7.0 ft, is sized so that the BOD loading is 30 lb/1000 cu ft/day and the wastewater is recirculated to provide a recirculation ratio of 1.0. What is the calculated maximum allowable BOD concentration in the raw wastewater at a temperature of 16° C to produce an effluent BOD of 30 mg/l?

11-16   Calculate the BOD efficiency in a two-stage trickling filter system following primary sedimentation at a wastewater temperature of 16° C. The first filter is loaded at 90 lb/1000 cu ft/day and $R = 0.5$. The second filter is the same size as the first and operates at $R = 0.5$. (*Answer* 80 percent)

11-17   In Example 11-4, the sizes of treatment units for a two-stage trickling filter plant are calculated and an effluent BOD of 26 mg/l is determined based on an influent BOD of 280 mg/l and wastewater temperature of 17° C. Using filters of the same size and the same hydraulic flow pattern, what is the maximum raw wastewater BOD concentration that will produce an effluent BOD of 30 mg/l at a wastewater temperature of 20° C?

11-18   The flow diagram of a proposed two-stage stone-media trickling filter plant is shown in Figure 11-65. The design wastewater flow is 2.0 mgd with a BOD of 250 mg/l. The effluent standards are 30 mg/l of BOD and 30 mg/l of suspended solids when the wastewater temperature is 16° C. Determine the dimensions for the primary clarifiers, first-stage filters, intermediate clarifiers, second-stage filters, and final clarifiers. (*Answers* 50-ft diameter and 8.0-ft depth, 52-ft diameter and 7.0-ft depth of media, 40-ft diameter and 7.0-ft depth, 52-ft diameter and 7.0-ft depth of media, and 40-ft diameter and 7.0-ft depth)

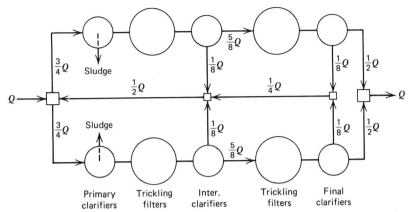

Figure 11-65   Flow diagram for Problem 11-18.

11-19   A biological tower is filled with random packing to a depth of 18 ft. The wastewater being treated contains 90 mg/l soluble BOD and 130 mg/l total BOD. Based on studies using an experimental filter filled with the random packing, the $n = 0.39$ and the $K = 0.090$ per min at a temperature of 20° C. The desired settled effluent quality at a wastewater temperature of 16° C is a soluble BOD of 8.0 mg/l, which is equivalent to a total BOD of 25 mg/l. What is the calculated allowable hydraulic loading with a recirculation ratio of 2.0 to meet the desired effluent BOD? (*Answer* 0.89 gpm/sq ft)

11-20   A treatment plant processes domestic wastewater by primary sedimentation, rotating biological contactors, and final clarification. Each RBC shaft has a length of 17 ft 4 in. with 12-ft-diameter disks for a nominal surface area of 60,000 sq ft. The installation has 16 shafts arranged with 4 rows of shafts of four stages each. Assuming a raw wastewater BOD of 200 mg/l and required effluent BOD of 30 mg/l at a wastewater temperature of 50° F, what is the calculated hydraulic capacity of the plant?

11-21   The aeration tank of the laboratory activated-sludge unit shown in Figure 9-3 has a liquid volume of 4.0 liters. Domestic wastewater with a BOD of 200 mg/l is fed to the tank at a rate of 12 l/d. The MLSS concentration is held at 2000 mg/l by wasting 200 ml of mixed liquor from the aeration tank each day. The clarifier efflu-

ent averages a BOD of 15 mg/l and suspended solids concentration of 20 mg/l. Calculate the BOD loading, F/M ratio, sludge age, aeration period, and BOD removal efficiency. (*Answers* 600 g/m³·d, 0.30 g/d·g, 12.5 d, 8.0 h, 92.5 percent)

11-22   A conventional activated-sludge tank is 24 ft wide, 98 ft long, and has a 13-ft liquid depth. The settled wastewater being treated is 0.91 mgd with a BOD of 130 mg/l. Compute the BOD loading and aeration period. The operating MLSS concentration is 2200 mg/l, and the settled sludge volume in the SVI test is 230 ml/l. Calculate the F/M ratio, SVI, recommended sludge recirculation rate, and solids concentration in the return sludge.

11-23   A step-aeration secondary is operating under the following conditions: settled influent wastewater flow = 5.5 mgd, average influent BOD = 126 mg/l, volume of aeration tanks = 150,000 cu ft, and the MLSS concentration is 2500 mg/l Determine the following: BOD loading, aeration period, F/M ratio, and estimated effluent BOD assuming good operation. (*Answers* 38.5 lb BOD/1000 cu ft/day, 4.9 hr, 0.25 lb BOD/day/lb MLSS, 15 to 20 mg/l BOD)

11-24   A design wastewater flow for a conventional activated-sludge plant is 3.0 mgd containing 280 mg/l of BOD and 300 mg/l of suspended solids. Calculate the diameter and side water depth for two primary clarifiers.

For the loading and operational parameters listed in Table 11-2, recommend the required volume for the aeration tanks. Based on the *GLUMRB Standards*, recommend the air requirement for a diffused air system. Calculate the diameter and side water depth for two final clarifiers.

11-25   An extended aeration process, without primary settling, treats a dairy wastewater of 0.30 mgd at 1000 mg/l BOD. For the aeration tank with a volume of 62,500 cu ft, calculate the BOD loading and aeration period. If the MLSS concentration in the aeration tank is 2500 mg/l, calculate the F/M ratio. Referring to Table 11-2, why is the aeration period indicative of extended aeration while the BOD loading and F/M values are in the range of conventional and step-aeration processes? If the settled sludge volume is 200 ml, what is the SVI value? If the return sludge rate is 100 percent, what is the concentration of suspended solids in the recirculated activated sludge?

11-26   A step-aeration plant consists of three identical flow streams as diagramed in Figure 11-31*b*. Each one third of the plant treats an average of 2000 m³/d of wastewater containing 250 mg/l of BOD and 250 mg/l of suspended solids. The primary clarifier has a surface area of 80 m² and a side water depth of 2.3 m. The aeration tank is 27 m long and 4 m wide with a liquid depth of 5.0 m. The mixed liquor in the aeration tank has a suspended solids concentration of 3000 mg/l, and the measured sludge volume index is normally 90 ml/g. Calculate the following: overflow rate and detention time in the primary clarifiers; BOD loading, F/M ratio and aeration period in the aeration tanks; and recommended minimum return sludge rate. What design parameters are recommended for sizing the final clarifiers?

11-27   Estimate the design flow and BOD loading for an extended aeration system with floating mechanical aerators. The units sizes are volume of aeration tanks = 2.50 mil gal, oxygen transfer capability of aerators = 340 lb/hr at 2.0 mg/l DO, and 10-ft deep final clarifiers with a surface area of 3400 sq ft. (*Answers* 2.7 mgd and 24 lb BOD/1000 cu ft/day)

11-28   A laboratory study was performed on a soluble wastewater to determine the kinetic constants for the mathematical model discussed in Section 11-8. The aeration chamber had a volume of 12 liters mixed by diffused aeration. The wastewater was fed at a constant rate of 41.2 l/d to provide an aeration period of 7.0 h. A measured volume of mixed liquor was wasted once a day from the aeration tank. The following data were collected for four different sludge ages. The symbols refer to those used in Figure 11-40.

Influent Flow Q (l/d)	Influent Soluble BOD S_0 (mg/l)	Wasted Mixed Liquor Q_w (l/d)	Aeration Volatile MLSS X (mg/l)	Effluent Soluble BOD S_e (mg/l)	Effluent Volatile SS X_e (mg/l)
41.2	126	0.30	1730	5.2	9.4
41.2	126	0.42	1500	7.3	8.0
41.2	126	0.96	968	10.5	8.4
41.2	126	1.08	848	11.5	7.9

Determine the values for Y, k_d, k and K_S. From the data given for each run, calculate θ_c, $1/\theta_c$, r_{su}, U, $1/U$, and $1/S_e$. Plot these values as shown in Figure 11-41 to determine the kinetic constants. (*Answers* $Y = 0.35$ mg VSS/mg BOD, $k_d = 0.04$ per day, $k = 4.0$ per day, $K_S = 90$ mg/l of soluble BOD)

11-29   A completely mixed activated-sludge process is being designed for a wastewater flow of 10,000 m³/d (2.64 mgd) using kinetics equations. The soluble raw wastewater BOD = 90 mg/l (total BOD = 130 mg/l), and the desired effluent soluble BOD = 10 mg/l (total BOD = 30 mg/l). For determining the aeration tank volume, the sludge age is selected to be 20 d and the volatile MLSS 2000 mg/l (total MLSS = 2800 mg/l). The kinetics constants determined from a laboratory study of the wastewater are $Y = 0.60$ mg VSS/mg BOD, $k_d = 0.06$ per day, $K_S = 60$ mg/l of BOD, and $k = 5.0$ per day. Calculate the volume required for the aeration tanks, aeration period, and F/M ratio. (*Answers* 2200 m³, 5.3 h, 0.20 g/d soluble BOD/g volatile MLSS)

11-30   Using the data and answers given in Problem 11-29, calculate the BOD loading and F/M ratio in the units listed in Table 11-2.

11-31 Stabilization ponds with a total surface area of 14.8 acres receive a wastewater flow of 140,000 gpd with a BOD of 320 mg/l. Calculate the BOD loading and winter storage available between the 2-ft and 5-ft depths assuming a water loss of 0.11 in./day during the winter months. (*Answers* 25 lb BOD/day/acre, 150 days)

11-32 Based on a loading of 25 lb BOD/day/acre, what stabilization pond area is needed for an average daily flow of 0.84 mgd with 200 mg/l BOD? Calculate the minimum water depth for a retention time of 90 days assuming that the difference between evaporation and seepage and precipitation is 1.0 in./week.

11-33 Stabilization ponds for a town are constructed with two cells for series operation. The first cell has a surface area of 27,000 m^2 and the second is smaller with 15,000 m^2. Water-level control in both ponds permits operation between 0.6 m and 1.5 m of water depth. Water loss from the ponds averages 2.1 mm/d. The average influent wastewater is 300 m^3/d with 250 mg/l of BOD. Calculate the daily BOD loading per unit area on the first cell. No discharge is allowed for a time period of 120 d. If the ponds are full, what is the calculated minimum lowering of the water level (in both ponds) required to ensure zero discharge for 120 days?

11-34 The proposed stabilization pond arrangement for a small community consists of three cells; the primary ponds provide a total surface area of 12 acres in two 6-acre cells, and the polishing cell, which receives no raw wastewater, has a water area of 2 acres. The anticipated wastewater loading is 160,000 gpd with a BOD of 280 mg/l. Based on available data, the net water loss (evaporation plus seepage minus precipitation) during the winter is 1.0 in./month. Design criteria specified for stabilization ponds in this area include (a) the primary cells should not be loaded at greater than 25 lb BOD/acre/day; (b) the minimum total water volume, based on influent flow and all cells at a water depth of 5 ft, should not be less than 120 days; and (c) the volume for winter storage, between the liquid depths of 1.5 and 5.0 ft, should be sufficient so that no discharge is necessary for the six-month period from November 1 to May 1. Does the proposed design meet these standards?

11-35 A food-processing wastewater is to be pretreated in a separate aerated lagoon prior to being discharged into municipal stabilization ponds. The design specification is to reduce the BOD from 600 mg/l to 150 mg/l at 15°C. Based on a laboratory study, the k_{20} of the wastewater is 0.70 per day and the θ is 1.047. What is the calculated aeration period for the aerated lagoon?

11-36 What is the primary reason for disinfecting wastewater discharges? Write the chemical reaction that takes place when hypochlorous acid solution from a chlorine feeder is mixed with wastewater effluent. Is chlorination an effective means of inactivating viruses? Why is residual chlorine detrimental to a receiving stream?

11-37 For each of the following treatment plants, estimate the quantity of sludge solids and volume of waste sludge produced per person. Assume a per capita contribution of 120 gpd of wastewater containing 0.20 lb of BOD and 0.24 lb of SS. (a) A two-stage trickling filter plant, with a flow diagram as in Figure 11-19, producing a sludge with 5.0 percent solids withdrawn from the primary clarifier. (b) A conventional activated-sludge plant with a flow diagram as in Figure 11-52. The sludge is a blend of primary sludge with 6.0 percent solids and waste activated sludge thickened to 4.0 percent solids by flotation. (c) An extended aeration plant without primary sedimentation. The sludge wasted from the bottom of the final clarifier contains 1.0 percent solids. [*Answers* (a) 0.16 lb, 0.38 gal, (b) 0.18 lb, 0.42 gal, (c) 0.14 lb, 1.6 gal]

11-38 Estimate the quantity of sludge produced by a two-stage trickling filter plant treating 10,000 m^3/d with a suspended solids concentration of 220 mg/l and a BOD of 200 mg/l. The water content of the sludge is 96.0 percent.

11-39 Estimate the waste activated sludge from a step-aeration process treating 10,000 m^3/d with a settled BOD of 130 mg/l operating at a F/M of 0.24 g BOD per day/g MLSS. The suspended solids in the waste sludge withdrawn from the return sludge pipe are 9500 mg/l. (*Answers* 590 kg/d, 62 m^3/d)

11-40 A gravity thickener processes 10,000 gpd of waste sludge, increasing the solids content from 3.5 to 7.0 percent with 85 percent solids recovery. Calculate the quantity of thickened sludge. (*Answer* 4300 gpd)

11-41   A dissolved air flotation thickener with a surface area of 400 sq ft processes 230,000 gal of waste activated sludge containing 1.0 percent solids in a 24-hr period. Calculate the solids loading in pounds per square foot per hour. If the solids content of the float is 4.0 percent, what is the volume of the thickened waste assuming 98 percent solids capture?

11-42   An activated-sludge plant has a flow scheme as shown in Figure 11-31. The step-aeration tank with a volume of 33,000 cu ft treats a settled wastewater of 1.1 mgd with a BOD of 160 mg/l. The operating MLSS concentration is 3000 mg/l, and the SVI is 80 ml/l. (a) Calculate the BOD loading, aeration period, and F/M ratio. (b) If the recirculated sludge is 0.45 mgd, what is the suspended solids concentration in the waste sludge? (c) What volume of recirculated sludge should be wasted each day? [*Answers* (a) 44 lb BOD/1000 cu ft/day, 5.4 hr, 0.24 lb BOD/day/lb MLSS, (b) 10,000 mg/l, (c) 7900 gal]

11-43   Raw sludge applied to a high-rate digester with a volume of 9600 cu ft is 4800 gpd with 1600 lb of solids that are 70 percent volatile. Calculate the concentration of solids in the raw sludge and in the digested sludge assuming 50 percent volatile solids destruction. Calculate the VS loading and detention time and compare these values to the general loading conditions given in Table 11-5. (*Answers* 4.0 percent, 2.6 percent, 0.12 lb VS/cu ft/day, 15 days)

11-44   A single-stage anaerobic digester has a capacity of 390 m^3 of which 230 m^3 is below the landing brackets. The following conditions prevail: average raw sludge solids = 260 kg/d, raw sludge solids content = 4.0 percent, digestion period = 30 days, total solids reduction during digestion = 45 percent, digested sludge water content = 94 percent, and digested sludge storage required = 90 days. Calculate the digester loading with the cover in the lowered position assuming 70 percent of the solids are volatile. Determine the digester capacity based on Eq. 11-40. Can this digester provide 90 days storage of digested sludge?

11-45   The raw-sludge yield from a trickling filter plant is 9000 gal/day containing 3800 lb of solids that are 70 percent volatile. Capacity of the single-stage, heated,

anaerobic digesters is 100,000 cu ft with the covers fully raised with 65,000 cu ft below the landing brackets. (a) Does the volatile solids loading with the covers resting on the landing brackets appear to be reasonable? (b) Assuming a volatile solids reduction of 65 percent during digestion and 5500 gallons of clear supernatant withdrawn daily calculate the solids concentration in the digested sludge. (c) Estimate the digestion gas production per day. (d) If this plant was located in an agricultural region, what is the most probable method used to dispose of the digested sludge?

11-46   The raw sludge pumped to an anaerobic digester contains 5.0 percent solids that are 65 percent volatile. The digested sludge withdrawn from the tank has 6.5 percent solids that are 40 percent volatile. Based on these data, calculate the percentage of volatile solids converted to gas during digestion, and the quantity of supernatant withdrawn for every 1000 gal of raw sludge fed to the digester. (*Answers* 65 percent, 550 gal)

11-47   The operator of a trickling filter plant is having problems with sludge stabilization in his single-stage, anaerobic digester. The gas production is erratic and low in methane, the supernatant is turbid and odorous, and the digested sludge is too dilute. From the operating records available, which were very incomplete, the following data were determined: The wastewater loading on the plant varied from 500,000 to 600,000 gpd with BOD contents in the range of 180 to 240 mg/l; the town had a sewered population of 5500 persons and several small manufacturing plants; raw sludge was pumped twice a day for a total of 50 min at a rate of 60 gpm from the primary tank into the digester; the temperature of the digesting sludge was maintained between 85 and 90° F; and the digestion tank had a total volume of 38,000 cu ft of which 28,000 cu ft was below the landing brackets. The operator claims his problems are because the digester is too small. What is your comment relative to this claim? Justify your statement with numerial data.

11-48   Anaerobic digestion systems perform the following functions: biological reduction of volatile solids, thickening of digested sludge by withdrawing supernatant, and storage of stabilized sludge. Which of these functions is applicable to aerobic digestion?

11-49 An aerobic digester is operated on a continuous basis by applying sludge and withdrawing supernatant and digested sludge daily. The volume of the tank is 5000 cu ft. After aerating for 20 hr, the aerators are turned off and the suspended solids are allowed to settle for 4 hr. A 1500-gal volume of the clear supernatant is decanted from the surface and a 1500-gal volume of thickened sludge is withdrawn from the bottom. Waste activated sludge is then pumped into the aerobic digester to replace the 3000 gal. The applied waste sludge contains 10,000 mg/l of suspended solids that are 70 percent volatile. The MLSS in the digesting sludge during aeration remains at a concentration of 8300 mg/l. Calculate the VS loading, the solids concentration in the thickened digested sludge withdrawn from the digester based on a volatile solids reduction of 50 percent, and the aeration period in degree-days for an operating temperature of 15° C. (*Answers* 0.035 lb VS/cu ft/day, 13,000 mg/l, 240 degree-days)

11-50 Figure 11-1 is a schematic of conventional wastewater treatment; the text under the diagram describes the processes. Starting with 120 gpcd of average sanitary wastewater (Table 9-2), show in step-by-step calculations how this is processed to one-half pint of filter cake with 30 percent solids. Your final answer need not be precisely one-half pint. List your assumptions. (1 gal = 8 pints)

11-51 A trickling filter plant produces a digested waste sludge with a solids content of 7.0 percent containing 5600 lb of dry solids daily. What area of vacuum filter surface would you recommend assuming a 7-hr operating period per day? Estimate the weight of cake produced assuming chemical additions of 7 percent lime and 3 percent ferric chloride and a water content of 75 percent. (*Answers* 130 sq ft, 25,000 lb/day)

11-52 A vacuum filter with an area of 20 m² dewaters 60 m³ of sludge containing 6.5 percent solids in 6 h of operation. Calculate filter yield and mass of cake with 20 percent solids content. Estimate the volume of filtrate assuming a specific gravity of 1.06 for the cake. Since a low dosage of polyelectrolyte is used in conditioning, the mass and volume of chemical feed can be ignored. (*Answers* 33 kg/m²·h, 20,000 kg/d, 42 m³)

11-53 A waste sludge with a solids content of 5.0 percent was evaluated for filterability on a pilot-scale vacuum filter. Test data for various dosages of conditioning polymer were:

Polymer Dosage, (percent)	Filter Yield, (lb/sq ft/hr)	Cake Solids, (percent)	Filtrate Suspended Solids, (mg/l)
0.2	3.4	22	4200
0.4	5.0	26	2300
0.6	5.8	28	1100
0.8	7.3	29	800
1.0	7.7	31	500

Plot the test data. For a filter yield of 6.0 lb/sq ft/hr, determine the following: polymer dosage in lb/ton of dry solids, weight of cake produced per 1000 gal of sludge processed, volume of filtrate per 1000 gal of sludge processed, and percentage of solids capture.

11-54 A plant that processes wastewater by primary sedimentation and activated-sludge aeration produces 20,000 gpd of anaerobically digested sludge containing 5.0 percent solids. What size belt filter press is recommended for dewatering this sludge in 7 or fewer hours of operation per day? What is the estimated cake production in tons per day?

11-55 A belt filter press housed in a mobile trailer was used to test the dewaterability of an anaerobically digested sludge. The effective width of the belt was 0.50 m. The operating data were sludge feed rate of 20 gpm, polymer dosage of 2.0 gpm containing 0.20 percent powdered polymer by weight, and wash-water flow of 17 gpm. Based on laboratory analyses, total solids in the feed sludge equaled 3.5 percent, total solids in the cake equaled 32 percent, wastewater from belt washing contained 2600 mg/l suspended solids, and the filtrate was 19 gpm with 500 mg/l of suspended solids. Calculate the hydraulic loading, solids loading, polymer dosage, and solids recovery. (*Answers* 40 gpm/m, 700 lb/m/hr, 11.4 lb/ton, and 92 percent)

11-56 At an existing treatment plant, the vacuum filters are being replaced by belt filter presses to dewater anaerobically digested sludge prior to stockpiling and spreading on agricultural land. The design sludge production, which is the average quantity during the maximum month of the year plus 25 percent for future plant expansion, equals 90,000 gpd with an average solids concentration of 5.0 percent. Using a trailer mounted press, the performance data for dewatering the 5.0-percent sludge were hydraulic loading of 45 gpm/m, cake solids content of 24 percent, polymer dosage of 4.0 gpm/m with a concentration of 0.20 percent powdered polymer, wash-water usage of 32 gpm/m, and wastewater production of 64 gpm/m containing 2200 mg/l of suspended solids. Calculate the following: solids loading, polymer dosage, wash-water usage and wastewater production per 1000 gal of sludge dewatered, solids recovery, and daily cake production. For a design operating schedule of a maximum of 12 hr/day with two presses, recommend the size of the presses for installation.

11-57 Why is a bulking agent mixed with dewatered sludge prior to placing in compost piles?

11-58 An anaerobically digested sludge is being spread on agricultural land. The crop is corn with an annual nitrogen uptake of 185 lb/ac. The maximum permissible addition of cadmium in sludge to agricultural land is 0.45 lb/ac/yr. If the sludge contains 80 lb of nitrogen and 0.08 lb of cadmium per ton of dry solids, what is the recommended sludge application rate in tons of dry solids per acre per year?

chapter 12
Operation of Wastewater Systems

Sewer maintenance requires a thorough knowledge of the layout and appurtenances used in collection systems covered in Chapter 10. Wastewater flows, infiltration, and inflow are discussed in Chapter 9. Prerequisite information regarding the function of various unit operations and how they relate to each other is presented in Chapter 11. The following sections are restricted to cleaning and inspection of sewer systems, infiltration and inflow surveys, regulation of sewer use, performance evaluation of treatment plants, and energy conservation.

12-1 SEWER MAINTENANCE AND CLEANING

Serious and expensive sewer problems can result from improper design or poor construction. Adequate slopes to maintain self-cleaning velocities are essential to minimizing maintenance. Selection of a suitable pipe joint is vital to prevent penetration of roots and excessive infiltration. Cutting of tree roots from sewer lines can be an expensive and recurring cleaning process. Groundwater entering joints carries with it soil from around the pipe, which ultimately causes structural failure. In addition to review of new design and supervision of construction, building permits should require careful inspection of all service connections before backfilling. It is important to make certain that unused service lines are properly capped when buildings are demolished.

A successful maintenance program operates on a planned schedule and requires keeping effective records. Maps are used to show location of manholes, flushing inlets, service connections, and other appurtenances. Records should be kept on maintenance performed with particular stress on troublesome lines that are known to require more frequent inspection or cleaning. While large sewers on adequate slopes may never require flushing or cleaning, others must be placed on a regular schedule that may range from every month to once a year. The number of emergency sewer blockages can be materially reduced by such preventative maintenance.

Sewer stoppages are caused chiefly by sand, greasy materials, sticks, stones, and tree roots. Common cleaning techniques are hydraulic scouring with high-pressure jets, scraping with mechanical tools, and addition of chemicals. When a clogged sewer line is reported, the manholes are inspected to determine the location. With few exceptions, the sewer pipe is cleaned in the up-slope direction by working in the first dry manhole downstream from the stoppage. This procedure allows working in a clear manhole, and the wastewater flow from upstream helps to flush the sewer line after the blockage is removed. If roots are suspected as the major cause, operating from the upstream manhole may have an advantage since roots grow in a downstream direction. The large tap root is therefore more easily reached from the upstream side, eliminating the necessity of working through all the fine feeder roots.

Sewer cleaning using a high-pressure water jet is performed as shown in Figure 12-1. The truck is positioned with the hose reel over the opening of the downstream manhole. After a shoe with rollers is placed into the manhole as a hose guide, the hose with a high-pressure

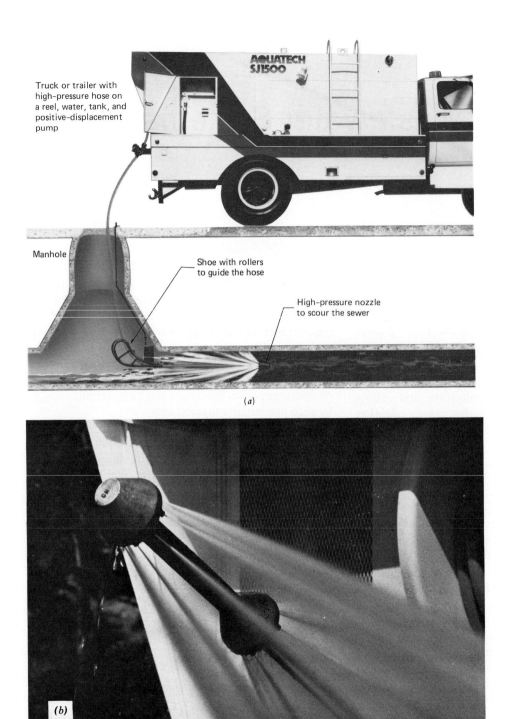

Truck or trailer with
high-pressure hose on
a reel, water, tank, and
positive-displacement
pump

Manhole

Shoe with rollers
to guide the hose

High-pressure nozzle
to scour the sewer

(a)

(b)

Figure 12-1 Sewer cleaning using a high-pressure water jet. (*a*) Schematic diagram of the system. (*b*) Detail of the hose nozzle. (Courtesy of Aquatech Inc., Cleveland, Ohio.)

(a)

Figure 12-2 Sewer cleaning using a high-pressure water jet and vacuum to remove debris. (*a*) The nozzle flushes solids back to a manhole with a temporary dam where the collected solids are removed by a vacuum tube. (*b*) The truck has two tanks. The first one stores solids separated from the wastewater by a cyclone separator. The second tank holds clarified wastewater for continuous cleaning using the high-pressure jet. (Courtesy of Aquatech, Inc., Cleveland, Ohio.)

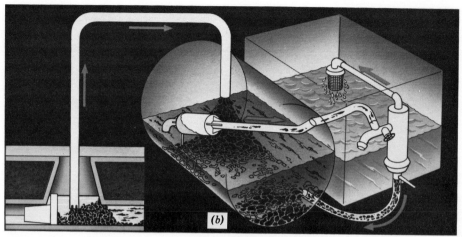

Figure 12-2   (*continued*)

nozzle on the end is placed down through the manhole and into the sewer pipe. Water pumped through the hose is ejected out through orifices on the rear of the nozzle at a pressure up to 2000 psi at a rate up to 60 gpm with a 400-ft hose (Figure 12-1*b*). The resulting force thrusts the nozzle through clogging materials as the jets of water break up the solids. By slowly withdrawing the nozzle, pipe cleaning is completed by washing the dislodged materials back into the manhole. A dam can be placed in the manhole to retain heavy solids for manual removal, so they are not carried into the downstream sewer pipe.

The sewer cleaning vehicle illustrated in Figure 12-2 jet cleans sewers with vacuum removal of debris using recycled water. Before cleaning is started, a plug is placed in the opening of the downstream pipe. The hose with a high-pressure nozzle is placed in the clogged sewer pipe, and the vacuum tube hung down into the manhole. As the water from jet cleaning flows into the manhole, the wastewater and debris are drawn out into an 850-gal cylindrical tank on the truck. The water is transferred to a second 1000-gal tank through a cyclone separator to clarify the wastewater by returning solids to the debris tank. From this tank, the water is recycled through the high-pressure cleaning nozzle. The cylindrical debris tank can be emptied by hydraulically opening a hinged door and tilting the tank to drain out the wastewater and accumulated solids.

Although hydraulic cleaning with high-pressure jets is preferred for most stoppages, a power rodding machine may be advantageous for cleaning out heavy root penetrations or dense solids. The danger in mechanical cleaning is damage to the sewer pipe, for example, cutting into a pipe at an offset joint. The power rodding machine, shown schematically in Figure 12-3, uses sections of flexible steel rod coupled together to hold the cleaning tool. A trailer unit, with a reel for holding several hundred feet of rod, has an engine-driven mechanism with variable-speed control. The rods, fed in through a hose guide braced in the manhole, can be pushed or pulled with the reel rotating or stationary. A number of different tools can be used depending on the reason for the stoppage. A rotating blade cutter is pulled through a sewer to cut grease and roots, or a root saw is pushed through the pipe to accomplish the same thing. An auger or corkscrew may be used to remove large objects. In addition, porcupine scrapers, spearheads, hole openers, and other appurtenances are available for special applications.

Tree roots are extremely troublesome in some sewer systems. Fine roots can be removed by high-pressure jetting. For heavier roots, treatment with chemicals like copper sulfate can be effective in removing and controlling recurrence of root intrusion. After the chemicals are allowed to kill and partially dissolve the roots, hydraulic cleaning can be used to clear the pipe. Power rodders, in

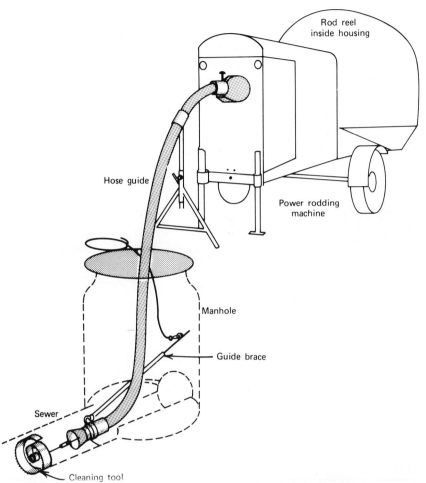

Figure 12-3 Power rodding machine for mechanical sewer cleaning. Tools attached to flexible steel rods can be rotated while being pushed or pulled through the pipeline.

combination with chemicals and flushing, are effective for removing severe root infestations.

12-2 TELEVISION INSPECTION OF SEWERS

Collection systems can be inspected for structural soundness by several techniques. Low-pressure smoke detection is an easy method of locating openings that connect to the atmosphere, such as service connections left unplugged from houses that have been removed, cross-connections with storm sewers, and unauthorized drainage connections. Since rats need access to a dry area above a sewer opening, smoke testing is a good way to locate holes accessible to rodents. A large volume of air combined with smoke is blown into a sewer line through a manhole by a mechanical air-smoke blowing machine. Flowing in the reverse direction of water flow, the smoke follows the sewer pipes to an opening and then rises to the surface through the drain line or cracks in the earth.

Smoke detection simultaneously tests the sewers, house connections, and house plumbing. The smoke is nontoxic and leaves no residue to damage the interior of buildings, although it is irritating to breathe. Before testing, the residents of buildings along a sewer line should be notified of smoke testing procedures to reduce the possibility of alarming people. If smoke enters a building, the occupants should be evacuated and the interior ventilated while locating the point of entry.

The best method for inspecting the inside of small-diameter pipes is by closed-circuit television. After a sewer line has been cleaned, the skid-mounted video camera is

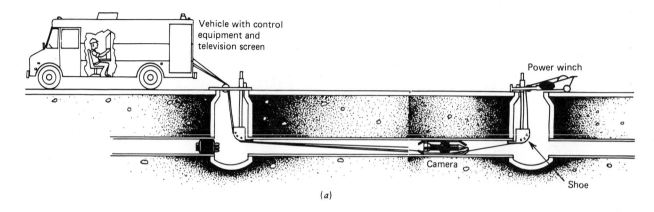

(a)

(b)

(c)

Figure 12-4   Television inspection of sewers. (a) Layout of the system. (b) Skid-mounted television camera. (c) Remote-controlled power winch. (Courtesy of CUES, Inc.) (d) Pictures taken from video tapes. (Lincoln, Nebraska.)

pulled through the pipe and a continuous picture is transmitted to a receiver in the service truck. This technique allows visual inspection and location of structural defects. The video tape can be replayed for further examination and retained for a permanent record. If tapes are reused, photographs can be taken of the video picture as a record of defective conditions in a sewer line.

A diagram of the procedure and pictures of major pieces of equipment used in television inspection are shown in Figure 12-4. The inspection vehicle is parked over the upstream manhole, and a power winch is located at the downstream manhole. The cable from the winch must be pulled through the sewer line to attach the camera at the upstream manhole. This is accomplished by floating a ball with a nylon rope attached down the sewer to the lower manhole. Water from a tank truck can be poured into the manhole to help float the ball downstream. After the nylon rope has been retrieved from the manhole, the stainless steel cable from the winch is attached to the rope

and pulled through the sewer to the television vehicle. The power winch (Figure 12-4c) is laid over the manhole opening so that the cable feeds over a pulley and through a shoe with rollers to guide the cable into the sewer pipe. The camera attached to armored video cables from a reel in the inspection vehicle is connected to the cable from the power winch and lowered into the manhole. The cables are guided into the manhole over a pulley and into the sewer pipe through a shoe with rollers.

The skid-mounted television camera (Figure 12-4b) is designed with a protective case for pulling through a sewer. Either black and white or color cameras are available. The control panel in the vehicle allows the operator to focus the camera and to adjust the light intensity. The camera is pulled forward by remote operation of the power winch and pulled backward by the power cable reel in the vehicle. The television screen on the control panel shows the picture from the camera. Displayed on the picture and recorded on the video tape

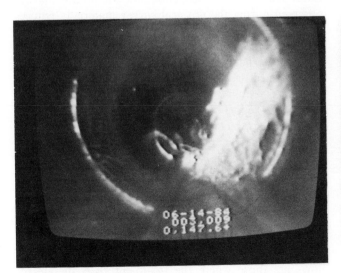
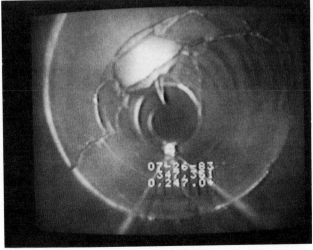

(d)

Figure 12-4 (*continued*)

are the date, manhole numbers, and distance from the sewer entrance. If a repair is to be made on a sewer line, the exact location of the defect is very important so that the excavation for repair will be made in the proper location. The two photographs in Figure 12-4d were taken from video tapes. The one on the left shows roots entering through a cracked joint in the crown of a sewer. In the right photograph, the pipe from a poorly constructed house connection extends into the sewer. The two lines on the bottom are the camera tow cables.

12-3 INFILTRATION AND INFLOW SURVEYS

Entrance of extraneous waters into sewer systems is of concern for several reasons. These include sewer surcharging during periods of intensive rainfall resulting in flooded basements, overloading of treatment plant facilities, overtaxing of pumping stations, excessive costs in processing diluted wastewater flows, and health hazards from discharging raw wastewater. In the past, problems of excessive infiltration and inflow were often handled by construction of relief sewers that by-passed treatment works and flowed directly to surface watercourses. This practice allowed adequate wastewater treatment 90 to 95 percent of the time and appeared at the time to be the best economical design choice. Currently, the goal is to eliminate these pollutional discharges by requiring treatment of all wastewater flows even during peak periods.

Infiltration results from groundwater entering sewer lines through poor joints and cracks in manholes and sewer pipe; the sources are widespread, and the flow is relatively steady during times of high groundwater levels. Inflow comes from direct connections, such as roof drains, and results in sudden high rates of flow of short duration. The quantity of infiltration-inflow is considered to be the maximum wastewater flow minus the peak domestic and industrial discharge; in other words, it is the difference between peak flow measurements during wet weather and dry weather.

A systematic approach to sewer system evaluation includes identifying the quantity and nature of infiltration-inflow, isolating problem areas, and then evaluating the most economical corrective measures. The basic alternatives are either to rehabilitate the sewer system to reduce extraneous flows, or to extend treatment facilities to handle peak wet-weather flows.

The diagram in Figure 12-5 outlines the general approach applied in infiltration-inflow surveys. The first step is to analyze and identify the magnitude of the problem. The sewer system is divided into areas that drain to key points, such as a manhole or pumping station. Flow tests are then conducted to determine peak discharges. Obviously, measurements must be taken at the proper times during both dry weather, and high groundwater and rainfall situations. Evaluation of these data reveals whether infiltration and inflow are reasonable. Detailed field investigations are then conducted in those areas where flows are considered excessive. Field investigations start with a survey of physical conditions, which requires descending every manhole in the study area and visually inspecting the sewer lines. This often uncovers deficiencies that are not obvious by merely looking into manholes from the street level. Cross-connections from storm sewers and unauthorized drains attached to sanitary sewers may be detected by using low pressure smoke or tracer dyes in water. Often elimination of inflows that are observed during the physical survey materially reduces peak discharges. For this reason, additional flow measurements may be required. Finally, those sewer lines identified as the source of excessive infiltration and inflow are inspected by using inspection television. Preparatory cleaning is required prior to televising.

The final report of the field survey presents data from investigations, recommendations for a sewer rehabilitation program, and, most important, an economic evaluation of the problem. Expenses of rehabilitation are compared to costs of expanding treatment works to process extraneous water. Decisions are made regarding the feasibility of repairing different sections of the collection system. Where it is more economical to treat than to repair, the extraneous water flows are considered to be nonexcessive, that is, in an economic sense. Excessive flows are those where the treatment costs dictate rehabilitation. Sewer repair may involve replacing pipelines, repairing structural deterioration or defects, or sealing openings by external grouting or pipe relining. The most

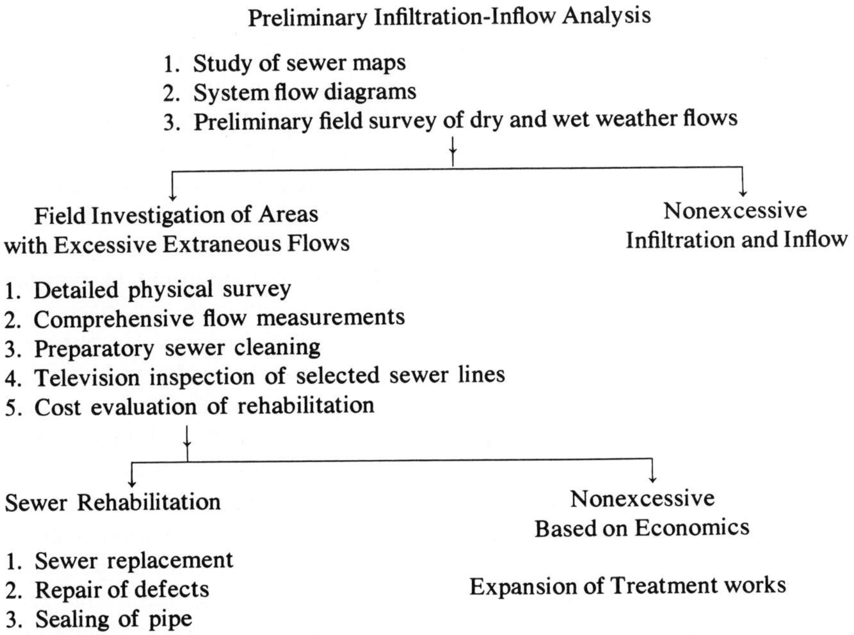

Preliminary Infiltration-Inflow Analysis

1. Study of sewer maps
2. System flow diagrams
3. Preliminary field survey of dry and wet weather flows

Field Investigation of Areas
with Excessive Extraneous Flows

Nonexcessive
Infiltration and Inflow

1. Detailed physical survey
2. Comprehensive flow measurements
3. Preparatory sewer cleaning
4. Television inspection of selected sewer lines
5. Cost evaluation of rehabilitation

Sewer Rehabilitation

Nonexcessive
Based on Economics

1. Sewer replacement
2. Repair of defects
3. Sealing of pipe

Expansion of Treatment works

Figure 12-5   Diagram outlining the general steps that are included in infiltration and inflow evaluations of sanitary sewer collection systems.

economic and best techniques must be determined from local conditions and experience. For example, the success of grouting with cement or polymers depends to a considerable extent on the type of soil. Care must be taken in external grouting to ensure that pipes are not fractured by grout injection. The economics of alternative repair methods are also considered; the cost to test and seal every joint in a manhole reach may be more than for complete sewer replacement. With manually accessible lines, hand-mortaring, coatings and guniting are primary repair methods. Internal sealing of small-diameter sewers commonly involves forcing sealants through pipe openings to gel in the surrounding soil, thus sealing the leaks.

12-4   REGULATION OF SEWER USE

The key purposes of a sewer ordinance are to control discharges to the sanitary sewer, to ensure that water quality standards of the receiving watercourse can be achieved, and to establish equitable customer charges for wastewater service. A comprehensive code contains regulations requiring use of public sewers where available, control of private waste disposal in the absence of public sewers, construction of service connections, control of the quantity and character of wastewaters admissible to municipal sewers, procedures for wastewater sampling and analyses, provisions for the powers and authority of inspectors, an enforcement (penalty) clause, and other legal clauses and signatures to validate the document. A model ordinance in the *Regulation of Sewer Use*[1] can be used as a guide in developing a regulation in a municipality where none has previously existed. Although several items included are common to all ordinances, each must be modified to incorporate local concerns of water pollution control and existing city documents, for example, the plumbing code.

Proper use and separation of sanitary and storm sewers are essential in protecting public health and reducing the quantity of inflow. Privies and septic tanks should not be permitted where sewer service is available. In the absence of public sewers, regulations should be established regarding construction, installation, and monitoring of private household systems to ensure that surface water and groundwater are not contaminated. Unpolluted waters must be excluded from the sanitary collection system and must be directed to storm sewers or a natural drainage outlet where feasible. These include inflow from downspouts, footing drains, surface-water inlets, swimming pools, and cooling water from air conditioners and industrial refrigeration units.

Certain wastes cannot be admitted to sanitary sewers, while others must be carefully controlled by specifying discharge limitations. These can be considered in the following four categories: (1) wastes that create a fire or explosion hazard; (2) substances that impair hydraulic capacity; (3) contaminants that create a hazard to people, the physical sewer system, or the biological treatment process; and (4) the refractory wastes that pass through treatment and result in degradation of the receiving watercourse. Examples of flammable liquids are gasoline, fuel oil, and cleaning solvents. Solids and viscous liquids that create sewer stoppages include ashes, sand, metal shavings, paunch manure, unshredded garbage, grease, and oil. The most common sewer stoppage, however, is related to root growth in sewer lines. This can be minimized by discouraging the planting of certain trees like elm, poplar, willow, sycamore, and soft maple near sewer lines. An alternate preventative measure is to use special jointing methods and materials when sewers are installed in areas of root influence.

Uncontrolled industrial wastes may contain corrosive or toxic compounds. For example, sulfur compounds and high temperature wastewaters can promote bacterial formation of sulfuric acid and lead to sewer crown corrosion. Acid wastewaters cause invert corrosion and, if not sufficiently diluted, may interfere with treatment processes. Toxic metal ions, such as chromium and zinc, and organic chemicals in rather small concentrations can lead to inhibition of the biological activity in aeration and anaerobic digestion. Dissolved salts, color, and odor-producing substances are only partially removed by conventional treatment. For these, protection of the

receiving watercourse is best achieved by separate disposal at industrial sites rather than by discharge to the sewer system. Examples are spent brine solutions, dye wastes, and phenols. Where industrial wastes are highly variable, it may be necessary to install equalizing tanks to prevent shock loads on the treatment works. In addition to controlling quality variations by neutralization and dilution, pretreatment by equalization can also provide discharge control to smooth out flow and to prevent shock hydraulic loads.

Flow measuring and sampling of wastewaters entering the sewer system are essential for enforcing a sewer ordinance and establishing equitable fees for sewer use. Each industry is required to install and maintain a suitable sampling station that may range from a simple manhole for grab samples to a structure that includes a flow recorder and automatic sampler. Inspectors must have the authority to enter all properties for the purposes of observation, measurement, sampling, and testing pertinent to wastewater discharge. Facilities for sampling and flow measurement are absolutely essential for instituting a sewer ordinance, and therefore their construction must be given primary consideration. For a small enterprise, the cost may involve only the installation of a manhole for sampling. Additional water meters may be needed to determine wastewater discharge if the wastewater is assumed to equal the water supplied minus the consumption for lawn watering, or discharge to the storm sewer. Automated sampling stations for large industries, particularly with more than one outlet sewer, may be a substantial investment.

Violation of an ordinance is declared a misdemeanor, or sometimes disorderly conduct. Penalties usually involve a fine for each violation plus costs of prosecution. If a user fails to correct an unauthorized discharge, the city is commonly authorized to discontinue sewer or water service or both. Repeated violations may result in revoking such an industry's discharge permit. Any of these actions requires written notice prior to institution. A hearing board may be defined in an ordinance to arbitrate differences between the city and an institution aggrieved by a penalty.

An ordinance also includes a section on charges for

wastewater service. Sometimes, this is put in a separate document apart from the regulations governing sewer use so that the two subjects can be handled independently at public hearings and during enactment. This approach is particularly practical when instituting a new code, since fees cannot be collected on the basis of wastewater flow or excessive strength until after installation of sampling stations.

Annual revenues of a wastewater system include costs associated with operation and maintenance of the collection system, treatment plant, and other facilities; principal and interest payments of providing debt financing for major capital improvements, as well as required reserve payments for general obligation or revenue bonds; and costs of capital additions not debt-financed, such as interim replacements, betterments, and minor extensions to physical facilities that are paid for from current revenues. The basis for allocating revenue requirements must give consideration to all costs of providing service. These vary with each municipality. Payment for service from each customer should be in proportion to use and benefits received. Arriving at a fair, proper, and practical system for raising annual revenue from users according to their responsibility for costs is a complex subject (see *Financing and Charges for Wastewater Systems*).[2]

Financial support may be obtained by ad valorem property taxes; however, these do not collect payments in proportion to customers' use or benefits received. For example, a disproportionately large charge may be placed on industries that produce little or no industrial waste. Wastewater service charges based on volume and strength reflect the actual physical use of the system to a considerable degree, but raising all revenues on the basis of service does not provide for payment by undeveloped property, for having facilities available whether they are used or not, or for the general community benefit that results from having adequate wastewater facilities. Yet a service charge system yields a relatively stable source of funds that allows for orderly planning, upgrading, and expansion. A major advantage in handling industrial wastes is that volume and strength charges encourage reduction of wastewater loads by in-plant modifications, improved housekeeping, by-product recovery, and water reuse. For most cities, the fairest system for raising revenues is a combination of property taxes and service charges.

Sewer collection systems are designed for peak hourly flows, while treatment facilities are sized on average daily flow. Although ideally desirable, it is impractical to base volume charges on discharge rates. Therefore, user charges are keyed to volume of flow without regard to time pattern of release, as long as excessive hydraulic shock loads are avoided. An exception is seasonal industries that should be assessed some charges during the nonuse season. The most equitable charge for flow is a uniform rate for all users regardless of quantity, with no tapered schedule of charges. In many cities, household production is considered to be the winter water consumption. This assumes that the water drawn is disposed of through the house drains and that none is applied for lawn watering. Water meters can also be used to determine wastewater flows from small industries and commercial establishments. Where uncontaminated water is discharged to a storm sewer, the flow is metered and this volume is subtracted from the metered water supply to yield the quantity of sanitary wastewater. Large industries install flow recorders that directly measure the quantity discharged to the municipal sewer.

The second component of a service charge is related to wastewater strength. This is generally applied as a surcharge for concentrations of pollutants that are greater than those found in domestic wastewater. The strength parameters most frequently employed are BOD and suspended solids. These must be determined by analyzing composite samples of industrial discharges in accordance with guidelines that are specified in the ordinance. Equation 12-1 is a typical formula for calculating service charges. The surcharge portion is for BOD in excess of 250 mg/l and suspended solids greater than 300 mg/l. Example 12-1 illustrates application of this formula.

$$\begin{aligned} \text{Wastewater} \atop \text{service charge} &= \text{charge for volume} + \atop \text{surcharge for strength} \\ &= VR_V + V[(\text{BOD} - 250)R_B \\ &\quad + (\text{SS} - 300)R_S]8.34 \qquad (12\text{-}1) \end{aligned}$$

where V = wastewater volume, million gallons

BOD = average BOD, milligrams per liter

SS = average suspended solids, milligrams per liter

R_V = charge rate for volume, dollars per million gallon

R_B = charge rate for BOD, dollars per pound

R_S = charge rate for suspended solids, dollars per pound

A sewer fee schedule should stipulate a minimum monthly charge for each property. The amount is often set on a graduated scale that increases with the size of water service to the property. One reason for a minimum bill is to cover the costs for maintaining capacity in the system even though it is not used. This is referred to occasionally as the right-to-service factor. Cities may also have a fee for making service connections to the municipal system. The charge may be small, only offsetting the costs of inspection, or much larger to pay a share of the cost involved in extending the sewer system.

EXAMPLE 12-1

Calculate the service charge for a dairy wastewater by using Eq. 12-1 based on the following: daily flow = 150,000 gal, average BOD = 910 mg/l, average suspended solids = 320 mg/l, service charge for flow = $450.00/mil gal, and surcharges of 2.38¢/lb excess BOD and 1.83¢/lb excess SS.

Solution

Substituting into Eq. 12-1,

$$SC = 0.15 \times 450 + 0.15[(910 - 250)0.0238$$
$$+ (320 - 300)0.0183]8.34$$
$$= 67.50 + 20.10 = \$87.60 \text{ per day}$$

12-5 PERFORMANCE EVALUATION OF TREATMENT PLANTS

The evaluation of a treatment plant involves examining each unit process to analyze its operation and how the process functions in the overall treatment scheme. Unfor-

tunately, analyses are often limited to influent and effluent testing as required by the National Pollutant Discharge Elimination System (NPDES) discharge permit. Studies of individual unit processes are frequently neglected. As a result, supervising personnel do not have data on the interrelationships of unit processes needed to provide overall optimum operation.

The first step in performance evaluation is to sketch a process flow diagram showing normal operation. Changes in processing to meet unusual flow and load conditions should be noted. Second, all treatment units should be dimensioned from field measurements and design drawings. Sampling points must be carefully selected to ensure collection of representative portions of flow for compositing. If access ports and flow-measuring devices are not available for all major in-plant flows, physical modifications must be made to permit isolation of each unit process for study. Finally, a laboratory facility is needed to perform routine tests in accordance with established procedures. Special studies and specific processes may dictate analyses of chemical oxygen demand (COD), grease, alkalinity, phosphate, various forms of nitrogen, sulfide, volatile acids, gas analyses, sludge filtrability, biological oxygen uptake, heavy metals, or total organic carbon (TOC).

The process diagram for a typical activated-sludge plant in Figure 12-6 suggests a minimum testing program for plant evaluation. Influent and effluent monitoring routinely requires testing for average biochemical oxygen demand (BOD), suspended solids (SS) concentrations, and pH. In some locations, tests for fecal coliform count, chlorine residual, phosphorus, ammonium nitrogen, and the presence of heavy metals may be specified by regulatory agencies. Also crossing the boundary of the treatment plant are conditioning chemicals imported for thickening and dewatering of sludge, and the export of grit and sludge cake to disposal.

Examination of Individual Processes

Examination of individual unit operations within a treatment plant requires understanding the following: (1) the purpose of each unit process and its function in the overall treatment scheme, (2) the proper procedures

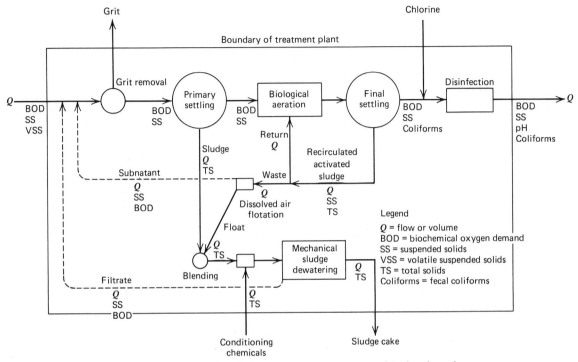

Figure 12-6 Typical process diagram for wastewater treatment by conventional activated sludge showing interrelated unit processes.

for operation, and (3) the performance expected under all possible loading conditions, including anticipated minimum and maximum values. Data collection must include composite testing of all influent and effluent flows and documenting the actual method of operation. Of course, the unit should be examined for any physical defects that would prevent satisfactory operation.

The measurements for evaluation of primary sedimentation include flow, BOD and SS of the influent, BOD and SS of the effluent, volume of sludge withdrawn, and its total solids content. Overflow rate, weir loading, and detention time can be calculated from the wastewater flow, and these values compared with design parameters to indicate the loading status of the tanks. Although efficiency of primary settling is most often related to BOD and SS removal, the quantity and solids content of sludge withdrawn are equally important as measures of satisfactory performance. Poor sludge

thickening in a sedimentation tank can be the result of either hydraulic overload or poor operating procedure. For example, a voluminous amount of thin sludge results if high pumping rates cause withdrawal of water from above the settled sludge layer.

Secondary biological aeration is examined on the basis of both organic matter removal and characteristics of the waste activated sludge. To view the aeration-clarification functions only in terms of BOD and SS removals, without regard for concentration of solids in the waste sludge, ignores the thickening function of biological flocculation. Air supply, rate of activated-sludge recirculation and sludge wasting can be adjusted to provide the thickest waste sludge possible while maintaining a clear wastewater effluent. Operation at various dissolved oxygen levels, mixed liquor suspended solids concentrations, and food-to-microorganism ratios with careful surveillance testing can define optimum performance.

Effluent disinfection requires careful analysis to ensure efficient use of chlorine. Poorly operated disinfection units can result in excessive chlorine residual being carried into the receiving watercourse. Chlorine residuals at different dosages can be measured under varying flow conditions and compared with laboratory tests on wastewater samples held for the same contact time. An efficient chlorination system provides rapid initial mixing of the chlorine solution in the wastewater followed by an adequate contact time in a plug-flow basin. Often, automatic residual monitoring and feedback control units are necessary to prevent either inadequate disinfection or excessive dosing.

Operation of a flotation thickener is judged on solids loading, float concentration, and clarity of the subnatant. For a given loading of waste activated sludge, operating variables for dissolved air flotation are air pressure, recycle ratio, air/solids ratio, and application of chemical aids, usually polyelectrolytes. This process has proven to be an effective method of thickening waste sludge with excellent solids capture commonly greater than 95 percent. Obviously, higher solids concentration in the influent requires less energy and reduced chemical dosage for separation of the suspended solids. Therefore, operation of the aeration system directly affects the sludge-thickening process. Excess activated sludge is normally discharged from the recirculation line, returning from the bottom of the final clarifier to the head of the aeration basin. If recycled flow is greater than that necessary for the settled sludge volume in the final clarifier, the settled solids are diluted by pumping overlying clarified wastewater, resulting in a thinner recirculated flow. For overall optimum plant operation, the final clarifier should be used to gravity thicken an activated sludge in order to reduce recirculation pumping and increase the solids content of the return flow and waste sludge.

Careful consideration must be given to all unit operations that discharge flow back to the head of the plant. Sludge processing may return supernatant from anaerobic or aerobic digesters, overflow from gravity thickeners, underflow from flotation thickeners, centrate from centrifuges, or filtrate from belt, pressure, or vacuum filters. Excessive suspended solids return can result in recycling of fine solids within the treatment plant. For

example, if insufficient conditioning chemical is applied for sludge dewatering, the filtrate may carry a large amount of solids back to the plant influent. Being primarily colloidal in nature, these solids pass through primary sedimentation for capture in biological aeration. They are then returned in the waste activated sludge for thickening and dewatering again. Cycling solids can lead to overloading and upset of all systems. However, their presence is normally first noticed in thinning of the primary sludge and increased oxygen demand in biological aeration.

Evaluating the operation of mechanical sludge dewatering normally requires extensive studies because it involves economic considerations related to operation time, chemical dosage, and disposal of sludge cake. Because the quantity of chemical used is related to the solids concentration in the sludge feed, satisfactory dewatering is keyed to the previous unit operations of wastewater treatment and sludge thickening. Performance assessment too often stops with sludge yield and cake solids without regard to the efficiency of solids capture or the quality of filtrate, which is frequently hard to sample. Solids processing is an integral part of plant evaluation frequently ignored.

Estimating Solids Capture in Sludge Processes

A relatively easy method of estimating solids capture by a sludge thickening or dewatering unit is to measure solids concentrations in the process flows. Figure 12-7 is a hypothetical processing unit for which the following relationships are apparent:

Solids mass balance is

$$M_S = M_F + M_C \qquad (12\text{-}2)$$

where M = mass of solids.

Liquid volumetric balance is

$$V_S = V_F + V_C \qquad (12\text{-}3)$$

where V = volume of liquid.

Without introducing significant error, the specific gravity of all flows can be assumed to be 1.0. (For example, a

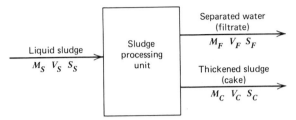

Figure 12-7 Flow diagram of a sludge-processing unit for Eqs. 12-2 through 12-5; M is mass of solids, V is volume of liquid, and S is solids concentration.

wastewater sludge with 10 percent organic solids has a specific gravity of approximately 1.02.) Hence,

$$M = S \times V \qquad (12\text{-}4)$$

Combining these equations results in the following:

$$\text{Fraction of solids removal} = \frac{M_C}{M_S} = \frac{S_C(S_S - S_F)}{S_S(S_C - S_F)} \qquad (12\text{-}5)$$

where S = solids concentration. This equation is easy to apply because no volumetric quantities are necessary.

The solids concentration S can be either total solids (residue upon evaporation) or SS (nonfiltrable residue) with the latter being more common. Testing for SS by the standard laboratory technique of filtration through a glass filter is feasible for dilute suspensions. Dissolved solids may be a major proportion of the total solids in supernatant or filtrate because of the low SS concentration. Waste flows with high solids contents are, however, difficult to test accurately by laboratory filtration. Therefore, SS analyses of sludge samples are performed by total solids tests and then corrected by subtracting an estimated dissolved solids concentration. This procedure does not create significant error since the filtrable solids content usually amounts to less than 5 percent of the total solids in sludge samples. Example 12-2 illustrates the use of this computational technique.

Record Keeping

Complete and accurate records of all phases of plant operation and maintenance are essential. Too often, extensive testing and flow records are filed daily and are

not reviewed until the plant is having serious operational problems or requires expansion. A rigid testing program without a directed purpose wastes labor and, worst yet, does not achieve its objectives. An effective sampling and testing program is one of continuous evaluation. Data collected may be applied to calculate existing hydraulic and organic loadings on all unit processes and to pinpoint problem areas in operations that are related to industrial wastes or excessive infiltration and inflow. Errors or omissions in laboratory testing will be recognized in the process of analyzing accumulated records. One procedure of value in compiling data is to trace suspended solids through the treatment processes, namely, removal from suspension in primary and secondary units, thickening, dewatering, and ultimate disposal. In-plant data on sludge production, thickening efficiency, dewatering operation, and others are extremely valuable to an engineer in modifying or expanding an existing facility. Effluent quality is of value as a measure of overall plant performance.

EXAMPLE 12-2

Performance of a vacuum filter dewatering waste sludge was evaluated to determine solids capture. Composite samples of the liquid sludge, filter cake, and filtrate were tested for total solids concentration. The filtrate was also tested for suspended solids concentration. Results of the laboratory analyses were

Total solids in sludge = 36,600 mg/l (3.66 percent)
Total solids in cake = 158,000 mg/l (15.80 percent)
Total solids in filtrate = 10,800 mg/l (1.08 percent)
Suspended solids in filtrate = 8,700 mg/l (0.87 percent)

The presence of conditioning chemicals was ignored since the polyelectrolyte addition was only 3 percent of the solids concentration in the sludge.

Solution

Dissolved solids in filtrate

$$= 10,800 - 8,700 = 2,100 \text{ mg/l}$$

Assuming the sludge and cake have the same concentration of dissolved solids, the estimated suspended solids concentrations are

$$S_S = 36,600 - 2,100 = 34,500 \text{ mg/l}$$

$$S_C = 158,000 - 2,100 = 155,900 \text{ mg/l}$$

From laboratory testing, the suspended solids in the filtrate are

$$S_F = 8,700 \text{ mg/l}$$

Substituting these values into Eq. 12-5,

$$\frac{M_C}{M_S} = \frac{155,900(34,500 - 8,700)}{34,500(155,900 - 8,700)} = 0.79$$

Therefore, the suspended solids capture in vacuum filtration was 79 percent.

12-6 ENERGY CONSERVATION

The sources of energy in wastewater treatment include electricity, fuel oil, natural gas, methane, and gasoline. Electricity accounts for 50 to 70 percent of the energy consumption, primarily to operate electric motors. Fuel oil and gas are used for building heating and sludge processing, and gasoline is fuel for vehicles. Since the majority of energy is used in treatment, the pattern of consumption in each plant varies with the methods of processing the wastewater and sludge. Pumping is a common primary use of electrical power. In activated-sludge plants, air blowers or mechanical aerators are major users of power. Sludge conditioning, dewatering, and disposal may have either a high or low energy demand. Incineration including dewatering and combustion is very energy intensive, whereas anaerobic digestion of gravity-thickened sludge can yield a surplus of energy in the form of methane gas. In order to determine the most energy efficient operation of any treatment plant, an energy audit is performed.

Electricity Rate Schedules

Most power companies establish an electricity pricing schedule incorporating the three components of energy usage, demand, and power factor. The energy usage charge is based on the amount of electricity consumed in kilowatt hours. The demand charge is determined by the maximum power demand in kilowatts for a duration of 15 to 30 min during a specified time period, for example, one year. The power factor charge is to compensate for the increased cost of supplying power to operate electrical equipment with a reactive load, such as an induction motor.

Energy usage is the measured kilowatt hours consumed during a billing period. The rate schedule is in blocks of usage commonly priced on a declining-charge scale. The first block is a specified amount of power at a given price per kilowatt hour, the second is an additional amount at a second lower price, and so forth. The demand charge is a penalty, in addition to the usage charge, for high peak consumption of electricity during a short period of time. Power companies discourage high instantaneous demands in order to reduce the required capacity of the power system and lessen the probability of overloading the system. Often, a high demand during one month results in a penalty charge for each of the following eleven months.

The power factor is the ratio of active power to apparent power. When the alternating current wave in a circuit coincides exactly in time with the voltage wave, the product of current times voltage is the magnitude of both apparent power and active power, and the power factor is equal to 1.00. When the current wave lags behind the voltage wave because of inductive reactance, or leads because of capacitive reactance, the peaks of the current and voltage waves do not coincide. Then, the product of effective current and voltage yields active power, which is less than apparent power. Consider the phase diagram in Figure 12-8a. The current wave leads the voltage wave as shown by the dashed lines. The solid line is a kilovolt-ampere (kVA) curve. The kVA value at any instant is the product of kilovolts of voltage times amperes of current occurring at the same time. The positive area under the kVA curve is active power, which is power that can perform work. The negative area is reactive power that cannot perform work. The power triangle in Figure 12-8b is a vector diagram of active power, reactive power, and apparent power. The power factor is numerically equal to the cosine of the phase angle, and the apparent power times the power factor equals the active power. Consider

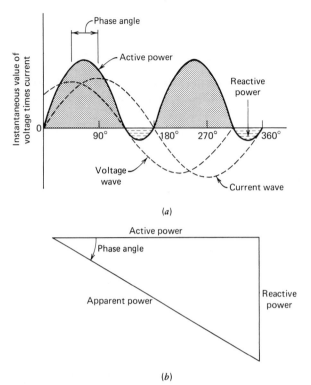

Figure 12-8 Power factor is the ratio of active power to apparent power. (*a*) When the current wave leads the voltage wave, the active power is the area under the product curve of instantaneous voltage times current. (*b*) Apparent power is the product of in-phase current times voltage. In a power triangle, the power factor is the cosine of the phase angle.

the example of an induction motor. The active power drawn during operation is 420 kW, while the calculated product of peak voltage times peak amperage in the electrical supply is 600 kVA, the apparent power drawn. The power factor is 420/600 = 0.70, or 70 percent. The difference between apparent power and active power occurs because of the fluctuating power draw that provides energy for the change in magnetic flux in the motor. All polyphase induction motors operate at lagging power factors in the range of 70 to 90 percent. In a purely resistive circuit, such as an incandescent light, the power factor equals 1.0 (phase angle equals 0 degrees).

Generation and power transmission capacities are sized for apparent power loads. In other words, for the power triangle in Figure 12-8*b*, the power company must supply the apparent power so that the active power is available to perform work. Therefore, many power companies impose a surcharge for power factors below 85 percent that result from the magnetic power draw of induction motors. The power factor change may increase electrical power costs by as much as 5 to 10 percent.

Reducing Power Charges

The energy usage charge can be reduced by simply consuming less electricity. Examples in conservation are adjusting thermostats to save energy for heating and air conditioning, turning off unneeded lights, and improving the operational efficiency of equipment. Automatic timers can be used to control pumps, air blowers, and mechanical equipment during periods of low wastewater loadings. The scraper in a grit chamber can be operated intermittently. Air supply to activated-sludge tanks and aerobic digesters can be minimized to reduce dissolved oxygen levels to the minimum necessary for biological metabolism and mixing of the liquid.

The demand charge can be reduced by scheduling the operation of selected equipment, lengthening the period between the starting of motors, and use of an emergency generator for peak demands. For example, some energy-consuming processes in sludge dewatering and incineration can be rescheduled to off-peak periods. Timing relays and interlocks can be used to prevent starting of several motors within too short a time period. Another method is to stop the least necessary operation for a period of time to prevent exceeding a maximum demand. Although expensive, an automatic device is available that can be connected to a demand meter to automatically stop electrical equipment as the maximum demand is approached. Peak demand may also be lowered by using the plant's emergency generator. A potential application is for wastewater pumping during peak flows that occur for a short period of time, for instance, high inflow resulting from rainfall.

The power factor charge can be reduced by changing motor loading, installing a different kind of motor, or use of capacitors. The improvement of the power factor is variable depending on the motor installation. An electric motor operates efficiently within a range of approximately 50 to 125 percent of rated output. As the load decreases below 50 percent, the power factor decreases rapidly. While reducing the load on a motor has the advantage of increasing service life, oversizing a motor may be economically offset by the reduction in power factor. Premium efficiency motors have high power factors that can result in cost savings over a standard motor. Synchronous motors are available with a power factor of 1.0 or a leading power factor of 0.8. The latter may be used to counteract the effect of a lagging power factor from induction motors.

The power factor of an induction motor circuit can be improved by a capacitor installation, after determining the best location. Capacitors may be placed on the primary side of the transformer, on the secondary side at the motor control center, or at the motor itself. The motor manufacturer should be consulted to determine the capacitance that can be used with a particular motor.

Energy Audit

The objectives of an energy audit are to assess energy uses, identify and evaluate energy conservation options, and implement an energy management plan. An energy audit is carried out in the following five steps: (1) develop baseline information on energy consumption and costs, (2) conduct an on-site facility survey, (3) identify alternative energy conservation measures, (4) perform an economic analysis of each alternative, and (5) develop an energy management plan.[3]

Baseline data on energy use are gathered in preparation for the on-site review. Energy consumption and costs are assembled for previous years, including electricity billing records, available internal electric metering records, and purchased fuels. Power company rate schedules are studied to see if modifications can be made to obtain a better rate. Information about future rate changes is also important for assessing energy management alternatives. Design data are examined to determine the accuracy of equipment specifications and electrical ratings. In order to evaluate how equipment is being operated and maintained, plant records are reviewed to see if performance can be improved. From the available data, energy con-

sumption profiles are developed by unit process and plant loadings.

The on-site facility survey includes reviewing the operating procedures for each unit process, inspecting equipment and reviewing maintenance records, observing building energy systems, and reviewing operational flexibility of treatment processes. Alternative energy conservation measures are grouped into three categories. Housekeeping measures are those that can be implemented immediately by modifying operations, for example, changing an operating sequence to reduce the demand for electricity. Minor process improvements, which require a low capital investment with a short payback period, are generally financed from current maintenance budgets. An example is the installation of timers to cycle operations automatically. Major capital improvements are those that require substantial investment and may affect process operations. For instance, a belt filter press may be installed to dewater sludge.

An economic analysis is performed to justify the cost of implementing each energy conservation measure. For minor process improvements, a simple payback period analysis that identifies the number of years required to recover initial costs is considered adequate. The computation for payback period is the initial investment in dollars divided by the annual savings in dollars per year. Although this is an easy comparison technique, its simplicity results in several limitations. Payback period analysis does not distinguish between alternatives with different useful lives; neither does it consider the value of money over time nor the impact of inflation. For major capital improvements, a life-cycle cost analysis is performed considering the initial capital investment, useful life of the equipment, cost of maintenance, and annual energy savings over the useful life. If the present value of the annual savings is greater than the present value of the cost, the improvement is economically justified. Based on these economic analyses, energy conservation options should be ranked according to cost effectiveness and feasibility of implementation.

An energy management plan ensures the adoption of energy conservation measures through planning, implementing, and monitoring. The planning phase includes revising housekeeping and operational procedures, initiating work orders for minor process improvements, and assembling budget documentation to support major capital improvements. Implementing involves assigning responsibility to individual employees who have authority to allocate resources, resolve conflicts, and establish implementation schedules. Monitoring requires periodic evaluation of the performance goals—for example, the assessment of energy consumption by a given treatment process on a monthly basis.

FURTHER READINGS

1. *Operation and Maintenance of Wastewater Collection Systems*, WPCF Manual of Practice No. 7, Water Pollution Control Federation, Washington, D.C., 1980.
2. *Prevention and Correction of Excessive Infiltration and Inflow into Sewer Systems*, Water Pollution Control Research Series, 11022 EFF 01/71, Environmental Protection Agency, Water Quality Office, 1971.

REFERENCES

1. *Regulation of Sewer Use*, WPCF Manual of Practice No. 3, Water Pollution Control Federation, Washington, D.C., 1975.
2. *Financing and Charges for Wastewater Systems*, A Joint Committee Report, American Public Works Association, American Society of Civil Engineers, and Water Pollution Control Federation, 1973.
3. *Wastewater Utility Management Manual*, Municipal Operations Branch, Municipal Construction Division, U.S. Environmental Protection Agency, 1981.

PROBLEMS

12-1 How is television inspection of sewers used in conducting infiltration and inflow surveys?

12-2 In an industrial city, why is a sewer ordinance essential for quality control of the municipal treatment plant effluent?

12-3 Using Eq. 12-1, calculate the wastewater service charge per month for a meat-processing industry with a discharge of 1.20 mil gal each day containing 1300 mg/l of BOD and 960 mg/l of suspended solids. The charge rate for volume is $450.00 per mil gal, $2.38 per lb of excess BOD, and $1.83 per lb of excess SS.

12-4 During the performance evaluation of a treatment plant, with a process diagram as shown in Figure 12-6, the following solids data were collected:

Waste (recirculated) activated sludge = 4000 mg/l of SS
Subnatant from flotation thickener = 500 mg/l of SS
Float from flotation thickener = 3.2 percent solids
Primary sludge from settling tank = 4.5 percent solids
Blended primary sludge and float = 4.0 percent solids
Cake from vacuum filter = 14.8 percent solids
Filtrate from vacuum filter = 12,700 mg/l of SS

Evaluate these solids data and comment on the performance of sludge handling processes.

12-5 Explain why a power company imposes a surcharge for power usage at a low power factor. How can the power factor charge be reduced?

chapter 13

Advanced Wastewater Treatment

Advanced wastewater treatment refers to methods and processes that remove more contaminants from wastewater than are taken out by conventional biological treatment. The term may be applied to any system that follows secondary treatment or that modifies or replaces a step in conventional processing. The expression tertiary treatment is often used as a synonym, but the two terms do not have precisely the same meaning. Tertiary suggests a third step that is applied after primary and secondary processing. Wastewater reclamation consists of a combination of conventional and advanced treatment processes employed to return a wastewater to its original quality, reclaiming the water.

The common advanced wastewater treatment processes are filtration, phosphorus precipitation, and nitrification. The primary purpose of effluent filtration is to remove suspended solids that do not settle out by gravity. Phosphorus removal is to reduce eutrophication of lakes receiving wastewater effluents. Oxidizing ammonia to nitrate reduces ammonia toxicity and nitrogen oxygen demand in the receiving stream. When planned or accidental reuse of a wastewater effluent involves human contact, virus inactivation is recommended for protection of human health. In water reclamation, the primary concern is removal of heavy metals, organic chemicals, inorganic salts, and pathogens.

13-1 PHOSPHORUS IN WASTEWATERS

The common forms of phosphorus in wastewater are orthophosphate ($PO_4^{\equiv}$), polyphosphates (polymers of phosphoric acid), and organically bound phosphates. Polyphosphates, such as hexametaphosphate, gradually hydrolyze in water to the soluble ortho form, and bacterial decomposition of organic compounds also releases orthophosphate. With the majority of compounds in wastewater soluble, phosphorus is removed only sparingly by plain sedimentation. Secondary biological treatment removes phosphorus by biological uptake; however, relative to the quantities of nitrogen and carbon, the amount of phosphorus is greater than necessary for biological synthesis. Consequently, conventional treatment removes only about 20 to 30 percent of the influent phosphorus.

The total phosphorus contribution to domestic wastewater is about 3.5 lb per capita per year (1.6 kg/person·y), resulting in an average concentration of 10 mg/l. Thirty to 50 percent of the phosphorus is from sanitary wastes, while the remaining 50 to 70 percent is from phosphate builders used in household detergents. One of the earliest concepts of phosphorus control was to substitute other chemical compounds for phosphate builders in laundry detergents. While this appears to be a reasonable approach, no suitable substance was found. Caustic additives did not clean as well, were irritating to the skin, and certain brands were sufficiently strong to damage eyes and mucus membranes if inhaled or eaten. Although the quantity of phosphate in detergents has been reduced in many brands, complete substitution has not been achieved for safety reasons.

Ecological Effects

The primary pollutional effect of phosphorus in surface waters is eutrophication (Section 5-8). Since phosphorus is the growth-limiting plant nutrient in natural waters, discharge of wastewater high in soluble phosphates leads to accelerated fertilization. The attendant results in lakes and reservoirs are excessive growth of algae causing reduced water transparency, depletion of dissolved oxygen, release of foul odors, loss of finer fish species, and dense growths of aquatic weeds in shallow bays. Flowing waters with turbidity that blocks the sunlight necessary for photosynthesis are not subject to eutrophication; however, estuaries and slow-moving rivers with a clear

water can be adversely affected in the same way as impounded waters.

In the United States, wastewaters from approximately 55 percent of the population are discharged into the ocean or into major river systems that flow to the ocean. Another 30 percent of the population lives in rural unsewered communities. This leaves only 15 percent contributing wastewater effluents to inland lakes in danger of eutrophication. Therefore, remedial action for phosphorus pollution is the treatment of wastewaters that discharge directly into lakes and rivers or streams that flow into lakes. Several states have adopted effluent standards for phosphorus. These standards have taken the form of an effluent concentration limit, and a requirement for a specified percentage reduction during treatment. Effluent limits range from 0.1 to 2.0 mg/l as P, with many established at 1.0 mg/l, while treatment efficiency requirements range from 80 to 95 percent.

Phosphorus Removal in Conventional Treatment

The nutrient composition of an average sanitary wastewater based on 120 gpcd (450 l/person·d) is listed in Table 13-1. The majority of phosphorus in wastewater is soluble; therefore, phosphorus is only sparingly removed by plain sedimentation. Based on the tabulated values, the total phosphorus is reduced from 10 mg/l to 9 mg/l by sedimentation. Secondary biological treatment removes phosphorus by biological uptake; however, the amount of phosphorus is surplus relative to the quantity of nitrogen and carbon necessary for synthesis (Section 9-1). In general, the amount of phosphorus in the excess biological floc produced in activated-sludge treatment of a wastewater is equal to about 1 percent of the BOD applied. Based on this, the total phosphorus is further reduced from 9 mg/l to approximately 8 mg/l. As a result, in Table 13-1, the total phosphorus of 10 mg/l in the raw wastewater is reduced to 8 mg/l in the biologically treated effluent.

Figure 13-1 is a diagram tracing phosphorus through a hypothetical treatment plant. The phosphorus concentration in the influent wastewater is assigned a value of

Table 13-1. Approximate Nutrient Composition of Average Sanitary Wastewater Based on 120 gpcd (450 l/person · d)

Parameter	Raw	After Settling	Biologically Treated
Organic Content (mg/l)			
Suspended solids	240	120	30
Biochemical oxygen demand	200	130	30
Nitrogen Content (mg/l as N)			
Inorganic nitrogen	15	15	24
Organic nitrogen	20	15	2
Total nitrogen	35	30	26
Phosphorus Content (mg/l as P)			
Inorganic phosphorus	7	7	7
Organic phosphorus	3	2	1
Total phosphorus	10	9	8

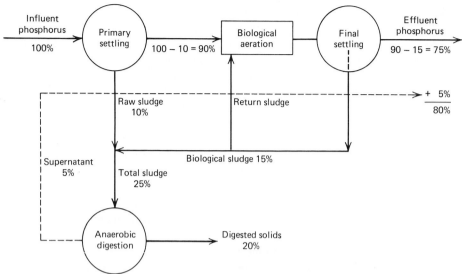

Figure 13-1 Diagram tracing wastewater phosphorus through a hypothetical treatment plant. Of the 100 percent influent phosphorus, 80 percent is in the treated wastewater and 20 percent is in the digested sludge solids.

100 percent. Phosphorus removal with the settled primary sludge is taken as 10 percent, and phosphorus synthesized in secondary biological aeration is assumed to be an additional 15 percent. Thus, the quantity in the raw sludge is 25 percent of the influent phosphorus. Yet the only portion of phosphorus extracted is that portion ultimately disposed of with the sludge solids. Stabilization and dewatering of the raw sludge may return some of the phosphorus back to wastewater processing. Mechanical dewatering of raw waste sludge followed by land burial or incineration results in little or no recycle of phosphorus. Anaerobic or aerobic digestion, on the other hand, returns supernatant containing soluble phosphorus to the influent of the treatment plant. Figure 13-1 assumes that anaerobic digestion returns 5 percent of the influent phosphorus, which then passes through the treatment system increasing the effluent phosphorus content to 80 percent of the influent. Of course, variations in characteristics of the wastewater and different methods of treatment can result in higher or lower removal efficiencies. Still, phosphorus removal in conventional biological treatment systems is generally within the range of 10 to 40 percent.

EXAMPLE 13-1

Using the values given in Table 13-1, calculate the concentration of total phosphorus in the effluent from a conventional activated-sludge plant. Assume the following: primary suspended solids removal of 50 percent containing 0.9 percent phosphorus; primary BOD removal of 35 percent; waste activated-sludge solids from biological aeration, based on a food-to-microorganism ratio of 0.40, containing 2.0 percent phosphorus; effluent BOD of 30 mg/l; effluent suspended solids of 30 mg/l; and no phosphorus recycling from sludge processing.

Solution

From Table 13-1, influent SS = 240 mg/l and BOD = 200 mg/l. At 50 percent SS removal, primary solids are 120 mg.
Phosphorus in the primary sludge equals

$$0.009 \times 120 = 1.1 \text{ mg of organic phosphorus}$$

Settled wastewater has a BOD equal to 130 mg/l.
From Figure 11-48 at an F/M of 0.40, K equals 0.50. Therefore,

$$W_s = 0.50 \times 130 = 65 \text{ mg}$$

Phosphorus in the secondary sludge equals

$$0.02 \times 65 = 1.3 \text{ mg of organic phosphorus}$$

Suspended solids in the effluent are 30 mg/l.
Organic phosphorus in the effluent equals

$$0.02 \times 30 = 0.6 \text{ mg/l of organic phosphorus}$$

Inorganic (soluble) phosphorus in the effluent is the influent minus the organic phosphorus contents of the primary sludge, secondary sludge, and effluent suspended solids. Therefore, inorganic phosphorus in the effluent equals

$$10.0 - 1.1 - 1.3 - 0.6 = 7.0 \text{ mg/l}$$

Total effluent phosphorus equals

$$0.6 + 7.0 = 7.6 \text{ mg/l}$$

These data are summarized in Figure 13-2. The total sludge solids produced per liter of wastewater treated equal 185 mg containing 2.4 mg of phosphorus. The influent phosphorus is 10.0 mg/l and effluent is 7.6 mg/l for a removal of 24 percent.

13-2 CHEMICAL-BIOLOGICAL PHOSPHORUS REMOVAL

Chemical precipitation using aluminum or iron coagulants is effective in phosphate removal. Although coagulation reactions are complex and only partially understood, the primary action appears to be the combining of orthophosphate with the metal cation. Polyphosphates and organic phosphorus compounds are probably removed by being entrapped, or adsorbed, in the floc particles. Aluminum ions combine with phosphate ions as follows:

$$Al_2(SO_4)_3 \cdot 14.3H_2O + 2PO_4^{\equiv}$$
$$= 2AlPO_4\downarrow + 3SO_4^{\equiv} + 14.3H_2O \quad (13\text{-}1)$$

The molar ratio for Al to P is 1 to 1 and the weight ratio of commercial alum to phosphorus is 9.7 to 1.0. Coagulation studies have shown that greater than this alum dosage is necessary to precipitate phosphorus from wastewater. One of the competing reactions, which accounts in part for the excess alum requirement, is with natural alkalinity

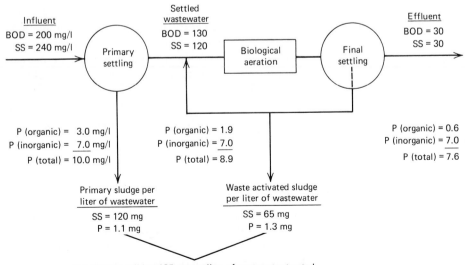

Figure 13-2 Summary of results from Example 13-1.

as follows:

$$Al_2(SO_4)_3 \cdot 14.3H_2O + 6HCO_3^-$$
$$= 2Al(OH)_3\downarrow + 3SO_4^= + 6CO_2 + 14.3H_2O \quad (13\text{-}2)$$

As a result, phosphorus reductions of 75 percent, 85 percent, and 95 percent require alum to phosphorus weight ratios of about 13 to 1, 16 to 1, and 22 to 1, respectively. For example, to achieve 85 percent phosphorus removal from a wastewater containing 10 mg/l of P, the alum dosage needed is approximately $16 \times 10 = 160$ mg/l, which is substantially greater than the $9.7 \times 10 = 97$ mg/l stoichiometric quantity of alum based on Eq. 13-1.

Iron coagulants precipitate orthophosphate by combining with the ferric ion as shown in Eq. 13-3 at a molar ratio of 1 to 1.

$$FeCl_3 \cdot 6H_2O + PO_4^=$$
$$= FePO_4\downarrow + 3Cl^- + 6H_2O \quad (13\text{-}3)$$

Just as with aluminum, a greater amount of iron is required in actual coagulation than this chemical reaction predicts. One of the competing reactions with natural alkalinity is

$$FeCl_3 \cdot 6H_2O + 3HCO_3^-$$
$$= Fe(OH)_3\downarrow + 3CO_2 + 3Cl^- + 6H_2O \quad (13\text{-}4)$$

Provided that the wastewater has sufficient natural alkalinity, ferric salts applied without coagulant aids result in phosphorus removal at Fe to P dosages of 1.8 to 1.0 or greater. This is equivalent to an application of approximately 150 mg/l of commercial ferric chloride for treatment of a wastewater containing 10 mg/l of P. Since the reaction of ferric chloride with natural alkalinity is relatively slow, lime or some other alkali may be applied to raise the pH and supply the hydroxyl ion for coagulation as follows:

$$2FeCl_3 \cdot 6H_2O + 3Ca(OH)_2$$
$$= 2Fe(OH)_3\downarrow + 3CaCl_2 + 6H_2O \quad (13\text{-}5)$$

Ferrous sulfate also forms a phosphate precipitate with an Fe to P molar ratio of 1 to 1, and the dosages for coagulation are similar to ferric salts.

Commercially available iron salts are ferric chloride, ferric sulfate, ferrous sulfate, and waste pickle liquor from the steel industry. The latter is the least expensive and most common source of iron coagulants for wastewater treatment in industrial regions. Pickle liquor is variable in composition depending on the metal treatment process. Ferrous sulfate from pickling with sulfuric acid and ferrous chloride from pickling with hydrochloric acid are the two common waste liquors from metal finishing. Waste liquors have an iron content from 5 to 10 percent and free acid ranges from a low of 0.5 percent to a high of 15 percent. Preparation prior to use includes neutralization and pH adjustment of the liquors with lime or sodium hydroxide.

Chemical-biological treatment combines chemical precipitation of phosphorus with biological removal of organic matter. Alum or iron salts are added either directly to the biological process or to the process effluent prior to final clarification. In activated-sludge aeration, the coagulant can be added to the aerating mixed liquor. Although the resulting chemical-biological floc has fewer protozoa, BOD removal efficiency is not adversely influenced. In trickling filtration or rotating biological contactors, the coagulant is usually mixed with the process effluent just prior to final sedimentation. Depending on coagulant dosage, the production of sludge solids in chemical-biological treatment is generally in the range of 30 to 60 percent more than the biological solids produced without chemical addition. In part, this increase in solids production is the result of improved effluent clarification because of chemical coagulation. The volume of sludge, on the other hand, is usually a smaller percentage increase because of the higher density of the chemical-biological sludge.

EXAMPLE 13-2

Alum is applied in the aeration tank of a conventional activated-sludge system (diagramed in Figure 13-2) to reduce the total phosphorus in the effluent to 1.0 mg/l. The commercial alum dosage is 130 mg/l. Calculate the alum dosage in terms of the weight ratio of alum applied to the phosphorus content of the wastewater and the molar ratio of aluminum to phosphorus. Calculate the anticipated sludge-solids production assuming the following: raw and settled wastewater characteristics as given in Figure 13-2, an operating F/M ratio in biological aeration of 0.40, and an effluent SS concentration of 15 mg/l. The reduction in effluent SS from 30 to 15 mg/l is the result of increased settleability of the chemical-biological floc as compared to the biological floc without alum addition. How much has the production of sludge solids increased?

Solution

Alum to phosphorus and aluminum to phosphorus ratios

Alum dosage = 130 mg/l

$$\text{Aluminum dosage} = 130\,\frac{2 \times 27}{600} = 11.7 \text{ mg/l}$$

Phosphorus in the settled wastewater = 8.9 mg/l

$$\frac{\text{Alum applied}}{\text{Phosphorus in wastewater}} = \frac{130}{8.9} = 14.6$$

Molar ratio of Al to P = 11.7/8.9 = 1.3

Sludge-solids production

SS removal in primary = 0.5 × 240 = 120 mg

The K from Figure 11-48 is 0.50 for an F/M of 0.40; therefore,

SS in waste activated sludge = 0.50 × 130 = 65 mg

Since the K from Figure 11-48 assumes an effluent SS of 30 mg/l and the actual effluent is 15 mg/l,

SS from improved settleability = 30 − 15 = 15 mg

Total biological solids removal = 65 + 15 = 80 mg

Organic phosphorus in biological solids is 2.0 percent by weight; therefore,

Phosphorus in biological solids removed

$$= 0.02 \times 80 = 1.6 \text{ mg}$$

Phosphorus removed by alum precipitation is the phosphorus in the wastewater after primary sedimentation minus the phosphorus in the biological solids removed minus the phosphorus in the process effluent, which is

$$8.9 - 1.6 - 1.0 = 6.3 \text{ mg}$$

From Eq. 13-1,

AlPO$_4$ precipitate

$$= \frac{(\text{P precipitated})(\text{molecular weight of AlPO}_4)}{(\text{molecular weight of P})}$$

$$= \frac{(6.3 \text{ mg})(122)}{(31)} = 25 \text{ mg}$$

From Eq. 13-1,

Unused alum

$$= \frac{\text{alum}}{\text{dosage}} - \frac{(\text{P precipitated})(\text{molecular weight of alum})}{(2 \times \text{molecular weight of P})}$$

$$= 130 \text{ mg/l} - \frac{(6.3)(600)}{(2 \times 31)}$$

$$= 69 \text{ mg/l}$$

The alum not used in phosphorus precipitation reacts with natural alkalinity to precipitate Al(OH)$_3$. From Eq. 13-2,

Al(OH)$_3$ precipitate

$$= \frac{(\text{unused alum})(2 \times \text{molecular weight of Al(OH)}_3)}{(\text{molecular weight of alum})}$$

$$= \frac{(69)(2 \times 78)}{600} = 18 \text{ mg}$$

The total sludge-solids production equals the sum of the SS removal in primary sedimentation, SS in waste activated sludge, SS from improved settleability in final clarification, AlPO$_4$ precipitate, and Al(OH)$_3$ precipitate. Therefore, solids production per liter of wastewater treated equals

$$120 + 65 + 15 + 25 + 18 = 243 \text{ mg}$$

Increase in sludge-solids production

The solids production by chemical-biological processing relative to conventional activated-sludge processing, as given in Figure 13-2, is 243 − 185 = 58 mg per liter of wastewater treated, which is an increase of 31 percent.

13-3   CHEMICAL PHOSPHORUS REMOVAL

The concentration of phosphorus in a wastewater can be reduced by chemical coagulation in either primary or tertiary treatment. In primary clarification, the chemicals added are waste pickle liquor or lime, whereas in tertiary treatment commercial coagulants or lime are common. Lime precipitation differs significantly from the action of metal coagulants. When added to wastewater, lime increase pH and reacts with carbonate alkalinity to precipitate calcium carbonate (Eq. 13-6). The calcium ion also combines with orthophosphate in the presence of hydroxyl ion to form gelatinous calcium hydroxyapatite as shown in Eq. 13-7.

$$Ca^{++} + OH^- + HCO_3^- = CaCO_3\downarrow + H_2O \quad (13\text{-}6)$$

$$5Ca^{++} + 4OH^- + 3HPO_4^=$$
$$= Ca_5(OH)(PO_4)_3\downarrow + 3H_2O \quad (13\text{-}7)$$

If sufficient lime is added, precipitation softening continues with the formation of magnesium hydroxide. A pH in the range of 9.5 to 11.5 is required to remove the major fraction of phosphorus. Lime dosages of 150 to 300 mg/l as CaO are commonly needed to remove 80 to 90 percent of the phosphorus from domestic wastewater. The actual amount required depends on the alkalinity, phosphorus concentration, degree of removal desired, and prior wastewater treatment.

In primary clarification, lime precipitation of a raw wastewater extracts approximately 60 percent of the BOD and up to 80 percent of the phosphorus without adversely affecting secondary biological treatment, if the pH is not raised above about 9.5. Microbial production of carbon dioxide in an activated-sludge secondary is usually sufficient to maintain a neutral pH in the aeration compartment.

In tertiary lime treatment following biological processing, the treatment scheme includes mixing, sedimentation, and filtration. Recarbonation is required prior to filtration for stabilization of the wastewater. Recarbonation in two stages is recommended. The first stage is to remove the excess lime as calcium carbonate by reducing the pH to approximately 9.5 followed by sedimentation. The second-stage addition of carbon dioxide converts a portion of the remaining carbonate ion to bicarbonate ion for stabilization. Tertiary lime treatment (a recommended process in wastewater reclamation) also precipitates heavy metals, removes some dissolved organic compounds, and provides disinfection.

13-4   NITROGEN IN WASTEWATERS

The common forms of nitrogen are organic, ammonia, nitrate, nitrite, and gaseous nitrogen. Decomposition of nitrogenous organic matter releases ammonia to solution, Eq. 13-8. Under aerobic conditions, nitrifying bacteria perform Eq. 13-9 to oxidize ammonia to nitrite and subsequently to nitrate. Bacterial denitrification, Eq. 13-10, occurs under anaerobic or anoxic conditions when organic matter (AH_2) is oxidized and nitrate is used as a hydrogen acceptor releasing nitrogen gas.

$$\text{Organic nitrogen compounds} \xrightarrow[\text{decomposition}]{\text{bacterial}} NH_3 \text{ (ammonia)} \quad (13\text{-}8)$$

$$NH_3 + O_2 \xrightarrow[\text{nitrification}]{\text{aerobic}} NO_3^- \text{ (nitrate)} \quad (13\text{-}9)$$

$$NO_3^- + AH_2 \xrightarrow[\text{denitrification}]{\text{anaerobic}} A + H_2O + N_2\uparrow \quad (13\text{-}10)$$

Nitrogen in municipal wastewater results from human excreta, ground garbage, and industrial wastes, particularly from food processing. Approximately 40 percent is in the form of ammonia and 60 percent is bound in organic matter, with negligible nitrate. The total nitrogen contribution is in the range of 8 to 12 lb per capita per year (4 to 6 kg/person·y), and the average concentration in domestic wastewater is 35 mg/l.

Ecological Effects

Un-ionized ammonia is toxic to fish and other aquatic animals. The amount of un-ionized ammonia is based on the pH of the water, since ammonia is converted to the nontoxic ammonium ion with decreasing pH. The criterion for salomonid fish is 0.02 mg/l of un-ionized ammonia, and for tolerant fish species is 0.08 mg/l. These values are equivalent to total ammonia-nitrogen concentrations of approximately 0.5 mg/l and 5.0 mg/l at pH 8, respectively. As a result of nitrification, ammonia can also result in dissolved oxygen uptake in flowing and impounded waters. Theoretically, 1.0 mg of ammonia nitrogen can exert an oxygen demand of 4.6 mg when converted to nitrate nitrogen. Nitrification of ammonia from a wastewater discharge rarely occurs to this extent in natural waters because of competing reactions, such as algal photosynthesis, and environmental conditions adverse to nitrifying bacteria. In oligotrophic waters, the contribution of nitrogen in wastewaters can increase the rate of eutrophication.

The maximum contaminant level for nitrate nitrogen in drinking water is 10 mg/l. Since organic and ammonia nitrogen are converted biologically to nitrate, wastewater effluents discharged to surface waters without adequate dilution can result in high nitrate concentrations. However, this is likely to occur only during drought flow in a river receiving large quantities of wastewater. A high concentration in a water supply is of concern since ordinary treatment processes do not remove the nitrate ion.

Nitrogen Removal in Conventional Treatment

The nitrogen compounds in an average sanitary wastewater are listed in Table 13-1. With most of the nitrogen in soluble and colloidal organic forms, the amount removed by primary sedimentation is limited to about 15 percent. Based on the tabulated values, the uptake in subsequent biological treatment is only another 10 percent. In general, the amount of nitrogen in excess biological floc produced in activated-sludge treatment of a wastewater is equal to about 4 percent of the BOD applied. With a total reduction of only 25 percent, the effluent contains 26 mg/l of the original 35 mg/l. Approximately 2 mg/l is organic nitrogen bound in the effluent suspended solids. The remaining 24 mg/l is in the form of ammonia, except when nitrification occurs during aeration. Oxidation of a portion of the nitrogen to nitrate is most likely to occur in an activated-sludge process treating a warm wastewater at a low BOD loading.

Figure 13-3 is a diagram tracing nitrogen through a hypothetical treatment plant. The nitrogen concentration in the influent wastewater is assigned a value of 100

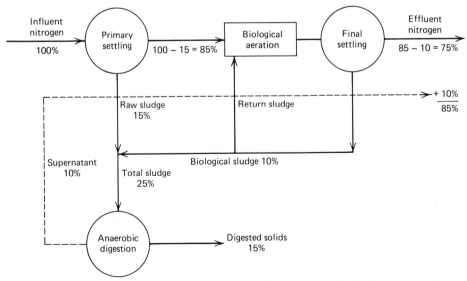

Figure 13-3 Diagram tracing wastewater nitrogen through a hypothetical treatment plant. Of the 100 percent influent nitrogen, 85 percent is in the treated wastewater and 15 percent is in the digested sludge solids.

percent. Primary removal by sedimentation of raw organic matter is 15 percent, and removal by biological synthesis in secondary aeration is assumed to be an additional 10 percent. Processing of the raw sludge can release some of the nitrogen extracted from the wastewater. In this treatment scheme, the sludge is stabilized by anaerobic digestion and the ammonia released from decomposition of the sludge solids is returned to the influent of the treatment plant in supernatant from the digester. Assuming 40 percent of the organic nitrogen in the sludge is converted to ammonia, 10 percent of the original 25 percent is recycled to the treatment plant and appears in the effluent, which then contains 85 percent of the influent nitrogen. Depending on variations in nitrogen content of the wastewater and methods of wastewater and sludge processing, nitrogen removal in conventional biological treatment systems ranges from nearly zero up to 40 percent.

EXAMPLE 13-3

Using the values given in Table 13-1, calculate the concentration of total nitrogen in the effluent from a conventional activated-sludge plant. Assume the following: primary suspended solids removal of 50 percent containing 4.2 percent nitrogen; primary BOD removal of 35 percent; waste activated-sludge solids from biological aeration, based on a food-to-microorganism ratio of 0.40, containing 6.2 percent nitrogen; effluent BOD of 30 mg/l; effluent suspended solids of 30 mg/l; and no nitrogen recycling from sludge processing.

Solution

From Table 13-1, influent SS = 240 mg/l and BOD = 200 mg/l. At 50 percent SS removal, primary solids are 120 mg. Nitrogen in the primary sludge equals

$$0.042 \times 120 = 5.0 \text{ mg of organic nitrogen}$$

Settled wastewater has a BOD equal to 130 mg/l. From Figure 11-48 at an F/M of 0.40, K equals 0.50. Therefore,

$$W_s = 0.50 \times 130 = 65 \text{ mg}$$

Nitrogen in the secondary sludge equals

$$0.062 \times 65 = 4.0 \text{ mg of organic nitrogen}$$

Suspended solids in the effluent are 30 mg/l.

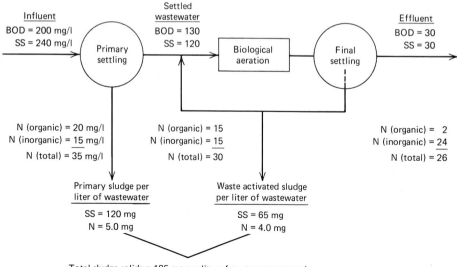

Figure 13-4 Summary of results from Example 13-3.

Organic nitrogen in the effluent equals

$$0.062 \times 30 = 2.0 \text{ mg/l of organic nitrogen}$$

Inorganic (ammonia + nitrate) nitrogen in the effluent is the influent minus the organic nitrogen contents of the primary sludge, secondary sludge, and effluent suspended solids. Therefore, inorganic nitrogen in the effluent equals

$$35 - 5.0 - 4.0 - 2.0 = 24 \text{ mg/l}$$

Total effluent nitrogen is $2.0 + 24 = 26$ mg/l.
These data are summarized in Figure 13-4. The total sludge solids produced per liter of wastewater treated equal 185 mg containing 9.0 mg of nitrogen. The influent nitrogen is 35 mg/l and the effluent is 26 mg/l for a removal of 35 percent.

13-5 BIOLOGICAL NITRIFICATION AND DENITRIFICATION

Nitrification of a wastewater is practiced where the ammonia content of the effluent causes pollution of the receiving watercourse. The process does not remove the nitrogen but converts it to the nitrate form (Eq. 13-9). Nitrification-denitrification, which reduces the total nitrogen content, includes conversion of the nitrate to gaseous nitrogen (Eq. 13-10). Since this is a costly process, it is generally performed only where the receiving watercourse is used as a source for public water supply and the dilution is not adequate to reduce the nitrate concentration to less than 10 mg/l.

Nitrification

Nitrification is usually a separate process following conventional biological treatment. Although nitrification may be possible to perform along with organic matter removal in a single-stage extended aeration unit in a warm climate, two-step treatment is necessary for reliable operation at a reduced wastewater temperature. The conventional biological treatment removes BOD, without oxidation of ammonia nitrogen, to produce a suitable effluent for nitrification. The high ammonia content and low BOD provide greater growth potential for the nitrifiers relative to the heterotrophs. This allows operation of the nitrification process at an increased sludge age

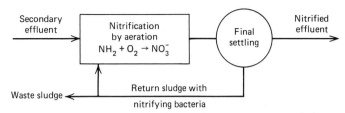

Figure 13-5 Flow diagram for nitrification by suspended-growth aeration following conventional biological treatment.

to compensate for lower operating temperature and to ensure that the growth rate of nitrifying bacteria is rapid enough to replace those lost through washout in the plant effluent. Although synthetic-media filtration and rotating biological contactor processes can perform nitrification, the most reliable system is suspended-growth aeration.

Figure 13-5 is the common flow scheme for nitrification following biological treatment of a wastewater for organic matter reduction. After the nitrifying bacteria oxidize the ammonia in the aeration tank, the activated sludge containing high populations of nitrifiers is settled in the final clarifier for return to the aeration tank. Sludge can be wasted, if necessary, to remove excess bacterial growth from the system. Because the rate of oxidation of ammonia is nearly linear, the tank configuration is plug flow to minimize short-circuiting. The important parameters in bacterial nitrification kinetics are temperature, pH, and dissolved oxygen concentration. Reaction rate is decreased markedly at reduced temperatures with about 8° C being the reasonable minimum value. Optimum pH is near 8.4, and the dissolved oxygen level should be greater than 1.0 mg/l. Since biological nitrification destroys alkalinity, lime may be needed to raise the pH to the optimum level in the nitrification tank. Ammonia nitrogen loadings applied to the aeration tank are 10 to 20 lb/1000 cu ft/day (160 to 320 g/m$^3 \cdot$ d) with corresponding wastewater temperatures of 10 to 20° C, respectively. For an average wastewater effluent, this is an aeration period of 4 to 6 hr.

Denitrification

Nitrate can be reduced to nitrogen gas by facultative heterotrophic bacteria in an anoxic environment. An organic carbon source, AH_2 in Eq. 13-10, is needed to act as a hydrogen donor and to supply carbon for biological synthesis. Although any biodegradable organic substance can serve as a carbon source, methanol is common because of its availability, ease of application, and ability to be applied without leaving a residual BOD in the process effluent. As with all other chemical carbon sources, methanol is expensive. In fact, the cost of methanol makes the widespread application of denitrification unrealistic in wastewater treatment.

The denitrification reaction between methanol and nitrate is as follows:

$$5CH_3OH + 6NO_3^-$$
$$= 3N_2\uparrow + 5CO_2\uparrow + 7H_2O + 6OH^- \quad (13\text{-}11)$$

Since the process effluent from nitrification also contains dissolved oxygen and nitrite, total methanol required as a hydrogen donor in denitrification is given in Eq. 13-12. In addition, methanol is used as a carbon source in bacterial synthesis. Based on approximately 30 percent excess methanol needed for synthesis, the total methanol demand is calculated using Eq. 13-13.

$$CH_3OH = 0.7DO + 1.1NO_2\text{-}N + 2.0NO_3\text{-}N \quad (13\text{-}12)$$

$$CH_3OH = 0.9DO + 1.5NO_2\text{-}N + 2.5NO_3\text{-}N \quad (13\text{-}13)$$

where CH_3OH = methanol, milligrams per liter
DO = dissolved oxygen, milligrams per liter

$NO_2\text{-}N$ = nitrite nitrogen, milligrams per liter
$NO_3\text{-}N$ = nitrate nitrogen, milligrams per liter

The recommended denitrification system consists of a plug-flow tank with underwater mixers followed by a clarifier for sludge separation and return, Figure 13-6. The level of agitation in the denitrification chambers must keep the microbial floc in suspension, but controlled to prevent undue aeration. Since nitrogen gas released from solution can float the biological floc, the last chamber must strip the nitrogen gas from solution for efficient final clarification. This may be done using an aeration chamber, which also has the advantage of oxygenating the plant effluent, or by a degasifier. The detention time required for denitrification of a domestic wastewater is usually in the range of 2 to 4 hr depending on nitrate loading and temperature.

Biological Nitrification-Denitrification

Unoxidized organic matter can be used as an oxygen acceptor (hydrogen donor) for conversion of nitrate to nitrogen gas. This reaction satisfies a portion of the BOD in a wastewater.

$$\begin{array}{ccc} \text{Unoxidized} & & \text{Oxidized} \\ \text{organic matter} + NO_3^- \rightarrow & \text{organic matter} & + N_2\uparrow \\ \text{(BOD)} & & \text{(reduced BOD)} \end{array}$$

$$(13\text{-}14)$$

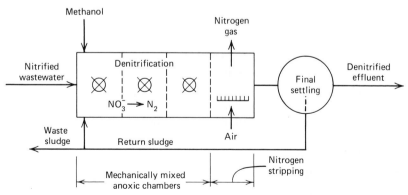

Figure 13-6   Flow diagram for denitrification by anoxic (anaerobic) suspended growth.

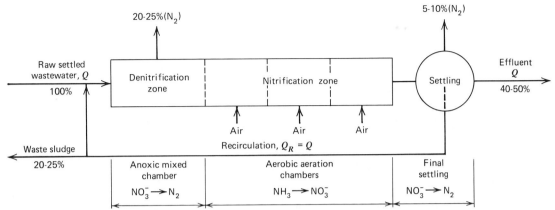

Figure 13-7   Diagram tracing wastewater nitrogen through a hypothetical biological nitrification-denitrification system. Of the 100 percent influent nitrogen, 30 to 35 percent is released as nitrogen gas, 20 to 25 percent is organic nitrogen in the waste sludge, and 40 to 50 percent appears in the effluent primarily as nitrate nitrogen.

The flow scheme of a biological nitrification-denitrification process requires mixing of raw organic matter with nitrified wastewater; hence, an aerobic zone is needed for nitrification and an anoxic zone for denitrification. (The term anoxic means low or lacking in oxygen.)

The plug-flow activated-sludge system, diagramed in Figure 13-7, blends nitrified recirculation flow with raw settled wastewater (primary effluent) in an anoxic zone. In this mechanically mixed chamber, or segregated zone of a long narrow tank, the biological floc in the return activated sludge uses recirculated nitrate as an oxygen source, releasing nitrogen gas. In the subsequent chambers mixed by diffused or mechanical aeration, dissolved oxygen is taken up by the biological floc for nitrification of the ammonia in the wastewater. Both the anoxic and aerobic chambers reduce the wastewater BOD—the anoxic zone by denitrification and the aerobic zone by uptake of dissolved oxygen. The final clarifier contributes to denitrification as the bacterial floc extract the oxygen from nitrate in metabolism of the organic solids during sedimentation. The majority of nitrogen removal, however, occurs in the anoxic chamber, and the degree of nitrogen removal is controlled by the rate of recirculated flow. The amount of return nitrate that can be reduced

depends on the maximum rate of denitrification possible in the anoxic zone. Too great a recirculation carries nitrate through the anoxic zone into the aerobic chambers, and, similarly, too short an aeration period can reduce the degree of nitrification. Other factors, such as the relative concentration of nitrogen to BOD in the wastewater and temperature, also influence the method of process operation. Figure 13-7 illustrates typical treatment of a settled domestic wastewater at a moderate temperature with a total retention time of approximately 8 hr in the biological chambers. Influent nitrogen is assigned a value of 100 percent. Of this, 30 to 35 percent is converted to nitrogen gas, 20 to 25 percent appears in the waste sludge as organic nitrogen, and 40 to 50 percent is in the plant effluent primarily as nitrate nitrogen. Thus, this biological nitrification-denitrification process removes 50 to 60 percent of the influent nitrogen.

Another process scheme for biological nitrogen removal is diagramed in Figure 13-8. The plant is a *Carrousel* system that has a deep, long, oval aeration tank with vertical walls and a mechanical surface aerator to propel the wastewater around in the channel. A similar system is a deep, vertical wall, oxidation ditch with horizontal rotor aerators. When treating an unsettled domestic wastewater, the aeration period is generally 24 hr

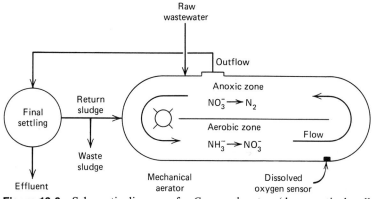

Figure 13-8   Schematic diagram of a *Carrousel* system (deep, vertical wall, oxidation ditch) operating as an extended aeration process with concurrent nitrification and denitrification.

and the mixed liquor suspended solids maintained at 3000 to 5000 mg/l to operate at a long sludge age. Under these conditions, viable populations of nitrifying bacteria can be maintained in the activated sludge for nitrification. Nevertheless, cold-weather operation that decreases the temperature of the aerating liquid below 10° C can adversely affect nitrification.

Denitrification in a *Carrousel* system is possible with proper operational control. By reducing the degree of aeration, the aerobic zone along the channel can be shortened to create an anoxic zone before the mixed liquor recirculates back to the aerator. A dissolved oxygen sensor located midway along the channel can be used to control the aerator. Consequently, the first half of the channel can be aerobic enough for nitrification, and the second half deficient so that oxygen is biologically removed from nitrate to satisfy wastewater BOD. Total nitrogen removal in the range of 50 percent to over 70 percent is possible in this combined nitrification-denitrification process.

EXAMPLE 13-4

A *Carrousel* extended aeration plant is operated as a biological nitrification-denitrification process. The aeration tank has a volume of 5600 m³. Based on several months operating data, the average wastewater flow was 4300 m³/d with the following characteristics: 260 mg/l BOD, 300 mg/l SS, 51 mg/l total N, and 31 mg/l NH_3-N. The average effluent quality was 5 mg/l BOD, 8 mg/l SS, 7 mg/l total N, 4 mg/l NH_3-N, and 2 mg/l NO_3-N. The operating MLSS concentration averaged 4700 mg/l and the temperature was in the range of 16 to 18° C. Calculate the volumetric BOD loading, F/M ratio, aeration period, BOD removal, SS removal, and total N removal.

Solution

$$\text{BOD loading} = \frac{(260 \text{ mg/l})(4300 \text{ m}^3/\text{d})}{5600 \text{ m}^3} = 200 \text{ g/m}^3 \cdot \text{d}$$

$$\text{F/M} = \frac{(260 \text{ mg/l})(4300 \text{ m}^3/\text{d})}{(4700 \text{ mg/l})(5600 \text{ m}^3)} = 0.042 \frac{\text{g BOD/d}}{\text{g MLSS}}$$

$$t = \frac{5600}{4300} 24 = 31 \text{ h}$$

$$\text{BOD removal} = \frac{260 - 5}{260} 100 = 98\%$$

$$\text{SS removal} = \frac{300 - 8}{300} 100 = 97\%$$

$$\text{Total N removal} = \frac{51 - 7}{51} 100 = 86\%$$

13-6 CHEMICAL NITROGEN REMOVAL

Air Stripping

Ammonia can be air stripped from solution at an alkaline pH of about 11, Eq. 13-15. After lime clarification, the wastewater is pumped to the top of a cooling tower and distributed over the packing, Figure 13-9. Forced air is drawn up through the packing to extract ammonia from the water droplets. A smooth plastic packing is used to reduce the accumulation of calcium carbonate. During warm weather, air stripping can achieve 90 to 95 percent removal of ammonia nitrogen at a pH of about 11.5 by using 400 cu ft of air per gallon of wastewater. Air stripping also removes some volatile organics. The simplicity of this process makes it the least expensive method of denitrification if the wastewater is pretreated with lime for phosphorus precipitation or as the first step in water reclamation.

$$NH_4^+ + OH^- \xrightleftharpoons[\text{acidic}]{\text{basic}} NH_4OH \xrightarrow[\text{stripping}]{\text{air}} NH_3\uparrow + H_2O$$

$$(13\text{-}15)$$

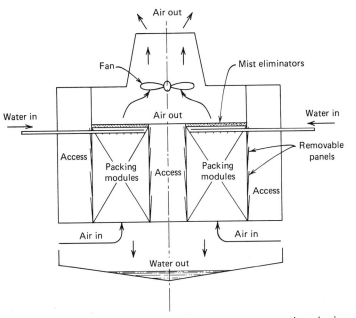

Figure 13-9 Cross section of a countercurrent air-stripping tower for ammonia removal.

Air stripping has several technical difficulties that limit its application. Experience with packed cooling towers has revealed the following operational problems: reduced ammonia removal efficiency at low wastewater temperature because of increased ammonia solubility; scale formation on the tower packing that may have to be removed frequently; ice formation in the tower during cold weather; and the failure of air stripping to remove nitrate nitrogen inadvertently produced in biological pretreatment. The inability to operate at a low ambient air temperature is a serious disadvantage during winter operation in cold climates. In southern regions during short periods of cool weather, breakpoint chlorination can be applied for supplemental nitrogen removal.

Breakpoint Chlorination

The chemistry of breakpoint reactions between ammonia and chlorine are not well defined. Equation 13-16 is an unbalanced chemical reaction to illustrate the possible products formed in the oxidation of ammonia; in order of importance, these are nitrogen gas, nitrous oxide, and nitrite-nitrate nitrogen.

$$NH_3 + HOCl \rightarrow N_2\uparrow + N_2O\uparrow + NO_2^- + NO_3^- + Cl^-$$

$$(13\text{-}16)$$

Weight ratios of chlorine to ammonia nitrogen required for breakpoint chlorination of wastewaters range from 8 to 1 to 10 to 1 as Cl_2 to N; the lower value is applicable for the most highly pretreated wastewater. Breakpoint chlorination in the pH range of 6.5 to 7.5 can yield up to 95 percent ammonia removal, and for initial ammonia nitrogen concentrations of 8 to 15 mg/l, the nitrate and nitrogen trichloride residuals do not exceed 0.5 mg/l.

The process is relatively inexpensive and easy to operate and control. One disadvantage of heavy chlorination is that essentially all of the chlorine added is reduced to chloride ion, thus contributing to the dissolved solids concentration in the treated wastewater. For example, at an 8 to 1 weight ratio of Cl_2 to N, the oxidation of 20 mg/l of ammonia nitrogen contributes 160 mg/l of chloride ion. In many instances, less than complete ammonia removal may be sufficient to meet water quality objectives; however, at sub-breakpoint oxidation the production of chloramines may be too high for discharge to a receiving watercourse. Dechlorination with sulfur dioxide is one solution, and another is adsorption on granular activated carbon.

Ion Exchange

The ammonium ion is present in a low concentration relative to other ions in wastewater. Consequently, it is difficult to selectively extract by ion exchange. In order for denitrogenation by ion exchange to be an economical process, the resin must have a relatively high selectivity for the ammonium ion, since removal of additional cations reduces process efficiency. The natural inorganic zeolite clinoptilolite has unusual selectivity for the ammonium ion and is currently the ion exchange medium of choice.

Pretreatment prior to ammonium cation exchange involves clarification by chemical coagulation and filtration. Adjustment to a pH of 6.5 is necessary to convert ammonia to ammonium ion, since free ammonia does not absorb or enter into ion exchange reactions with clinoptilolite. Spent medium is regenerated with a salt solution. The ammonia is recovered as ammonium sulfate by means of a closed-loop stripping-absorption process.

Regardless of the apparent advantages and need for an ion-exchange method for nitrogen removal, the process of ammonium ion extraction and regeneration of the exchange resin are still experimental. Future application in the nitrogen removal of wastewater is difficult to predict.

13-7   SUSPENDED SOLIDS REMOVAL

The inability of gravity sedimentation in final clarifiers to remove small particles is the major limitation of suspended solids and BOD removal by conventional wastewater treatment. Where needed, tertiary granular-media filtration can be employed to upgrade effluent quality. The filters are similar to those used in water treatment (Section 7-4). Design must take into account, however, the higher suspended solids content and fluctuating rates of wastewater flow not common to water processing. The preferred filters are coal-sand dual media or mixed media containing anthracite coal, garnet, and sand. A sand medium alone cannot accommodate the quantity of coarse solids in wastewater because of surface plugging. Multimedia beds allow in-depth filtration and greater solids holding capacity, resulting in longer filter runs. Efficient backwashing requires auxiliary scour since hydraulic suspension of the media, by upward flow of water through the bed, does not provide adequate cleaning. Either air scrubbing or a rotating agitator improves scouring action. The filters may be either gravity or pressure units depending on size of the treatment plant.

Gravity filters in wastewater processing usually have deep filter boxes with the water applied to a series of filters by influent flow splitting. Figure 13-10 shows that the influent is divided equally to all operating filters from a common channel or pipe through valve 1. When filtration starts, the influent weir box discharges into a wash trough that overflows to the wastewater above the filter. The media are not disturbed by the water entering the filter box because the static water level is above the surface of the bed. To prevent accidental dewatering of the media, the filtered wastewater passes over an effluent weir in the discharge chamber that is higher in elevation than the surface of the media. After the wash troughs are submerged, the influent weir discharges directly into the wastewater above the bed. After passing through the filter media, the wastewater enters a discharge tank, which is usually the inlet end of a chlorination chamber, through valve 2. Head loss through the clean filter is h_c, which is the minimum head loss. The maximum head loss available for filtration is at the maximum water level in the filter box. When this level is reached, the filter is backwashed.

Cleaning of a filter is started by closing the influent valve 1 and opening the drain valve 3 to the wash-water outlet channel. This drains out the water overlying the filter and lowers the level to the bottom of the wash troughs. After valve 2 is closed to the discharge tank, the wash-water valve 4 is opened allowing filtered wastewater to flush impurities out of the fluidized bed. Mechanical agitation or air scouring for dislodging the impurities from the media precedes hydraulic backwashing and may continue as the backwash water rises in the bed. The clean media are allowed to restratify in quiescent water before restarting filtration.

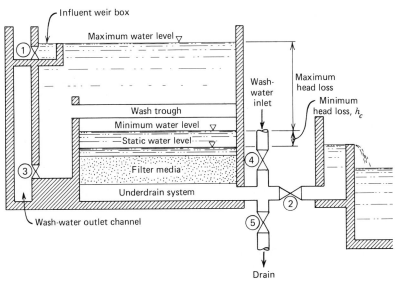

Figure 13-10 Cross-sectional view of a gravity filter with a deep filter box.

A deep filter box, rather than the traditional pressure-suction water filtration system, prevents the possibility of causing air binding in the bed as a result of creating a negative pressure in the filter. Another advantage of a deep filter box is eliminating the installation of costly rate of flow controllers and effluent piping. In flow splitting, effluent flow meters are not necessary since the influent to each filter is the total plant influent divided by the number of units in operation. Also, head loss in a filter can be easily observed by the water level in the filter box.

Hydraulic loading and time of filtration before a bed requires backwashing are interrelated. Design rates in wastewater filtration are generally in the range of 3 to 4 gpm/sq ft (2.0 to 2.7 l/m²·s). Proper design of a filtration system depends on providing sufficient bed area. Although removal efficiency is relatively independent of the hydraulic loading rate, too small a surface area causes short filter runs. Excess backwash water cycled to the head of the plant and filter downtime result in inefficient operation. A minimum filtration time of 24 hr between backwashes is desirable under normal conditions of wastewater flow and suspended solids concentration. Effective suspended solids removal relies on the design of the installation and the characteristics of the applied

wastewater. In general, the best effluent quality achievable by plain filtration is 5 to 10 mg/l of suspended solids. If further reduction is desired, chemical coagulation must precede filtration to flocculate the colloidal solids and reduce the effluent suspended solids to less than 5 mg/l.

13-8 VIRUS REMOVAL

The presence of enteric viruses in wastewater effluent is of major concern where reuse of the treated wastewater involves human contact. This reuse may be un-intentional—for instance, discharge to a dry streambed where children can play, or intentional as in the case of landscape watering. In the effluent from a conventional treatment plant, suspended biological floc can trap viruses inside, protecting them against inactivation during chlorination. Although biological treatment, particularly activated-sludge processes, greatly reduce the populations of viruses, disinfection is still necessary to ensure a safe effluent for human contact.

The Pomona Virus Study[1] investigated virus removal efficiency of several tertiary treatment systems on a pilot-plant scale. The results demonstrated that direct filtration with a low alum dosage of 5 mg/l with polymer aid was as effective as alum coagulation at a dosage of 150 mg/l with conventional inline mixing, flocculation, and sedimentation units prior to filtration. The rate of filtration was 5 gpm/sq ft and chlorination was in a plug-flow tank with a residual of 5 to 10 mg/l after a 2-hr contact period. The concentration of coliforms in the tertiary effluent was reduced to less than 2.2 per 100 ml. The efficiencies of the tertiary systems were based on removal of poliovirus added to the wastewater before processing. The major conclusions of the study were that the majority of virus inactivation occurred during disinfection and the main function of preceding treatment was to remove suspended solids that interfere with effluent disinfection. Also, the concentration of the chlorine residual directly affected the effluent virus concentration.

Tertiary treatment by direct filtration has been implemented at several wastewater plants in California as diagramed in Figure 13-11. For example, the Whittier Narrows Water Reclamation Plant has chemical pretreatment using alum (5 to 7 mg/l) and anionic polymer (0.01 mg/l), dual-media filtration at 3.4 gpm/sq ft at average flow rate and 5.4 gpm/sq ft at peak flow, and plug-flow chlorination with a contact time of about 1.5 hr at peak flow. The effluent water-quality standard is a seven-day median coliform concentration less than or equal to 2.2 per 100 ml. Because of the difficulties in monitoring for the presence of viruses, a standard for virus concentration is not used. The filters are constant rate controlled by influent flow splitting, and the backwash system is combined air-water with wash-water recovery.

A chlorination chamber must be properly designed for effective disinfection. Rapid mixing is employed to blend the chlorine solution with the wastewater, followed by plug flow through a contact tank that is a long, narrow channel. Chlorine dosage is maintained by automatic residual monitoring and feedback control. If dechlorination of the effluent is necessary, a solution of sulfur dioxide can be added at the discharge end of the chlorination chamber.

13-9   WASTEWATER RECLAMATION

Reclaiming a wastewater to near potable water quality requires removal of heavy metals, organic chemicals, viruses, and inorganic salts. Reclamation involves biological treatment followed by chemical-physical processes uniquely designed for removal of these contaminants. Pretreatment of industrial wastewaters is essential in the control of wastewater quality. Since many of the hazardous contaminants in industrial wastes are difficult to remove, reductions in their concentrations in the wastewater are essential to effective operation of a water reclamation plant. Reliability of reclaimed water quality is directly related to the quality variations of the influent wastewater.

The preferred initial treatment is conventional activated-sludge processing designed and operated to yield the best possible effluent. Flow equalization is necessary, either before or after biological treatment, to reduce peak pollutant concentrations and to provide a uniform flow rate for the subsequent chemical-physical processes. If the wastewater cannot be discharged to a watercourse, emergency storage is needed to collect and

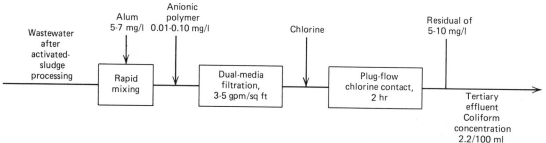

Figure 13-11   Tertiary treatment scheme for virus removal that is equivalent to potable water treatment by alum coagulation, sedimentation, filtration, and chlorination.

recycle wastewater of poor quality prior to tertiary treatment.

Lime precipitation of the biological effluent is the first step in chemical-physical processing. High-lime treatment is effective in precipitation of heavy metals, suspended organics, some dissolved organics, and phosphates. An alkaline pH of about 11 also kills bacteria and inactivates viruses. Although flocculator-clarifiers have been used, the common scheme is to separate rapid mixing, flocculation, and sedimentation in a series of tanks. The lime sludge can be gravity thickened and mechanically dewatered for disposal or for purification and recalcination for reuse of the lime. The carbon dioxide produced in the recalcining process can be used in recarbonation of the wastewater.

After sedimentation and before recarbonation, the alkaline wastewater can be aerated to strip volatile organics and ammonia, commonly in a cooling tower. Air stripping, however, cannot be performed at a cold air temperature because of excessive cooling of the wastewater. One process as a possible substitute for air stripping is selective ion exchange for removal of the ammonium ion, following granular-media filtration.

Recarbonation after high-lime treatment is necessary to stabilize the wastewater to prevent precipitation of calcium carbonate scale in subsequent processes. Two stages are preferred in recarbonation to remove some of the calcium from the wastewater and to recover calcium carbonate for recalcination. In the first stage, only sufficient carbon dioxide is applied to convert the hydroxide to carbonate, at about pH 8.4, and precipitate calcium carbonate. In the second stage, the pH is further reduced to about 7.5 in order to convert the carbonate remaining in the settled wastewater to bicarbonate. Granular-media filtration is used to remove nonsettleable solids prior to subsequent treatment.

Granular activated-carbon adsorption following filtration removes trace concentrations of a variety of nonpolar organic compounds. Disinfection by chlorination or ozonation is also integrated with carbon adsorption; for example, the sequence may be carbon adsorption–ozonation–carbon adsorption.

Finally, a portion of the wastewater flow is processed for reduction of dissolved salts. Reverse osmosis is the common demineralization process. Besides reducing the content of the undesirable ions of sodium, chloride, nitrate, and trace metals, passage of water through a membrane removes nonspecific dissolved organic compounds. Post treatment after blending the reverse-osmosis effluent with the by-pass flow is disinfection using chlorine or chlorine dioxide.

13-10 WATER RECLAMATION PLANT, ORANGE COUNTY, CALIFORNIA

Water Factory 21 in southern California reclaims sanitary wastewater after activated-sludge treatment and equalization of flow.[2] The design capacity is 15 mgd (57,000 m³/d) at a uniform flow rate. The final process of demineralization by reverse osmosis has a design flow rate of only 5 mgd, since removal of dissolved solids from one third of the water is adequate to provide a blended effluent of suitable quality. As diagrammed in Figure 13-12, the processing consists of lime precipitation with sludge recalcination, air stripping, recarbonation, granular activated-carbon adsorption with carbon regeneration, disinfection by chlorination, and reduction of dissolved solids by reverse osmosis. An aerial view of the plant showing the orientation of treatment units is in Figure 13-13.

Lime treatment is performed in separate rapid mixing, flocculation, and sedimentation tanks at a controlled pH of 11.0 by application of 350 to 400 mg/l of CaO with a small amount of anionic polymer to aid coagulation. The primary objectives in lime precipitation are extraction of heavy metals and reduction of organic solids. The concentrations of heavy metals are significantly reduced. Occasionally, cadmium and chromium have exceeded the maximum contaminant levels for drinking water, but only during periods of high influent concentrations. Suspended solids reduction results in an average COD removal of 50 percent and turbidity reduction of over 90 percent. Another benefit of high-lime treatment is disinfection resulting in very high coliform reductions and virus removals greater than 98 percent.

The sludge from lime precipitation of the wastewater and calcium carbonate from first-stage recarbonation are concentrated in a gravity thickener to 8 to 15 percent

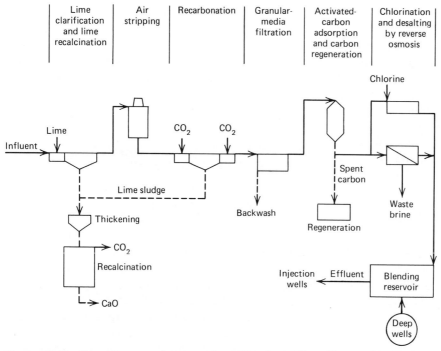

Figure 13-12 Flow diagram of water reclamation plant, Water Factory 21, Orange County, California.

solids. After dewatering by centrifugation, approximately 50 percent of the solids are recalcined in a multiple-hearth furnace for recovery of lime for reuse. By wasting one half of the precipitate, the buildup of impurities is controlled in the recalcined lime to prevent recycling of contaminants to the wastewater. Storage is provided in the lime-processing building for both recalcined lime and unused lime. Carbon dioxide from the recalcination process is used for recarbonation.

Air stripping is done in countercurrent induced-draft towers with a design aeration capacity of 400 cu ft per gallon of wastewater. Currently, the wastewater cascades over the tower packing with only natural ventilation. Because of improved biological treatment of the wastewater, the ammonia concentration is reduced to less than 5 mg/l and forced-air ventilation is not necessary. Nevertheless, air stripping by natural ventilation is still beneficial for removing a variety of volatile organic compounds. After air stripping, the wastewater is stabilized by two-stage recarbonation and filtered through gravity granular-media beds for suspended solids removal prior to activated-carbon adsorption.

The granular activated-carbon columns are operated in a downflow direction at an empty-bed contact time of approximately 30 min. In upflow operation, as originally designed, carbon fines smaller than 25 μm were carried out of the columns and accumulated on the membranes in the reverse-osmosis modules. Efficiency of COD removal by carbon adsorption is generally in the range of 50 to 60 percent by the adsorption of a wide variety of organic compounds. Chlorinated compounds with one or two carbons, however, were not readily adsorbed. Some heavy metals, such as chromium and copper, are partially removed.

One third of the effluent from the carbon columns is processed by reverse osmosis to remove dissolved inorganic solids, Figure 13-14. Typical operation is 90 percent reduction of dissolved solids to a level of about

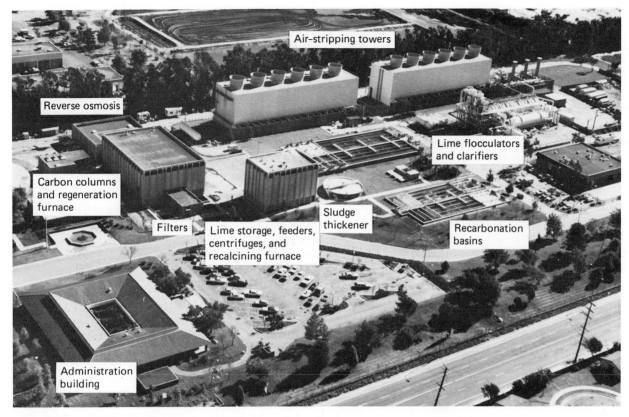

Figure 13-13 Aerial view of Water Factory 21. (Courtesy of Orange County Water District.)

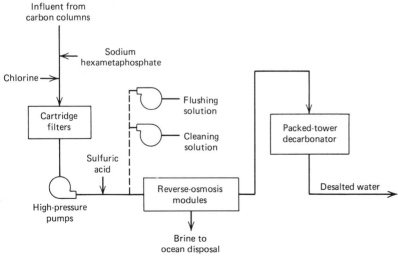

Figure 13-14 Flow diagram of the reverse-osmosis plant at Water Factory 21, Orange County, California.

100 mg/l with a recovery of approximately 85 percent of the feed water. In addition, passage of the water through membranes also removes high-molecular-weight organic substances. The influent is treated with sodium hexameta-phosphate as a scale inhibitor and chlorine to control biological growth on the membranes. Cartridge filters with replaceable polypropylene elements with an effective size of 25 μm are used to remove particulate matter. After pressurizing by feed pumps to 550 psi, sulfuric acid is injected to adjust the pH to 5.5 to minimize membrane hydrolysis and scale formation, primarily calcium carbonate and calcium sulfate.

Each reverse osmosis module consists of six spiral-wound cellulose acetate elements enclosed in a 23-ft-long, 8-in.-diameter, fiberglass cylinder. The water passes through 3 stages consisting of 42 modules. First, the feed water flows in parallel through 24 first-stage modules; the concentrate (brine) from these units is passed through 12 second-stage modules; and the concentrate from the second pass flows through the 6 remaining modules. The reject brine, which is approximately 15 percent of the feed water, is discharged to an outfall sewer for ocean disposal.

The product water is air stripped in a packed tower to remove carbon dioxide, which resulted from prior acidic pH adjustment.

Loss of productivity is a major problem in reverse-osmosis treatment of wastewater. The membranes must be periodically cleaned to restore the rate of water passage. The inhibiting layer that forms on the membrane surface consists primarily of bacteria that attach to the surface of the cellulose acetate. Common cleaning solutions contain detergents and enzymes. After cleaning, a flushing solution is used to remove the cleaning chemicals; if the process is to be shutdown for several days, the flushing solution is used to displace wastewater from the modules.

Water Factory 21 reliably produces a treated water with contaminant concentrations less than the maximum levels dictated by drinking water standards. Therefore, the reclaimed water is considered satisfactory for underground injection to prevent saltwater intrusion by forming a water mound blocking infiltration of seawater from the Pacific Ocean. The injected water is required to meet drinking water standards since it can flow through the

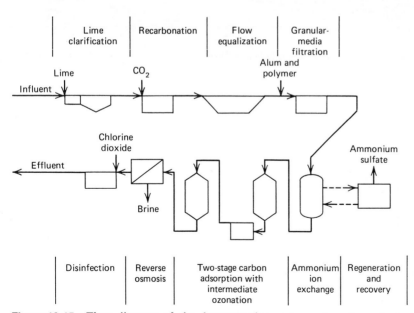

Figure 13-15 Flow diagram of the demonstration water reclamation plant, Denver, Colorado.

aquifer toward inland wells used for public water supplies. Heavy metals are removed, viruses inactivated, and trace organic compounds reduced to extremely low levels. The safety of reusing reclaimed wastewater for groundwater recharge is increased by dilution with natural groundwater and filtration as the water moves through the aquifer.

13-11   DEMONSTRATION WATER RECLAMATION PLANT, DENVER, COLORADO

The Denver water department has constructed a demonstration plant with a capacity of 1 mgd. The purpose is to evaluate the technical and economic feasibilities of reclaiming wastewater to potable quality for unrestricted reuse. The processing scheme, shown in Figure 13-15, is similar to that of Water Factory 21 except nitrogen removal is by selective ion exchange and ozonation is incorporated with two-stage carbon adsorption. The influent is unchlorinated treated wastewater from secondary biological processing.

The first step in treatment is lime precipitation at a pH of 11 or greater for removal of suspended solids, heavy metals, phosphorus, bacteria, and viruses. After clarification, recarbonation is used to stabilize the wastewater before it flows into a ballast pond. After storage for flow equalization, nonsettleable solids are removed by direct filtration in dual-media pressure filters following chemical additions of alum and polymer.

The ammonium ion is removed by selective ion exchange after interfering substances are removed by the previous physical-chemical treatment, particularly suspended and colloidal solids that can foul the zeolite (clinoptilolite). By regenerating the zeolite with a salt solution, the ammonia can be recovered as ammonium sulfate in a stripping process for use as a fertilizer. Two-stage carbon adsorption with intermediate ozonation removes refractory organic compounds. Spent carbon is transferred to holding tanks prior to regeneration in a fluidized bed furnace. Ten percent of the carbon column effluent (0.1 mgd) is treated by reverse osmosis for reduction of dissolved solids and subsequent aeration to remove dissolved gases and strip volatile organic com-

pounds. The final step is disinfection using chlorine dioxide.

Existing drinking water standards are founded on raw-water supplies being drawn from sources of natural flowing or impounded water and groundwater. These sources are protected from excessive contamination by adequate environmental control. Since water reclamation starts with wastewater (a grossly polluted source) and the product water may be reused with little or no dilution, existing drinking water standards do not provide adequate consumer protection. Because of the lack of quality standards for potable reuse of reclaimed wastewater, the Denver demonstration project is conducting special analyses on the chemical composition of the product water and health effects on animals. Animal toxicity testing over a period of several years is intended to identify potential health problems in consumption of reclaimed water.

REFERENCES

1. *Pomona Virus Study*, Sanitation Districts of Los Angeles County, California State Water Resources Control Board, Sacramento, CA, 1977.
2. *Wastewater Contaminant Removal for Groundwater Recharge at Water Factory 21*, U.S. Environmental Research Laboratory, EPA 600/2-80-144, August, 1980.

FURTHER READING

Viessman, W., Jr., and M. J. Hammer. *Water Supply and Pollution Control*. New York: Harper & Row, 1985.

PROBLEMS

13-1   For Eq. 13-1, verify by appropriate calculations the molar ratio for Al to P of 1 to 1 and the weight ratio of commercial alum to phosphorus of 9.7 to 1.0.

13-2   A conventional activated-sludge plant treats a domestic wastewater with 220 mg/l of SS, 180 mg/l of BOD, and 9.0 mg/l of total P. After primary settling, the characteristics are 110 mg/l of SS, 120 mg/l of BOD, and 7.5 mg/l of total P. Based on a pilot-plant study, a dosage of 116 mg/l of alum is required to reduce the effluent total P to 1.0 mg/l with the aeration tank operating at an F/M of 0.20 lb BOD/day/lb MLSS. The effluent SS concentration averages 10 mg/l. Calculate the estimated sludge-solids production per million gallons of wastewater treated. If the primary sludge has a solids content of 5.0 percent and the waste sludge from the chemical-biological aeration system has a concentration of 1.5 percent, calculate the volume of combined sludge produced per million gallons of wastewater treated.

13-3 A laboratory activated-sludge system as shown in Figure 9-3 was used to evaluate chemical-biological phosphorus removal by applying alum to the aeration chamber. Twelve liters of settled sanitary wastewater were applied daily at a constant rate to the aeration chamber, which had a volume of 3.6 l. The aeration tank MLSS concentration was held near 2000 mg/l by wasting 200 to 250 ml of mixed liquor each day. The temperature was 22 to 24° C, pH 7.3 to 7.7, and SVI varied from 90 to 130 ml/g. The alum solution feed to the aeration tank had a strength of 10.0 mg of commercial alum per milliliter. The following data were collected at various alum feed rates after arriving at steady-state conditions.

Wastewater Feed (l/d)	Alum Applied (ml/d)	Influent			
		BOD (mg/l)	SS (mg/l)	P (mg/l)	Alk (mg/l)
12.0	0	150	94	10.3	350
12.0	70	158	104	10.9	350
12.0	140	169	114	10.6	340
12.0	210	173	135	10.4	360
12.0	390	173	123	9.3	320

Wastewater Feed (l/d)	Alum Applied (ml/d)	Effluent			
		BOD (mg/l)	SS (mg/l)	P (mg/l)	Alk (mg/l)
12.0	0	6	7	7.9	230
12.0	70	6	8	5.4	200
12.0	140	5	11	2.4	160
12.0	210	10	10	1.0	140
12.0	390	8	17	0.5	80

(a) Calculate the phosphorus removal and the weight ratio of alum applied to total phosphorus in the influent wastewater for each run. Plot a graph of percentage of phosphorus removal versus the weight ratio of alum to phosphorus. On the same graph, plot the phosphorus removal and alum dosage values given in the text under Eq. 13-2. (b) Calculate the average BOD and SS removal efficiencies. Suggest some reasons why the effluent BOD and SS values are lower and resulting efficiencies higher than are normally achieved by a full-scale treatment plant. (c) Why did the concentration of alkalinity in the effluent decrease with increasing alum dosage?

13-4 A settled wastewater is treated for phosphorus removal by chemical-biological aeration with the addition of waste pickle liquor (ferrous sulfate). The influent wastewater has 110 mg/l of SS, 120 mg/l of BOD, and 7.5 mg/l of total P. The effluent contains 26 mg/l of SS, 20 mg/l of BOD, 1.0 mg/l of P, and 1.4 mg/l of Fe. The waste pickle liquor containing 8.8 percent iron by weight is fed at a rate of 160 gal per million gallons of wastewater. Calculate the weight ratios of iron applied to influent phosphorus and iron precipitated to phosphorus removed. Estimate the chemical-biological sludge-solids production per million gallons of wastewater treated.

13-5 A domestic wastewater with 220 mg/l of SS, 180 mg/l of BOD, and 9.0 mg/l of P is treated by lime precipitation in primary settling followed by conventional activated-sludge processing. A dosage of 200 mg/l of CaO removes 80 percent of the influent phosphorus. Of the remaining 1.8 mg/l of P, 0.8 mg/l is removed by biological uptake during aeration and 1.0 mg/l appears in the plant effluent. Lime precipitation in the primary also removes 80 percent of the SS and 60 percent of the BOD. The plant effluent SS concentration averages 20 mg/l. Calculate the estimated sludge solids production per million gallons of wastewater treated based on the following data. The chemical precipitate produced is 2.0 mg/l (calcium hydroxyapatite plus calcium carbonate) per milligram of applied CaO. The aeration process operates at an F/M of 0.20 lb BOD/day/lb MLSS. How does this sludge production compare to the sludge solids generated by the biological-chemical treatment in Problem 13-2?

13-6 Modify the nitrogen data in Figure 13-4 for Example 13-3 assuming that the waste sludge is anaerobically digested with the supernatant being returned to the plant influent.

13-7 Describe the process of nitrification and how it is usually performed.

13-8 Why is methanol, or some other carbon source, needed in biological denitrification?

13-9 Calculate the methanol demand for denitrifying one million gallons of nitrified effluent with a nitrate nitrogen concentration of 25 mg/l and dissolved oxygen concentration of 8 mg/l. Express answers in pounds of methanol and gallons of 95 percent methanol with a specific gravity of 0.82.

13-10 Convert the percentage data in Figure 13-7 to concentration values based on an influent nitrogen concentration of 25 mg/l.

13-11 Why is air stripping to remove ammonia from wastewater restricted to a warm climate?

13-12 One disadvantage of breakpoint chlorination for removal of ammonia is the addition of chloride ion to the water. Verify by calculation that the oxidation of 20 mg/l of ammonia contributes 160 mg/l of chloride ion.

13-13 What are the major differences between the gravity filters shown in Figure 13-10 and Figure 7-12?

13-14 Effluent filtration is added to a conventional activated sludge plant (primary sedimentation and secondary aeration) that treats 5.0 mgd. The characteristics of the raw, settled, and biologically treated wastewaters are as listed in Table 9-2. For a nominal filtration rate (all filters operating with average design flow entering the plant) of 3.0 gpm/sq ft, calculate the area of granular-media bed required for effluent filtration. If the volume of backwash water needed is 200 gal/sq ft of filter area and all of the beds are cleaned each day, how many gallons of wash water are returned daily to the head of the plant? What is this volume expressed as a percentage of the daily wastewater flow? Before filtration, the SS in the effluent averaged 30 mg/l; after filtration, the SS averages 5 mg/l. Calculate the waste solids produced in primary sedimentation, activated-sludge aeration, and filtration. What is the percentage increase in solids production resulting from effluent filtration?

13-15 In tertiary treatment for virus removal, why does filtration precede chlorination?

13-16 How should municipal wastewater be pretreated prior to chemical-physical reclamation treatment?

13-17 What are the main purposes for each of the following processes: lime precipitation, air stripping, recarbonation, granular activated-carbon adsorption, and reverse osmosis?

13-18 What is the primary problem in operation of the reverse-osmosis process at Water Factory 21? What maintenance procedures are used to reduce the problem?

13-19 How does the processing scheme of the demonstration water reclamation plant at Denver differ from the treatment scheme at Water Factory 21?

chapter 14

Water Reuse and Land Disposal

Indirect reuse of wastewater occurs when effluents that have been discharged and diluted in a river are withdrawn downstream for irrigation or as a public water supply. In some instances, the treated wastewater in a stream is a significant portion of the drought flow. The practice of discharging to surface waters and withdrawal for reuse provides dilution and separation in time and space, thus allowing the processes of natural purification to take place. In contrast to this, planned direct reuse of reclaimed waters is practiced for several applications without dilution in natural waters. In the United States, the major direct reuse is land disposal of wastewater by agricultural irrigation for growing fodder, fiber, and seed crops. Other significant reuse applications are industrial cooling and landscape irrigation. Only about one tenth of the total reuse of reclaimed waters is for groundwater recharge through percolation basins, fish and wildlife habitat, and recreational impoundments.

14-1 WATER QUALITY AND REUSE APPLICATIONS

Protection of human health is the main consideration in the reuse of wastewater; hence, the primary water-quality parameter for reclaimed water is the concentration of coliform bacteria. Since treatment processes vary in effectiveness in removing different types of pathogens, the quality of water for reuse is also judged by the sequence of unit processes used in treatment. The coliform standard

and specified treatment processes for a given application take into account both the risk involved in the wastewater reuse and the degree of treatment that is feasible. Quality criteria have been adopted by several states, and in some cases a permit system is included to ensure operational control. For example, a permit may specify the kinds of crops that can be irrigated at a particular site for the quality of reclaimed water available. The purpose is to reduce the temptation, in order to have greater profits, to irrigate food crops sensitive to contamination by pathogens with an inadequately treated wastewater. Figure 14-1 relates treatment processes and coliform standards to various applications of water reuse; the coliform concentrations listed are typical values applied in the United States.

Agricultural Irrigation

The potential advantages of reuse by irrigation are the low cost of the water, beneficial use of the plant nutrients in the wastewater, use of land application as tertiary treatment, and income gained from growing cash crops. In arid regions, wastewater may be the only source of water available for irrigation. Where advanced wastewater treatment is required, irrigation may be used to renovate the wastewater by percolation through the soil profile, or the wastewater may be disposed of by evaporation and transpiration with no discharge from the irrigation site.

The potential disadvantages are related to the large storage volume that may be required to balance the supply and demand for irrigation water, and the restriction on application rates because of the effects of contaminants in the wastewater on groundwater, soils, and crops. While the supply of wastewater is continuous, the use for irrigation water depends on the growing season and climatic conditions. Hence, the volume for storage is greatest in humid northern regions requiring winter storage and least in dry southern regions with a long growing season. The major contamination problems are percolation of nitrate to groundwater and retention of heavy metals in the soil.

Contamination of crops with pathogens and the hazard to farm workers are the major health risks. For irrigation of fodder, fiber, and seed crops, the common wastewater treatment is biological processing and several

weeks of storage in open basins for natural purification. Actually, the length of time in storage ranges from months in a northern climate to a few days in an arid region. If the detention time in storage is not adequate, chlorination of the water is required for disinfection. The common coliform standard for restricted irrigation of nonedible crops is a maximum geometric mean of 5000 coliforms per 100 ml with a maximum concentration of 10,000 per 100 ml in a single sample (Figure 14-1). Although raw wastewater after plain sedimentation has been used for decades in surface irrigation of seed and fiber crops, the current requirements of biological treatment and partial disinfection were established primarily to protect farm workers and reduce the risk if edible crops are deliberately or inadvertently irrigated. Public access to irrigation sites is limited by fencing and warning signs around the perimeter. If the watering is done by spray irrigators,

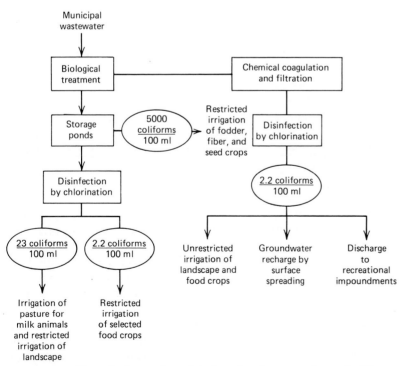

Figure 14-1   General scheme of municipal wastewater processing and coliform standards for various applications of water reuse.

buffer zones are established along the boundaries of the site to prevent aerosols from drifting into adjacent public access areas.

Chemical chlorination of the treated wastewater is required prior to irrigation of pasture for milk animals as well as of restricted landscapes, such as golf courses and grassed areas along highways. Contact of the irrigation water by people is limited by controlling the watering times. A common standard for these reuse applications is a median coliform concentration of less than 23 per 100 ml. Selected food crops that are eaten cooked may be irrigated with reclaimed water provided the median coliform concentration does not exceed 2.2 per 100 ml with no sample exceeding 23 per 100 ml. In order to achieve such a low coliform count, treatment of the wastewater involves extensive processing, for example, a high degree of biological treatment, a long period of storage in ponds, and chlorine disinfection.

Restricted Irrigation, Infiltration, and Recreational Impoundments

Unrestricted irrigation of such landscape as parks and lawns results in public contact with the reclaimed water. In order to ensure that the reclaimed water is free of resistant pathogens, the specified treatment of the wastewater includes chemical coagulation and filtration through granular media prior to disinfection by chlorination (Figure 14-1). The microbiological standard is a median concentration of 2.2 or fewer coliforms per 100 ml. The filtration process removes protozoa cysts, worm ova, and suspended solids that can protect bacteria and viruses from the oxidizing action of chlorine. To reduce the probability of people drinking the irrigation water, warning signs are posted and watering is done only at night using pop-up nozzles.

Wastewater treatment plants in arid regions often discharge into watercourses during periods of little or no dilutional flow. If the streambed is porous, the effluent can percolate to groundwater naturally or impoundments can be constructed for planned recharge to replenish groundwater withdrawn by wells. These impoundments may also be open to the public for recreation, including water sports. Because of groundwater recharge and recreational

reuse, the recommended processing of the wastewater following conventional biological treatment is coagulation and filtration followed by chlorination for virus removal (refer to Section 13-8).

Treatment to reclaim wastewater for reuse, beyond ordinary biological processing and disinfection, requires a high degree of reliability. Confidence that effluent quality will continuously meet the design standard is necessary since failure can lead to contamination, resulting in transmission of an infectious disease. Dependability of a plant can be improved by installation of standby treatment units, alarms to announce the failure of a process, an alternate power supply, and provisions for emergency storage or disposal if the reclaimed water does not meet the quality standards. Of course, adequate funds for equipment maintenance and operator training are essential.

Groundwater Recharge and Potable Supply

Groundwater recharge with reclaimed water is being performed in coastal regions by direct injection through wells to form a barrier against underground intrusion of saltwater from the ocean. Because a portion of this injected water can seep toward inland potable groundwater supplies, it is reclaimed to meet drinking water standards. This is the highest degree of wastewater renovation required in the United States, since direct potable reuse is not practiced. Primary pollutants of concern in underground injection are heavy metals, organic chemicals, inorganic salts, and pathogens. In order to achieve drinking water quality, water reclamation requires biological, chemical, and physical processes in a series of stages. The advanced treatment processes used in Water Factory 21 in southern California to treat wastewater for groundwater injection are lime precipitation, air stripping, recarbonation, granular-media filtration, granular-carbon adsorption, chlorination, and reverse osmosis. (Refer to Sections 13-9 and 13-10.) A high degree of treatment reliability is achieved by requiring pretreatment of industrial wastewaters, installation of duplicate processing units, and discharge of unsatisfactory reclaimed water to a surface watercourse rather than pumping it into the recharge wells.

Direct reuse of reclaimed wastewater for a domestic water supply—even when diluted with a natural water—is not considered safe. Wastewaters contain a variety of refractory organic and inorganic compounds and unidentified viruses. Until the effectiveness of fail-safe advanced treatment can be clearly demonstrated, the risk of wastewater contaminants being present in the reclaimed water is too great to justify domestic reuse, particularly when natural water sources are available.

Drinking water standards are based on the assumption that the source is a groundwater or a relatively unpolluted surface water so that contaminants can be removed by conventional water treatment. Refractory organic and inorganic compounds are in low concentration in the treated water since they are present in only limited quantities in the raw water. In contrast, wastewater contains many unknown and unidentified pollutants. Therefore, current drinking water standards are not appropriate for evaluating the quality of reclaimed water for a drinking supply.

14-2 METHODS OF LAND DISPOSAL

The use of land for wastewater treatment or disposal depends on climate and availability of a suitable site. In a humid climate, land application of wastewater is often considered infeasible treatment because the wastewater is not considered a valuable resource and flowing surface waters can be used to dilute and convey away treated wastewater. Another major factor is the general lack of available sites for wastewater irrigation since nearly all of the land in a humid region is already productive farmland. In arid and semiarid regions, irrigation has been practiced for decades for disposal of wastewater rather than discharge to a dry streambed. Irrigation has the advantage of controlled reuse of the water to grow crops for an economic gain, and at the same time undiluted wastewater does not flow along a streambed where people and animals might mistake it for fresh water. Where land disposal is applicable, the three kinds of systems in use are irrigation, rapid infiltration, and overland flow. A primary consideration in system selection is soil permeability. Porous sands are required for rapid infiltration, moder-

ately permeable soils are appropriate for irrigation, and impermeable clays are needed for overland flow.

Irrigation is the best of the three land disposal systems for predictability of the renovated water quality, beneficial reuse of the water, and ease of operation and maintenance. Water application rates range from 0.5 to 4 in./week (1 to 10 cm/week) during the growing season. The allowable application rate depends on type of soils, topography, infiltration capacity, weather conditions, kind of crop, and wastewater characteristics. A conservative estimate appears to be an average of 2 in./week (54,000 gal/acre) at a rate of 0.25 in./hr for an 8-hr period when site and climatic conditions are suitable for irrigation. Based on site topography and crop, the method of watering can be by either spray or surface flow. Common crops include wheat, corn, alfalfa, cotton, and a variety of grasses for animal forage. Corn generally has a high market value, but also requires more effort in annual cultivation and planting. Perennial grasses require only harvest and have the advantage of an extensive root system to take up wastewater nutrients throughout the irrigation season. Annual crops, in contrast, can allow significant nitrogen losses by infiltration during the period of early growth.

Rapid infiltration, usually for the purpose of groundwater recharge, is performed by spreading or flooding wastewater onto land capable of a percolation rate of several feet per week. Generally water is applied to a series of ponds for 10 to 14 days followed by a drying period of 10 to 20 days, depending on the season of the year. Percolation rates during the application period range from 1 to 4 ft/day (0.3 to 1.2 m/d) with a maximum of about 330 ft/yr (100 m/y). The drying cycle is necessary for oxidation of organic matter to restore permeability, since anaerobic conditions that occur during flooding can result in reduced porosity of the soil. The bottom of the basins can be barren or grassed. Grasses may be beneficial to prevent clogging of the soil pores and retain high infiltration rates. The degree of water renovation by rapid infiltration is difficult to predict. Implications are that coarse-grained soils under high hydraulic loadings remove very limited amounts of dissolved solids.

Overland flow is applicable only for sloping sites with relatively impervious soils. Wastewater is discharged by spraying or spreading to produce sheet flow down the slope, rather than infiltration. Since wastewater treatment is achieved by movement of the water through grass and decaying debris, a continuous vegetative cover is essential for filtering action. Sites must be leveled and graded to provide slopes of 2 to 6 percent. For each slope length of 200 to 300 ft, a cutoff trench is constructed to intercept the flow in the soil mantle. Application points and collection ditches are located to provide the necessary contact time for treatment. Depending on the degree of pretreatment, wastewater application rates vary from 2 to 16 in./week (5 to 40 cm/week) during a daily application period of 4 to 6 hr. Sites should have at least 6 to 8 in. of good topsoil to support grass growth. The two major restrictions to overland flow are difficulty of maintaining consistent quality in the runoff and site preparation costs. Since wastewater flowing over the ground surface is exposed to weather, biological activity and consequently the degree of treatment are adversely effected by lack of sunshine and cold temperature, so year-around operation is practical only in mild climates.

14-3 DESIGN OF IRRIGATION SYSTEMS

Processing of municipal wastewater by biological treatment and disinfection prior to irrigation oxidizes the organic matter, greatly reduces the numbers of bacteria and viruses, and converts nitrogen to inorganic forms without significant removal. The concentrations of heavy metals are controlled by pretreatment of industrial wastewaters prior to discharge to the municipal sewer system. Of the pollutants in reclaimed water, only dissolved solids are likely to percolate to groundwater. The primary ion of concern is nitrate because it is readily carried downward through the soil profile by infiltration.

The field area required for irrigation is determined by either the hydraulic loading or nitrogen loading. The water balance is calculated by the relationship

water loading + precipitation

$$= \text{evapotranspiration} + \text{percolation} + \text{runoff} \quad (14\text{-}1)$$

The value for precipitation is the calculated maximum for a ten-year return period. Evapotranspiration is estimated based on climatic conditions and crops, and percolation is

determined by soil conditions. This relationship is used to calculate water balances for any desired length of time, usually by the week, month, and year.

The inorganic nitrogen in reclaimed water is readily oxidized to the nitrate form in storage ponds or during spray irrigation. In a properly operated system, the majority of the nitrogen is removed from the percolating water by crop synthesis. Under anoxic conditions created in wet soils, some of the nitrate may be denitrified to gaseous nitrogen and lost to the atmosphere. Nitrate ions can also be carried through the soil profile by percolating water and eventually appear in the groundwater. The relationship for nitrogen balance is

applied nitrogen = crop uptake + loss by denitrification

$$+ \text{loss by percolation} \quad (14\text{-}2)$$

The greatest nitrogen uptake is in the approximate range of 200 to 400 lb/ac/yr (220 to 450 kg/ha·y) by forage crops such as alfalfa, clover, and grasses. Much less nitrogen, in the range of 70 to 170 lb/ac/yr (80 to 190 kg/ha·y) during one growing season, is taken up by field crops like corn, soybeans, and cotton. Denitrification is difficult to determine under field conditions; therefore, the nitrogen loss is generally assumed to be 15 to 25 percent of the applied nitrogen. The loss by percolation should be limited to 10 mg/l, which is the maximum contaminant level for drinking water.

Water storage is required in irrigation systems because of the imbalance between the reclaimed water supply and the application rate determined by crop growth and climatic conditions. In a cold climate, the operation of an irrigation site is stopped for several months each year. The length of time is determined by the growing season of either annual crops or perennial grasses when the ground is frozen. In a warm climate, the required storage is calculated by determining the difference between the application rate established by either hydraulic or nitrogen loading and the water supply. The method of numerical analysis is performed by calculating the surplus or deficit of water each month and then algebraically adding these values to determine the storage needed to hold the maximum amount of surplus water.

Water distribution for irrigation can be either surface or sprinkler. Surface distribution, which is best suited for soils with moderate to low permeability, uses gravity flow from a network of pipes and ditches to periodically flood an area with several inches of water. In border strip flooding, crops are planted in constructed shallow rectangular basins that are flooded with water entering from the borders. Ridge and furrow irrigation consists of running streams of water along small channels (furrows) bordered by raised beds (ridges) upon which the crops are grown.

Sprinkler distribution simulates rainfall and can be used for watering soils with a wide range of permeabilities and a variety of crops. The choice of sprinkler equipment depends on site topography, kinds of crops, and the cost of purchase and installation. For portable application, irrigation pipes can be placed on the ground with attached risers and spray nozzles. The location of the pipes in the field can be moved by hand or by towing from the end with a tractor. For large, permanent systems the pipe network is buried with only risers and spray nozzles appearing above ground. Mechanical systems are side-wheel roll and center pivot. A side-wheel roll irrigator consists of a wheel-supported pipe that is connected to a hydrant with a flexible hose. The irrigator is moved from one position to the next across a field by a centrally located drive unit powered by a gasoline engine. A center pivot irrigator is a wheel-supported spray boom that rotates around a fixed pivot in the center of the field. For a square field, the irrigator can be equipped with traveling end sprinklers to water the corners. The main advantages of central pivots are the continuous self-propelled operation, the adaptability to uneven terrain, and the unrestricted crop height.

Irrigated vegetation removes nutrients (primarily nitrogen and phosphorus) from the wastewater, maintains soil permeability, and reduces erosion of the soil. Selection of a crop is determined by the reclaimed water quality, required nitrogen uptake, and profitability. Grasses can be irrigated with a poorer quality of water, have a higher nutrient uptake over a longer growing season, and require less maintenance and skill to grow. Nevertheless, field crops like corn are generally preferred in large irrigation systems because of greater monetary gain, even though watering must be scheduled more carefully to prevent the application of either too much water or nitrogen.

EXAMPLE 14-1

Calculate the water balance for the month of July for an irrigation site of 120 acres based on the following data. The reclaimed water supply averages 1.0 mgd, and precipitation is 2.1 in./mon. Evapotranspiration for the crop is assumed to be 6.0 in./mon, and the allowable percolation is 10 in./mon with a nitrogen concentration in the percolate of 10 mg/l.

Solution

Water supply

$$= \frac{1.0 \text{ mil gal/day} \times 31 \text{ days/mon} \times 36.8 \text{ ac-in./mil gal}}{120 \text{ ac}}$$

$$= 9.5 \text{ in./mon}$$

Substituting into Eq. 14-1, assuming zero runoff,
Water loading + 2.1 in./mon

$$= 6.0 \text{ in./mon} + 10 \text{ in./mon}$$

$$\text{Water loading} = 16.0 - 2.1$$

$$= 13.9 \text{ in./mon}$$

Water drawn from storage

$$\frac{(13.9 - 9.5) \text{ in./mon} \times 120 \text{ ac}}{31 \text{ days/mon} \times 36.8 \text{ ac-in./mil gal}} = 0.46 \text{ mgd}$$

EXAMPLE 14-2

Calculate the nitrogen crop uptake required for the water balance data given in Example 14-1. Assume a concentration of 10 mg/l in the percolate and a denitrification loss of 20 percent. The average nitrogen content of the reclaimed water is 25 mg/l.

Solution

$$\text{Applied nitrogen} = \frac{\begin{array}{c} 1.0 \text{ mil gal/day} \times \\ 31 \text{ days/mon} \times 25 \text{ mg/l} \times 8.34 \end{array}}{120 \text{ ac}}$$

$$= 54 \text{ lb/ac/mon}$$

Loss by denitrification $= 0.20 \times 54$

$$= 11 \text{ lb/ac/mon}$$

Loss by percolation

$$= \frac{10 \text{ in./mon} \times 1.0 \text{ ac} \times 10 \text{ mg/l} \times 8.34}{36.8 \text{ ac-in./mil gal}}$$

$$= 23 \text{ lb/ac/mon}$$

Substituting into Eq. 14-2,

54 lb/ac/mon

$$= \text{crop uptake} + 11 \text{ lb/ac/mon} + 23 \text{ lb/ac/mon}$$

Required crop uptake

$$= 20 \text{ lb/ac/mon}$$

Surface Irrigation at Bakersfield, California

Land application has been practiced historically for wastewater disposal in the semiarid San Joaquin Valley of central California. Since the annual rainfall at Bakersfield is only 5 to 6 in./yr, settled municipal wastewater has been a valuable resource for irrigation of forage, fiber, and seed crops for decades. Because of the long, hot summers and mild winters, crops can be selected for year-round irrigation. Evapotranspiration accounts for 70 to 80 percent of the water loss and percolation is 20 to 30 percent. The soils are composed of poorly drained, alkaline loams underlain by dense clayey strata. Although accumulation of salt results from wastewater irrigation, unavoidable salt buildup in agricultural soils is a regional problem and reuse of wastewater does not have a significantly greater impact than irrigation with imported water. The groundwater is unsuitable for potable use because of naturally occurring high total dissolved solids and nitrate concentrations; hence, additional groundwater pollution resulting from irrigation is not considered a serious problem.

The wastewater processing and irrigation system is diagramed in Figure 14-2. The treatment plant designed for 19 mgd consists of primary clarification, with anaerobic digestion of the sludge, and secondary biological aeration in two-stage aerated ponds, with a detention time of seven days. Effluent is pumped into two equalizing storage reservoirs with a total capacity of 1710 mil gal, which is equivalent to 90 days of storage at design flow.

These reservoirs allow storage of reclaimed water during periods of low irrigation demand. The water is pumped to the fields through pipelines and distributed by surface irrigation using both flooding and ridge-and-furrow methods. Tailwater (runoff) is collected and recycled to the fields.

The crops grown include alfalfa, barley, grasses, sugar beets, cotton, and corn. Seasonal crops like cotton are irrigated only 6 to 8 mon/yr and use less water while year-round crops like alfalfa, which is irrigated 12 mon of the year, have greater water consumption. The average irrigation demand for design of the system was 60 in./yr. The irrigation rates for various crops are based on both water needs and nitrogen synthesis rates.

Spray Irrigation at Muskegon, Michigan

The crop irrigation and soil filtration system at Muskegon is for tertiary treatment of the municipal wastewater. After biological treatment and retention in large storage basins, the reclaimed water is applied to corn by spray irrigation during the growing season. The soils are pervious sands and sandy loams resulting in rapid percolation (2 to 10 in./hr). Subsurface drainage tiles in the fields, drainage wells, and ditches surrounding the site collect the subsurface water for discharge to surface watercourses that flow to Lake Michigan. An irrigation system was chosen for tertiary treatment because it was less costly than advanced wastewater treatment to remove phosphorus and other contaminants to prevent eutrophication and pollution of recreational waters.

The treatment and irrigation system has six major components: collection and transportation piping, aerated stabilization ponds, storage basins, crop land with a natural filter of sandy soil, and a drainage network (Figure 14-3). The design wastewater flow is 42 mgd composed of a combination of domestic and industrial (papermill) wastewaters. After collection in the city, it is conveyed by gravity flow and pumping in force mains to the irrigation site located 11 miles east of the City of Muskegon. After biological treatment in three aerated ponds (labeled A), the flow passes through a settling pond (B) to storage basins that are sealed on the bottom to prevent exfiltration. The large storage volume of 5300 mil

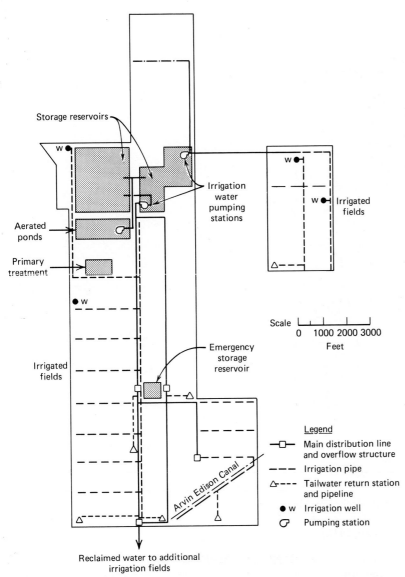

Figure 14-2 Diagram of the surface irrigation system at Bakersfield, California. (From *Process Design Manual for Land Treatment of Municipal Wastewater*, U.S. Environmental Protection Agency, EPA 625/1-77-008, 1977.)

gal is needed to accumulate the wastewater flow for at least four months of the year when spray irrigation is not possible because of frozen ground. Reclaimed water leaving through the outlet pond (*C*) can be chlorinated if

necessary and then pumped through a piping system to 54 center pivot spray irrigators.

The total field area is 5350 ac and the average application rate of irrigation water is 3.0 in./week. The

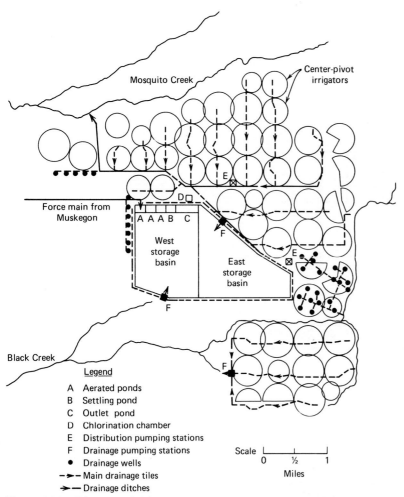

Figure 14-3   Diagram of the spray irrigation system at Muskegon, Michigan. (From *Process Design Manual for Land Treatment of Municipal Wastewater*, U.S. Environmental Protection Agency, EPA 625/1-77-008, 1977.)

wastewater provides adequate amounts of phosphorus and potassium for the corn crop; however, the low concentration of nitrogen in the wastewater must be supplemented when the corn is in rapid growth. Nitrogen fertilizer is added based on the amount of wastewater applied and crop requirements for synthesis. The pollutants in the renovated water removed by the underground drainage system are commonly less than 3 mg/l of BOD, 2 mg/l of inorganic nitrogen, and less than 0.1 mg/l of phosphorus.

FURTHER READINGS

1. Middlebrooks, E. J. *Water Reuse.* Ann Arbor: Ann Arbor Science Publishers, 1982.
2. *Process Design Manual for Land Treatment of Municipal Wastewater*, U.S. Environmental Protection Agency, EPA 625/1-77-008; U.S. Army Corps of Engineers, COE EM 1110-1-501 (October 1977).

Appendix

COMMON ABBREVIATIONS USED IN TEXT

ac	acres	kN	kilonewton
BOD	biochemical oxygen demand	kPa	kilopascal
°C	degrees Celsius	kW	kilowatt
cfm	cubic feet per minute	l	liters
cm	centimeters	lb	pounds
cu ft	cubic feet	m	meters
d	day	meq	milliequivalents
DO	dissolved oxygen	meq/l	milliequivalents per liter
°F	degrees Fahrenheit	mg	milligrams
F/M	food-to-microorganism ratio	mg/l	milligrams per liter
fps	feet per second	mgad	million gallons per acre per day
ft	feet	mgd	million gallons per day
g	grams	mi	miles
gal	gallons	mil gal	million gallons
gpcd	gallons per capita per day	min	minutes
gpg	grains per gallon	mm	millimeters
gpm	gallons per minute	MPN	most probable number
gr	grains	ppm	parts per million
h	hour	psi	pounds per square inch
ha	hectare	s	second
hr	hour	sec	second
in.	inches	sq ft	square feet
kg	kilogram	y	year
km	kilometer	yr	year

COMMON CONVERSION FACTORS

Multiply	By	To Obtain
acres	43,560	square feet
acres	0.001,56	square miles
acre-feet	43,560	cubic feet
acre-feet	325,850	gallons
atmospheres	76.0	centimeters of mercury
atmospheres	29.92	inches of mercury
atmospheres	33.90	feet of water
atmospheres	14.7	pounds per square inch
centimeters	0.3937	inches
cubic centimeters	0.061,02	cubic inches
cubic feet	1728	cubic inches
cubic feet	7.481	gallons
cubic feet	28.32	liters
cubic feet per second	0.6463	million gallons per day
cubic feet per second	448.8	gallons per minute
cubic meters	35.31	cubic feet
cubic meters	264.2	gallons
cubic meters	1000	liters
cubic meters per second	2,120	cubic feet per minute
day	1440	minutes
day	86,400	seconds
feet	30.48	centimeters
feet	12	inches
feet	0.3048	meters
feet of water	0.4335	pounds per square inch
gallons	3785	cubic centimeters
gallons	0.1337	cubic feet
gallons	231	cubic inches
gallons	3.785	liters
gallons of water	8.345	pounds of water
gallons per minute	0.002,23	cubic feet per second
gallons per minute	8.021	cubic feet per hour
grains per gallon	17.12	milligrams per liter
grains per gallon	142.9	pounds per million gallons
grams	0.035,27	ounces
grams	0.002,21	pounds
inches	2.540	centimeters
kilograms	2.205	pounds
kilograms per cubic meter	0.062,43	pounds per cubic foot
kilograms per square centimeter	14.22	pounds per square inch
kilometers	0.6214	miles
liters	1000	cubic centimeters

Multiply	By	To Obtain
liters	0.035,31	cubic feet
liters	0.2642	gallons
liters per second	15.85	gallons per minute
$\log_{10}$	2.303	$\log_e$
$\log_e(e = 2.7183)$	0.4343	$\log_{10}$
meters	100	centimeters
meters	3.281	feet
meters	39.37	inches
meters per second	3.281	feet per second
microns	0.001	millimeters
milligrams per liter	1	parts per million
milligrams per liter	0.0584	grains per gallon
milligrams per liter	8.345	pounds per million gallons
pounds	7000	grains
pounds	453.6	grams
pounds per square foot	0.006,95	pounds per square inch
square centimeters	0.1550	square inches
square meters	10.76	square feet
square miles	640	acres
temperature		
°C	9/5	+32 = °F
°F − 32	5/9	°C
tons	2000	pounds

SELECTED SI UNITS (Système International d'Unités)

Unit Type	Quantity	Unit	Symbol	Expression in Other Units
Base Units	Length	meter	m	
	Mass	kilogram[a]	kg	
	Time	second	s	
	Thermodynamic temperature	kelvin[b]	K	
	Amount of substance	mole	mol	
Supplementary Unit	Plane angle	radian	rad	

Derived Units Having Special Names	Force	newton	N	$kg \cdot m/s^2$
	Pressure	pascal	Pa	N/m^2
	Energy, work	joule	J	$N \cdot m$
	Power	watt	W	J/s
	Area	hectare	ha	$10^4 \ m^2$
	Volume	liter	l	$10^{-3} \ m^3$
Derived Units	Area	square meter		m^2
	Volume	cubic meter		m^3
	Linear velocity	meter per second		m/s
	Linear acceleration	meter per second squared		m/s^2
	Density	kilogram per cubic meter		kg/m^3

[a] The kilogram is the only base unit with a prefix.

[b] The Celsius temperature scale (previously called centigrade) is commonly used for measurements even though it is not part of the SI system. A difference of one degree on the Celsius scale (°C) equals one kelvin (K), and zero on the thermodynamic scale is 273.15 kelvin below zero degrees Celsius.

SI UNIT PREFIXES

Amount	Multiples and Submultiples	Prefixes	Symbols
1 000 000 000 000	10^{12}	tera	T
1 000 000 000	10^9	giga	G
1 000 000	10^6	mega	M
1 000	10^3	kilo	k
100	10^2	hecto	h
10	10	deka	da
0.1	10^{-1}	deci	d
0.01	10^{-2}	centi	c
0.001	10^{-3}	milli	m
0.000 001	10^{-6}	micro	μ
0.000 000 001	10^{-9}	nano	n
0.000 000 000 001	10^{-12}	pico	p
0.000 000 000 000 001	10^{-15}	femto	f
0.000 000 000 000 000 001	10^{-18}	atto	a

CONVERSION FACTORS ENGLISH TO SI METRIC

Multiply	By	To Obtain
Length		
mile, mi	1.609	kilometer, km
yard, yd	0.914 4	meter, m
foot, ft	0.304 8	meter, m
inch, in.	25.40	millimeter, mm
Area		
square mile, sq mi	2.590	square kilometer, km^2
acre, ac	0.404 7	hectare, ha
	4047.	square meter, m^2
square yard, sq yd	0.836 1	square meter, m^2
square foot, sq ft	0.092 9	square meter, m^2
square inch, sq in.	645.2	square millimeter, mm^2
Volume		
acre foot, ac ft	1234.	cubic meter, m^3
cubic yard, cu yd	0.764 9	cubic meter, m^3
cubic foot, cu ft	0.028 32	cubic meter, m^3
	28.32	liter, l
US gallon, gal	3.785	liter, l
Velocity		
foot/second, ft/sec	0.304 8	meter/second, m/s
foot/minute, ft/min	0.005 08	meter/second, m/s
Flow		
million US gallons/day, mgd	3785.	cubic meter/day, m^3/d
	43.81	liter/second, l/s
US gallons/minute, gpm	5.450	cubic meter/day, m^3/d
	0.063 09	liter/second, l/s
cubic foot/second, cu ft/sec	0.028 32	cubic meter/second, m^3/s
Mass		
ton (2000 lb)	0.907 2	tonne (1000 kg), t
	907.2	kilogram, kg
pound, lb	0.453 6	kilogram, kg
	453.6	gram, g
Density		
pound/cubic foot, lb/cu ft	16.02	kilogram/cubic meter, kg/m^3
BOD loading rate		
pound/1000 cubic foot/day, lb/1000 cu ft/day	16.02	gram/cubic meter·day, $g/m^3 \cdot d$
pound/acre/day, lb/ac/day	0.1121	gram/square meter·day, $g/m^2 \cdot d$
	1.121	kilogram/hectare·day, $kg/ha \cdot d$

Multiply	By	To Obtain
Solids loading rate		
pound/cubic foot/day, lb/cu ft/day	16.02	kilogram/cubic meter·day, $kg/m^3 \cdot d$
pound/square foot/day, lb/sq ft/day	4.883	kilogram/square meter·day, $kg/m^2 \cdot d$
Hydraulic loading rate		
million US gallons/acre/day, mgad	0.935 3	cubic meter/square meter·day, $m^3/m^2 \cdot d$
	9353.	cubic meter/hectare·day, $m^3/ha \cdot d$
US gallon/square foot/day, gal/sq ft/day	0.040 75	cubic meter/square meter·day, $m^3/m^2 \cdot d$
US gallon/square foot/minute, gpm/sq ft	0.679 0	litre/square meter·second $l/m^2 \cdot s$
US gallon/cubic foot/day, gal/cu ft/day	0.133 7	cubic meter/cubic meter·day, $m^3/m^3 \cdot d$
Concentration		
pound/million US gallons, lb/mil gal	0.119 8	milligram/liter, mg/l
Force		
pound force, lb	4.448	newton, N
Pressure		
pound/square foot, psf	0.047 88	kilopascal, kPa
pound/square inch, psi	6.895	kilopascal, kPa
Power		
horsepower, hp	1.341	kilowatt, kW
Aeration		
cubic foot (air)/pound (BOD), cu ft/lb	0.062 43	cubic meter/kilogram, m^3/kg

CONVERSION FACTORS SI METRIC TO ENGLISH

Multiply	By	To Obtain
Length		
kilometer, km	0.621 4	mile, mi
meter, m	3.281	foot, ft
	1.094	yard, yd
millimeter, mm	0.039 37	inch, in.

Multiply	By	To Obtain
Area		
square kilometer, km^2	0.386 1	square mile, sq mi
hectare, ha	2.471	acre, ac
square meter, m^2	10.76	square foot, sq ft
	1.196	square yard, sq yd
square millimeter, mm^2	0.001 55	square inch, sq in.
Volume		
cubic meter, m^3	264.2	US gallon, gal
	35.31	cubic foot, cu ft
	1.308	cubic yard, cu yd
liter, l	0.264 2	US gallon, gal
	0.035 31	cubic foot, cu ft
Velocity		
meter/second, m/s	3.281	foot/second, ft/sec
	196.8	foot/minute, ft/min
Flow		
cubic meter/day, m^3/d	0.183 5	US gallon/minute, gpm
cubic meter/second, m^3/s	35.31	cubic foot/second, cu ft/sec
liter/second, l/s	15.85	US gallon/minute, gpm
	0.022 83	million US gallons/day, mgd
Mass		
tonne (1000 kg), t	1.102	ton (2000 lb)
kilogram, kg	2.205	pound, lb
Density		
kilogram/cubic meter, kg/m^3	0.624 2	pound/cubic foot, lb/cu ft
BOD loading rate		
gram/cubic meter · day, g/m^3 · d	0.062 43	pound/1000 cubic foot/day, lb/1000 cu ft/day
gram/square meter · day, g/m^2 · d	8.921	pound/acre/day, lb/ac/day
kilogram/hectare · day, kg/ha · d	0.892 1	pound/acre/day, lb/ac/day
Solids loading rate		
kilogram/cubic meter · day, kg/m^3 · d	0.062 43	pound/cubic foot/day lb/cu ft/day
kilogram/square meter · day, kg/m^2 · d	0.204 8	pound/square foot/day, lb/sq ft/day
Hydraulic loading rate		
cubic meter/square meter · day, m^3/m^2 · d	24.54	US gallon/square foot/day, gal/sq ft/day
	1.069	million US gallons/acre/day, mgad

Multiply	By	To Obtain
cubic meter/cubic meter · day, m³/m³ · d	7.481	US gallon/cubic foot/day, gal/cu ft/day
liter/square meter · second, l/m² · s	1.473	US gallon/square foot/minute, gpm/sq ft
Concentration		
milligram/liter, mg/l	8.345	pound/millionUS gallons, lb/mil gal
Force		
newton, N	0.224 8	pound force, lb
Pressure		
kilopascal, kPa	0.145 0	pound/square inch, psi
Power		
kilowatt, kW	0.745 7	horsepower, hp
Aeration		
cubic meter (air)/kilogram (BOD), m³/kg	16.02	cubic foot/pound, cu ft/lb

TABLE OF CHEMICAL ELEMENTS

Name	Symbol	Atomic Number	Atomic Weight	Name	Symbol	Atomic Number	Atomic Weight
Actinium	Ac	89	—	Mercury	Hg	80	200.59
Aluminum	Al	13	26.9815	Molybdenum	Mo	42	95.94
Americium	Am	95	—	Neodymium	Nd	60	144.24
Antimony	Sb	51	121.75	Neon	Ne	10	20.183
Argon	Ar	18	39.948	Neptunium	Np	93	—
Arsenic	As	33	74.9216	Nickel	Ni	28	58.71
Astatine	At	85	—	Niobium	Nb	41	92.906
Barium	Ba	56	137.34	Nitrogen	N	7	14.0067
Berkelium	Bk	97	—	Nobelium	No	102	—
Beryllium	Be	4	9.0122	Osmium	Os	76	190.2
Bismuth	Bi	83	208.980	Oxygen	O	8	15.9994
Boron	B	5	10.811	Palladium	Pd	46	106.4
Bromine	Br	35	79.904	Phosphorus	P	15	30.9738
Cadmium	Cd	48	112.40	Platinum	Pt	78	195.09
Calcium	Ca	20	40.08	Plutonium	Pu	94	—
Californium	Cf	98	—	Polonium	Po	84	—
Carbon	C	6	12.01115	Potassium	K	19	39.102
Cerium	Ce	58	140.12	Praseodymium	Pr	59	140.907
Cesium	Cs	55	132.905	Promethium	Pm	61	—

Name	Symbol	Atomic Number	Atomic Weight	Name	Symbol	Atomic Number	Atomic Weight
Chlorine	Cl	17	35.453	Protactinium	Pa	91	—
Chromium	Cr	24	51.996	Radium	Ra	88	—
Cobalt	Co	27	58.9332	Radon	Rn	86	—
Copper	Cu	29	63.546	Rhenium	Re	75	186.2
Curium	Cm	96	—	Rhodium	Rh	45	102.905
Dysprosium	Dy	66	162.50	Rubidium	Rb	37	85.47
Einsteinium	Es	99	—	Ruthenium	Ru	44	101.07
Erbium	Er	68	167.26	Samarium	Sm	62	150.35
Europium	Eu	63	151.96	Scandium	Sc	21	44.956
Fermium	Fm	100	—	Selenium	Se	34	78.96
Fluorine	F	9	18.9984	Silicon	Si	14	28.086
Francium	Fr	87	—	Silver	Ag	47	107.868
Gadolinium	Gd	64	157.25	Sodium	Na	11	22.9898
Gallium	Ga	31	69.72	Strontium	Sr	38	87.62
Germanium	Ge	32	72.59	Sulfur	S	16	32.064
Gold	Au	79	196.967	Tantalum	Ta	73	189.948
Hafnium	Hf	72	178.49	Technetium	Tc	43	—
Helium	He	2	4.0026	Tellurium	Te	52	127.60
Holmium	Ho	67	164.930	Terbium	Tb	65	158.924
Hydrogen	H	1	1.00797	Thallium	Tl	81	204.37
Indium	In	49	114.82	Thorium	Th	90	232.038
Iodine	I	53	126.9044	Thulium	Tm	69	168.934
Iridium	Ir	77	192.2	Tin	Sn	50	118.69
Iron	Fe	26	55.847	Titanium	Ti	22	47.90
Krypton	Kr	36	83.80	Tungsten	W	74	183.85
Lanthanum	La	57	138.91	Uranium	U	92	238.03
Lead	Pb	82	207.19	Vanadium	V	23	50.942
Lithium	Li	3	6.939	Xenon	Xe	54	131.30
Lutetium	Lu	71	174.97	Ytterbium	Yb	70	173.04
Magnesium	Mg	12	24.312	Yttrium	Y	39	88.905
Manganese	Mn	25	54.9380	Zinc	Zn	30	65.37
Mendelevium	Md	101	—	Zirconium	Zr	40	91.22

GEOMETRIC FORMULAS FOR BASINS OF VARIOUS SHAPES

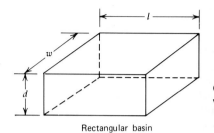

Surface area = $w \times l$
Circumference = $2w + 2l$
Volume = $w \times l \times d$

Rectangular basin

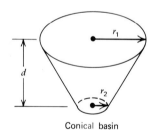

Surface area = $\pi \times r^2$ = 3.14 $\times r \times r$
Circumference = $2 \times \pi \times r$ = 6.28 $\times r$
Volume = $\pi \times r^2 \times d$

Circular basin

$Area_1 = \pi \times r_1$
Volume = $\frac{d}{3} \times (A_1 + A_2 + \sqrt{A_1 \times A_2})$
$Area_2 = \pi \times r_2^2$

Conical basin

Volume = $\frac{d}{6} \times (A_1 + 4A_2 + A_3)$

where A_1 = surface area

A_2 = area of midsection

A_3 = bottom area

d = depth

Prismoidal basin

TWO-CYCLE LOGARITHMIC PROBABILITY PAPER FOR USE IN SOLVING PROBLEM 4-48

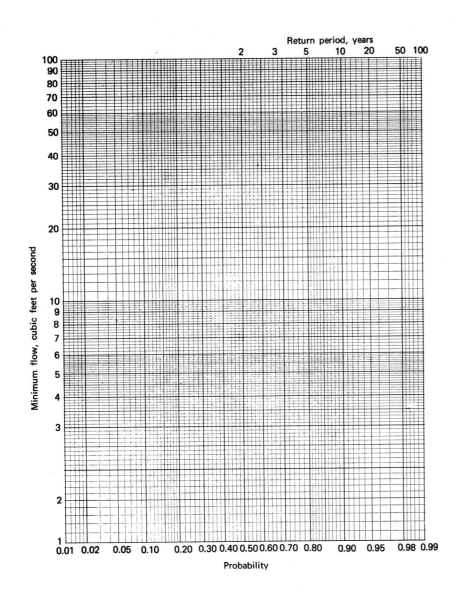

Index